Microbial Resilience

We owe profound thanks to our chapter contributors for their rigorous scholarship, from metagenomic dissections to field-validated models. Gratitude extends to Springer Nature for shepherding this vision, and to our institutions—Swami Rama Himalayan University, ICAR-Vivekananda Parvatiya Krishi Anusandhan Sansthan, Graphic Era University, Sri Lanka Institute of Information Technology, and the University of Arkansas—for nurturing inquiry. To families enduring our obsessions, your steadfastness grounds us.

As tipping points loom, *Microbial Resilience* urges a paradigm shift: view microbes as allies, not footnotes. May these pages spark alliances, ignite discoveries, and instill optimism—that in adaptation's quiet forge, humanity finds its path forward.

Dehradun, Uttarakhand, India — Geeta Bhandari
Almora, Uttarakhand, India — Priyanka Khati
Dehradun, Uttarakhand, India — Saurabh Gangola
Malabe, Sri Lanka — Dhanushka Udayanga
Fayetteville, AR, USA — Neelendra K. Joshi

December 2025

Contents

harsh terrains. She has published over 40 articles and reviews in renowned journals, co-authored 16 book chapters with Elsevier, Springer, and Taylor & Francis, and co-edited volumes like *Advances in Nanotechnology for Smart Agriculture* (Taylor & Francis) and commemorative works on ICAR-VPKAS history. She contributes to in-house and external projects, edits publications such as the Hindi magazine *Haritima*, and is a life member of the Association of Microbiologists of India, bridging lab innovations with field impacts for sustainable agriculture.

Saurabh Gangola is an Assistant Professor in the Department of Microbiology at Graphic Era University, Dehradun, India. He completed his Ph.D. in Microbiology and boasts over 6 years of combined teaching and research experience. Specializing in environmental microbiology, his investigations target the biodegradation and bioremediation of xenobiotic pollutants—including pesticides, dyes, heavy metals, and polyaromatic hydrocarbons—offering eco-friendly cleanup solutions amid climate-induced contamination surges. Honored with the Governor's Research Award-2019 from Uttarakhand, he has contributed 70 research papers and reviews to high-impact journals, 38 book chapters, 3 edited books with Springer and Elsevier, and 5 patents. His work underscores microbial roles in restoring polluted ecosystems, aligning with global sustainability goals. Through rigorous experimentation and interdisciplinary collaborations, he advances knowledge on resilient microbial communities, equipping future scientists to tackle environmental crises with innovative, biology-driven approaches.

Dhanushka Udayanga is a distinguished Professor in the Department of Applied Sciences at the Faculty of Humanities and Sciences, Sri Lanka Institute of Information Technology (SLIIT). With a Ph.D. from Mae Fah Luang University, Thailand, and experience as a visiting scientist at the United States Department of Agriculture's Agriculture Research Service (USDA-ARS) in Beltsville, Maryland, USA, he excels in biotechnology and mycology. His research delves into molecular systematics of fungi, their agricultural applications, and environmental remediation, particularly in tropical contexts vulnerable to climate shifts. He has authored over 50 articles in high-impact journals and serves as an editor and reviewer for leading publications. His contributions illuminate fungal adaptations that bolster ecosystem resilience, from pest control to soil health enhancement. By fostering global collaborations, he champions interdisciplinary strategies to harness microbial diversity for sustainable development in changing climates.

Neelendra K. Joshi is an Associate Professor in the Department of Entomology and Plant Pathology at the University of Arkansas, Fayetteville, AR, USA. His research integrates pesticide toxicology, ecotoxicology, risk assessment, biodiversity conservation, and beneficial species protection across ecosystems. A dual Ph.D. holder in Entomology and Comparative/International Education from Pennsylvania State University, where he also completed postdoctoral training, he earned his M.Sc. from GBPUAT Pantnagar and B.Sc. from Kumaun University, India. A Fellow of the Royal Entomological Society (UK) and Ford Foundation International Fellow, he has garnered numerous awards. With over 125 publications—including articles, book chapters, and 250+ presentations—he edits *Environmental Entomology* (Oxford) and serves on boards for *Scientific Reports* (Nature) and *Frontiers in Ecology and Evolution*. His work critiques climate-exacerbated risks, advocating microbe-entomology synergies for resilient agroecosystems.

Contributors

Anju Arora CCUBGA, Division of Microbiology, ICAR-Indian Agricultural Research Institute, New Delhi, India

Narendra Singh Bhandari Faculty of Agricultural Sciences, Shree Guru Gobind Singh Tricentenary University, Gurugram, Haryana, India

Mahendra Singh Bhinda ICAR- Vivekananda Parvatiya Krishi Anusandhan Sansthan, Almora, Uttarakhand, India

Shubhranil Brahma Department of Zoology, Iswar Chandra Vidyasagar College, Belonia, Tripura, India

Somnath Chatterjee Department of Zoology, Dr. Bhupendra Nath Dutta Smriti Mahavidyalaya, Burdwan, West Bengal, India

Anuj Chaudhary Department of Biotechnology, Graphic Era Deemed to be University, Dehradun, Uttarakhand, India

Parul Chaudhary School of Agriculture, Graphic Era Hill University, Dehradun, Uttarakhand, India

Shaohua Chen College of Plant Protection and the Integrative Microbiology Research Centre, South China Agricultural University (SCAU), Guangzhou, China

Mahipal Choudhary ICAR-Central Arid Zone Research Institute, Jodhpur, Rajasthan, India

Hemant Dasila Department of Microbiology, Akal College of Basic Sciences, Eternal University, Baru Sahib, Rajgarh, Himachal Pradesh, India

Prasenjit Debbarma Department of Botany, Iswar Chandra Vidyasagar College, Belonia, Tripura, India

Anant Deogaonkar Symbiosis Institute of Business Management, Nagpur, Symbiosis International (Deemed) University, Pune, Maharashtra, India

Atrayee Dey Department of Zoology, Banwarilal Bhalotia College, Asansol, West Bengal, India

Saurabh Gangola Department of Microbiology, Graphic Era (Deemed to be University), Dehradun, Uttarakhand, India

Minakshi Grover ICAR-Indian Agricultural Research Institute, New Delhi, India

Durgesh Kumar Jaiswal Department of Biosciences, Graphic Era (Deemed to be University), Dehradun, Uttarakhand, India

Pranita Jaiswal CCUBGA, Division of Microbiology, ICAR-Indian Agricultural Research Institute, New Delhi, India

Hem Ch. Joshi Department of Botany, D.S.B. Campus, Kumaun University, Nainital, Uttarakhand, India

Neelendra K. Joshi Department of Entomology and Plant Pathology, University of Arkansas, Fayetteville, AR, USA

Samiksha Joshi School of Biosciences, Swami Rama Himalayan University, Dehradun, Uttarakhand, India

Pooja Kala School of Agriculture, Graphic Era Hill University, Dehradun, Uttarakhand, India

Himani Karki Department of Botany, D.S.B. Campus, Kumaun University, Nainital, Uttarakhand, India

Sunil Kashyap Punjab Agriculture University, Regional Research Station, Gurdaspur, Punjab, India

Rajeev Kaushik Division of Microbiology, ICAR-Indian Agricultural Research Institute, New Delhi, India

Shivani Kaushik Division of Microbiology, ICAR-Indian Agricultural Research Institute, New Delhi, India

Amir Khan CSIR-Central Institute of Medicinal & Aromatic Plants (CIMAP), Research Centre, Pantnagar, Uttarakhand, India

Priyanka Khati Crop Production Division, ICAR VPKAS, Almora, Uttarakhand, India

Vijay Krishna ICAR-Indian Agricultural Research Institute, Hazaribag, Jharkhand, India

Bharti Kukreti Department of Microbiology, College of Basic Sciences and Humanities, Govind Ballabh Pant University of Agriculture and Technology, Pantnagar, Uttarakhand, India

Dipender Kumar CSIR-Central Institute of Medicinal & Aromatic Plants (CIMAP), Research Centre, Pantnagar, Uttarakhand, India

Pradeep Kumar Agricultural and Biosystems Engineering, South Dakota State University, Brookings, SD, USA

Prashant Kumar ICAR-Indian Agricultural Research Institute, Hazaribag, Jharkhand, India

Asha Kumari ICAR-Indian Agricultural Research Institute, Hazaribag, Jharkhand, India

Yama Santhoshi Lavanya Professor Jayashankar Telangana State Agricultural University Press, Hyderabad, Telangana, India

Rajeshwari Lekhwar Department of Microbiology, Graphic Era (Deemed to be University), Dehradun, Uttarakhand, India

Swati Lohani ICAR-VPKAS, Almora, Uttarakhand, India

Himanshu Mahawar Directorate of Weed Research (ICAR), Jabalpur, India

Jyotsana Maura DIBER, Kumaun University Nainital, Haldwani, Uttarakhand, India

Innocent Ojeba Musa Department of Microbiology, Skyline University Nigeria, Kano, Kano State, Nigeria

Neetika Naudiyal Department of Biochemistry, CBSH, G. B. Pant University of Agriculture and Technology, Pantnagar, Uttarakhand, India

Ayush Om ICAR-Indian Agricultural Research Institute, Hazaribag, Jharkhand, India

Prabha Pant Department of Botany, D.S.B. Campus, Kumaun University, Nainital, Uttarakhand, India

Department of Botany, DSB Campus, Kumaun University, Nainital, Uttarakhand, India

Babita Patni High Altitude Plant Physiology Research Centre, Hemvati Nandan Bahuguna Garhwal University, Srinagar, Uttarakhand, India

Arjita Punetha G. B. Pant National Institute of Himalayan Environment, Almora, Uttarakhand, India

Kamali Raja ICAR-Indian Agricultural Research Institute, New Delhi, India

Vijaya Rani ICAR-IARI, Division of Agricultural Microbiology, New Delhi, India

Meena Rathore Division of Microbiology, ICAR-Indian Agricultural Research Institute, New Delhi, India

Thottiyan Saeed ICAR-Indian Agricultural Research Institute, New Delhi, India

Sanjay Kumar Sanadya CSK HP Agricultural University, Palampur, Himachal Pradesh, India

S. Santhiya Crop Production Division, ICAR-VPKAS, Almora, Uttarakhand, India

M. Saritha ICAR-Central Arid Zone Research Institute, Jodhpur, Rajasthan, India

Pushpendra Sharma Division of Microbiology, ICAR-Indian Agricultural Research Institute, New Delhi, India

Raj Shekhar Sharma Department of Microbiology, School of Basic and Applied Science, Sri Guru Ram Rai University, Dehradun, Uttarakhand, India

Ayisha Shibin ICAR-Indian Agricultural Research Institute, New Delhi, India

Ajay Veer Singh Department of Microbiology, College of Basic Sciences and Humanities, Govind Ballabh Pant University of Agriculture and Technology, Pantnagar, Uttarakhand, India

Devendra Singh ICAR-Central Arid Zone Research Institute, Jodhpur, Rajasthan, India

Pragati Srivastava Department of Molecular biology and Biotechnology, ICAR-National Institute for Plant Biotechnology, IARI, PUSA, New Delhi, India

Kimi Tiwari Department of Microbiology, Graphic Era (Deemed to be University), Dehradun, Uttarakhand, India

Dhanushka Udayanga Department of Applied Sciences, Faculty of Humanities and Sciences, Sri Lanka Institute of Information Technology, Malabe, Sri Lanka

Abhiruchi Uniyal Wildlife Institute of India, Dehradun, Uttarakhand, India

Kumar Vishal ICAR-Indian Agricultural Research Institute, Hazaribag, Jharkhand, India

The Intersection of Microbiology and Climate Change

Rajeshwari Lekhwar, Kimi Tiwari, Pooja Kala, Narendra Singh Bhandari, Saurabh Gangola, Neelendra K. Joshi, and Anant Deogaonkar

Abstract

Microbe-environment interaction is how microbes interact with their environment, and these interactions are increasing in prominence due to climate change. The effects of microbes on climate change can be positive or negative. Warming and shifting precipitation patterns are fragmenting microbial assemblages, with consequences for ecosystem and global biogeochemical cycling. For example, as permafrost melts, methane-producing microbes are released, and some microbes can also capture carbon dioxide. Ocean microbes such as phytoplankton and heterotrophic bacteria are also central players in the carbon cycle. Effects of climate change on microbial processes, including denitrification and nitrogen fixation, have the potential to impact both agricultural practices and water quality. Predicting ecosystem responses to climate change and identifying sustainable mitigation strategies depend on understanding these intricate relationships. This

R. Lekhwar · K. Tiwari · S. Gangola (✉)
Department of Microbiology, Graphic Era (Deemed to be University), Dehradun, Uttarakhand, India

P. Kala
School of Agriculture, Graphic Era Hill University, Dehradun, Uttarakhand, India

N. S. Bhandari
Faculty of Agricultural Sciences, Shree Guru Gobind Singh Tricentenary University, Gurugram, Haryana, India

N. K. Joshi
Department of Entomology and Plant Pathology, University of Arkansas, Fayetteville, AR, USA
e-mail: nkjoshi@uark.edu

A. Deogaonkar
Symbiosis Institute of Business Management, Nagpur, Symbiosis International (Deemed) University, Pune, Maharashtra, India

G. Bhandari et al. (eds.), *Microbial Resilience*,
https://doi.org/10.1007/978-981-95-8883-1_1

suggests that further research into the connection between microbiology and climate change could drive more transformative solutions and lead to more ecological resilience.

Keywords

Microbiology · Climate change · Microbial ecology · Carbon cycle · Ecosystem functioning · Sustainability

1 Introduction

The geochemist Wallace Broecker is credited with popularizing the terms "climate change" and "global warming" in the 1970s; however, it is unclear whether he was the first to use these terms (Broecker, 1975). Just 4 years later, the National Academy of Sciences stated, "the study group has found no basis for concluding that climate changes will not occur, or that the changes will be so slight as to be of no consequence, if carbon dioxide continues to grow" (National Academy of Sciences: Carbon Dioxide and Climate (1979), 2017). Since then, the scientists' initial hypothesis that human activities releasing CO_2 into the atmosphere have altered the Earth's climate and, consequently, the way of life for everyone on the planet has been validated by the effects of CO_2 on global temperature and climate. So, on Earth, there are a lot of microbes (Lennon & Locey, 2016). These submicroscopic organisms that live in soil, city streets, the atmosphere, beneath the surface of the Earth, or in layers of water include viruses, bacteria, archaea, fungi, algae, and protozoa. Microorganisms are tiny in size, yet they can influence the climate at a global scale because they have an incredibly high abundance. The microbes are the essential symbionts of world crops, significant producers and consumers of greenhouse gases, and key players in the global geochemical cycle. Few people outside of the microbial research community discuss or incorporate microbial activities into their climate change predictions, even though bacteria have a significant influence on climate change (Lennon & Locey, 2016). The microbes that live on Earth will shift along with changes in the climate. Microbes are robust and will change in response to changes in the Earth's climate because they have done so for more than 3 billion years. Humans respond more slowly. To address this deadly problem, we need to understand the links among the health of humans, animals, the environment, and microbes (Nguyen & Casadevall, 2021). We need new ideas, unconventional approaches, and out-of-the-box solutions, as climate change presents an unprecedented challenge to mankind. Advancing on our current knowledge, increasing our understanding of microbes, and putting our knowledge into practice by developing sustainable and microbe-based solutions are crucial steps to help stem climate change, fight this crisis of such magnitude, and promote the health and well-being of humans across the globe.

2 Role of Microorganisms in Earth's Climate System

The connection between bacteria and their environments has been studied for decades. The abundance of microorganisms found in aquatic, terrestrial, and urban settings has been shown by the development of sequencing and metagenomics. Researchers may know which bacteria are found in the environment, but they do not know much about what those microbes are doing in each one or how microbial communities and their ecosystem functions will be affected by climate change. Climate change is indicated by variations in temperature, precipitation, and season duration (Singh et al., 2010). So, the primary expression of its mode of action is in the temperature and moisture contents. The Intergovernmental Panel on Climate Change (IPCC) states that climate change is a pressing global threat. Their analysis suggests that between 3.3 and 3.6 billion people live in regions that are highly vulnerable to the impacts of climate change. Many factors, such as poverty, inequality, and unsustainable means of development, exacerbate this vulnerability. According to the report, populations and ecosystems are becoming more vulnerable to climate change because of present development patterns that increase exposure to climate hazards (Microbes and climate change—Science, people & impacts, 2022)

Climate change (CC) is referred to as the change in climate of a particular region due to human activities and natural phenomena, such as the ozone layer and the greenhouse effect (Shivanna, 2022). Global climate change is expected to bring about dramatic changes in food security, agricultural production, and biodiversity (Vermeulen et al., 2012). The global climate has been relatively stable for over 12,000 years because our species, humans, has been dependent on this kind of stability to survive (Smith et al., 2018). Global warming is a result of certain gases such as carbon dioxide, methane, and nitrous oxide, which absorb the infrared radiation and then send it back toward the Earth (Bapuji Rao et al., 2014). Over the last 10 years, there has been a general recognition of human-induced climate change and consequent calls for technological and legal solutions to reduce greenhouse gas (GHG) emissions by governments, the scientific community, and international financial markets (Eskander & Fankhauser, 2020).

3 Climate-Induced Shifts in Pathogenic Microorganisms and Their Adaptations

Changes in temperature, precipitation, and season length are indicators of climate change (Singh et al., 2010). Therefore, variations in temperature and moisture are the main ways its effects manifest. The Intergovernmental Panel on Climate Change (IPCC) has emphasized that climate change is a major global challenge. Their report states that about 3.3–3.6 billion people live in regions highly vulnerable to climate impacts. It is expected that rainfall patterns will shift, and flooding and heat waves will intensify under global warming and climate change, which will likely influence the Earth's ecology. Food and human security are at risk with rising sea levels and severe weather; the threats escalate significantly at 1.5° and beyond due to increased greenhouse gas emissions, mainly CO_2. Furthermore, food-borne diseases are more closely linked to climate change and cause millions of illnesses and deaths each year, especially in Africa. Warmer temperatures encourage and spread of food-borne parasites, leading to more contamination and health risk. Since agricultural diseases are expected to spread faster because of climate change, knowledge of plant-pathogen interactions is important (Naga et al., 2025). The current practice of doing unsustainable development is posing a threat to climate change (Microbes and climate change—Science, people & impacts, 2022). Human activities mainly lead to climate change and emissions (Shivanna, 2022). Global climate change is expected to lead to significant changes in biodiversity, agriculture, and food security (Nguyen et al., 2021). The global climate has been stable for 12,000 years (Smith et al., 2018). Climate change is altering water bodies and their ecosystems, such as microbial communities, which are crucial to ecological functioning and human health. Perturbed microbial (e.g., warm water, lab stratification, simulate increased pollution, or extreme weather) communities may favor pathogens such as *Vibrio cholerae* or toxigenic cyanobacteria that contaminate food and water and cause significant health problems. Environmental factors related to climate change are also implicated in the spread of antibiotic resistance genes, posing further challenges to global healthcare. This highlights the need for solutions such as building global monitoring networks, improving wastewater treatment technologies, and establishing early warning systems for disease outbreaks. It shows that international cooperation is key to solving this urgent problem and protecting water sources from the impacts of climate change, particularly in developing countries with fragile infrastructure (Yu, 2025). Cyanobacteria, such as *Prochlorococcus* and *Synechococcus,* first appeared with the capacity to photosynthesize (Baumgartner et al., 2020), and about 2 billion years ago, Earth had a dramatic increase in the buildup of oxygen in the atmosphere, known as the Great Oxidation Event (Wang et al., 2025). The incident serves as a dramatic illustration of how microbiological life can alter our globe. Although the great oxidation event caused a vast extinction of anaerobic life, including many methanogenic archaea such as *Methanobacterium* and *Methanobrevibacter,* it also created a favorable environment for the growth of aerobic multicellular life (Berner et al., 2018). The first ice age resulted from the oxidation of atmospheric methane by increased oxygen, which in turn reduced the amount of this greenhouse gas and lowered global temperatures (Shen et al., 2021). Microbial metabolism and nutrient cycling will surely be impacted by changes in temperature, precipitation, and nutrient availability, which may in turn influence colonization and pathogenicity in microbes such as *Vibrio cholerae, Salmonella enterica,* and *Campylobacter jejuni* (Classen et al., 2015). It is necessary to do research on how infections' metabolic processes, enzyme activity, and community dynamics alter in response to temperature and nutritional changes. Additionally, as humans move into new natural settings, their exposure to disease increases. Surveillance studies enable the tracking of spillback events, in which humans reintroduce viruses to wildlife, as human-animal interaction increases. Zoonotic infections can use this as a reservoir to adapt, pick up new mutations, and develop novel traits that could affect their virulence (Tiedje et al., 2022)

4 Microbial Contribution to Greenhouse Gas Emissions

Climate change is highly influenced by greenhouse gases (Xavier, 2016). The heterotrophs, such as *Pseudomonas, Bacillus,* and *Actinobacteria*, emit many greenhouse gases. Microorganisms play a crucial role in the global biogeochemical cycle and influence atmospheric quantities of important greenhouse gases (GHGs), including nitrous oxide (N_2O), carbon dioxide (CO_2), and methane (CH_4) (Yi et al., 2025). These gases play an important role in climate change and radiative forcing on the Earth. In both terrestrial and aquatic environments, microbial processes such as respiration, methanogenesis, nitrification, denitrification, and anaerobic ammonium oxidation control the generation and uptake of these GHGs. A region's average weather, including its temperature, precipitation, and wind, is referred to as its climate (Das & Mandal, 2022). For example, to sustain prawn production, it is necessary to understand the microbial interactions with greenhouse gas (GHG) emissions. In a work, microbial communities and key functional genes (*mcrA, pmoA, amoA*, and *nosZ*) were investigated in relation to CO_2, CH_4, and N_2O fluxes in prawn farms in Bangladesh. It was found that microbial richness was positively associated with CO_2/N_2O emission but negatively associated with CH_4 emissions. Methanogenic archaea, prevalent in high salinity farms, were weakly associated with CO_2 and strongly with CH_4 flux. Anaerobic methanotrophs were negatively related to CH_4 emissions at the plot level, suggesting that their function in CH_4 oxidation might be important for reducing CH_4 emissions. Most microbial groups were negatively correlated with N_2O emissions, in contrast to Chloroflexi, a N_2O consumer, which was positively correlated. The correlation of the *mcrA* gene and CH_4 production in the study indicates the potential of manipulating microbial communities toward the mitigation of emissions in prawn farming (Bashar et al., 2025). The most complicated worldwide issue of our day is climate change, which has elements of science, economics, society, politics, morality, and ethics (Smith et al., 2018). These gases function similarly to a blanket. The Earth would be 20–30 ° C cooler and far less habitable without the blanket higher level of the four greenhouse gases carbon dioxide, methane, nitrous oxide, and chlorofluorocarbons in the atmosphere are the cause of it (Ibáñez et al., 2023). The Earth's atmosphere traps solar radiation, a process known as the greenhouse effect. It is caused by molecules like CO_2, water vapor, and CH_4 that enable incoming sunlight and absorb the heat reflected from the Earth while allowing sunlight to pass through. Climate trends have a significant impact on the breakdown of microbial organic matter (Crowther et al., 2015), 85% of global methane emissions are caused by microbes (Dutta & Dutta, 2016). When biogeochemical cycles and microbial populations are connected, they function as an effective mechanism to address climate change. The enzyme methyl coenzyme M reductase, which converts CO_2 into CH_4, is an example of this. Many of these local bacteria may produce more methane because of warming trends (Rossi et al., 2015). Methane-consuming microbes may extract atmospheric methane, even at extremely low concentrations (Abatenh et al., 2018) (Table 1).

Table 1 Microbial biogeochemical processes influencing greenhouse gas emission

References	Process	Climate impact
Conrad (2023)	Temperature dependence of methanogenesis	Demonstrated how warming shifts pathways from acetate to H_2/CO_2 and community composition, with methane production rates showing temperature compensation
Xu et al. (2025)	Soil denitrification under warming and acidification	Combined warming and low pH significantly increase N_2O emission via active microbial communities with transcriptomic insights
Doherty et al. (2025)	Permafrost microbial community coalescence	Thaw-induced mixing of active layer and permafrost microbes, affecting CO_2/CH_4 release
He et al. (2025)	Global importance of microbial denitrification and anammox	Quantified 79.8 ± 0.4% of nitrogen loss via denitrification; emphasizes aquatic vs terrestrial pathways
Weston et al. (2024)	Physiochemical triggers for N_2O via denitrification	Demonstrated how soil sediment perturbations enhance N_2O output

5 Impacts of Climate Change on Microbial Communities

Climate change responses of ecosystems without specifically mentioning microbial communities. However, soil microbial communities are the primary forces behind soil nutrient cycling due to their varied metabolic activities, and regulating climate change and ecosystem functioning should not be underestimated (Malchair et al., 2010). Numerous ecosystem scales, including fluctuations in climate, increased CO_2, and disturbance, might affect the metabolic activity of microorganisms (Matzarakis et al., 1999). Shifts in microbial activity can lead to changes in decomposition, nitrogen mineralization, organic carbon storage, and other ecosystem processes. The response of microbial communities to climate change and interactions between climate change and other global change factors may be highly dependent on specific ecosystems and historical adaptation of the community (Classen et al., 2015). Changes in soil microbial diversity and their processes can result from abiotic stress caused by climate change (Shade et al., 2012). Different microbial groups are distinct. Alter the temperature ranges they favor for growth and activity; higher temperatures may affect the microbial communities (Fierer et al., 2003). The human gut microbiota plays an important role in modulating immune responses, nutrient absorption, and even neurochemical signaling. Climate-induced factors such as malnutrition, poor water quality, altered food systems, and disease outbreaks disrupt microbiome composition, weakening host immunity and increasing disease susceptibility (Rio et al., 2024). For instance, frequent flooding and rising temperatures promote waterborne and vector-borne diseases, which are exacerbated when the gut microbiome is compromised (Vangay et al., 2018).

5.1 Terrestrial

Climate change is the primary cause of the soil rising in temperature. Changing weather patterns and temperature are the causes of the temperature variation in the soil. Numerous microorganisms and other organic substances can be found in the soil. The various soil layers alter the soil. The topsoil layer has the maximum number of organic materials (Oishy et al., 2025). As fast-growing bacteria are chosen for diversity may shift, changing the dynamics of the structure of the microbial community (Adamczyk et al., 2020). The effect on species alpha and beta diversity is also determined by the regions' changing climate. At different levels, different species are impacted. There is little information available regarding soil viromers, and most of the research focuses on the taxonomic variety of bacteria and fungi. Only soil between 15 and 20 cm has been the focus of the study (Antiqueira et al., 2018). There is extremely little data available and little research on subsurface bacteria. The microorganism that has been living beneath the surface has not yet been investigated. These microorganisms require study since they have been present in the soil for a very long time (Beaver & Neufeld, 2024). To answer these problems, a gene to ecosystems strategy will be used. Climate models and strategies that address the effect of climate change can be informed by these research priorities (Wang et al., 2023). In addition to altering temperature, climate change also affects soil characteristics like pH and porosity, and raises the likelihood of natural disasters like fires, droughts, and floods that physically and chemically disturb the soil. Even if bacteria have developed strategies to withstand drought stress, changes in plant life that are more resistant to fire or drought will ultimately change the microbial population associated with the roots (Reinsch et al., 2024). Additionally, hotter and drier soil results in less microbial variety and richness, which might lower total production. Increase nutrient leaching, soil erosion, and even anaerobiosis that can produce CH_4 and N_2O can all be caused by excessive rain brought on by more intense storms (Uzel et al., 2023).

5.2 Permafrost and Methanogenesis

Changes in soil microbial populations are brought about by climate change. Changes in temperature, precipitation, storms, soil organic carbon, and greenhouse gases all affect the bacteria (*Bacillus, Pseudomonas*), archaea (*Methanobacterium*), and fungal hyphae (*Aspergillus, Penicillium*) in the center, which leads to the availability of nutrients and their cycling by microbes can be impacted by soil warming and physical changes, which may have unanticipated environmental ripple consequences (Microbes and climate change—Science, people & impacts, 2022). The main greenhouse gases that cause global warming, CO_2, CH_4, and N_2O, are produced and consumed by microorganisms. Climate change-induced changes in soil microbial respiration or denitrification rates might impact microbial metabolism and nutrient feedback loops, resulting in inaccurate measurements of these key greenhouse gases that may impede mitigation efforts (Paranaíba et al., 2023). Although found that higher temperature is linked to higher metabolic rates, little is known about how changing metabolic kinetics affects microbial populations. To forecast future climate change implications on the Earth and people, it is essential to investigate how changes in nutrient and carbon availability affect the bacterial, fungal, protozoan, and viral ecosystems and how that contributes to climate change. The main greenhouse gases that cause global warming, CO_2, CH_4, and N_2O, are produced and consumed by microorganisms (Yu, 2025).

5.3 Nitrification and Denitrification

One of the most significant nutrient cycles in terrestrial ecosystems is the nitrogen cycle. Four microbial processes are involved in nitrogen cycling: nitrogen fixing, nitrification, denitrification, and mineralization (decay). Large-scale nitrogen fertilizer inputs to agricultural fields affect these processes, particularly nitrification and denitrification, and raise N_2O production (Pajares & Bohannan, 2016). Microbial activities like nitrification and denitrification play a major role in the emissions of nitrous oxide, N_2O, which is a powerful greenhouse gas that has around 300 times the potential to cause global warming. Ammonia-oxidizing bacteria (AOB) and archaea (AOA) are the driving forces behind nitrification, which transforms NH_4^+ into NO^{3-} and frequently yields N_2O as a byproduct. Under subtoxic circumstances, denitrifying bacteria such as *Paracoccus* and *Pseudomonas* convert nitrate to N_2, which is then released as N_2O (Liang & Robertson, 2021). The plant creates a hollow thread that leads into the root in reaction to the bacterium attaching to one of its root hairs. Through this infection thread, bacteria proliferate and ultimately start the root nodule creation process, up to 30% of the nodules' weight could be bacteria. Through nitrogen fixation, the plant transforms atmospheric energy into a form that can be utilized, and symbiotic nitrogen fixation is exemplified by this (Tsyganova et al., 2021).

6 Microbial Activity and Ecology in Urban Environments

Climate change also affects how microorganisms affect urban areas. Microbial corrosion of civil infrastructure is accelerated by warming. Microbiologically influenced corrosion is caused by microbial adhesion and metabolic activity on surfaces. While methanogens or bacteria that reduce sulfate can result in corrosive metal loss, microbial biofilms adhere directly to the surface (Microbes and climate change—Science, people & impacts, 2022). Concrete is harmed by weak organic and inorganic acids made by bacteria, fungi, and algae, which also lower the pH of the surrounding environment to permit further microbial colonization (Jiang et al., 2022). Damage caused by microorganisms can cost millions to billions of dollars and potentially generate greenhouse gases that hasten climate change. Furthermore, microbial activity that alters soil conditions, temperature, precipitation, and humidity can cause direct harm to infrastructure. In light of climate change, further investigation is required into the function of microbes in the deterioration or restoration of man-made structures (Subramanian et al., 2023). Changes in precipitation and temperature also alter urban areas' groundwater chemistry and levels of dissolved organic carbon. The sewage wastewater systems are triggered by the changes in the climate (McDonough et al., 2020). A study found that climate factors are important for microbial growth. Elevated temperatures have also triggered *Legionella pneumophila* (agent of Legionnaires' disease) overgrowth and have caused significant changes. Warming can also cause the temperatures in the snow cap to rise (Han, 2021). Climate change will have a rigorous impact on different aspects. Vulnerable, minority, and lower-income communities are especially exposed to climate change-associated impacts because they tend to live in "more polluted, less secure, and high-risk environments" that drive inequitable microbial exposure. The poor communities were affected by Hurricane Katrina in Louisiana, USA, in 2005 (Microbes and climate change—Science, people & impacts, 2022). Higher rates of bacterial contamination in drinking water have also been found in immigrant and low-income communities, hurting community health. This "microbial injustice" stemming from the unjust and disproportionate microbial exposures and risks of vulnerable communities will worsen as climate change increases the frequency and/or intensity of storms (Seymour & McLellan, 2025). The marine carbon pump uses microbial processes to sequester large amounts of carbon in the deep ocean, helping regulate atmospheric CO_2 levels. Global warming can affect the carbon sequestration by the oceans. Actions of the microbial loop and viral shunt drive carbon cycling in the oceans, though there is a current lack of understanding about microorganisms in deep ocean systems (Zhang et al., 2018). Ocean warming impacts microbial communities, which, in turn, affect the overall balance between carbon sequestration and release of CO_2. Studies to uncover how microbial community composition and carbon cycling are impacted by ocean warming under different conditions and in diverse regions will be important to inform Earth climate models and environmental management strategies (Guan et al., 2025).

7 Incorporating Microbes into Earth System Climate Models

Climate models are crucial for forecasting changes in long-term trends in temperature, precipitation, and storm event frequency and intensity on the local and global scale. The interplay of the atmosphere, land surface, oceans, and sea ice is considered in Earth's system climate models (Sillmann et al., 2017). The ability to forecast land, ocean, and climatic changes that affect weather and storm conditions over both short and long durations is made possible by these models. Adding more details about microbial dynamics and activities, like the generation and consumption of greenhouse gases, can enhance model performance, even though current models have significantly improved over the past 50 years (Chen et al., 2023). Unfortunately, microbial activities that impact the cycle of carbon and nitrogen in terrestrial and aquatic ecosystems are currently only weakly explicitly included in most current Earth system climate models. It is difficult to include microbial activity in climate

Table 2 Key microbial groups and their roles in climate-related biogeochemical processes

Microbial group	Function	Habitat and pathways	Key reference
Methanogens	CH_4 production	Anoxic wetland soils, rumen, flooded paddies; hydrogenotrophic and acetolactic pathways	Conrad (2020)
Aerobic methanotrophs	CH_4 consumption	Oxic zones; type 1/2; nitrogen-fixing	Liu et al. (2025)
Anaerobic methanotrophs (ANME)	CH_4 consumption	Anoxic sediments with alternative electron acceptors via AOM (anaerobic oxidation of methane)	Zhao et al. (2024)
Nitrifiers	NH_3 and NO_2^- oxidation → NO_3^-	Soils, freshwater, marine systems, aerobic nitrification	Zhu et al. (2013)
Sulfate reducing bacteria (SRB)	Organic C oxidation + sulfate → H_2S	Anoxic marine and freshwater sediments	Tripathi et al. (2021)
Photosynthetic cyanobacteria	CO_2 fixation, O_2 production	Aquatic systems, soil crusts, wetlands	Mangan and Brenner (2014)
Plant growth-promoting rhizobacteria (PGPR)	Indirect climate role via C sequestration	Rhizosphere: produce ACC deaminase, fix N_2, solubilize P	Dutta et al. (2022)

models of the Earth system. Large amounts of carbon, nitrogen, and phosphorus are transformed annually by microbes, and many features of microbial populations and their responses seem to be random (Cavicchioli et al., 2019). It is challenging to model microbial activities and outputs at the planetary scale using laboratory or field studies. Microbial activity patterns are predicted to shift in a variety of ways due to climate change, which will increase complexity and unpredictability. The key to developing models will be predicting emerging features of altered communities and comprehending changes to microbial community dynamics and nutritional feedback loops when microbial populations transition and adapt to different temperatures, humidity, etc. One of the biggest challenges at the moment is the absence of trustworthy time series data about microbial communities and the kinetics of microbial processes for parameterizing models. Information on deep ocean and subterranean microorganisms (such as the viral shunt) and the adaptation of microbial communities to locate and regional climate changes is required (Fasaeiyan et al., 2024) (Table 2).

8 Microbial Taxa as Bioindicators of Environment

Metabarcoding combined with multivariate statistics and machine learning has proven to be a reliable method to identify bacterial bioindicators for the assessment of the biological status of water bodies. Therefore, although bacteria are known to have fundamental roles and predictable patterns of diversity, there remains a significant technology gap to use these organisms for biomonitoring due to a lack of data and models that are implementable. The research demonstrates a predictive relationship between environmental drivers, bacterial beta-diversity, and the biological condition of downstream water, as forecastable from upstream sites. Genus such as *Fluvicola, Acinetobacter, Flavobacterium*, and *Rhodoluna* were confirmed as reliable bioindicators by network analysis and showed also a taxonomically related redundancy among taxa (Fontaine et al., 2023). With the risk to coral so high, early detection of indicators of stress in the ecosystem is essential to allow for better management and restoration. Although it is recognized that bacteria, e.g., *Rhodobacteraceae, Vibrio corallilyticus, Verrucomicrobia*, and *Flavobacterium*, have a major role in the health of coral reefs, there is limited understanding about how different microbiomes respond to environmental perturbations, which constrains our capacity to foresee ecosystem stress. As changes in temperature affect microbial communities and are associated with coral bleaching events, monitoring microbial community data in conjunction with seawater quality monitoring may provide an integrated assessment of reef status (Glasl et al., 2019). It is vital to understand how abundant (AT) and rare (RT) taxa react when subjected to the same environmental forces to assess ecological mechanisms. Both subcommunities were dominated by the *Woeseia* genus. RT exhibited higher α- and β-diversity than AT. Taxa diversity and compositional differences were greatly affected by metal stress, and AT were more intolerant than salt stress. Stress also influences the biotic network by connecting more closely in RT but less lightly in AT. In addition, RT had more salt tolerance and a greater variety of genes for metal resistance than AT. Community assembly in the AT was predominantly deterministic under salinity stress, but in the RT, the proportion of stochastic processes increased with metal (Li et al., 2024). Plants, plankton, animals, and microbes (*Hylocomium splenders*) are bioindicators, living organisms that we can use to monitor the health of an ecosystem and changes in the environment. They provide information on the environmental condition, and plankton is a significant biomarker for pollution and the quality of water (Parmar et al., 2016).

9 Advances in Omics and Data Integration for Microbe-Climate Research

With the advent of advanced omics methodologies, the biological responses of microbial cells at native and artificial interfaces are now being profoundly investigated. These enable the separation and analysis of proteins, metabolites, and nucleic acids derived from complex mixtures, in fields such as material sciences, biological science, forensic science, biotechnology, and archaeology (González-Plaza et al., 2022). The sequencing of DNA from entire microbial communities has revolutionized microbial ecology by allowing researchers to assess taxonomic composition, functional potential, and stress responses of microbial populations in situ. This approach provides a comprehensive view of microbial ecosystem function and helps identify microbial shifts related to climate-driven variables such as warming, acidification, or salinization. In long-term soil warming experiments, metagenomics revealed reduced abundance of lignin-degrading genes and increased carbohydrate metabolism genes, indicating changes in decomposition processes under elevated temperatures (Pérez-Cobas et al., 2020). In oceanic environments, functional gene markers (amoA for nitrifiers, mcrA for methanogens) are used to monitor biogeochemical cycling under acidification and warming. These biomarkers serve as tools to assess microbial adaptation, carbon use efficiency, and nutrient cycling under climate perturbations (Speth & Orphan, 2018). Despite their strength, omics technologies need to be used effectively so that data can be shared and used by other parties. In addition to accounting for local influences, this continuity of research strategy and data type should allow data comparison across various habitats and microbial populations. Funding must be allocated by agencies to the development and upkeep of these data systems. To comprehend the dynamics of marine ecosystems, the Darwin Project serves as an example of how to combine environmental data with information about the physiology of marine microorganisms and community structure (Chervitz et al., 2011). To assist in anticipating future ocean changes and global nutrient cycling, these models especially look at how plankton diversity and better data coordination can help infectious disease models as well as Earth climate models by enabling the incorporation of real-time air and water surveillance data. Vibrio Map Viewer uses the real-time data to predict the sea surface temperature. The modeling can help in the sustenance of sea surface temperature. The need to improve the datasets in relation to the climate change model is required (Henson et al., 2021). It is important to conserve the microbial communities and conserve them. While these developments are promising, microbiome engineering must be approached with caution. It is essential to consider ecological safety, the potential for horizontal gene transfer, and the risk of unintended interactions with native microbiota. These areas require thorough risk assessments and robust regulatory frameworks (Amen et al., 2025). Furthermore, it is necessary to consider the scaling of microbial processes from controlled studies to climatic models. Microbiological data must therefore be consistent with the input and software requirements of climate modeling, and vice versa. The incorporation of microbiological processes into models also requires improved data coordination. Broad data use requires better data coordination between the top level and bottom level, even while national databases ("top level") and individual research groups ("bottom level") each have well-organized data management systems. Additionally, cross-field data synchronization is required. To establish agreement on research and data collection procedures that provide proper data interpretation and model parameterization, systematic collaboration among microbiologists, geochemists, ecologists, computational scientists, and modelers will be required.

10 Conclusion

Microbes are important players in the climate-change equation. There is a microbial community all around us. These microbiomes are crucial for the survival of microbial populations. The future of microbial communities will be in the convergence of multiple platforms, such as biotechnology. To identify microbial solutions, cooperative microbiology research is needed. And there is a need to identify the tools and data sets to enable collaborative research. There should be concerted efforts at the global level to identify solutions for the menace of bacteria due to climate change. We must exploit microbial potential and microbial communities as key means to sustain the ecosystem and prevent its breakdown. Microbial communities are the bricks that build the soil ecosystem, and their protection will lead to the protection of many more species. Collaborative efforts are required to fill these gaps and understand the microbial ecosystem.

References

Abatenh, E., Gizaw, B., Tsegaye, Z., & Tefera, G. (2018). Microbial function on climate change—A review. *Environment Pollution and Climate Change, 2*(1). https://doi.org/10.4172/2573-458X.1000147

Adamczyk, M., Perez-Mon, C., Gunz, S., & Frey, B. (2020). Strong shifts in microbial community structure are associated with increased litter input rather than temperature in high Arctic soils. *Soil Biology and Biochemistry, 151*, 108054. https://doi.org/10.1016/j.soilbio.2020.108054

Amen, R. A., Hassan, Y. M., Essmat, R. A., Ahmed, R. H., Azab, M. M., Shehata, N. R., Elgazzar, M. M., & El-Sayed, W. M. (2025). Harnessing the microbiome: CRISPR-based gene editing and antimicrobial peptides in combating antibiotic resistance and cancer. *Probiotics and Antimicrobial Proteins, 17*(4), 1938–1968. https://doi.org/10.1007/s12602-025-10573-8

Antiqueira, P. A. P., Petchey, O. L., dos Santos, V. P., de Oliveira, V. M., & Romero, G. Q. (2018). Environmental change and predator diversity drive alpha and beta diversity in freshwater macro and microorganisms. *Global Change Biology, 24*(8), 3715–3728. https://doi.org/10.1111/gcb.14314

Bapuji Rao, B., Santhibhushan Chowdary, P., Sandeep, V. M., Rao, V. U. M., & Venkateswarlu, B. (2014). Rising minimum temperature trends over India in recent decades: Implications for agricultural production. *Global and Planetary Change, 117*, 1–8. https://doi.org/10.1016/j.gloplacha.2014.03.001

Bashar, A., Djurhuus, A. M., Browne, P. D., Raju Podduturi, M., Hasan, Z., Al Rakib, A., Hooper, C., et al. (2025). Identification of microbial groups and functional genes associated with greenhouse gas emissions in prawn farms. *Scientific Reports, 15*(1), 38450. https://doi.org/10.1038/s41598-025-22768-8

Baumgartner, M., Bayer, F., Pfrunder-Cardozo, K. R., Buckling, A., & Hall, A. R. (2020). Resident microbial communities inhibit growth and antibiotic-resistance evolution of Escherichia coli in human gut microbiome samples. *PLoS Biology, 18*(4), e3000465. https://doi.org/10.1371/journal.pbio.3000465

Beaver, R. C., & Neufeld, J. D. (2024). Microbial ecology of the deep terrestrial subsurface. *The ISME Journal, 18*(1). https://doi.org/10.1093/ismejo/wrae091

Berner, C., Bertos-Fortis, M., Pinhassi, J., & Legrand, C. (2018). Response of microbial communities to changing climate conditions during summer cyanobacterial blooms in the Baltic Sea. *Frontiers in Microbiology, 9*. https://doi.org/10.3389/fmicb.2018.01562

Broecker, W. S. (1975). Climatic change: Are we on the brink of a pronounced global warming? *Science, 189*(4201), 460–463. https://doi.org/10.1126/science.189.4201.460

Cavicchioli, R., Ripple, W. J., Timmis, K. N., Azam, F., Bakken, L. R., Baylis, M., Behrenfeld, M. J., et al. (2019). Scientists' warning to humanity: Microorganisms and climate change. *Nature Reviews. Microbiology, 17*(9), 569–586. https://doi.org/10.1038/s41579-019-0222-5

Chen, L., Han, B., Wang, X., Zhao, J., Yang, W., & Yang, Z. (2023). Machine learning methods in weather and climate applications: A survey. *Applied Sciences, 13*(21), 12019. https://doi.org/10.3390/app132112019

Chervitz, S. A., Deutsch, E. W., Field, D., Parkinson, H., Quackenbush, J., Rocca-Serra, P., Sansone, S.-A., et al. (2011). Data standards for omics data: The basis of data sharing and reuse. *Methods in Molecular Biology (Clifton, N.J.), 719*, 31–69. https://doi.org/10.1007/978-1-61779-027-0_2

Classen, A. T., Sundqvist, M. K., Henning, J. A., Newman, G. S., Moore, J. A. M., Cregger, M. A., Moorhead, L. C., & Patterson, C. M. (2015). Direct and indirect effects of climate change on soil microbial and soil microbial-plant interactions: What lies ahead? *Ecosphere, 6*(8), 1–21. https://doi.org/10.1890/ES15-00217.1

Conrad, R. (2020). Importance of hydrogenotrophic, aceticlastic and methylotrophic methanogenesis for methane production in terrestrial, aquatic and other anoxic environments: A mini review. *Pedosphere, 30*(1), 25–39. https://doi.org/10.1016/S1002-0160(18)60052-9

Conrad, R. (2023). Complexity of temperature dependence in methanogenic microbial environments. *Frontiers in Microbiology, 14*. https://doi.org/10.3389/fmicb.2023.1232946

Crowther, T. W., Glick, H. B., Covey, K. R., Bettigole, C., Maynard, D. S., Thomas, S. M., Smith, J. R., et al. (2015). Mapping tree density at a global scale. *Nature, 525*(7568), 201–205. https://doi.org/10.1038/nature14967

Das, N., & Mandal, S. (2022). Microbial populations regulate greenhouse gas emissions in Sundarban Mangrove Ecosystem, India. *Acta Ecologica Sinica, 42*(6), 641–652. https://doi.org/10.1016/j.chnaes.2021.07.011

Doherty, S. J., Thurston, A. K., & Barbato, R. A. (2025). Active layer and permafrost microbial community coalescence increases soil activity and diversity in mixed communities compared to permafrost alone. *Frontiers in Microbiology, 16*. https://doi.org/10.3389/fmicb.2025.1579156

Dutta, H., & Dutta, A. (2016). The microbial aspect of climate change. *Energy, Ecology and Environment, 1*(4), 209–232. https://doi.org/10.1007/s40974-016-0034-7

Dutta, P., Muthukrishnan, G., Gopalasubramaiam, S. K., Dharmaraj, R., Karuppaiah, A., Loganathan, K., Periyasamy, K., et al. (2022). Plant Growth-Promoting Rhizobacteria (PGPR) and its mechanisms against plant diseases for sustainable agriculture and better productivity. *Biocell, 46*(8), 1843–1859. https://doi.org/10.32604/biocell.2022.019291

Eskander, S. M. S. U., & Fankhauser, S. (2020). Reduction in greenhouse gas emissions from national climate legislation. *Nature Climate Change, 10*(8), 750–756. https://doi.org/10.1038/s41558-020-0831-z

Fasaeiyan, N., Jung, S., Boudreault, R., Arenson, L. U., & Maghoul, P. (2024). A review on mathematical modeling of microbial and plant induced permafrost carbon feedback. *Science of the Total Environment, 939*, 173144. https://doi.org/10.1016/j.scitotenv.2024.173144

Fierer, N., Schimel, J. P., & Holden, P. A. (2003). Influence of drying-rewetting frequency on soil bacterial community structure. *Microbial Ecology, 45*(1), 63–71. https://doi.org/10.1007/s00248-002-1007-2

Fontaine, L., Pin, L., Savio, D., Friberg, N., Kirschner, A. K. T., Farnleitner, A. H., & Eiler, A. (2023). Bacterial bioindicators enable biological status classification along the continental Danube River. *Communications Biology, 6*(1), 862. https://doi.org/10.1038/s42003-023-05237-8

Glasl, B., Bourne, D. G., Frade, P. R., Thomas, T., Schaffelke, B., & Webster, N. S. (2019). Microbial indicators of environmental perturbations in coral reef ecosystems. *Microbiome, 7*(1), 94. https://doi.org/10.1186/s40168-019-0705-7

González-Plaza, J. J., Furlan, C., Rijavec, T., Lapanje, A., Barros, R., Tamayo-Ramos, J. A., & Suarez-Diez, M. (2022). Advances in experimental and computational methodologies for the study of microbial-surface interactions at different omics levels. *Frontiers in Microbiology, 13*, 1006946. https://doi.org/10.3389/fmicb.2022.1006946

Guan, Y., Jin, S., Zhang, Z., Chen, S., Chen, Y., Liu, S., Fan, M., Ding, Y., & Yuan, X. (2025). A comprehensive review of climate warming and carbon dynamics in wetland ecosystems. *Discover Sustainability, 6*(1), 672. https://doi.org/10.1007/s43621-025-01579-x

Han, X. Y. (2021). Effects of climate changes and road exposure on the rapidly rising legionellosis incidence rates in the United States. Pausata, F. S. R. (Ed.). *PLoS ONE, 16*(4), e0250364. https://doi.org/10.1371/journal.pone.0250364

He, G., Deng, D., Delgado-Baquerizo, M., Liu, W., & Zhang, Q. (2025). Global relative importance of denitrification and anammox in microbial nitrogen loss across terrestrial and aquatic ecosystems. *Advancement of Science, 12*(8). https://doi.org/10.1002/advs.202406857

Henson, S. A., Cael, B. B., Allen, S. R., & Dutkiewicz, S. (2021). Future phytoplankton diversity in a changing climate. *Nature Communications, 12*(1), 5372. https://doi.org/10.1038/s41467-021-25699-w

Ibáñez, A., Garrido-Chamorro, S., & Barreiro, C. (2023). Microorganisms and climate change: A not so invisible effect. *Microbiological Research, 14*(3), 918–947. https://doi.org/10.3390/microbiolres14030064

Jiang, L., Pettitt, T. R., Buenfeld, N., & Smith, S. R. (2022). A critical review of the physiological, ecological, physical and chemical factors influencing the microbial degradation of concrete by fungi. *Building and Environment, 214*, 108925. https://doi.org/10.1016/j.buildenv.2022.108925

Lennon, J. T., & Locey, K. J. (2016). The underestimation of global microbial diversity. *MBio, 7*(5). https://doi.org/10.1128/mBio.01298-16

Li, Y., Ma, G., Xi, Y., Wang, S., Zeng, X., & Jia, Y. (2024). Divergent adaptation strategies of abundant and rare bacteria to salinity stress and metal stress in polluted Jinzhou Bay. *Environmental Research, 245*, 118030. https://doi.org/10.1016/j.envres.2023.118030

Liang, D., & Robertson, G. P. (2021). Nitrification is a minor source of nitrous oxide (N_2O) in an agricultural landscape and declines with increasing management intensity. *Global Change Biology, 27*(21), 5599–5613. https://doi.org/10.1111/gcb.15833

Liu, J., Cao, J., Rina, S., Yan, L., Wang, K., Haiyang, H., & Bao, Z. (2025). Variations in the N_2 fixation and CH_4 oxidation activities of type I methanotrophs in the rice roots in saline-alkali paddy field under nitrogen fertilization. *Rice, 18*(1), 17. https://doi.org/10.1186/s12284-025-00766-8

Malchair, S., De Boeck, H. J., Lemmens, C. M. H. M., Merckx, R., Nijs, I., Ceulemans, R., & Carnol, M. (2010). Do climate warming and plant species richness affect potential nitrification, basal respiration and ammonia-oxidizing bacteria in experimental grasslands? *Soil Biology and Biochemistry, 42*(11), 1944–1951. https://doi.org/10.1016/j.soilbio.2010.07.006

Mangan, N., & Brenner, M. (2014). Systems analysis of the CO_2 concentrating mechanism in cyanobacteria. *eLife, 3*, e02043. https://doi.org/10.7554/eLife.02043

Matzarakis, A., Mayer, H., & Iziomon, M. G. (1999). Applications of a Universal Thermal Index: Physiological equivalent temperature. *International Journal of Biometeorology, 43*(2), 76–84. https://doi.org/10.1007/s004840050119

McDonough, L. K., Santos, I. R., Andersen, M. S., O'Carroll, D. M., Rutlidge, H., Meredith, K., Oudone, P., et al. (2020). Changes in global groundwater organic carbon driven by climate change and urbanization. *Nature Communications, 11*(1), 1279. https://doi.org/10.1038/s41467-020-14946-1

Microbes and climate change—Science, people & impacts. (2022). https://doi.org/10.1128/AAMCol.Nov.2021

Naga, N. G., Taha, R. M., Hamed, E. A., Nawar, E. A., Jaheen, H. O., Mobarak, A. A., Radwan, Y. M., Faramawy, A. G., & Arayes, M. A. (2025). The silent microbial shift: Climate change amplifies pathogen evolution, microbiome dysbiosis, and antimicrobial resistance. *Tropical*

Diseases, Travel Medicine and Vaccines, 11(1), 43. https://doi.org/10.1186/s40794-025-00275-y
National Academy of Sciences: Carbon Dioxide and Climate (1979). (2017). Making climate change history (pp. 175–179). University of Washington Press. https://doi.org/10.1515/9780295741406-035
Nguyen, N., & Casadevall, A. (2021). Our health, our action, our planet-a call to action for microbiologists to engage in climate research. *MBio, 12*(5), e0250221. https://doi.org/10.1128/mBio.02502-21
Nguyen, J., Lara-Gutiérrez, J., & Stocker, R. (2021). Environmental fluctuations and their effects on microbial communities, populations and individuals. *FEMS Microbiology Reviews, 45*(4). https://doi.org/10.1093/femsre/fuaa068
Oishy, M. N., Shemonty, N. A., Fatema, S. I., Mahbub, S., Mim, E. L., Raisa, M. B. H., & Anik, A. H. (2025). Unravelling the effects of climate change on the soil-plant-atmosphere interactions: A critical review. *Soil & Environmental Health, 3*(1), 100130. https://doi.org/10.1016/j.seh.2025.100130
Pajares, S., & Bohannan, B. J. M. (2016). Ecology of nitrogen fixing, nitrifying, and denitrifying microorganisms in tropical forest soils. *Frontiers in Microbiology, 7*. https://doi.org/10.3389/fmicb.2016.01045
Paranaíba, J. R., Struik, Q., Erdociain, M., van Dijk, G., Smolders, A. J. P., van der Knaap, J., Veraart, A. J., & Kosten, S. (2023). CO_2, CH_4, and N_2O emissions from dredged material exposed to drying and zeolite addition under field and laboratory conditions. *Environmental Pollution, 337*, 122627. https://doi.org/10.1016/j.envpol.2023.122627
Parmar, T. K., Rawtani, D., & Agrawal, Y. K. (2016). Bioindicators: The natural indicator of environmental pollution. *Frontiers in Life Science, 9*(2), 110–118. https://doi.org/10.1080/21553769.2016.1162753
Pérez-Cobas, A. E., Gomez-Valero, L., & Buchrieser, C. (2020). Metagenomic approaches in microbial ecology: An update on whole-genome and marker gene sequencing analyses. *Microbial Genomics, 6*(8). https://doi.org/10.1099/mgen.0.000409
Reinsch, S., Robinson, D. A., van Soest, M. A. J., Keith, A. M., Parry, S., & Tye, A. M. (2024). Temperate soils exposed to drought—Key processes, impacts, indicators, and unknowns. *Land, 13*(11), 1759. https://doi.org/10.3390/land13111759
Rio, P., Caldarelli, M., Gasbarrini, A., Gambassi, G., & Cianci, R. (2024). The impact of climate change on immunity and gut microbiota in the development of disease. *Diseases (Basel, Switzerland), 12*(6). https://doi.org/10.3390/diseases12060118
Rossi, F., Olguín, E. J., Diels, L., & De Philippis, R. (2015). Microbial fixation of CO_2 in water bodies and in drylands to combat climate change, soil loss and desertification. *New Biotechnology, 32*(1), 109–120. https://doi.org/10.1016/j.nbt.2013.12.002
Seymour, J. R., & McLellan, S. L. (2025). Climate change will amplify the impacts of harmful microorganisms in aquatic ecosystems. *Nature Microbiology, 10*(3), 615–626. https://doi.org/10.1038/s41564-025-01948-2
Shade, A., Peter, H., Allison, S. D., Baho, D. L., Berga, M., Bürgmann, H., Huber, D. H., et al. (2012). Fundamentals of microbial community resistance and resilience. *Frontiers in Microbiology, 3*. https://doi.org/10.3389/fmicb.2012.00417
Shen, J., Zhang, P., Chang, Y., Zhang, L., Hao, Y., Tang, S., & Xiong, X. (2021). The environmental performance of greenhouse versus open-field cherry production systems in China. *Sustainable Production and Consumption, 28*, 736–748. https://doi.org/10.1016/j.spc.2021.07.002
Shivanna, K. R. (2022). Climate change and its impact on biodiversity and human welfare. *Proceedings of the Indian National Science Academy, 88*(2), 160–171. https://doi.org/10.1007/s43538-022-00073-6
Sillmann, J., Thorarinsdottir, T., Keenlyside, N., Schaller, N., Alexander, L. V., Hegerl, G., Seneviratne, S. I., et al. (2017). Understanding, modeling and predicting weather and climate

extremes: Challenges and opportunities. *Weather and Climate Extremes, 18*, 65–74. https://doi.org/10.1016/j.wace.2017.10.003

Singh, B. K., Bardgett, R. D., Smith, P., & Reay, D. S. (2010). Microorganisms and climate change: Terrestrial feedbacks and mitigation options. *Nature Reviews. Microbiology, 8*(11), 779–790. https://doi.org/10.1038/nrmicro2439

Smith, D. J., Ravichandar, J. D., Jain, S., Griffin, D. W., Hongbin, Y., Tan, Q., Thissen, J., et al. (2018). Airborne bacteria in Earth's lower stratosphere resemble taxa detected in the troposphere: Results from a new NASA aircraft bioaerosol collector (ABC). *Frontiers in Microbiology, 9*, 1752. https://doi.org/10.3389/fmicb.2018.01752

Speth, D. R., & Orphan, V. J. (2018). Metabolic marker gene mining provides insight in global McrA diversity and, coupled with targeted genome reconstruction, sheds further light on metabolic potential of the methanomassiliicoccales. *PeerJ, 6*, e5614. https://doi.org/10.7717/peerj.5614

Subramanian, A., Nagarajan, A. M., Vinod, S., Chakraborty, S., Sivagami, K., Theodore, T., Sathyanarayanan, S. S., Tamizhdurai, P., & Mangesh, V. L. (2023). Long-term impacts of climate change on coastal and transitional eco-systems in India: An overview of its current status, future projections, solutions, and policies. *RSC Advances, 13*(18), 12204–12228. https://doi.org/10.1039/D2RA07448F

Tiedje, J. M., Bruns, M. A., Casadevall, A., Criddle, C. S., Eloe-Fadrosh, E., Karl, D. M., Nguyen, N. K., & Zhou, J. (2022). Microbes and climate change: A research prospectus for the future. *MBio, 13*(3), e0080022. https://doi.org/10.1128/mbio.00800-22

Tripathi, A. K., Thakur, P., Saxena, P., Rauniyar, S., Gopalakrishnan, V., Singh, R. N., Gadhamshetty, V., et al. (2021). Gene sets and mechanisms of sulfate-reducing bacteria biofilm formation and quorum sensing with impact on corrosion. *Frontiers in Microbiology, 12*. https://doi.org/10.3389/fmicb.2021.754140

Tsyganova, A. V., Brewin, N. J., & Tsyganov, V. E. (2021). Structure and development of the legume-rhizobial symbiotic interface in infection threads. *Cells, 10*(5), 1050. https://doi.org/10.3390/cells10051050

Uzel, B., Stanton, T. D., & Scott, N. A. (2023). Impact of a warmer cold season on soil nitrogen and carbon cycling in high Arctic wet Sedge Tundra. *Geoderma, 437*, 116570. https://doi.org/10.1016/j.geoderma.2023.116570

Vangay, P., Johnson, A. J., Ward, T. L., Al-Ghalith, G. A., Shields-Cutler, R. R., Hillmann, B. M., Lucas, S. K., et al. (2018). US immigration westernizes the human gut microbiome. *Cell, 175*(4), 962–972.e10. https://doi.org/10.1016/j.cell.2018.10.029

Vermeulen, S. J., Campbell, B. M., & Ingram, J. S. I. (2012). Climate change and food systems. *Annual Review of Environment and Resources, 37*(1), 195–222. https://doi.org/10.1146/annurev-environ-020411-130608

Wang, F., Harindintwali, J. D., Wei, K., Shan, Y., Mi, Z., Costello, M. J., Grunwald, S., et al. (2023). Climate change: Strategies for mitigation and adaptation. *The Innovation Geoscience, 1*(1), 100015. https://doi.org/10.59717/j.xinn-geo.2023.100015

Wang, J., Qiu, M., Shen, Z., Chen, L., & Ge, Y. (2025). Microbial modulation of soil PH: A self-benefiting mechanism exemplified by bacillus. *Soil Biology and Biochemistry, 210*, 109949. https://doi.org/10.1016/j.soilbio.2025.109949

Weston, N. B., Troy, C., Kearns, P. J., Bowen, J. L., Porubsky, W., Hyacinthe, C., Meile, C., Van Cappellen, P., & Joye, S. B. (2024). Physicochemical perturbation increases nitrous oxide production from denitrification in soils and sediments. *Biogeosciences, 21*(21), 4837–4851. https://doi.org/10.5194/bg-21-4837-2024

Xavier, J. B. (2016). Sociomicrobiology and pathogenic bacteria. Kudva, I. T. & Plummer, P. J. (Eds.). *Microbiology Spectrum, 4*(3). https://doi.org/10.1128/microbiolspec.VMBF-0019-2015

Xu, P., Gao, M., Li, Y., Ye, J., Jianqiang, S., & Li, H. (2025). Combined effects of acidification and warming on soil denitrification and microbial community. *Frontiers in Microbiology, 16*. https://doi.org/10.3389/fmicb.2025.1572497

Yi, X., Guo, H., Deng, J., Kang, F., Gong, H., Qiang, J., & Wang, Y. (2025). Impact of indigenous microbial metabolism on coal pore structure and CO_2 adsorption in coal mines. *Journal of Environmental Chemical Engineering, 13*(4), 117373. https://doi.org/10.1016/j.jece.2025.117373

Yu, H. (2025). Climate change unveils hidden microbial dangers. *Environmental Science and Ecotechnology, 24*, 100544. https://doi.org/10.1016/j.ese.2025.100544

Zhang, C., Dang, H., Azam, F., Benner, R., Legendre, L., Passow, U., Polimene, L., et al. (2018). Evolving paradigms in biological carbon cycling in the ocean. *National Science Review, 5*(4), 481–499. https://doi.org/10.1093/nsr/nwy074

Zhao, Y., Liu, Y., Cao, S., Hao, Q., Liu, C., & Li, Y. (2024). Anaerobic oxidation of methane driven by different electron acceptors: A review. *Sci Total Environ, 946*, 174287. https://doi.org/10.1016/j.scitotenv.2024.174287

Zhu, X., Burger, M., Doane, T. A., & Horwath, W. R. (2013). Ammonia oxidation pathways and nitrifier denitrification are significant sources of N_2O and NO under low oxygen availability. *Proceedings of the National Academy of Sciences, 110*(16), 6328–6333. https://doi.org/10.1073/pnas.1219993110

Microbial Contributions to Greenhouse Gas Emissions

Prabha Pant, Himani Karki, and Hem Ch. Joshi

Abstract

Microorganisms play a crucial role in the global carbon and nitrogen cycles, exerting a significant impact on greenhouse gas (GHG) emissions. These microorganisms, mainly bacteria and archaea, mediate key biogeochemical processes such as carbon fixation, methane production (methanogenesis), nitrification and denitrification, which contribute directly to the atmospheric concentration of GHGs. In terrestrial and aquatic environments, microbes can both mitigate and aggravate GHG emissions, depending on environmental conditions, substrate availability, and microbial community composition. By exploring their ecological roles in various environments present chapter highlights the importance of microbial activities in regulating atmospheric GHG levels.

Keywords

Microbes · Biogeochemical cycle · GHGs · Environment · Ecology

1 Introduction

Climate change is transforming the planet and poses a threat to humanity. Burning of fossil fuels, deforestation, and fast population growth have all contributed to climate change by increasing the amount of greenhouse gases (GHGs) in the atmosphere (USGRCP, 2017). Global temperatures have increased, and the climate has changed in every part of the world due to heat retention by greenhouse gases, which may have detrimental effects on agricultural production, human health, and the international

P. Pant (✉) · H. Karki · H. Ch. Joshi
Department of Botany, D.S.B. Campus, Kumaun University, Nainital, Uttarakhand, India

G. Bhandari et al. (eds.), *Microbial Resilience*,
https://doi.org/10.1007/978-981-95-8883-1_2

economy (IPCC, 2022). Microorganisms play a significant role in the natural biogeochemical cycles that produce and absorb greenhouse gases (GHGs), even though human activities are largely responsible for the increase in GHG emissions.

Greenhouse gases (GHGs) generally refer to gases that can absorb the Earth's thermal radiation and enhance the greenhouse effect. The greenhouse effect is one of the crucial environmental problems humans have so far faced in the twenty-first century (Wang et al., 2022). Carbon dioxide (CO_2), methane (CH_4), and nitrous oxide (N_2O) are the main GHGs, each with distinctive properties and sources. Other GHGs include hydrofluorocarbons (HFCs), perfluorocarbons (PFCs), sulfur hexafluoride (SF6), and nitrogen trifluoride (NF3). Human activities like fossil fuel combustion and deforestation are the main factors of GHG emissions. These anthropogenic activities have increased GHG concentrations, which, in turn, have increased the Earth's temperature and altered the climate globally (IPCC, 2021). The role of microorganisms in generating these gases is equally significant and often overlooked. This chapter discusses exclusively how microbial processes contribute to the emission of GHGs, highlighting their essential role in climate change.

Microorganisms are the most abundant and diverse organisms on Earth. These include fungi, bacteria, viruses, algae, archaea, and protozoa and are found in all areas of the planet, including terrestrial, urban, atmospheric, subsurface, and aquatic ecosystems. Microorganisms have a critical role within soil ecosystems, actively involved in the formation and development of soils (Margesin et al., 2009). These microbes also play an important role in the cycling of various elements on Earth, especially in the cycling of carbon and nitrogen (Bardgett et al., 2008)

The contribution of microorganisms to the Earth's climate is significant because of their sheer numbers. Microbes involved in various metabolic processes directly or indirectly influence the atmospheric concentrations of GHGs. Microbial communities in agricultural soils significantly contribute to greenhouse gas emissions, releasing approximately 1.5–3 billion tons of carbon annually. Key processes include microbial methanogenesis, nitrification, denitrification, and respiration, which influence the various biogeochemical cycles and overall greenhouse gas dynamics (Shakoor et al., 2020).

The diversity, abundance, and activity of soil microbial communities are dependent on climate change, leading to an interdependent relationship (Guo et al., 2020; Yuan et al., 2021; Tiedje et al., 2022). There are many factors, such as temperature, moisture, pH, and organic matter content, as well as other physical and chemical properties, affecting microbial activities, which change the rate of GHG emissions (Wang et al., 2022).

2 Major Microbial Pathways Associated with GHG Emissions

Microorganisms play a crucial role in the cycling of different elements and are essential parts of the Earth's biogeochemical cycles. These cycles are closely related to the regulation of greenhouse gases, which are significant contributors to climate change. While microbes are essential for the functioning of these cycles, their activities can both mitigate and exacerbate climate change, making them key drivers in the climate system (Singh et al., 2019).

Each greenhouse gas has distinct origins and characteristics. Greenhouse gas (GHG) emissions, especially those of methane (CH_4), carbon dioxide (CO_2), and nitrous oxide (N_2O), are largely caused by microbial metabolic pathways (Insam & Wett, 2008). These processes are highly dependent on environmental conditions, such as the temperature, moisture, pH, availability of oxygen, the type of organic matter, and the presence of other electron acceptors. The following are the main microbial pathways associated with microbial greenhouse gas emissions.

2.1 Nitrification and Denitrification (Nitrous Oxide Emissions)

Agricultural soils, heavily fertilized with nitrogen-based fertilizers, are hotspots for emission of highly potent greenhouse gas nitrous oxide (N_2O) (Zhu et al., 2024). In terrestrial ecosystem N_2O is produced from both natural and anthropogenic sources. The agricultural sector is thought to be responsible for about 70% of the annual increase in N_2O emissions, including direct and indirect, at a rate of 50 ppb (Hofstra & Boweman, 2005; Huang et al., 2007; Ramulu et al., 2008). It is the most important ozone-depleting substance (Ravishankara et al., 2009), with a global warming potential 273 times greater than that of CO_2 on a mass equivalent basis (Lee et al., 2023).

Microbial processes such as nitrification and denitrification are the main processes producing N_2O (Ussiri & Lal, 2013). Ammonia-oxidizing archaea (AOA) and ammonia-oxidizing bacteria (AOB) play a pivotal role in N_2O emission in many soil environments. The availability of nitrogen substrates, oxygen levels, and soil moisture content are key factors influencing the rates of these microbial processes.

Nitrification occurs when soil ammonium is available, and environmental conditions such as temperature and moisture are favorable for the microbes converting ammonia to nitrite and then to nitrate, a process vital for nitrogen cycling in ecosystems. These situations generally occur in many agronomic zones (Subbarao et al., 2009). In oxygen-deficient conditions, microorganisms use nitrate as the electron acceptor for the anaerobic oxidation process known as denitrification. The process of denitrification is common in wet soils where diffusion of oxygen is inhibited by saturated conditions inside the soil aggregate (Davidson, 1991).

Nitrification

Ammonia-oxidizing bacteria (AOB) and archaea (AOA) convert ammonium (NH^{4+}) into nitrite (NO_2^-) and then nitrate (NO_3^-), releasing nitrous oxide as a byproduct and an important step in the global nitrogen cycle. Some of the microbes involved in this process are *Nitrosomonas*, *Nitrosospira*, *Nitrosococcus*, *Nitrospira*, *Nitrobacter*, *Nitrococcus*, and *Nitrospina*.

Denitrification

Facultative anaerobic bacteria reduce nitrate to nitrogen gas (N_2) via intermediate forms, including nitrous oxide. Some denitrifying bacteria include species of *Bacillus*, *Paracoccus*, and *Pseudomonas* carried out the process of denitrification. There is recent evidence that some eukaryotes are also capable of denitrification (Risgaard-Petersen et al., 2006).

2.2 Methanogenesis (Methane Emission)

Methane is considered the second most important greenhouse gas in the atmosphere, and its concentration is directly correlated with anthropogenic activity (Whalen, 2005; Karakurt et al., 2012). Wetlands are a potent natural source of CH_4, accounting for 62% of the total methane emission into the atmosphere. Rice fields, ruminants, landfills, biomass burning, and fossil fuel combustion are anthropogenic sources for methane emission (Nazaries et al., 2013). Due to its higher efficiency in trapping radiation, CH_4 is 25 times more potent than CO_2 as a GHG over a 100-year horizon and on a mass basis, even though a short lifetime in the atmosphere of approximately 8 years (Shindell et al., 2009).

Methanogenesis is an ancient microbial metabolism that most likely played a significant role in the evolution of life on Earth (Ueno et al., 2006; Kasting & Catling, 2003; Sauterey et al., 2020). It is an anaerobic process carried out by methanogenic archaea, a unique group of microorganisms that thrive in anaerobic conditions (Houghton et al., 2001). These microbes utilize substrates such as carbon dioxide, acetate, and hydrogen to produce methane. Furthermore, various bacterial species in aerobic conditions participate in methane production by using methylated compounds such as phosphonates (Repeta et al., 2016; Wang et al., 2017), methylated-sulfur compounds (Damm & Helmke, 2009), and methylamines (Bižić-Ionescu et al., 2018). Freshwater and marine Cyanobacteria, Marine algae, could produce methane as a byproduct of photosynthesis (Lenhart et al., 2016; Klintzsch et al., 2019; Bižić et al., 2020).

2.2.1 Major Biochemical Pathways Involved in Methanogenesis

Acetoclastic Methanogenesis

Acetoclastic methanogenesis is considered a major pathway by which methane is produced in an anaerobic codigestion (ACoD) of organic substrates, such as fats, oils, and grease. Presently, only two methanogenic genera (*Methanosaeta* and *Methanosarcina*) are identified to perform acetoclastic methanogenesis (Kurade et al., 2019).

Methanosaeta is an obligatory acetate consumer that mainly performs at low acetate concentration, whereas *Methanosarcina* is the most metabolically and physiologically agile methanogen that can convert different substrates, such as acetate, hydrogen, and methyl-containing groups to methane (Conklin et al., 2006; Blume et al., 2010; De Vrieze et al., 2012).

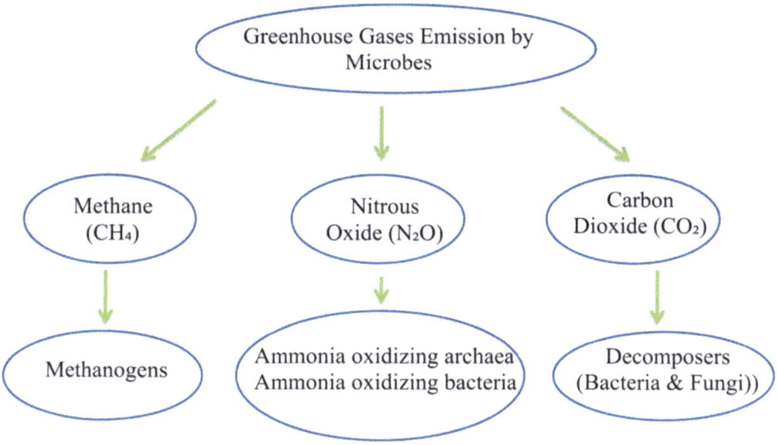

Fig. 1 Microbes involved in greenhouse gas emission

Hydrogenotrophic Methanogenesis

Hydrogenotrophic methanogenesis is another process in the anaerobic environment where methane is produced. According to Thauer et al. (2008), it is the most extensive pathway and has been suggested to represent the ancestral form of methane production (Bapteste et al., 2005). The metabolism of hydrogenotrophic methanogens involves Wolfe-cycle where hydrogen is used to reduce carbon dioxide (CO_2) to methane (CH_4) (Ray et al., 2023). Members of Methanobacteriales, Methanococcales, and Methanomicrobiales are mainly involved in this process.

Methylotrophic Methanogenesis

In this pathway, methane is produced by Archaea using methylated compounds like methanol and methylamines as substrates rather than hydrogen and carbon dioxide. Methylotrophic methanogenesis is mainly performed by members of the Methanosarcinales, Methanobacteriales, and Methanomassiliicoccales (Kurth et al., 2020; Vanwonterghem et al., 2016) (Fig. 1).

2.3 CO_2 Emission

Carbon dioxide is naturally present in the atmosphere as part of the Earth's carbon cycle, but human activities, mainly the burning of fossil fuels, significantly increase its concentration, leading to global warming. According to the IPCC (2021), CO_2 is a notable greenhouse gas, and an increase in its concentration causes the greenhouse effect, which is the predominant factor influencing climate change. CO_2 contributes approximately 70% of the challenges associated with climate change (Allan et al., 2021).

Soil constitutes the largest carbon reservoir in the terrestrial ecosystem, containing three times more carbon than vegetation and two times more than the metabolism typically experiences acceleration, resulting in the enhancement of microbial processes such as methanogenesis and denitrification. These metabolic activities are involved in the emission of methane and nitrous oxide, respectively.

Various studies have shown that temperature is the main factor affecting the production and emissions of GHG in soil. Increased soil temperature can enhance the root respiration of crops, accelerate the breakdown of organic matter in the soil, increase soil microbial activity, and ultimately, accelerate the diffusion of CO_2 in soil (Carey et al., 2016; Voigt et al., 2017). Zhang et al. indicated that a 2 °C rise in temperature was correlated with a 10–20% increase in methane emissions from rice paddies. In another study, Chen et al., 2020 documented that methanogens also have thermal adaptability, and microbial activity decreases after prolonged warming and increases after prolonged cooling. Davidson et al. (2000) proposed that temperature elevations in tropical soils facilitated increased N_2O emissions through intensified rates of nitrification and denitrification.

Moisture

Soil moisture levels have vital importance, as they significantly influence the availability of water essential for various microbial processes that affect GHG emissions in soils. Soil moisture is an important factor that affects N_2O and CH_4 emissions in soils. The water-filled pore space (WFPS), which offers extensive information on the water content and overall porosity of the soil system, is a common indicator of soil moisture (Adviento-Borbe & Linquist, 2016). Anaerobic conditions, characterized by a deficiency of oxygen, arising from high moisture content, facilitate the growth of methanogenic microorganisms and denitrifying bacteria, thus enhancing the production of methane and nitrous oxide (Hu et al., 2014, 2015). Soil carbon dioxide emissions exhibit a Birch effect in relation to soil moisture. Optimal moisture content serves to stimulate microbial activity and consequently elevates CO_2 emissions, while excessively high moisture levels may impede soil respiration (Li et al., 2020).

Soil moisture content is one of the predominant factors regulating N_2O emission from soils. However, it has been observed that changes in the soil water content due to wetting events like rainfall and irrigation can stimulate the generation of N_2O by the process of nitrification and denitrification (Kollah et al., 2018). Water-filled pore space (WFPS) and N_2O emission N_2O emission is highly correlated. The largest N_2O emissions in a calcareous fluvo-aquic soil under 70% WFPS were caused by both nitrification (35–53%) and denitrification (44–58%) (Huang et al., 2014). The favorable conditions for N_2O production from nitrification occur within the range of 30–70% WFPS (Hu et al., 2015), whereas denitrification dominates N_2O production in wet soils with >80–90% WFPS (Huang et al., 2014).

Nutrient Availability

The availability of essential nutrients, including carbon, nitrogen, phosphorus, and trace elements, is pivotal for microbial growth. The scarcity of specific nutrients may

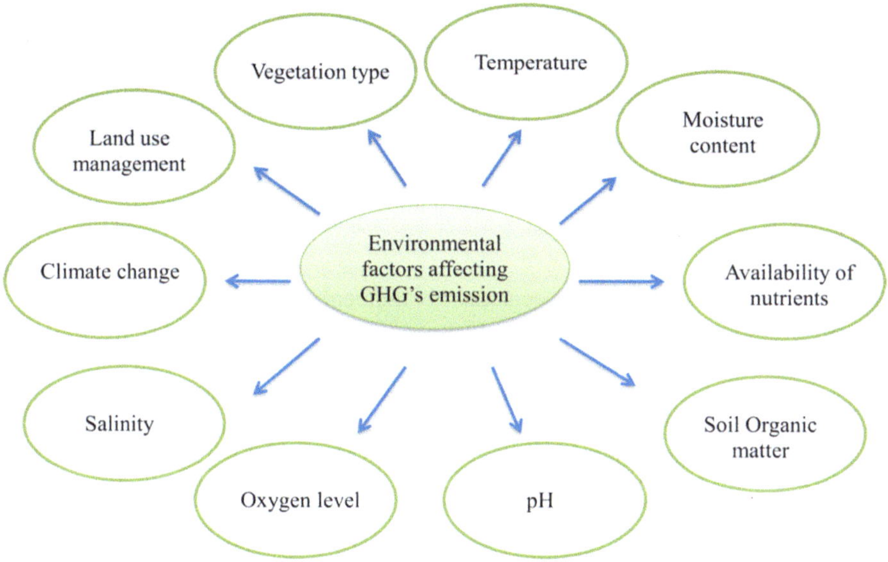

Fig. 2 Environmental factors affecting the rate of microbial GHG emissions

limit microbial communities, thereby affecting their growth rates and other physiological processes. Different microbes have distinct nutritional requirements, influencing their ability to flourish in diverse ecological niches (Porollo, 2023).

The availability of nutrients promotes microbial growth and activity, leading to higher GHG emissions. Use of nitrogen-containing fertilizers in soil triggers higher N_2O emissions by fueling microbial processes. One of the main factors influencing microbial activity and greenhouse gas production is the availability of carbon, frequently through organic materials. Excess carbon can increase soil organic matter decomposition, potentially leading to a net emission of carbon dioxide (CO_2).

Oxygen Level

Availability of oxygen significantly influences the type of microbes present and the types of GHGs produced. Methanogens thrive well in anaerobic (low-oxygen) conditions, thus elevating more methane production. On the other hand, aerobic (adequate oxygen) conditions promote the growth of microorganisms that mainly perform aerobic respiration and release more CO_2 (Fig. 2).

pH

Soil pH is one of the major factors that affect microbial activities and the rates of various biogeochemical processes, thus showing an influence on greenhouse gas emissions. The influence of soil pH on CO_2, CH_4, and N_2O emissions is complex. Different microbes thrive well at different pH levels, and the ideal pH range for most microorganisms in soil is 6–8. An environment that is too acidic or too alkaline will have an impact on microbial activity and change the rate of greenhouse gas emissions.

According to Awais et al. (2021), the ideal pH for the growth and reproduction of methanogens is about 7, and an acidic environment will decrease soil CH_4 emissions. An increase in pH level enhances the process of methanogenesis, leading to increased methane emission (Marquart et al., 2019).

Vegetation Type

Vegetation type significantly affects microbial greenhouse gas (GHG) emissions by influencing soil organic matter, moisture, and microbial community structure and functions (Hsieh et al., 2021). Soil GHG emissions can be impacted by changes in the quantity, quality, and components of organic matter inputs as well as net primary production due to variations in species composition and community structure, which also affect root biomass and organizational components (IPCC, 2021). The production of greenhouse gases is directly impacted by the different litter and root exudates produced by plants.

Soil Organic Matter

The quantity and type of organic matter within the soil constitute a significant factor for GHG emissions. Soil organic matter (SOM) is usually defined as a category of polymeric organic compounds characterized by complex constituents and enduring properties that are generated from organic residues in the soil through microbial activities or various physical and chemical processes (Sylvie & Katell, 2015). SOM serves as a principal carbon reservoir for soil respiration and plays a substantial role in influencing soil greenhouse gas emissions. The organic carbon activated by soil serves as a substrate for microbial growth, with its concentration directly affecting the microbial activity, which subsequently influences greenhouse gas emissions. The concentration of soluble organic matter within SOM is intricately linked to the production of CO_2 in the soil (Li et al., 2015; Chen et al., 2018).

Microbial carbon use efficiency was found to be directly correlated with soil organic carbon content, which in turn affected soil CO_2 and CH_4 emissions (Tao et al., 2023; Khan et al., 2024a, b). The emission of N_2O, associated with nitrification and denitrification processes, is also regulated by the quantity and quality of soil organic matter, as it provides electrons for these processes (Marín-Muñiz et al., 2015; Wang et al., 2019; Khan et al., 2024a, b).

Salinity

Soil salinity, defined as a high concentration of soluble salts in soil, is one of the major worldwide issues that severely impacts agricultural production and environmental sustainability. The availability of carbon (C) and nitrogen (N) in soil, as well as different microbial processes and activities, can be impacted by soil salinity (Poffenbarger et al., 2011; Shahariar et al., 2021). These changes can result in changes to soil organic carbon (SOC) and greenhouse gas emissions (Shahariar et al., 2021; She et al., 2021). According to earlier studies, soil salinity primarily reduces CO_2 and CH_4 emissions (She et al., 2021; Shahariar et al., 2021). However, the response of N_2O emissions to soil salinity varies, showing either a linear increase, a decrease, or no response (Thapa et al., 2017).

Land Use and Management

Land use practices, such as deforestation, intensive agriculture, and urbanization, can alter soil properties (like moisture, temperature, and nutrient availability), which, in turn, affect the diversity of soil microbes and processes (Bandyopadhyay & Lal, 2014). Changes in soil conditions affect the microbial communities and their activities, leading to changes in GHG emissions.

The decomposition of organic matter (OM) in the soil due to microbes and root respiration has been linked to soil CO_2 emissions. Increased CO_2 emissions may result from changes in soil properties brought about by changes in land use. Methanogenic bacteria produce CH_4 emissions from soil under anaerobic conditions; however, changes in land use, such as the conversion of forests to grassland, can increase CH_4 emissions because the concentration of methanogenic microbes in the soil is higher. However, microorganisms produce N_2O in the soil during the nitrification and denitrification processes, and nitrogen fertilizer in agricultural regions can enhance N_2O emissions, particularly when it is linked to soil moisture and the availability of organic carbon.

Climate Change

One of the most important global issues of the twenty-first century is climate change, which has a profound impact on ecosystems, biodiversity, and human civilization (Bardgett & Van Der Putten, 2014). Climate change, through several interrelated factors such as temperature, precipitation, and soil properties, influences microbial communities and biogeochemical processes, affecting GHG emissions (Singhal et al., 2021).

The impact of climate change on the biomass, diversity, and composition of microbes causes them to become extinct or change, which can have either positive or negative effects on their physiology and greenhouse gas emissions. Microbial community structures change when the temperature rises, and processes like respiration, fermentation, and methanogenesis also accelerate. Through the breakdown of organic matter, they accelerate the rate of global warming and ultimately raise the atmospheric flux of CO_2 (Tyagi et al., 2014; Weiman, 2015; Castro et al., 2010).

4 Conclusion and Outlook

Greenhouse gases are gases in Earth's atmosphere, such as Carbon dioxide (CO_2), methane (CH_4), nitrous oxide (N_2O), hydrofluorocarbons (HFCs), perfluorocarbons (PFCs), sulfur hexafluoride (SF_6), nitrogen trifluoride (NF_3), and water vapor. The major greenhouse gases are carbon dioxide, nitrous oxide, and methane, each with distinctive properties and sources. These gases trap heat and enhance the greenhouse effect, which is one of the most important environmental issues in the twenty-first century. The increase in GHG emissions in the environment also leads to global warming, changes in weather patterns, and other environmental issues.

Microorganisms are an integral part of the biogeochemical cycles that regulate the Earth's climate system. Their role in greenhouse gas emissions is very important because they serve as a natural catalyst for GHG emissions. Microorganisms, by breaking down organic matter through nutrient cycling, release greenhouse gases and accelerate global climate change. However, their activities are highly sensitive to various environmental factors that alter the rate of GHG emissions.

A deeper understanding of microbial ecology and metabolic processes, simultaneously with technological advancements and modeling techniques, is essential for addressing the microbial aspects of climate change. Future research should focus on (1) fully understanding the complex microbial processes and their impact on GHG emissions and (2) quantifying the contributions of microorganisms and formulating biotechnological interventions for the reduction of greenhouse gas emissions.

References

Adviento-Borbe, M. A., & Linquist, B. (2016). Assessing fertilizer N placement on CH_4 and N_2O emissions in irrigated rice systems. *Geoderma, 266*, 40–45.

Allan, R. P., Hawkins, E., Bellouin, N., & Collins, B. (2021). Summary for policymakers. In Intergovernmental Panel on Climate Change (IPCC) (Ed.), *Climate Change 2021: The physical science basis*. Cambridge University Press. https://doi.org/10.1017/9781009157896

Awais, S., Saba, S., Abdul, R., Fatima, A., Muhammad, A., Sher, M. S., Taimoor, H. F., Muhammad, A., Muhammad, A. M., Muhammad, M. A., et al. (2021). Effect of animal manure, crop type, climate zone, and soil attributes on greenhouse gas emissions from agricultural soils-a global meta-analysis. *Journal of Cleaner Production, 278*, 124019.

Bandyopadhyay, K. K., & Lal, R. (2014). Effect of land use management on greenhouse gas emissions from water stable aggregates. *Geoderma, 234*, 363–372. https://doi.org/10.1016/j.geoderma.2014.05.025

Bapteste, E., Brochier, C., & Boucher, Y. (2005). Higher-level classification of the archaea: Evolution of methanogenesis and methanogens. *Archaea, 1*, 353–363.

Bardgett, R. D., & Van Der Putten, W. H. (2014). Belowground biodiversity and ecosystem functioning. *Nature, 515*(7528), 505–511.

Bardgett, R. D., Freeman, C., & Ostle, N. J. (2008). Microbial contributions to climate change through carbon cycle feedbacks. *The ISME Journal, 2*(8), 805–814. https://doi.org/10.1038/ismej.2008.58

Batjes, N. H., & Sombroek, W. G. (1997). Possibility for carbon sequestration in tropical and subtropical soils. *Global Change Biology, 3*(2), 161–173. https://doi.org/10.1111/j.1365-2486.1997.gcb136.x

Bižić, M., Klintzsch, T., Ionescu, D., Hindiyeh, M. Y., Günthel, M., Muro-Pastor, A. M., Eckert, W., Urich, T., Keppler, F., & Grossart, H. P. (2020). Aquatic and terrestrial cyanobacteria produce methane. *Science Advances, 6*(5), eaax5343. https://doi.org/10.1126/sciadv.aax5343

Bižić-Ionescu, M., Ionescu, D., Günthel, M., Tang, K. W., & Grossart, H.-P. (2018). Oxic methane cycling: New evidence for methane formation in oxic lake water. In *Biogenesis of hydrocarbons* (pp. 1–22). Springer. https://doi.org/10.1007/978-3-319-53114-4_1

Blume, F., Bergmann, I., Nettmann, E., Schelle, H., Rehde, G., Mundt, K., & Klocke, M. (2010). Methanogenic population dynamics during semi-continuous biogas fermentation and acidification by overloading. *Journal of Applied Microbiology, 109*(2), 441–450. https://doi.org/10.1111/j.1365-2672.2010.04660.x

Carey, J. C., Tang, J. W., Templer, P. H., Kroeger, K. D., Crowther, T. W., Burton, A. J., Dukes, J. S., Emmett, B., Frey, S. D., Heskel, M. A., et al. (2016). Temperature response of soil respiration largely unaltered with experimental warming. *Proceedings of the National Academy of Sciences, USA, 113*, 13797–13802.

Castro, H. F., Classen, A. T., Austin, E. E., Norby, R. J., & Schadt, C. W. (2010). Soil microbial community responses to multiple experimental climate change drivers. *Applied and Environmental Microbiology, 76*, 999–1007.

Chen, H., Yang, Z., Chu, R. K., Tolic, N., Liang, L., Graham, D. E., Wullschleger, S. D., & Gu, B. (2018). Molecular insights into Arctic soil organic matter degradation under warming. *Environmental Science & Technology, 52*(8), 4555–4564. https://doi.org/10.1021/acs.est.7b05424

Chen, H., Zhu, T., Li, B., Fang, C., & Nie, M. (2020). The thermal response of soil microbial methanogenesis decreases in magnitude with changing temperature. *Nature Communications, 11*, 5733.

Conklin, A., Stensel, H. D., & Ferguson, J. (2006). Growth kinetics and competition between Methanosarcina and Methanosaeta in mesophilic anaerobic digestion. *Water Environment Research, 78*(5), 486–496. https://doi.org/10.2175/106143006X95393

Damm, E., & Helmke, E. (2009). Methane production in aerobic oligotrophic surface water in the central Arctic Ocean. *Biogeosciences Discussions, 6*(6), 10355–10379. https://doi.org/10.5194/bgd-6-10355-2009

Davidson, E. A. (1991). Fluxes of N_2O and nitric oxide from terrestrial ecosystems. In J. E. Rogers & W. B. Whitman (Eds.), *Microbial production and consumption of greenhouse gases: CH_4, nitrogen oxides and HaloCH_4S* (pp. 219–235). American Society of Microbiology.

Davidson, E. A., Verchot, L. V., Cattânio, J. H., Ackerman, I. L., & Carvalho, J. E. M. (2000). Effects of soil water content on soil respiration in forests and cattle pastures of eastern Amazonia. *Biogeochemistry, 48*(1), 53–69. https://doi.org/10.1023/A:1006204113917

De Vrieze, J., Hennebel, T., Boon, N., & Verstraete, W. (2012). Methanosarcina: The rediscovered methanogen for heavy-duty biomethanation. *Bioresource Technology, 112*, 1–9. https://doi.org/10.1016/j.biortech.2012.02.079

Deng, C. X., Li, R. R., Xie, B. G., Wan, Y. L., Li, Z. W., & Liu, C. C. (2021). Impacts of the integrated pattern of water and land resources use on agricultural greenhouse gas emissions in China during 2006–2017: A water-land-energy-emissions nexus analysis. *Journal of Cleaner Production, 308*, 127221. https://doi.org/10.1016/j.jclepro.2021.127221

Guo, X., Gao, Q., Yuan, M., Wang, G., Zhou, X., Feng, J., Shi, Z., Hale, L., Wu, L., Zhou, A., Tian, R., Liu, F., Wu, B., Chen, L., Jung, C. G., Niu, S., Li, D., Xu, X., Jiang, L., & Zhou, J. (2020). Gene-informed decomposition model predicts lower soil carbon loss due to persistent microbial adaptation to warming. *Nature Communications, 11*, 4897. https://doi.org/10.1038/s41467-020-18706-4

Haj-Amor, Z., Araya, T., Kim, D.-G., Bouri, S., Lee, J., Ghiloufi, W., Yang, Y., Kang, H., Jhariya, M. K., Banerjee, A., et al. (2022). Soil salinity and its associated effects on soil microorganisms, greenhouse gas emissions, crop yield, biodiversity and desertification: A review. *Science of the Total Environment, 843*, 156946.

Hofstra, N., & Boweman, A. F. (2005). Denitrification in agricultural soils: Summarizing published data and estimating global annual rates. *Nutrient Cycling in Agroecosystems, 72*, 267–278.

Houghton, J., Ding, Y., Griggs, D., Noguer, M., van der Linden, P., Dai, X., Maskell, K., & Johnson, C. (2001). *Climate change 2001: The scientific basis*. IPCC. ISBN: 0309075742.

Hsieh, S. H., Yuan, C. S., Ie, I. R., Yang, L., Lin, H. J., & Hsueh, M. L. (2021). In-situ measurement of greenhouse gas emissions from a coastal estuarine wetland using a novel continuous

monitoring technology: Comparison of indigenous and exotic plant species. *Journal of Environmental Management, 281*, 111–905.
Hu, H., Xu, Z., & He, J. (2014). Ammonia-oxidizing archaea play a predominant role in acid soil nitrification. *Advances in Agronomy, 125*, 261–302. https://doi.org/10.1016/B978-0-12-800137-0.00006-5
Hu, H., Chen, D., & He, J. Z. (2015). Microbial regulation of terrestrial nitrous oxide formation: Understanding the biological pathways for prediction of emission rates. *FEMS Microbiology Reviews, 39*(5), 729–749. https://doi.org/10.1093/femsre/fuv021
Huang, S., Pant, H. K., & Lu, J. (2007). Effects of water regimes on N_2O emission from soils. *Ecological Engineering, 31*, 9–15.
Huang, T., Gao, B., Hu, X.-K., Lu, X., Well, R., Christie, P., Bakken, L. R., & Ju, X.-T. (2014). Ammonia-oxidation as an engine to generate nitrous oxide in an intensively managed calcareous Fluvo-aquic soil. *Scientific Reports, 4*.
Huang, N., Wang, L., Song, X.-P., Black, T. A., Jassal, R. S., Myneni, R. B., Wu, C., Wang, L., Song, W., & Ji, D. (2020). Spatial and temporal variations in global soil respiration and their relationships with climate and land cover. *Science Advances, 6*(39), eabb8508. https://doi.org/10.1126/sciadv.abb8508
Insam, H., & Wett, B. (2008). Control of GHG emission at the microbial community level. *Waste Management, 28*, 699–706.
Intergovernmental Panel on Climate Change (IPCC). (2021). Climate change 2021: The physical science basis. Contribution of Working Group I to the sixth assessment report of the Intergovernmental Panel on Climate Change. Cambridge: Cambridge University Press. https://doi.org/10.1017/9781009157896
IPCC. (2022). *Global warming of 1.5°C: IPCC special report on impacts of global warming of 1.5°C above pre-industrial levels in context of strengthening response to climate change, sustainable development, and efforts to eradicate poverty* (1st ed.). Cambridge University Press.
IPCC: Climate Change. (2021). The physical science basis. In V. P. Zhai, A. Pirani, S. L. Connors, C. Péan, S. Berger, N. Caud, Y. Chen, L. Goldfarb, M. I. Gomis, M. Huang, et al. (Eds.), *Contribution of Working Group I to the sixth assessment report of the Intergovernmental Panel on Climate Change*. Cambridge University Press.
Karakurt, I., Aydin, G., & Aydiner, K. (2012). Sources and mitigation of methane emissions by sectors: A critical review. *Renewable Energy, 39*(1), 40–48. https://doi.org/10.1016/j.renene.2011.09.006
Kasting, J. F., & Catling, D. (2003). Evolution of a habitable planet. *Annual Review of Astronomy and Astrophysics, 41*, 429–463. https://doi.org/10.1146/annurev.astro.41.071601.170049
Khan, J., Tariq, M., Alabbosh, K., Rehman, A., Jalal, A., Khan, A., Farooq, M., Li, G., Iqbal, B., Ahmad, N., et al. (2024a). Soil microplastics: Impacts on greenhouse gasses emissions, carbon cycling, microbial diversity, and soil characteristics. *Applied Soil Ecology, 197*, 105–343.
Khan, K., Tang, Y., Cheng, P., Song, Y., Li, X., Lou, J., Iqbal, B., Zhao, X., Hameed, R., Li, G., et al. (2024b). Effects of degradable and non-degradable microplastics and oxytetracycline co-exposure on soil N_2O and CO_2 emissions. *Applied Soil Ecology, 197*, 105–331.
Klintzsch, T., Langer, G., Nehrke, G., Wieland, A., Lenhart, K., & Keppler, F. (2019). Methane production by three widespread marine phytoplankton species: Release rates, precursor compounds, and relevance for the environment. *Biogeosciences Discussions*, 1–25. https://doi.org/10.5194/bg-2019-116
Kollah, B., Patra, A. K., Mohanty, S. R., Adhya, K., et al. (Eds.). (2018). Microbial cycling of greenhouse gases and their impact on climate change. In: *Advances in soil microbiology: Recent trends and future prospects, microorganisms for sustainability* (Vol. 3). https://doi.org/10.1007/978-981-10-6178-3_7
Kurade, M. B., Saha, S., Salamaa, E.-S., Patil, S. M., Govindwar, S. P., & Jeon, B.-H. (2019). Acetoclastic methanogenesis led by Methanosarcina in anaerobic co-digestion of fats, oil, and grease for enhanced production of methane. *Bioresource Technology, 272*, 351–359. https://doi.org/10.1016/j.biortech.2018.10.051

Kurth, J. M., Op den Camp, H. J. M., & Welte, C. U. (2020). Several ways, one goal—Methanogenesis from unconventional substrates. *Applied Microbiology and Biotechnology, 104*(15), 6839–6854. https://doi.org/10.1007/s00253-020-10744-0

Lee, H., Calvin, K., Dasgupta, D., Krinner, G., Mukherji, A., Thorne, P., Trisos, C., Romero, J., Aldunce, P., & Barrett, K. (2023). *Climate change 2023: Synthesis report. Contribution of Working Groups I, II and III to the sixth assessment report of the Intergovernmental Panel on Climate Change*. Intergovernmental Panel on Climate Change (IPCC).

Lenhart, K., Klintzsch, T., Langer, G., Nehrke, G., Bunge, M., Schnell, S., & Keppler, F. (2016). Evidence for methane production by the marine algae Emilianiahuxleyi. *Biogeosciences, 13*(11), 3163–3174. https://doi.org/10.5194/bg-13-3163-2016

Li, Y., Dong, S., Liu, S., Zhou, H., Gao, Q., Cao, G., Wang, X., Su, X., Zhang, Y., & Tang, L. (2015). Seasonal changes of CO_2, CH_4 and N_2O fluxes in different types of alpine grassland in the Qinghai-Tibetan Plateau of China. *Soil Biology and Biochemistry, 80*, 306–314. https://doi.org/10.1016/j.soilbio.2014.10.026

Li, J., Junmin, P., Elise, P., Changming, F., & Ming, N. (2020). Spatial heterogeneity of temperature sensitivity of soil respiration: A global analysis of field observations. *Soil Biology and Biochemistry, 141*, 107675. https://doi.org/10.1016/j.soilbio.2019.107675

Margesin, R., Jud, M., Tscherko, D., & Schinner, F. (2009). Microbial communities and activities in alpine and subalpine soils. *FEMS Microbiology Ecology, 67*(2), 208–218. https://doi.org/10.1111/j.1574-6941.2008.00620.x

Marín-Muñiz, J. L., Hernandez, M. E., & Moreno-Casasola, P. (2015). Greenhouse gas emissions from coastal fresh water wetlands in Veracruz Mexico: Effect of plant community and seasonal dynamics. *Atmospheric Environment, 107*, 107–117.

Marquart, K. A., Haller, B. R., Paper, J. M., et al. (2019). Influence of pH on the balance between methanogenesis and iron reduction. *Geobiology, 17*(2), 185–198. https://doi.org/10.1111/gbi.12320

Martins, C. S., Nazaries, L., Macdonald, C. A., Anderson, I. C., & Singh, B. K. (2015). Water availability and abundance of microbial groups are key determinants of greenhouse gas fluxes in a dryland forest ecosystem. *Soil Biology and Biochemistry, 86*, 5–16.

Nazaries, L., Murrell, J. C., Millard, P., Baggs, L., & Singh, B. K. (2013). Methane, microbes and models: Fundamental understanding of the soil methane cycle for future predictions. *Environmental Microbiology, 15*(9), 2395–2417. https://doi.org/10.1111/1462-2920.12149

Oertel, C., Matschullat, J., Zurba, K., Zimmermann, F., & Erasmi, S. (2016). Greenhouse gas emissions from soils—A review. *Geochemistry, 76*(3), 327–352. https://doi.org/10.1016/j.chemer.2016.04.002

Phillips, C. L., Bond-Lamberty, B., Desai, A. R., Lavoie, M., Risk, D., Tang, J., Todd-Brown, K., & Vargas, R. (2017). The value of soil respiration measurements for interpreting and modeling terrestrial carbon cycling. *Plant and Soil, 413*, 1–25. https://doi.org/10.1007/s11104-016-3086-0

Poffenbarger, H. J., Needelman, B. A., & Megonigal, J. P. (2011). Salinity influence on methane emissions from tidal marshes. *Wetlands, 31*, 831–842.

Porollo, M. (2023). The role of environmental influence from microbial growth dynamics. *Applied Microbiology Open Access, 9*(6), 1000287.

Ramulu, S., Sahoo, S. K., Baral, S. S., Das, S. N., Swamy, Y. V., & Chaudhury, G. R. (2008). Estimation of nitrous oxide emission from pulse cultivation in rain fed uplands. *Current Science, 95*, 584–585.

Ravishankara, A. R., Daniel, J. S., & Portmann, R. W. (2009). Nitrous oxide (N_2O): The dominant ozone-depleting substance emitted in the 21st century. *Science, 326*(5949), 123–125. https://doi.org/10.1126/science.1176985

Ray, S., Kuppam, C., Pandit, S., & Kumar, P. (2023). Biogas upgrading by hydrogenotrophic methanogens: An overview. *Waste and Biomass Valorization, 14*(2), 537–552. https://doi.org/10.1007/s12649-022-01795-4

Repeta, D. J., Ferrón, S., Sosa, O. A., Johnson, C. G., Repeta, L. D., Acker, M., Delong, E. F., & Karl, D. M. (2016). Marine methane paradox explained by bacterial degradation of dissolved organic matter. *Nature Geoscience, 9*(12), 884–887. https://doi.org/10.1038/ngeo2837

Risgaard-Petersen, N., Revsbech, N. P., Rysgaard, S., Fossing, H., & Dalsgaard, T. (2006). Evidence for complete denitrification in a benthic foraminifer. *Nature, 443*(7107), 93–96. https://doi.org/10.1038/nature05070

Sauterey, B., Charnay, B., Affholder, A., Mazevet, S., & Ferrière, R. (2020). Co-evolution of primitive methane-cycling ecosystems and early Earth's atmosphere and climate. *Nature Communications, 11*, 2705. https://doi.org/10.1038/s41467-020-16374-7

Shahariar, S., Farrell, R., & Soolanayakanahally, R. (2021). Elevated salinity and water table drawdown significantly affect greenhouse gas emissions in soils from contrasting land-use practices in the prairie pothole region Angela Bedard-Haughn. *Biogeochemistry*. https://doi.org/10.1007/s10533-021-00818

Shakoor, A., Ashraf, F., Shakoor, S., Mustafa, A., Rehman, A., & Altaf, M. M. (2020). Biogeochemical transformation of greenhouse gas emissions from terrestrial to atmospheric environment and potential feedback to climate forcing. *Environmental Science and Pollution Research, 27*(31), 38513–38536. https://doi.org/10.1007/s11356-020-09820-1

She, R., Yu, Y., Ge, C., & Yao, H. (2021). Soil texture alters the impact of salinity on carbon mineralization. *Agronomy, 11*(1), 128.

Shindell, D. T., Faluvegi, G., Koch, D. M., Schmidt, G. A., Unger, N., & Bauer, S. E. (2009). Improved attribution of climate forcing to emissions. *Science, 326*(5953), 716–718. https://doi.org/10.1126/science.1174760

Singh, B. K., Bardgett, R. D., Smith, P., & Reay, D. S. (2019). Microorganisms and climate change: Terrestrial feedbacks and mitigation options. *Nature Reviews. Microbiology, 5*, 245–262.

Singhal, A., Pandey, S., Kumari, N., Chauhan, D. K., & Jha, P. K. (2021). Impact of climate change on soil microbes involved in biogeochemical cycling. In D. K. Choudhary, A. Mishra, & A. Varma (Eds.), *Climate change and the microbiome* (pp. 63–78). Springer. https://doi.org/10.1007/978-3-030-76863-8_5

Subbarao, G. V., Kishii, M., Nakahara, K., Ishikawa, T., Ban, T., Tsujimoto, H., George, T. S., Berry, W. L., Hash, C. T., & Ito, O. (2009). Biological nitrification inhibition—Is there potential for genetic interventions in the Triticeae? *Breeding Science, 59*(5), 529–545. https://doi.org/10.1270/jsbbs.59.529

Sylvie, D., & Katell, Q. (2015). Analytical pyrolysis as a tool to probe soil organic matter. *Journal of Analytical and Applied Pyrolysis, 111*, 108–120. https://doi.org/10.1016/j.jaap.2014.12.011

Tao, F., Huang, Y., Hungate, B., Manzoni, S., Frey, S., Schmidt, M., Reichstein, M., Carvalhais, N., Ciais, P., Jiang, L., et al. (2023). Microbial carbon use efficiency promotes global soil carbon storage. *Nature, 618*, 981–985.

Thapa, R., Chatterjee, A., Wick, A., & Utcher, K. B. (2017). Carbon dioxide and nitrous oxide emissions from naturally occurring sulfate-based saline soils at different moisture contents. *Pedosphere, 27*(5), 868–876.

Thauer, R. K., Kaster, A.-K., Seedorf, H., Buckel, W., & Hedderich, R. (2008). Methanogenic archaea: Ecologically relevant differences in energy conservation. *Nature Reviews. Microbiology, 6*, 579–591.

Tiedje, J. M., Bruns, M. A., Casadevall, A., Criddle, C. S., Eloe-Fadrosh, E., Karl, D. M., Nguyen, N. K., & Zhou, J. (2022). Microbes and climate change: A research prospectus for the future. *MBio, 13*(1), e03417–e03421. https://doi.org/10.1128/mbio.03417-21

Tyagi, S., Singh, R., & Javeria, S. (2014). Effect of climate change on plant-microbe interaction: An overview. *European Journal of Molecular Biotechnology, 5*, 149–156.

Ueno, Y., Yamada, K., Yoshida, N., Maruyama, S., & Isozaki, Y. (2006). Evidence from fluid inclusions for microbial methanogenesis in the early Archaean era. *Nature, 440*, 516–519.

USGRCP. (2017). Climate science special report. In D. J. Wuebbles, D. W. Fahey, K. A. Hibbard, D. J. Dokken, B. C. Stewart, & T. K. Maycock (Eds.), *Fourth national climate assessment* (Vol. I). U.S. Global Change Research Program.

Ussiri, D., & Lal, R. (2013). *Soil emission of nitrous oxide and its mitigation* (p. 378). Springer.

Vanwonterghem, I., Evans, P., Parks, D., et al. (2016). Methylotrophic methanogenesis discovered in the archaeal phylum Verstraetearchaeota. *Nature Microbiology, 1*, 16170. https://doi.org/10.1038/nmicrobiol.2016.170

Voigt, C., Lamprecht, R. E., Marushchak, M. E., Lind, S. E., Novakovskiy, A., Aurela, M., Martikainen, P. J., & Biasi, C. (2017). Warming of subarctic tundra increases emissions of all three important greenhouse gases-carbon dioxide, methane, and nitrous oxide. *Global Change Biology, 23*, 3121–3138.

Wang, Q., Dore, J. E., & McDermott, T. R. (2017). Methylphosphonate metabolism by Pseudomonas sp. populations contributes to the methane oversaturation paradox in an oxic freshwater lake. *Environmental Microbiology, 19*(6), 2366–2378. https://doi.org/10.1111/1462-2920.13750

Wang, C., Wu, B., Jiang, K., Wei, M., & Wang, S. (2019). Effects of different concentrations and types of Cu and Pb on soil N-fixing bacterial communities in the wheat rhizosphere. *Applied Soil Ecology, 144*, 51–59.

Wang, H., Zhao, R., Zhao, D., Liu, S., Fu, J., Zhang, Y., Dai, N., Song, D., & Ding, H. (2022). Microbial-mediated emissions of greenhouse gas from farmland soils: A review. *Process, 10*(11), 2361. https://doi.org/10.3390/pr10112361

Weiman, S. (2015). Microbes help to drive global carbon cycling and climate change. *Microbe Magazine, 10*, 233–238.

Whalen, S. C. (2005). Biogeochemistry of methane exchange between natural wetlands and the atmosphere. *Environmental Engineering Science, 22*(1), 73–94. https://doi.org/10.1089/ees.2005.22.73

Xin, Y., Ji, L., Wang, Z., Li, K., Xu, X., & Guo, D. (2022). Functional diversity and CO_2 emission characteristics of soil bacteria during the succession of halophyte vegetation in the Yellow River Delta. *International Journal of Environmental Research and Public Health, 19*(19), 12919. https://doi.org/10.3390/ijerph191912919

Yuan, M. M., Guo, X., Wu, L., Zhang, Y., Xiao, N., Ning, D., Shi, Z., Zhou, X., Wu, L., Yang, Y., Tiedje, J. M., & Zhou, J. (2021). Climate warming enhances microbial network complexity and stability. *Nature Climate Change, 11*, 343–348. https://doi.org/10.1038/s41558-021-00993-1

Zhu, B., Chen, Z., Hu, H., Andrei, A.-S., & Li, Y. (2024). Editorial: Greenhouse gas emissions and mitigation: Microbes, mechanisms and modeling. *Frontiers in Microbiology, 15*, 1363814. https://doi.org/10.3389/fmicb.2024.1363814

Keywords

Temperature changes · Metabolic processes · Stress response · Microbial adaptation · Microbial metabolism

1 Introduction

One essential physical factor affecting all biological systems—but especially microbes—is temperature. Microbial life has demonstrated an astounding ability to thrive across a vast range of thermal environments, from sizzling hydrothermal vents to the frigid Polar Regions. These organisms not only persist under extreme conditions but also perform vital ecological roles. Temperature influences nearly all cellular processes, affecting protein structure and function, membrane fluidity, enzyme activity, and the efficiency of energy-yielding biochemical pathways. As a result, it directly shapes the metabolic activity and ecological distribution of microbial communities.

In the era of accelerating climate change, the implications of even minor thermal shifts on microbial behavior and ecosystem functioning are profound. Changes in ambient temperature can disrupt global biogeochemical cycles, including carbon, nitrogen, and phosphorus fluxes, by altering microbial community composition and function. Thus, understanding microbial responses to temperature extremes is increasingly vital, not only from an ecological standpoint but also for their biotechnological and agricultural potential.

Microorganisms exhibit a remarkable capacity to adapt to both heat and cold stresses. These adaptations are expressed through a variety of mechanisms, such as protein refolding via chaperonins, membrane restructuring through changes in lipid composition, metabolic reprogramming, and the production of stress-responsive proteins like heat shock proteins (HSPs) and cold shock proteins (CSPs). These molecular and physiological responses ensure microbial survival while supporting essential processes like nutrient cycling, organic matter decomposition, and plant growth promotion, particularly in temperature-sensitive agroecosystems.

Recent studies have highlighted the functional diversity and resilience of microbial populations adapted to thermally challenging environments, such as those found in the Indian Himalayas. Here, temperature variability across elevation gradients significantly influences rhizospheric microbial communities. For instance, research by Parihar et al. (2025) has demonstrated how indigenous microbes from Himalayan soils adapt to cold temperatures while supporting plant productivity. Further investigations by Khati et al. (2024) and Tomer et al. (2024) have isolated plant growth-promoting rhizobacteria (PGPR) exhibiting both drought and temperature tolerance, capable of solubilizing phosphate, fixing nitrogen, and synthesizing growth-regulating phytohormones even under abiotic stress conditions.

Parallel to biological adaptations, the use of nanotechnology has opened new frontiers in managing soil microbial responses to temperature stress. Nanomaterials like nanozeolite, nanogypsum, and nanochitosan have been shown to modulate rhizospheric microbial communities, enhancing their resilience and promoting beneficial plant–microbe interactions (Chaudhary et al., 2021a, b, c, d; Khati et al., 2017, 2018, 2019a, b). These nanocompounds can influence microbial diversity and activity by modifying the soil microenvironment, thus contributing to thermal stress mitigation. Advanced molecular techniques such as metagenomics and transcriptomics further reveal that nanomaterials can induce significant shifts in microbial functional traits, improving nutrient availability and plant health (Chaudhary et al., 2021a, b, c, d; Khati et al., 2019a, b).

In addition, long-term agronomic practices involving organic amendments and biofertilizers have proven effective in enhancing microbial metabolic stability and the soil's thermal buffering capacity (Panday et al., 2024). This has direct implications for sustaining crop productivity under climate-induced temperature fluctuations. Meanwhile, thermotolerant microbes have shown promise in the bioremediation of pesticide-contaminated soils. For example, certain bacterial strains possess the enzymatic machinery—such as esterase and laccase—to degrade persistent agrochemicals like cypermethrin and sulfosulfuron, even under heat or cold stress conditions (Bhatt et al., 2015a, b, 2016, 2019; Gangola et al., 2015a, b). These metabolic capabilities emphasize the ecological importance of thermally adaptive microbes in sustaining both environmental safety and agricultural productivity.

The relevance of microbial thermotolerance extends far beyond the soil. Microbes have been discovered in some of Earth's coldest environments—permafrost, glacial ice, and polar sea ice—where they endure subzero temperatures by exploiting physicochemical microenvironments. For example, the presence of ion-rich liquid veins and thin films of unfrozen water along ice grain boundaries supports minimal microbial metabolism by allowing nutrient transport and waste removal (Price et al., 2000; Rivkina et al., 2000; Gilichinsky et al., 2008). Certain solutes, such as mineral acids and salts, can depress the freezing point of water to as low as -90 °C, sustaining microbial activity in these extreme niches (Deming, 2002). Studies by Bakermans (2008), Junge et al. (2004), and Miteva et al. (2007) provide direct and indirect evidence of viable microbial life in these cryoenvironments.

Even though laboratory conditions allow microbes to double rapidly under nutrient-rich settings, their growth in natural environments is generally constrained by energy limitation, nutrient scarcity, and biotic pressures. In such conditions, average doubling times may span millennia, punctuated by brief periods of activity (Hoehler & Jørgensen, 2013). To cope, microbes employ a variety of survival strategies when faced with nutrient depletion or metabolic stress. These include dormancy, sporulation, cannibalism, formation of ultramicrocells, and expression of starvation-induced genes and stress proteins. At the cellular level, microbes may alter membrane lipid composition, reduce cellular water content, increase the ordering of intracellular water, or even extract energy through redox reactions in thin aqueous films or ice veins. These adaptive strategies allow microorganisms to persist at the thermodynamic limits of life.

By exploring these microbial adaptations and survival mechanisms, this chapter aims to dissect the intricate interplay between temperature and microbial physiology. It also seeks to illuminate the broader ecological and biotechnological ramifications

4 Effect on Enzyme Kinetics and Metabolic Pathways

From nutrient cycling in ecosystems to biochemical processes within cells, temperature exerts a profound impact on biological functions at all levels of biological organization (Brown et al., 2004; Hochachka & Somero, 2002; Kleiber & Rogers, 1961; Schulte, 2015; Yvon-Durocher et al., 2012). Among the most basic thermal responses is the modulation of metabolic rate, which directly affects the temperature sensitivity of enzyme kinetics and many downstream biological processes (Dell et al., 2011; Gillooly et al., 2001). Understanding how microbial metabolic rates react to temperature is thus crucial to thermal ecology, especially in the context of global carbon cycling and climate change.

The soil is the largest reservoir of terrestrial carbon (Jobbágy & Jackson, 2000), and microbial metabolism plays a pivotal role in controlling carbon storage and release. Soil microbes utilize organic carbon for energy and biomass production, emitting CO_2 in the process and contributing to both short- and long-term carbon sequestration (Davidson & Janssens, 2006; Schimel et al., 2000; Schlesinger & Andrews, 2000). Consequently, temperature-mediated alterations in microbial enzyme function and energy-yield pathways can significantly impact atmospheric CO_2 levels and climate feedbacks (Billings et al., 2024; Hungate et al., 2009). Despite its critical role, the regulation of microbial energy metabolism—specifically ATP, NADH, and NADPH production and utilization in whole soil microbial communities—is still poorly understood (Allison et al., 2010; Chapman & Gray, 1986; Del Giorgio & Cole, 1998; Schimel & Weintraub, 2003). Microbial energy status governs key ecological processes, including nutrient immobilization versus mineralization, by determining whether carbon is allocated toward biomass, respiration, or storage. Gaining insight into these energy-regulated pathways is essential for modeling soil carbon (C) and nitrogen (N) cycling accurately.

Temperature fluctuations influence not only enzymatic reaction rates but also the flux of carbon through core metabolic networks. The majority of microbial ATP and biosynthetic precursors are produced through glycolysis, the tricarboxylic acid (Krebs) cycle, and the pentose phosphate pathway (PPP) (Almaas et al., 2004). The flow through these pathways varies depending on environmental conditions, substrate availability, and cellular energy and redox requirements (Gerstmeir et al., 2003). Temperature directly affects rate-limiting enzymes within these routes, thereby altering carbon flow and microbial physiology. At low temperatures, many microorganisms—including plant and animal cells—tend to shift away from glycolysis and NADH production toward increased PPP activity and NADPH generation (Chung et al., 1976; Wittmann et al., 2007; Malone et al., 2006). This metabolic reprogramming serves multiple purposes: meeting NADPH demands for biosynthesis, combating oxidative stress via NADPH-dependent antioxidant pathways (Ralser et al., 2009), or bypassing heat-labile enzymatic steps such as phosphofructokinase in glycolysis (Malone et al., 2006).

Carbon use efficiency (CUE)—the ratio of substrate carbon converted into microbial biomass—is another key parameter influenced by temperature. While previously considered a fixed value across ecosystems (Manzoni & Porporato,

2009; Steinweg et al., 2008), CUE is now recognized as variable depending on environmental conditions. Elevated temperatures often reduce CUE due to increased maintenance energy costs (Allison et al., 2010; Devêvre & Horwáth, 2000). Other factors like substrate quality (C:N ratio), moisture, oxygen levels, and microbial predation also affect CUE (Bremer & Kuikman, 1994; Shen & Bartha, 1996; Parsons & Smith, 1989). Variations in CUE and microbial energy metabolism ultimately affect CO_2 emissions, nitrogen mineralization, and the formation or decomposition of soil organic matter (De Graaff et al., 2006; Hungate et al., 2009; Van Groenigen et al., 2006). These processes contribute significantly to ecosystem-level feedbacks under warming conditions. Therefore, understanding how temperature influences enzyme kinetics, metabolic pathway fluxes, and energy allocation is essential for predicting the resilience and functionality of microbial communities in a rapidly changing climate (Heimann & Reichstein, 2008).

Temperature thus has a far-reaching impact on microbial enzyme activity and metabolic pathways, shaping microbial community function and global biogeochemical cycling. It alters not only enzymatic reaction rates but also redirects carbon flow through central metabolic routes like glycolysis, the TCA cycle, and the PPP. These changes affect the energy-to-biomass balance, carbon use efficiency, and soil organic matter dynamics. Given the central role of microbial metabolism in regulating nutrient and carbon cycles, a mechanistic understanding of temperature-sensitive processes is vital for projecting climate change impacts on ecosystems. Future research integrating microbial energetics, environmental physiology, and systems-level metabolic modeling will be essential for refining estimates of terrestrial carbon fluxes in a warming world.

5 Temperature-Induced Stress Responses in Microorganisms

Microorganisms are often subjected to variable environmental conditions, which include temperature extremes that lead to cellular stress. These stressors activate a variety of adaptive processes in terms of general and specific pathways associated with DNA damage, oxidative stress, envelope perturbation, heat and cold shock, nutrient starvation, and osmotic imbalance (Shima & Takagi, 2009; Mattanovich et al., 2004). These cellular responses are important in microbial adaptability, survival, virulence, and even antibiotic resistance. Stress-induced regulatory mechanisms profoundly affect bacterial physiology by inducing phenotypic variation, causing resistance mutations, increasing biofilm production, and affecting persistence mechanisms—all of which enhance reduced antibiotic susceptibility (Smits & Brul, 2005). Stress responses thus complicate the effectiveness of most antibiotics and are a major obstacle to effective clinical treatment (Jenkins et al., 1991). Additionally, in ecological microbial populations, where organisms are often exposed to nutrient limitation, poisonous substances, and host immunity, stress responses are critical for sustaining microbial community dynamics and function (Piper, 1995). Microbial stress responses are regulated by signal transduction pathways that perceive environmental signals and govern corresponding gene expression alterations. Prokaryotes and eukaryotes both employ these mechanisms to balance cellular homeostasis upon facing environmental perturbations like temperature changes, pH changes, osmotic stress, and oxidative conditions (Piper, 1995; Mattanovich et al., 2004). Interestingly, there are overlaps among the various pathways of stress responses to facilitate cross-protection—exposure to one stress type enhances tolerance to another (Jenkins et al., 1991).

5.1 Temperature Stress Responses

Temperature is one of the chief selective forces that impact microbial evolution and physiology. It influences essential biomolecules like DNA, RNA, proteins, and membranes. Microorganisms have developed intricate responses to survive both high and low temperature extremes (Fig. 1).

(a) *Heat Shock Response*

Heat can cause misfolding, denaturation, and aggregation of proteins. To mitigate this, microorganisms cause heat shock proteins (Hsps) to be induced, which function as molecular chaperones to refold denatured proteins and avoid aggregation (Piper, 1995; Mattanovich et al., 2004). The heat shock response is mainly regulated by sigma factor σ^{32}, which controls the transcription of Hsp-encoding genes in bacteria (Shima & Takagi, 2009). Thermosensors such as certain proteins, RNA, and DNA architectures sense changes in temperature and activate the heat shock response. For example, the RNA thermometers in the untranslated regions (UTRs) of genes *tviA* and *hsp* allow for temperature-dependent regulation of translation. Upon rise in temperature, structural changes in RNA reveal ribosome binding sites, enhancing translation of virulence and chaperone proteins (Smits & Brul, 2005). In addition,

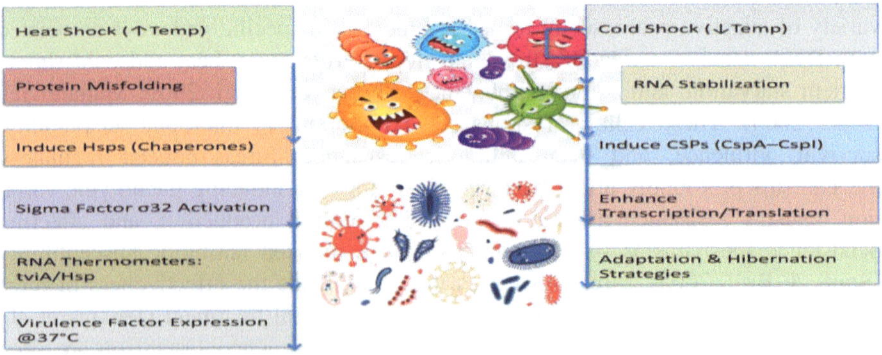

Fig. 1 Schematic representation of microbial responses to temperature extremes. The heat shock response (left) is triggered by elevated temperatures, leading to protein misfolding and the induction of heat shock proteins. The cold shock response (right) involves RNA stabilization and the synthesis of cold shock proteins (CSPs) such as CspA–CspI, which facilitate translation under cold stress

increased temperatures (e.g., 37 °C) in pathogenic bacteria tend to induce virulence factors that enable host colonization and evasion of the host's immune system (Piper, 1995).

(b) *Cold Shock Response*

At reduced temperatures, microbial metabolism slows, and nucleic acid rigidity increases. To counter cold stress, microbes trigger the expression of cold shock proteins (CSPs), such as CspA–CspI, which function as nucleic acid chaperones and facilitate transcription and translation during low-temperature stress (Shima & Takagi, 2009). CSPs prevent the formation of stable secondary RNA structures that could obstruct ribosome binding and translation initiation. The cold shock response is vital for maintaining viability and adaptive capability in cold environments, including refrigerated foods and natural cold habitats (Smits & Brul, 2005).

(c) *Cross-Protection and Adaptive Strategies*

Experimental studies in *Escherichia coli* provided early evidence of cross-protection, where prior exposure to one type of stress (e.g., heat) improved microbial survival during subsequent stresses such as oxidative or acid stress (Jenkins et al., 1991). This phenomenon underscores the complex interplay among stress response networks and highlights the metabolic plasticity of microorganisms. Cross-protection enables microbes to withstand sequential or overlapping environmental challenges by reprogramming gene expression and stress-related metabolic pathways.

6 Regulatory Networks and Temperature-Responsive Gene Expression

Temperature shifts impinge on virtually every cellular process, so bacteria have evolved molecular thermosensors that translate a change in temperature into precise transcriptional or translational outputs. Three mechanistic layers recur across disparate species and regulons (Table 2).

6.1 DNA Supercoiling as a Temperature Fuse

The ratio between gyrase (adding negative supercoils) and topoisomerase I (removing them) adjusts chromosomal superhelicity. Warm shifts increase metabolic ATP/ADP, condensing supercoils and initiating promoters needing negative superhelicity, while cold shifts do the opposite (Dorman, 1991).

Table 2 Mechanisms of thermosensing and temperature-dependent gene regulation in bacteria

Thermosensing principle	Typical sensor	Downstream effect	Representative examples
DNA supercoiling	Global DNA topology set by gyrase/ topoisomerase I	Alters promoter accessibility to RNA polymerase or transcription factors	*pap*, *vir*, *yop* virulence loci; heat-shock regulon (Dorman, 1991)
RNA structural switches	mRNA stem–loop in 5′-UTR or coding region	Melting/unfolding exposes the Shine–Dalgarno sequence and starts translation	*lcrF* in *Yersinia*, *rpoH* in *E. coli*, λ *cIII* in phage (Hoe & Goguen, 1993)
Protein conformational changes	Coiled-coil or oligomerization-prone domains	Temperature-dependent folding or multimerization switches activity	TlpA antiterminator, σ^32 activator RpoH, H-NS, Tar thermoreceptor (Lupas, 1996; Nishiyama et al., 1997)

- *Pap pili (E. coli)*—Thermal upshift to 37 °C enhances *papI* transcripts; PapI counteracts the nucleoid-associated repressor H-NS, relaxing repression of the pilus operon (Göransson et al., 1990).
- *VirF/VirB cascade (Shigella)*—VirF is only activated with supercoiled templates; H-NS bound at lower temperatures occludes the *virB* promoter (Tobe et al., 1993).
- *Yop regulon (Yersinia)*—The analogue of the H-NS, YmoA, represses the *yop* genes at 25 °C; temperature-induced supercoil shifts and VirF in *Y. enterocolitica* alleviate repression (Rohde et al., 1994).

These instances show the way in which temperature-regulated DNA architecture and occupancy of the H-NS together regulate transcription.

6.2 RNA Thermometers at the Translational Level

Some virulence activators are constitutively transcribed but translated only upon melting of an mRNA hairpin.

- *LcrF (Yersinia pestis)*—A stem–loop covering the Shine–Dalgarno region unfolds at ≥37 °C, inducing translation of the master activator independently of other proteins (Hoe & Goguen, 1993).
- *σ^32 mRNA (rpoH) (E. coli)*—Several 5′ UTR hairpins block ribosome entry at 30 °C; heat melts them, increasing σ^32 synthesis and heat shock regulon (Kamath-Loeb & Gross, 1991; Nagai et al., 1991).
- *Phage λ cIII*—Low temperature conformer ("structure B") stimulates translation, favoring lysogeny (Altuvia et al., 1989). RNA thermometers, therefore, function as direct, protein-independent temperature sensors.

6.3 Protein-Based Thermosensors: Coiled-Coil and Oligomeric Switches

The coiled-coil structure easily acquires or loses heptad repeat interactions with temperature fluctuations.

- *Salmonella typhimurium TlpA*—A dimeric repressor which unfolds at temperatures above 42 °C, derepressing downstream genes, refolds on cooling, functioning as a reversible "protein thermometer."
- *E. coli Tar aspartate chemoreceptor*—Flanking coiled-coil domains (MH1/MH2) are likely to modulate receptor signaling in a temperature-dependent manner (Nishiyama et al., 1997).
- *M protein (Group A streptococci)*—found as dimers at 25 °C but as monomers at 37 °C, suggesting an extracellular coiled-coil thermometer that has the potential to convey signals through the wall (Cedervall et al., 1997).
- *H-NS itself*—Binding affinity, oligomer size, and ability to span DNA depend on temperature; heat-induced dissociation can activate previously suppressed loci (White-Ziegler et al., 1998).

Pathogens seldom use temperature alone, but instead combine temperature signals with pH, osmolarity, redox status, and quorum signals (Mekalanos, 1992). Predictive mathematical models that pair experimentally measured promoter kinetics with supercoiling dynamics and folding thermodynamics of RNA are now tractable, offering quantitative insight into the optimal timing of virulence for host invasion.

7 Applications in the Biotechnology Industry

In industrial biotechnology, temperature is also an important parameter for determining the efficiency of microorganisms. Optimized production methods in fermentation, enzyme production, bioremediation, and synthetic biology have ensued with an understanding of the possibility of controlling microbial reactions to temperature changes.

7.1 Fermentation Process Optimization

Temperature significantly influences microbial fermentation. Each industrial strain is also specialized to a certain temperature at which it maximizes the production of the highest number of metabolites and biomass. During ethanol fermentation, for instance, mesophilic microbes like *Saccharomyces cerevisiae* are used; they are most active under temperatures ranging from 30 to 35 °C (Mattanovich et al., 2004). Alternatively, thermophilic organisms like *Thermus thermophilus* (Adams & Mitchell, 2002) are utilized in high-temperature fermentations, which lower contamination as well as enhance reaction kinetics.

7.2 Increased Enzyme Production

Temperature control is needed for industrial enzyme manufacturing. Overexpression of specific proteins may be achieved using stress-inducible promoters by exploiting heat-shock or cold-shock responses (Bhandari et al., 2024). Induction of high-level expression of recombinant proteins, for example, may be triggered by the heat-shock promoter (*groESL* or *dnaK* in *E. coli*) (De Marco, 2007). In addition, molecular biology techniques such as PCR depend upon thermostable enzymes from thermophiles, for example, Taq polymerase from *Thermus aquaticus* (Chien et al., 1976).

7.3 Stress Engineering to Make Production Strains Stronger

By incorporating chaperone systems, cold-shock proteins, or thermoprotective solutes such as trehalose, biotechnological strains are often engineered to resist heat stress (Goyal et al., 2005). In industrial-scale bioreactors where hot spots can occur, such adaptations help keep productivity at optimum levels under suboptimal or changing temperature conditions (Neubauer et al., 2013).

7.4 Synthetic Biology and Metabolic Engineering

Temperature-sensitive regulatory systems serve as inducible switches in synthetic biology. Temperature-sensitive repressors, like *cI857*, and RNA thermometers enable the precise control of gene expression upon heat induction (Kortmann & Narberhaus, 2012). Such systems are extremely useful in environmental biosensing, biopharmaceutical production, and the regulation of multistep metabolic pathways.

7.5 Scale-Up and Bioprocess Control

Reliable microbial performance in large-scale bioreactors is based on monitoring and controlling temperature. Batch-to-batch variation is reduced and yield stability increases by incorporating temperature-feedback loops, advanced thermoregulation equipment, and process analytics with digital computer systems. Thermotolerant strains also reduce cooling and sterilization costs, making bioprocesses more energy-efficient (Zeldes et al., 2015).

7.6 Bioremediation Through Heat

The functional traits of soil microorganisms also play a critical role in biodegradation and detoxification of xenobiotics, as evident in studies focusing on cypermethrin, sulfosulfuron, and petroleum hydrocarbon degradation by *Bacillus* spp. and microbial consortia (Bhatt et al., 2015a, b; Gangola et al., 2015a, b; Kumar et al., 2017). These insights contribute to understanding microbial resilience under chemical stress, especially in polluted agro-ecosystems. In addition, bioremediation technologies utilize thermophilic bacteria, particularly for the degradation of organic contaminants in extremely hot, high-temperature environments such as composting facilities and hot industrial waste streams. Due to their capacity to degrade complex molecules at high temperatures, they are a green waste management option. Microbial thermosensitivity is not only something that needs to be corrected but can also be used as a biotechnological tool. By controlling gene expression, strain resistance, and enzymatic function, temperature control provides enormous potential for enhancing industrial microbial processes.

Microbial metabolism is heavily influenced by temperature, which also influences gene expression, enzyme function, membrane fluidity, and stress response. Microorganisms have developed a range of strategies to survive across a broad temperature range. Some of these include adaptation of the membrane, production of stress proteins, and thermosensitive regulatory devices. These thermal responses have significant biotechnological applications as well as controlling microbial survival and ecological activity. Temperature control enhances microbial performance in industrial applications, ranging from optimizing fermentation and enzyme yields to enabling bioremediation and synthetic biology. Knowledge of microbial thermoadaptation is essential in predicting ecosystem response and optimizing biotechnological outcomes when climate change alters ambient temperatures. Temperature is thus both a challenge and an instrument for optimizing the potential of microorganisms.

8 Conclusion

Temperature is a fundamental abiotic factor that exerts a profound influence on microbial metabolism, physiology, gene regulation, and ecological function. Microorganisms have evolved a wide array of structural, biochemical, and genetic adaptations to survive and thrive across diverse thermal environments, from the icy extremes of polar regions to the heat of hydrothermal vents. These include modifications to membrane fluidity, enzyme kinetics, stress protein expression, and thermosensitive regulatory networks.

The interplay between temperature and microbial metabolism is especially critical in the context of global climate change. Even minor shifts in environmental temperatures can disrupt microbial community structure, alter biogeochemical cycling, and affect ecosystem resilience. Changes in microbial carbon use efficiency, enzyme-driven decomposition, and nutrient turnover have direct implications for soil health, greenhouse gas emissions, and climate feedback loops.

From an applied perspective, understanding microbial thermal responses is essential for optimizing industrial processes such as fermentation, enzyme production, and bioremediation. Advances in synthetic biology and metabolic engineering offer promising tools to harness thermosensitive elements for precise control of microbial behavior in industrial and environmental settings.

9 Future Prospects

Future research on temperature–microbe interactions should prioritize the integration of microbial thermal responses into predictive climate models to better assess ecosystem-level feedbacks under global warming. High-throughput omics approaches—such as transcriptomics, proteomics, and metabolomics—will be instrumental in identifying novel thermotolerant genes, stress regulators, and enzymes with industrial relevance. Advances in synthetic biology offer promising avenues to engineer microbial strains with enhanced thermal resilience, enabling more stable performance in bioreactors and environmental applications. Furthermore, the development of temperature-sensitive regulatory elements, such as RNA thermometers and protein-based bioswitches, can be harnessed for precise control of gene expression in biotechnology. Exploration of extremophiles from polar and hydrothermal environments may also yield unique biomolecules with potential applications in pharmaceuticals, agriculture, and green chemistry, expanding the biotechnological utility of microbes in a warming world.

References

Adams, M., & Mitchell, R. (2002). Fermentation and pathogen control: A risk assessment approach. *International Journal of Food Microbiology, 79*(1–2), 75–83.

Allison, S. D., Wallenstein, M. D., & Bradford, M. A. (2010). Soil-carbon response to warming dependent on microbial physiology. *Nature Geoscience, 3*(5), 336–340.

Almaas, E., Kovács, B., Vicsek, T., Oltvai, Z. N., & Barabási, A. L. (2004). Global organization of metabolic fluxes in the bacterium Escherichia coli. *Nature, 427*(6977), 839–843.

Altuvia, S., Kornitzer, D., Teff, D., & Oppenheim, A. B. (1989). Alternative mRNA structures of the cIII gene of bacteriophage λ determine the rate of its translation initiation. *Journal of Molecular Biology, 210*(2), 265–280.

Bakermans, C. (2008). Limits for microbial life at subzero temperatures. In *Psychrophiles: From biodiversity to biotechnology* (pp. 17–28). Springer.

Bhandari, G., Gangola, S., Bhatt, P., & Rafatullah, M. (2024). Potential of the plant rhizomicrobiome for bioremediation of contaminants in agroecosystems. *Frontiers in Plant Science, 15*, 1397360. https://doi.org/10.3389/fpls.2024.1397360

Bhatt, P., Negi, G., Gangola, S., Khati, P., Srivastava, A., & Sharma, A. (2015a). Optimization of sulfosulfuron biodegradation through response surface methodology using indigenous bacterial strain isolated from contaminated agriculture field. *International Journal of Current Microbiology and Applied Sciences, 4*(8), 105–112.

Bhatt, P., Sharma, A., Khati, P., Kumar, G., Gangola, S., & Srivastava, A. (2015b). Novel pathway of cypermethrin biodegradation in a *Bacillus* sp. strain SG2 isolated from cypermethrin-contaminated agriculture field. *3 Biotech, 6*, 45.

Bhatt, P., Negi, G., Gangola, S., Khati, P., Srivastava, A., & Sharma, A. (2016). Differential expression and characterization of cypermethrin degrading potential proteins in *Bacillus thuringiensis* strain SG4. *3 Biotech, 6*, 225.

Bhatt, P., Gangola, S., Chaudhary, P., Khati, P., Kumar, G., Sharma, A., & Srivastava, A. (2019). Pesticide induced up-regulation of esterase and aldehyde dehydrogenase in indigenous *Bacillus* spp. *Bioremediation Journal, 23*(1), 42–52.

Billings, S. A., Brecheisen, Z., Cherkinsky, A., Lehmeier, C., Cook, C. W., Markewitz, D., et al. (2024). Persistent biogeochemical signals of land use-driven, deep root losses illuminated by C and O isotopes of soil CO_2 and O_2. *Biogeochemistry, 167*(12), 1469–1489.

Bremer, E., & Kuikman, P. (1994). Microbial utilization of 14C [U] glucose in soil is affected by the amount and timing of glucose additions. *Soil Biology and Biochemistry, 26*(4), 511–517.

Brown, J. H., Gillooly, J. F., Allen, A. P., Savage, V. M., & West, G. B. (2004). Toward a metabolic theory of ecology. *Ecology, 85*(7), 1771–1789.

Cavicchioli, R., Thomas, T., & Curmi, P. M. (2000). Cold stress response in Archaea. *Extremophiles, 4*, 321–331.

Cedervall, T., Johansson, M. U., & Åkerström, B. (1997). Coiled-coil structure of Group A Streptococcal M Proteins. Different temperature stability of Class A and C proteins by hydrophobic—Nonhydrophobic amino acid substitutions at Heptad positions A and D. *Biochemistry, 36*(16), 4987–4994.

Chapman, S. J., & Gray, T. R. G. (1986). Importance of cryptic growth, yield factors and maintenance energy in models of microbial growth in soil. *Soil Biology and Biochemistry, 18*(1), 1–4.

Chaudhary, P., Khati, P., Chaudhary, A., Gangola, S., Kumar, R., & Sharma, A. (2021a). Bioinoculation using indigenous *Bacillus* spp. improves growth and yield of *Zea mays* under the influence of nanozeolite. *3 Biotech, 11*, 11.

Chaudhary, P., Khati, P., Chaudhary, A., Maithani, D., Kumar, G., & Sharma, A. (2021b). Cultivable and metagenomic approach to study the combined impact of nanogypsum and *Pseudomonas taiwanensis* on maize plant health and its rhizospheric microbiome. *PLoS ONE, 16*(4), e0250574.

Chaudhary, P., Khati, P., Gangola, S., Kumar, A., Kumar, R., & Sharma, A. (2021c). Impact of nanochitosan and *Bacillus* spp. on health, productivity and defence response in *Zea mays* under field condition. *3 Biotech, 11*(5), 237.

Chaudhary, P., Sharma, A., Chaudhary, A., Khati, P., Gangola, S., & Maithani, D. (2021d). Illumina-based high-throughput analysis of microbial diversity of maize rhizosphere treated with nanocompounds and *Bacillus* sp. *Applied Soil Ecology, 159*, 103386.

Chien, A., Edgar, D. B., & Trela, J. M. (1976). Deoxyribonucleic acid polymerase from the extreme thermophile Thermus aquaticus. *Journal of Bacteriology, 127*(3), 1550–1557.

Chung, B. H., Cannon, R. Y., & Smith, R. C. (1976). Influence of growth temperature on glucose metabolism of a psychotrophic strain of Bacillus cereus. *Applied and Environmental Microbiology, 31*(1), 39–45.

Davidson, E. A., & Janssens, I. A. (2006). Temperature sensitivity of soil carbon decomposition and feedbacks to climate change. *Nature, 440*(7081), 165–173.

De Graaff, M. A., Van Groenigen, K. J., Six, J., Hungate, B., & van Kessel, C. (2006). Interactions between plant growth and soil nutrient cycling under elevated CO_2: A meta-analysis. *Global Change Biology, 12*(11), 2077–2091.

De Marco, A. (2007). Protocol for preparing proteins with improved solubility by co-expressing with molecular chaperones in Escherichia coli. *Nature Protocols, 2*(10), 2632–2639.

Del Giorgio, P. A., & Cole, J. J. (1998). Bacterial growth efficiency in natural aquatic systems. *Annual Review of Ecology and Systematics, 29*(1), 503–541.

Dell, A. I., Pawar, S., & Savage, V. M. (2011). Systematic variation in the temperature dependence of physiological and ecological traits. *Proceedings of the National Academy of Sciences, 108*(26), 10591–10596.

Deming, J. W. (2002). Psychrophiles and polar regions. *Current Opinion in Microbiology, 5*(3), 301–309.

Devêvre, O. C., & Horwáth, W. R. (2000). Decomposition of rice straw and microbial carbon use efficiency under different soil temperatures and moistures. *Soil Biology and Biochemistry, 32*(11–12), 1773–1785.

Dorman, C. J. (1991). DNA supercoiling and environmental regulation of gene expression in pathogenic bacteria. *Infection and Immunity, 59*(3), 745–749.

Gangola, S., Bhatt, P., Khati, P., & Sharma, A. (2015a). Mycoremediation of imidacloprid in the presence of different soil amendments using *Trichoderma longibrachiatum* and *Aspergillus oryzae* isolated from pesticide-contaminated agricultural field of Uttarakhand. *Journal of Bioremediation & Biodegradation, 6*, 310.

Gangola, S., Sharma, A., Bhatt, P., Khati, P., & Chaudhary, P. (2015b). Presence of esterase and laccase in *Bacillus subtilis* facilitates biodegradation and detoxification of cypermethrin. *Scientific Reports, 8*, 12755.

Gangola, S., Chaube, S., Bayram, A., Joshi, S., Bhandari, G., Malik, S., & Khan, A. A. (2024). Optimizing microbial strain selection for pyrethroid biodegradation in contaminated environments through a TOPSIS-based decision-making system. *Scientific Reports, 14*(1), 14928. https://doi.org/10.1038/s41598-024-59223-z

Gerstmeir, R., Wendisch, V. F., Schnicke, S., Ruan, H., Farwick, M., Reinscheid, D., & Eikmanns, B. J. (2003). Acetate metabolism and its regulation in Corynebacterium glutamicum. *Journal of Biotechnology, 104*(1–3), 99–122.

Gilichinsky, D., Vishnivetskaya, T., Petrova, M., Spirina, E., Mamykin, V., & Rivkina, E. (2008). Bacteria in permafrost. In *Psychrophiles: From biodiversity to biotechnology* (pp. 83–102). Springer.

Gillooly, J. F., Brown, J. H., West, G. B., Savage, V. M., & Charnov, E. L. (2001). Effects of size and temperature on metabolic rate. *Science, 293*(5538), 2248–2251.

Göransson, M., Sondén, B., Nilsson, P., Dagberg, B., Foreman, K., Emanuelsson, K., & Uhlin, B. E. (1990). Transcriptional silencing and thermoregulation of gene expression in Escherichia coli. *Nature, 344*(6267), 682–685.

Goyal, K., Walton, L. J., & Tunnacliffe, A. (2005). LEA proteins prevent protein aggregation due to water stress. *Biochemical Journal, 388*(1), 151–157.

Heimann, M., & Reichstein, M. (2008). Terrestrial ecosystem carbon dynamics and climate feedbacks. *Nature, 451*(7176), 289–292.

Hochachka, P. W., & Somero, G. N. (2002). *Biochemical adaptation: Mechanism and process in physiological evolution*. Oxford University Press.

Hoe, N. P., & Goguen, J. D. (1993). Temperature sensing in Yersinia pestis: Translation of the LcrF activator protein is thermally regulated. *Journal of Bacteriology, 175*(24), 7901–7909.

Hoehler, T. M., & Jørgensen, B. B. (2013). Microbial life under extreme energy limitation. *Nature Reviews. Microbiology, 11*(2), 83–94.

Hungate, B. A., Van Groenigen, K. J., Six, J., Jastrow, J. D., Luo, Y., De Graaff, M. A., et al. (2009). Assessing the effect of elevated carbon dioxide on soil carbon: A comparison of four meta-analyses. *Global Change Biology, 15*(8), 2020–2034.

Jenkins, D. E., Auger, E. A., & Matin, A. (1991). Role of RpoH, a heat shock regulator protein, in Escherichia coli carbon starvation protein synthesis and survival. *Journal of Bacteriology, 173*(6), 1992–1996.

Jobbágy, E. G., & Jackson, R. B. (2000). The vertical distribution of soil organic carbon and its relation to climate and vegetation. *Ecological Applications, 10*(2), 423–436.

Junge, K., Eicken, H., & Deming, J. W. (2004). Bacterial activity at −2 to −20°C in Arctic wintertime sea ice. *Applied and Environmental Microbiology, 70*(1), 550–557.

Kamath-Loeb, A. S., & Gross, C. A. (1991). Translational regulation of sigma 32 synthesis: Requirement for an internal control element. *Journal of Bacteriology, 173*(12), 3904–3906.

Khati, P., Chaudhary, P., Gangola, S., & Sharma, A. (2017). Nanochitosan-induced growth of *Zea mays* with soil health maintenance. *3 Biotech, 7*(81), 1–9.

Khati, P., Chaudhary, P., Bhatt, P., Nisha, Kumar, R., & Sharma, A. (2018). Effect of nanozeolite and plant growth promoting rhizobacteria on maize. *3 Biotech, 8*(141), 1–12.

Khati, P., Chaudhary, P., Gangola, S., & Sharma, A. (2019a). Influence of nanozeolite on plant growth promotory bacterial isolates recovered from nanocompound-infested agriculture field. *Environment and Ecology, 37*(2), 521–527.

Khati, P., Sharma, A., Chaudhary, P., Singh, A. K., Gangola, S., & Kumar, R. (2019b). High-throughput sequencing approach to assess the impact of nanozeolite treatment on species richness and evenness of soil metagenome. *Biocatalysis and Agricultural Biotechnology, 20*, 101249.

Khati, P., Mishra, P. K., & Kant, L. (2024). Isolation, screening and characterization of drought tolerant plant growth promoting bacteria from Indian Himalayas. *Pantnagar Journal of Research, 22*(2).

Kleiber, M., & Rogers, T. A. (1961). Energy metabolism. *Annual Review of Physiology, 23*(1), 15–36.

Kortmann, J., & Narberhaus, F. (2012). Bacterial RNA thermometers: Molecular zippers and switches. *Nature Reviews. Microbiology, 10*(4), 255–265.

Kumar, G., Arya, K., Verma, A., Bhatt, P., Khati, P., Gangola, S., Kumar, R., Sharma, A., & Singh, H. (2017). Bioremediation of petrol engine oil polluted soil using microbial consortium and wheat crop. *Journal of Pure and Applied Microbiology, 11*(3), 1583–1588.

Lupas, A. (1996). [30] Prediction and analysis of coiled-coil structures. In *Methods in enzymology* (Vol. 266, pp. 513–525). Academic.

Malone, J. G., Mittova, V., Ratcliffe, R. G., & Kruger, N. J. (2006). The response of carbohydrate metabolism in potato tubers to low temperature. *Plant & Cell Physiology, 47*(9), 1309–1322.

Manzoni, S., & Porporato, A. (2009). Soil carbon and nitrogen mineralization: Theory and models across scales. *Soil Biology and Biochemistry, 41*(7), 1355–1379.

Mattanovich, D., Gasser, B., Hohenblum, H., & Sauer, M. (2004). Stress in recombinant protein producing yeasts. *Journal of Biotechnology, 113*(1–3), 121–135.

Mekalanos, J. J. (1992). Environmental signals controlling expression of virulence determinants in bacteria. *Journal of Bacteriology, 174*(1), 1–7.

Miteva, V., Sowers, T., & Brenchley, J. (2007). Production of N_2O by ammonia oxidizing bacteria at subfreezing temperatures as a model for assessing the N_2O anomalies in the Vostok ice core. *Geomicrobiology Journal, 24*(5), 451–459.

Nagai, H., Yuzawa, H., & Yura, T. (1991). Interplay of two cis-acting mRNA regions in translational control of sigma 32 synthesis during the heat shock response of Escherichia coli. *Proceedings of the National Academy of Sciences, 88*(23), 10515–10519.

Neubauer, P., Cruz, N., Glauche, F., Junne, S., Knepper, A., & Raven, M. (2013). Consistent development of bioprocesses from microliter cultures to the industrial scale. *Engineering in Life Sciences, 13*(3), 224–238.

Nishiyama, S. I., Nara, T., Homma, M., Imae, Y., & Kawagishi, I. (1997). Thermosensing properties of mutant aspartate chemoreceptors with methyl-accepting sites replaced singly or multiply by alanine. *Journal of Bacteriology, 179*(21), 6573–6580.

Panday, S. C., Parihar, M., Meena, R. P., Chaudhary, M., Meena, V. S., Mondal, T., Khati, P., Singh, A. K., Bisht, J. K., Kant, L., & Singh, R. D. (2024). Long-term fertilization strategies for improving productivity, profitability and water use efficiency of soybean–wheat cropping system. *The Journal of Agricultural Science*, 1–14.

Parihar, M., Mishra, P. K., Paschapur, A., Kumar, A., Khati, P., Joshi, D., Bisht, J., & Kant, L. (2025). Elevation gradients and soil characteristics shape arbuscular mycorrhizal fungal communities in the Indian mid-Himalaya. *Nordic Journal of Botany, 1–13*, e04760.

Parsons, L. L., & Smith, M. S. (1989). Microbial utilization of carbon-14-glucose in aerobic vs. anaerobic denitrifying soils. *Soil Science Society of America Journal, 53*(4), 1082–1085.

Piper, P. W. (1995). The heat shock and ethanol stress responses of yeast exhibit extensive similarity and functional overlap. *FEMS Microbiology Letters, 134*(2–3), 121–127.

Price, G. D., Ruffell, A. H., Jones, C. E., Kalin, R. M., & Mutterlose, J. (2000). Isotopic evidence for temperature variation during the early Cretaceous (late Ryazanian–mid-Hauterivian). *Journal of the Geological Society, 157*(2), 335–343.

Ralser, M., Wamelink, M. M., Latkolik, S., Jansen, E. E., Lehrach, H., & Jakobs, C. (2009). Metabolic reconfiguration precedes transcriptional regulation in the antioxidant response. *Nature Biotechnology, 27*(7), 604–605.

Rivkina, E. M., Friedmann, E. I., McKay, C. P., & Gilichinsky, D. A. (2000). Metabolic activity of permafrost bacteria below the freezing point. *Applied and Environmental Microbiology, 66*(8), 3230–3233.

Rohde, J. R., Fox, J. M., & Minnich, S. A. (1994). Thermoregulation in Yersinia enterocolitica is coincident with changes in DNA supercoiling. *Molecular Microbiology, 12*(2), 187–199.

Santos, H., & Da Costa, M. S. (2002). Compatible solutes of organisms that live in hot saline environments. *Environmental Microbiology, 4*(9), 501–509.

Schimel, J. P., & Weintraub, M. N. (2003). The implications of exoenzyme activity on microbial carbon and nitrogen limitation in soil: A theoretical model. *Soil Biology and Biochemistry, 35*(4), 549–563.

Schimel, D., Melillo, J., Tian, H., McGuire, A. D., Kicklighter, D., Kittel, T., et al. (2000). Contribution of increasing CO_2 and climate to carbon storage by ecosystems in the United States. *Science, 287*(5460), 2004–2006.

Schlesinger, W. H., & Andrews, J. A. (2000). Soil respiration and the global carbon cycle. *Biogeochemistry, 48*, 7–20.

Schulte, P. M. (2015). The effects of temperature on aerobic metabolism: Towards a mechanistic understanding of the responses of ectotherms to a changing environment. *The Journal of Experimental Biology, 218*(12), 1856–1866.

Shen, J., & Bartha, R. (1996). Metabolic efficiency and turnover of soil microbial communities in biodegradation tests. *Applied and Environmental Microbiology, 62*(7), 2411–2415.

Shima, J., & Takagi, H. (2009). Stress-tolerance of baker's-yeast (Saccharomyces cerevisiae) cells: Stress-protective molecules and genes involved in stress tolerance. *Biotechnology and Applied Biochemistry, 53*(3), 155–164.

Smits, G. J., & Brul, S. (2005). Stress tolerance in fungi—To kill a spoilage yeast. *Current Opinion in Biotechnology, 16*(2), 225–230.

Steinweg, J. M., Plante, A. F., Conant, R. T., Paul, E. A., & Tanaka, D. L. (2008). Patterns of substrate utilization during long-term incubations at different temperatures. *Soil Biology and Biochemistry, 40*(11), 2722–2728.

Stetter, K. O. (2006). Hyperthermophiles in the history of life. *Philosophical Transactions of the Royal Society B: Biological Sciences, 361*(1474), 1837–1843.

Tobe, T., Yoshikawa, M., Mizuno, T., & Sasakawa, C. (1993). Transcriptional control of the invasion regulatory gene virB of Shigella flexneri: Activation by virF and repression by H-NS. *Journal of Bacteriology, 175*(19), 6142–6149.

Tomer, S., Khati, P., Suyal, D. C., Khan, F., Perveen, K., & Barasarathi, J. (2024). Relative multi-beneficial effect of microorganisms on plant health of chickpea (*Cicer arietinum* L. var. PG-186). *Frontiers in Microbiology, 15*, 1452553.

Van de Vossenberg, J. L., Ubbink-Kok, T., Elferink, M. G., Driessen, A. J., & Konings, W. N. (1995). Ion permeability of the cytoplasmic membrane limits the maximum growth temperature of bacteria and archaea. *Molecular Microbiology, 18*(5), 925–932.

Van Groenigen, K. J., Six, J., Hungate, B. A., de Graaff, M. A., Van Breemen, N., & Van Kessel, C. (2006). Element interactions limit soil carbon storage. *Proceedings of the National Academy of Sciences, 103*(17), 6571–6574.

White-Ziegler, C. A., Angus Hill, M. L., Braaten, B. A., Van Der Woude, M. W., & Low, D. A. (1998). Thermoregulation of Escherichia coli pap transcription: H-NS is a temperature-dependent DNA methylation blocking factor. *Molecular Microbiology, 28*(6), 1121–1137.

Wittmann, C., Weber, J., Betiku, E., Krömer, J., Böhm, D., & Rinas, U. (2007). Response of fluxome and metabolome to temperature-induced recombinant protein synthesis in Escherichia coli. *Journal of Biotechnology, 132*(4), 375–384.

Yvon-Durocher, G., Caffrey, J. M., Cescatti, A., Dossena, M., Giorgio, P. D., Gasol, J. M., et al. (2012). Reconciling the temperature dependence of respiration across timescales and ecosystem types. *Nature, 487*(7408), 472–476.

Zeldes, B. M., Keller, M. W., Loder, A. J., Straub, C. T., Adams, M. W., & Kelly, R. M. (2015). Extremely thermophilic microorganisms as metabolic engineering platforms for production of fuels and industrial chemicals. *Frontiers in Microbiology, 6*, 1209.

Adaptations to the Extremes: Survival Mechanisms of Microbes in Heat and Cold Environments

Amir Khan, Arjita Punetha, Raj Shekhar Sharma, Bharti Kukreti, Ajay Veer Singh, Hemant Dasila, Dipender Kumar, and Sunil Kashyap

Abstract

The exceptional adaptations of extremophilic microbes permit them to thrive in a broad range of temperatures, which include hot hydrothermal vents to freezing polar regions. The genetic and metabolic adaptations of extremophilic microorganisms allow their survival in such extreme conditions. Thermophilic microbes are located in hot environments like hydrothermal vents and hot springs and have adaptations to endure temperatures above 45 °C and, in some circumstances, above 100 °C. They hold a variety of molecular and physiological adaptations to survive, such as a dedicated lipid membrane system that is resistant to heat and thermostable enzymes, etc. On the other hand, subzero ecosystems such as permafrost, deep-sea habitats, and polar ice caps are examples of the

A. Khan (✉) · D. Kumar
CSIR-Central Institute of Medicinal & Aromatic Plants (CIMAP), Research Centre, Pantnagar, Uttarakhand, India

A. Punetha
G. B. Pant National Institute of Himalayan Environment, Almora, Uttarakhand, India

R. S. Sharma
Department of Microbiology, School of Basic and Applied Science, Sri Guru Ram Rai University, Dehradun, Uttarakhand, India

B. Kukreti · A. V. Singh
Department of Microbiology, College of Basic Sciences and Humanities, Govind Ballabh Pant University of Agriculture and Technology, Pantnagar, Uttarakhand, India

H. Dasila
Department of Microbiology, Akal College of Basic Sciences, Eternal University, Baru Sahib, Rajgarh, Himachal Pradesh, India

S. Kashyap
Punjab Agriculture University, Regional Research Station, Gurdaspur, Punjab, India

G. Bhandari et al. (eds.), *Microbial Resilience*,
https://doi.org/10.1007/978-981-95-8883-1_4

home of psychrophilic microbes. These psychrophilic microbes produce cold-active enzymes, antifreezing proteins that prevent the production of ice crystals, and cryo-protectants like trehalose and glycerol that protect cells from any harm during cycles of freezing and thawing under low temperatures. Moreover, unsaturated fatty acids present in the membrane also preserve membrane fluidity. Such amazing adaptations of thermophilic and psychrophilic microorganisms are discussed in this chapter, with an emphasis on the tactics that allow them to survive and proliferate in both high- and low-temperature environments.

Keywords

Adaptations · Microbes · Thermophile · Psychrophiles

1 Introduction

Many different types of environmental conditions exist on the surface of the Earth. Extreme conditions, including the polar regions, iced waters, acidic hot springs, arid rock surfaces, deserts, and even deep-sea habitats, are ideal for a variety of bacteria. The prevailing temperature among these places has a significant variation. Temperature immediately impacts the interior of the cells, making it a significant component in microbial growth and function (Marzban & Tesei, 2025). Microorganisms can be divided into three major groups based on how well they tolerate different temperatures: thermophiles, mesophiles, and psychrophiles. High-altitude environments, characterized by extremely low temperatures, reduced oxygen availability, intense ultraviolet radiation, and nutrient-deficient soils, these harsh conditions impose severe constraints on every forms of life, including plants and microbes. Even in an iced region, microorganisms that have acclimated to the cold can flourish. The chemical and physical characteristics of live cells are significantly changed by cold stress (Khan et al., 2023a). These microorganisms are further divided into two groups: psychrophiles sensu stricto, which grow best at temperatures below 15 °C with a maximum of 20 °C, and psychrotrophs (also known as psychrotolerants), which can divide at temperatures as low as 0 °C and thrive at temperatures between 20 and 25 °C. Thermophiles and hyperthermophiles, on the other hand, are incredible species that thrive in extremely hot surroundings. They have amazing adaptations that enable them to not only survive but also thrive in extremely hot environments. Temperatures between 45 °C and above 100 °C are ideal for the growth and reproduction of the majority of thermophiles and hyperthermophiles. Despite facing similar thermal restrictions, thermophiles and psychrophiles are members of separate genera and use distinct molecular processes to adapt to severe temperatures. Maintaining critical biological processes at both high and low temperatures requires modifications. Microorganisms adapt to these temperature variations by changing several physical and biochemical characteristics, such as protein structure, membrane fluidity, and stability, and gene expression. Protein genetic alterations affect the translation system and enzyme activity, two processes essential for temperature adaptation. Lipid changes are also essential for controlling the permeability and fluidity of membranes. Microorganisms experience a variety of metabolic modifications that improve their survival in environments with high temperatures (Kochhar et al., 2022). Microbes also create molecular chaperones, heat shock proteins, cold shock proteins, and suitable solutes to assist them in surviving in extremely hot or cold environments (Storey & Storey, 2023). Given that the majority of terrestrial and aquatic ecosystems on Earth are either permanently or seasonally exposed to freezing temperatures, extremophilic microbes are essential to world ecology. 90% of the oceans' volume is kept below 5 °C, and they make up 71% of the planet's surface. These microbes contribute to nitrogen cycling and crop growth in the polar areas, which comprise 14% of the Earth's surface (Hébraud & Potier, 1999). However, severe heat settings, including fumaroles, geysers, volcanic regions, hot springs, mudpots, coastal thermal springs, and deep-sea hydrothermal vents, are ideal for thermophiles and hyperthermophiles. Microbes and the bioproducts they generate are useful in a variety of industrial and medical fields in addition to their ecological significance. For example, food processing, baking, textile production, bioremediation, and biotransformation all depend on enzymes that are cold-adapted and thermostable (Khan et al., 2020; Mesbah, 2022). The ecological and industrial significance and microbial adaptations to temperature changes are discussed in this chapter.

2 Microbial Adaptations in High-Temperature Conditions

In order to survive in the high temperature environment, thermophiles and hyperthermophiles acquire unique genomic and biochemical features, along with specialized structural modifications, that enhance their cellular firmness and functionality in such extreme conditions. Their cellular membranes, nucleic acids, and proteins are uniquely adapted to sustain their integrity and operational efficiency, ensuring that they can carry out essential biological processes even in environments that would be detrimental to most other forms of life. This remarkable ability to endure and excel in high-temperature habitats makes thermophiles and hyperthermophiles fascinating subjects of study in microbiology and biotechnology.

2.1 Thermophiles: Mechanism of Adaptations

Thermophiles are the microorganisms that flourish at moderately high temperatures, typically ranging from 45 to 80 °C. They survive in such extreme circumstances by employing various molecular and structural adaptations. Their proteins are highly hydrophobic, which minimizes interaction with water and helps avert protein unfolding at elevated temperatures. Additionally, salt bridges and electrostatic interactions between charged amino acids such as arginine, glutamate, and lysine stabilize the protein structure, which minimizes the risk of denaturation. To manage protein misfolding, thermophiles utilize heat shock proteins (Hsp60 and Hsp70)

(Wang et al., 2015). HSPs or heat shock proteins consist of a class of pervasive proteins, which are believed to protect the organism from thermal stress as well as assist in refolding of several important proteins and obstruct aggregation. Membrane lucidity is attained by increasing the wedge of saturated fatty acids in the lipid bilayer, which reduces fluidity of the membrane and prevents leakage (Hu et al., 2022). These unique adaptations enable thermophiles to thrive in geothermal environments, including hot springs and industrial reactors.

2.1.1 Protein Thermostability

Temperature has an impact to cell with no difference between outer and inner cellular boundaries. The considerable provocation for thermophiles is to support the cellular protein in its native configuration. Thermophilic proteins maintain higher stability at high temperatures because of multiple structural characteristics that enhance stability. Not all proteins in thermophiles are thermally stable. Only those proteins that perform some specific functions, like regulator proteins and catalysts, are thermally stable. These heat-resistant proteins include thioredoxin reductase, proteasome-related proteins, triosephosphate isomerase, translation initiation factors, etc. In addition to this, thermophiles possess an elevated amount of disulfide bonds in their proteins with respect to the mesophiles and use them as a major mechanism for the stabilization of proteins (Wang et al., 2015). Quantification of disulfide bond revealed that *Pyrobaculum aerophilum* contains about 47% of cysteine residues as compared to *E. coli*, which only contains 8% of cystine residue in cell lysate (Boutz et al., 2007). In addition to this, the thermostability of protein also depends on some additional factors like increased ionic interactions, hydrophobic core enhancement, and secondary structure reinforcement. Increased ionic interactions contain enhanced ionic connections that produce enhanced salt bridges, which strengthen the protein framework. Strengthening the hydrophobic core and enhancing hydrophobic interactions among core residues of protein contributes to stabilizing the protein structure and reduces its susceptibility to temperature-induced denaturation. Secondary Structure Reinforcement, the stringency and steadiness of proteins increase because of higher concentrations of α-helices and β-sheets (Kohli et al., 2020).

2.1.2 Genomic Features: Adaptation at the DNA Level

Thermal denaturation of DNA is affected by nucleotide composition. Organisms inhabiting at higher temperature contain the high level of G+C content. The thermal stability of DNA increases with high G+C content because G–C base pair formation creates stronger triple hydrogen bonds. There is no obvious relation between the G +C content of a genome and the optimal environmental temperature for growth of an organism. Many thermophilic bacteria, like *Aquifex aeolicus* and *Pyrococcus abyssi*, have genomic G+C content >50%, while several mesophilic bacteria, like Mycobacterium, have much more G+C content in their genomes (Hickey & Singer, 2004). Hence, it is concluded that thermophiles have several other mechanisms instead of G +C content for sustaining the double-stranded DNA structure at high temperature. Thermophilic enzymes and dinucleotide selection also contribute to DNA thermostability (Nakashima et al., 2003). DNA supercoiling also plays an interesting role in DNA thermostability. DNA supercoiling becomes positive through reverse gyrase activity, thus raising DNA melting temperatures, which helps safeguard DNA from thermal denaturation (Hickey & Singer, 2004).

2.1.3 Membrane Adaptations

The job of the biological membrane in thermophiles is essential; it acts as a barrier for solute diffusion, performs a central or key role in energy storage and processing through an ion gradient. Therophilic and hyperthermophilic Archaea possess bipolar membrane-spanning tetraether lipids, and hence, it forms lipid monolayer membranes instead of bilayer membranes. The advantage of bipolar lipids is it packs more than archaeol-based bilayer membranes, which increases membrane rigidity in addition to checking little proton leakage at high temperature and acidic pH (Cario et al., 2015). According to Charlier and Droogamans (2005), for the actual functioning of thermophilic membranes at high temperatures, the liquid-crystalline state of the membrane should be maintained, sufficiently impermeable for small molecules or ions, properly generate proton motive force, and perform pH homeostasis. To cope with high temperatures, membrane adaptation includes modification in phospholipids for fluidity and integrity. Major adaptations in thermophilic bacteria involve extension in acyl chain length, level of fatty acids saturation, cyclization, and branching (Min et al., 2020). Thermophile organisms contain ether-linked lipids as part of their membranes since these chemical structures withstand heat-induced hydrolysis better than the ester-linked lipids found in other cells (Wani et al., 2022). Furthermore, the detrimental effects of increased proton permeability of the membrane at high temperatures are limited by the nonstoichiometric proton sodium antiport system, which generates a secondary sodium gradient, and its maintenance is less temperature dependent (Albers et al., 2001).

2.1.4 Heat Shock Proteins and Chaperonins

Heat shock proteins or HSPs are diverse sets of proteins majorly classified on the basis of molecular weight. Despite thermophilic bacteria have its own ideal growth temperatures, responses connected to heat shock are still induced at temperatures greater than optimal temperatures. During heat shock stress, distinct heat shock proteins are induced, such as Hsp70, Hsp90, Hsp104, Hsp60, Hsp27, Hsp40, and Hsp10, according to their molecular weight (Sahlan & Yohda, 2013). Several of these proteins function as molecular chaperones, which catalyze the refolding of denatured proteins, also assist in the turning up of newly synthesized proteins, preventing protein aggregation, and in the folding of nascent or unfolded polypeptides into mature conformations (Laksanalamai et al., 2004). Molecular chaperon systems of several thermophilic bacteria are quite distinct from those of thermophilic archaea. Acidothermophilic archaea, i.e., *Sulfobolus* sp. strain B12, exhibit an acquired temperature tolerance response (Trent et al., 1991). TF55 and TCP1 were identified as a new class of molecular chaperons; TF55 is mainly present in archea while TCP1 is mainly present in eukaryotic cytosol. Later, these archaeal and cytosolic chaperones were classified as group II chaperonins (Sahlan & Yohda, 2013).

2.2 Hyperthermophile: Mechanism of Adaptation

Hyperthermophiles thrive in extreme conditions, with ideal living temperatures ranging from above 80 °C to as high as 122 °C. To thrive in harsh environments, they have developed remarkable adaptations. Their proteins are stabilized by a greater standard of hydrophobic packing in the core and enhanced ionic interactions, which significantly minimize protein unfolding at high temperatures. Additionally, disulfide bonds established linking the cysteine residues further strengthen protein structures. Small heat shock proteins and chaperonins, such as GroEL and GroES, assist in maintaining proper protein folding. Hyperthermophilic archaea's molecular chaperone system is very similar to those of other organisms. In hyperthermophilic archaeal system only six types of chaperones have been discovered so far, i.e., peptidyl-prolyl cis-trans isomerases, ATPase associated with various cellular activities (AAA proteins), small heat-shock proteins, group II chaperonins, and NAC chaperonins (Sahlan & Yohda, 2013). These hyperthermophilic archaea lack Hsp70, Hsp90, and Hsp100 chaperone systems (Laksanalamai et al., 2004). In the hyperthermophilic archaea group II chaperonins are highly induced during tensity or stress response mechanisms and at rising temperatures (Sahlan & Yohda, 2013). Hyperthermophiles also contain unique tetraether lipid monolayers composed mainly of diglycerol tetraethers and cyclopentane rings. These structures provide exceptional rigidity to membranes, preventing dissipation under high thermal stress. A distinctive adaptation found only in hyperthermophiles is the presence of reverse DNA gyrase, which introduces positive supercoiling into DNA, thereby preventing denaturation at extreme temperatures. Furthermore, histone-like proteins compact and stabilize DNA, while highly efficient DNA repair mechanisms help mitigate heat-induced DNA damage. The metabolic pathways of hyperthermophiles are optimized with heat-tolerant enzymes and modified electron transport chains, allowing them to function effectively even at high temperatures and allowing hyperthermophiles to thrive in hydrothermal vents, volcanic environments, and deep-sea geothermal ecosystems.

2.2.1 Thermostability of Protein and Enzymes

Recently, the completion of more and more genomic sequences of hyperthermophiles has led to an overwhelming quantity of comparative studies on protein and enzyme thermostability. The proteins of hyperthermophilic organisms achieve stability under extreme conditions since they contain particular molecular structures that maintain functionality at high temperatures. The amino acid makeup of hyperthermophilic proteins incorporates elevated concentrations of arginine and tyrosine amino acids because these amino acids generate stabilizing molecular bonds (Kohli et al., 2020). Proteins extracted from thermophilic organism contains high number of bulky hydrophobic and charged residues, mainly arginine, for stabilizing the core and ion networks, along with fewer nonpolar residues for stabilizing the periphery (Wang et al., 2015). Some other commonly used strategies for protein thermostability include the presence of proline residues in the β sheet, the presence of negatively charged amino acids at the N-terminus of the α-helix, and metal introduction at the binding site (Hawwa et al., 2009). Proteins find extraordinary stability through their numerous internal forces, which include ionic bonds, together with hydrogen bonds and hydrophobic contacts. *Homotetrameric methyltransferase* TrmI extracted from *Pyrococcus abyssi* is stabilized by intersubunit bridges formed due to disulfide bonding (disulfide bridges). Other stabilized proteins known as chaperone proteins facilitate protein stabilization. Thermal stress causes protein aggregation, but heat shock proteins and other chaperone proteins help fold or refold proteins properly to prevent this aggregation (Hickey & Singer, 2004).

2.2.2 Genomic Features

Thermal stability or resistance of double helical DNA appears to be improved in hyperthermophiles by an enzyme called reverse gyrase. It's a unique type of DNA topoisomerase I, which causes positive supercoiling for stabilization of DNA for thermal resistance. All hyperthermophiles (archaea and bacteria) contains reverse gyrase enzyme. This reverse gyrase is an exclusive feature of hyperthermophiles. Hyperthermophilic Archaea need to relax the slightly positive supercoiled DNA; in spite of these, Bacteria, mesophilic Archaea, and Eukarya have negatively supercoiled DNA. This topological state of hyperthermophilic Archaean DNA is possible due to the combined action of reverse gyrase as well as the presence of type II topoisomerase activity, which is responsible for the removal of negative supercoiling and boosts DNA thermal resistance (Charlier & Droogmans, 2005). In spite of this, archaeal hyperthermophiles contain histones which are phylogenetically related to eukaryotic histones, e.g., H2A, H3, H2B, and H4. Some special proteins which tie to DNA protect the DNA from heat denaturation and help maintain its structural integrity (Wang et al., 2015).

2.2.3 Membrane Adaptations

Components of cell-like proteins, nucleic acids, and lipids are usually heat-sensitive. Hyperthermophilic bacterial membrane lipids of *Thermotoga maritima* contain an unusual glycerol ether lipid, i.e., 15,16—dimethyl-30-glyceryloxy—triacontanedioic acid, which significantly helps in increasing the membrane stability against hydrolysis at high temperatures with respect to the ester lipids present in the mesophiles (Stetter, 1999). Some hyperthermophilic cells deploy lipids in monolayers built from tetraether lipids, which create remarkable thermal stability for their membranes (Wani et al., 2022). Cario et al. (2015) reported that *Thermococcus barophilus*, one of the utmost members of hyperthermophilic piezophiles, was able to thrive and grow up to the temperature 103 °C and pressure up to 80 MPa with the help of a major membrane-spanning tetraether lipid glycerol-dibiphytanyl glycerol tetraether (GDGT-0). Cyclopentane ring structures or cyclized lipids also enhance the membrane stability (Driessen & Albers, 2007). Hyperthermoacidophilic Archea like *Sulfolobus* contains C40 biphytanyl chains, which undergo further modification, i.e., internal cyclization (cyclopentane rings) as a result of triggered response to increasing temperatures.

2.2.4 Metabolic Adaptations

The high-temperature conditions force hyperthermophiles to follow chemolithoautotrophic metabolism by utilizing inorganic energy resources due to the scarce availability of organic nutrients. Hyperthermophilic Archaea, which thrives the temperature above 80 °C, possess a modified central metabolic pathway, often with promiscuous enzyme which help these organisms thermoadaptation, additionally to maintaining and thriving in very harsh environments (Schmerling et al., 2022). According to Amend and Shock (2001) judgment of the energy budget of hyperthermophiles quantitatively at appropriate pressure, chemical composition, and temperature of the system of interest is an important prerequisite for the determination of a large array of metabolic reactions that might be energy yielding (exergonic). Mandelli et al. (2017) reported that the bacterium *Thermus filiformis*, during heat shock, influenced its carotenoid profile, which results in favoring the synthesis of thermo-bis-zeaxanthins and thermo-zeaxanthins, which can be directly or indirectly involved in membrane stabilization at high temperatures. High temperatures trigger oxidative stress, which leads to the gene repression involved in the tricarboxylic acid cycle and glycolysis. Hence, in heat stress, glucose metabolism occurs through the pentose phosphate pathway instead of the glycolysis pathway.

3 Ecological and Industrial Significance of Thermophiles and Hyperthermophiles

As thermophiles and hyperthermophiles reside in harsh environment so bioproducts which are producing by them have unique features and therefore they are important in the sectors of industry, agriculture, and medicine. These organisms produce enzymes that are stable at temperatures between 90 and 105 °C, including amylase, cellulase, glucosidase, galactosidase, protease, pullulanase, and xylanase (Ajeje et al., 2021). These thermoenzymes, also known as extremozymes, are especially useful in high-temperature-operating industries like feed production, food processing, pulp and paper, and brewing. Furthermore, it may be possible to find novel medicinal chemicals from extremophilic bacteria. For instance, pyochelin, an iron-binding antifungal drug, has been isolated from a novel strain of Pseudomonas, and extracts from these organisms have shown antifungal activity against *Aspergillus* and *Candida* species (Combie et al., 2001). The applications of thermophilic and hyperthermophilic microorganisms are vast and continue to expand across various fields, as discussed below:

3.1 Industrial Applications of Thermophiles and Hyperthermophiles

3.1.1 Liquid biofuels

Biofuels are made by fermenting materials like lignocellulosic biomass and starch or by removing lipids from different sources, including microbes and plants. In addition to being necessary in the current oil crisis, the production of bioethanol also contributes significantly to the reduction of greenhouse gas emissions (Bala & Singh, 2019). Particularly, bioethanol produced from lignocellulose is eco-friendly, which is produced by pretreatment, enzymatic conversion of polysaccharides into fermentable monosaccharides, and fermentation into bioethanol. One crucial process that has a big influence on total production costs is the enzymatic saccharification of biomass. One essential characteristic that improves biomass saccharification efficiency is thermostatability (Candido et al., 2019). Thermostable enzymes from thermophilic microbes can be used to optimize this process (Kuhad et al., 2016). For the generation of liquid biofuel, thermophiles and hyperthermophiles can produce thermostable enzymes that efficiently break down lignocellulosic biomass components such as xylan, cellulose, and hemicellulose. For instance, cellulases and xylanases from *Clostridium* spp., *Caldanaerobius*, and *Caldicellulosiruptor* have high lignocellulosic degradation activity, which makes them useful for the production of different liquid biofuels (Sun et al., 2015). Bhandiwad et al. (2013) conducted a noteworthy study on the synthesis of biobutanol and found that overexpressing certain genes in the bcs operon of *Thermoanaerobacterium thermosaccharolyticum*, which are in charge of generating butyryl CoA, increased n-butanol production by 180%. Microorganisms such as cyanobacteria can produce biohydrogen by photosynthesizing sunlight and water (Parmar et al., 2011). Extremozymes like hydrogenase are better suited than conventional enzymes because they convert glucose to hydrogen more effectively at greater temperatures. For example, one of only two enzymes known to use NADPH electrons for hydrogen generation is the hyperthermophile *Pyrococcus furiosus*, which is found in deep-sea hydrothermal vents. It produces a hydrogenase that functions best at 85 °C (Raven et al., 1992). By secreting enzymes, some thermophilic fungi, such as *Thielavia terrestris*, *Thermoascus aurantiacus*, *Scytalidium thermophilum*, and *Sporotrichum thermophile*, are also efficient at breaking down lignocellulosic biomass (Ajeje et al., 2021).

3.1.2 Biosurfactants

Microorganisms produce amphiphilic molecules called biosurfactants, which are essential for enhancing the emulsification and dispersion of water-insoluble materials. Glycolipids, lipopolysaccharides, lipoproteins, phospholipids, hydroxylated and cross-linked fatty acids, and even entire cells are examples of these substances. Because of their benefits, which include reduced toxicity, increased biodegradability, improved environmental compatibility, increased foam formation, increased activity under extreme temperature, pH, and salinity conditions, and the potential to be derived from renewable resources, their use has expanded dramatically across a variety of industries, including food, agriculture, pharmaceuticals, petroleum, and paper, pulp (Sharafi et al., 2014). Because they are more stable at high temperatures and resistant to adverse physicochemical conditions, such as excessive pH and salinity, thermophiles have demonstrated special promise in the creation of biosurfactants. The genera *Aneurinibacillus*, *Geobacillus*, *Alcaligenes*, *Bacillus*, and *Brevibacillus* are home to the promising species that have been found thus far (Urbieta et al., 2015). For example, *Bacillus licheniformis* was recovered from the Zilaei oil reservoir in southwest Iran by Daryasafar et al. (2016). At 50°C, this strain flourished and produced a biosurfactant that decreased interfacial tension from 36.8 to 0.93 mN m^{-1} and surface tension from 72 to 23.8 mN m^{-1}. Comparably, in a different investigation, Coronel-León et al. (2015) isolated *Bacillus licheniformis* AL 1.1. This strain developed a biosurfactant that showed stability throughout a wide pH range (6.0–11.0), high temperatures (up to 120 °C), and different salt concentrations (up to 20% NaCl). It also obtained a critical micelle concentration of 15 mg L^{-1} and reduced water surface tension to 28.5 mN m^{-1}. These results demonstrate how thermophilic microbes can produce strong biosurfactants that are appropriate for a range of industrial uses.

3.1.3 Pulp and paper industry

The hydrolysis of hemicellulose by xylanases is essential to the enzymatic treatment of wood for the formation of pulp. One of the most prevalent organic compounds on Earth and the main component of hemicellulose, xylan, is broken down by these enzymes (de Miguel Bouzas et al., 2006). High temperatures and an alkaline pH are used during the pulp-making process to efficiently break down the cell wall structure. High thermostability and activity at neutral to alkaline pH are hence requirements for industrially viable xylanases. To stop cellulose fibers from breaking down, these enzymes should also not have cellulolytic activity. Therefore, finding enzymes that maintain their stability and activity in the face of significant pH and temperature changes is essential for their use in the pulp and paper sectors. In the paper business, thermostable enzymes are superior to thermostable ones in a number of ways. There is a lot of promise for using xylanases and cellulases that are stable at high temperatures and resistant to high pH in the pulp and paper industry. It has been demonstrated that these enzymes improve the concentration, purity, brightness, and fiber surface permeability of dissolved pulp, which increases paper strength and makes it easier for bleaching agents to diffuse (Yang et al., 2020). *Streptomyces griseorubens* LH-3 is the source of a highly thermostable endo-xylanase that functions best at 60 °C and pH 5.0. When eucalyptus Kraft pulp was biobleached using this purified xylanase, the brightness increased by 14.5% and the kappa numbers decreased by 24.5%. It is regarded as a promising biobleaching agent for the production of pulp and paper because of its advantageous industrial characteristics (Wu et al., 2018). Similarly, the biobleaching effect of a thermo-alkali-stable xylanase generated by *Bacillus tequilensis* strain UD-3, which was isolated from a hot spring, was assessed on rice straw pulp. During the Xyl + Zn treatment, this xylanase, which is most active at 50 °C and pH 8.0, dramatically decreased the amounts of reducing sugars (50%), phenolic compounds (29.19%), hydrophobic compounds (33.20%), and lignin (35.86% and 40.48%) as compared to xylanase alone (Patel & Dudhagara, 2020). This highlights how thermostable and alkali-stable enzymes might improve the sustainability and efficiency of pulp and paper manufacturing processes.

3.1.4 Food industries

Every group of thermophiles has distinct qualities that can be used to create enzymes with a variety of industrial uses. For instance, cellulases produced from bacteria and fungi show great potential for use in food production. Cellulases are employed in conjunction with other enzymes in the juice business to increase output, optimize extraction effectiveness, and stabilize and clarify juices (Chakraborty et al., 2016). Additionally, these enzymes help extract flavonoids from floral nectars and seeds and effectively lower the viscosity of nectars and purees made from fruits, including apricot, mango, pomegranate, pear, and peach (Uzuner & Cekmecelioglu, 2019). Specifically, β-glucosidases can alter glycosylated precursors to improve the aroma of the wine (Chakraborty et al., 2016). Cellulases, when combined with other enzymes, may effectively extract olive oil, which is well-known for its health benefits, such as being a high source of monounsaturated fats, vitamin E, phenolic compounds, and antioxidants (Bisht et al., 2015). Furthermore, *Rhodothermus marinus* and other thermophilic bacteria are prized for their thermally stable α-amylase enzymes, which are frequently employed in fermentation (Gimbi & Kitabatake, 2002). Together with other enzymes, xylanases are used in baking to increase yields and enhance bread volume, quality, and baking efficiency (Tekkol et al., 2017). Xylanases improve water absorption and produce a softer, more elastic dough by breaking down hemicellulose in wheat flour (Dahiya & Singh, 2019).

3.2 Ecological Significance of Thermophiles and Hyperthermophiles

3.2.1 Bioremediation

A possible method for treating metal-contaminated locations is the use of extremophilic bacteria, especially thermophiles. By means of metabolic oxidation processes involving sulfur and/or iron, thermophiles improve metal solubilization (Sen et al., 2014). They can immobilize hazardous metals during bioremediation, even in heated wastewater or radioactive waste disposal sites, because they can reduce a variety of metals, such as Mn(IV), Cr(VI), U(VI), Tc(VII), Co(III), Mo(VI), Au(I, III), and Hg(II). Species such as *Geobacillus*, *Anoxybacillus flavithermus*, *Thermus thermophilus*, and the archaeon *Thermococcus zilligii* have been used to study the biosorption of metals, including Cd, Cu, Ni, Zn, and Mn (Chatterjee et al., 2010; Sar et al., 2013). In the treatment of heated nuclear waste streams, thermophiles provide notable benefits for the precipitation and immobilization of radionuclides and hazardous metals (such as U, Cr, Tc, and Co). *Thermus scotoductus*, *Pyrobaculum islandicum*, *Thermoanaerobacter* sp., and *Thermoterrabacterium ferrireducens* are among the species that have been shown to reduce uranium and technetium enzymatically (Chernyh et al., 2007). Furthermore, it has been reported that thermophilic species reduce Cr(VI), Se(IV), and Te (IV) (Sar et al., 2013). Since higher temperatures make hydrocarbons more soluble and bioavailable, thermophiles are also useful for the biodegradation of hydrocarbons with limited water solubility. Using an upflow anaerobic sludge blanket inoculated with sulfidogenic sludge from hydrothermal vent sediments, Guerrero-Barajas et al. (2014) reported a 74% biodegradation of trichloroethylene (TCE) in the case of haloorganic compounds. Genera including *Dehalobacter*, *Desulfotomaculum*, *Sulfospirillum*, *Desulfitobacterium*, *Desulfovibrio*, and *Clostridium* were among the microorganisms found during TCE biodegradation. Additionally, thermophiles have the capacity to break down extremely harmful organic substances. For example, it has been shown that thermophilic bacteria belonging to the genera *Aeribacillus* and *Geobacillus* can biodegrade a variety of resistant hydrocarbons and aromatic chemicals (Mnif et al., 2014). Additionally, azo dyes, which make up around 50% of industrial colors and are extremely dangerous, have been degraded by *Anoxybacillus* sp. Thermophilic biofiltration has the benefit of not requiring gas cooling, which lowers operating expenses. The removal of volatile organic chemicals, including ethanol, benzene, ethyl acetate, and gases containing sulfur, has shown encouraging results (Ryu et al., 2009). These uses demonstrate the adaptability and promise of thermophiles in tackling a range of environmental issues.

3.2.2 Soil functioning

Since many vital activities of the soil are mostly or completely driven by microbes, microbial activity is strongly linked to the health and usefulness of soil. Extracellular enzymes, which are necessary for breaking down organic debris, can be produced by thermophiles in temperate habitats during recurrent or sporadic high-temperature episodes. These enzymes have the ability to last in the environment and become active during hot weather, which aids in the breakdown of organic matter in the soil (Gomez et al., 2020). According to Rosa et al. (2023), microorganisms play a crucial role in maintaining soil activities, such as the biogeochemical cycles of components that promote plant growth and ease interactions between microbes and plants. In harsh environments, such as high temperatures and desiccation, soil thermophiles are able to carry out essential tasks that other microbes are unable to. For example, substantial amounts of nitrogen are found in soil organic matter, such as humic acids (Yan et al., 2021). Nitrogen is released as ammonium by thermophiles at rates similar to those of mesophilic super-ammonifiers. Through the breakdown of soil organic matter, thermophiles are able to preserve or even increase the availability of inorganic nitrogen during times of intense heat and dryness. Due to its restricted availability of sulfur in soils, sulfur is frequently supplied as a fertilizer in agriculture; nevertheless, the majority of soil sulfur is bound in organic matter, including proteinaceous compounds and humic acids (Yan et al., 2021). According to studies, thermophiles are crucial for the sulfur cycle in severe environments because they emit sulfate at significantly higher rates than mesophiles (Santana et al., 2021). Despite being abundant in soils, phosphorus, another necessary nutrient, is frequently scarce because it is usually bonded in minerals and organic matter. Phosphates can be dissolved by a variety of microbes, including thermophiles, phosphatases enzymes generated by some soil thermophiles, such as strains linked to *Geobacillus*, play a critical role in phosphate solubilization during high-temperature and desiccation periods (Margalef et al., 2021). These results highlight how important thermophiles are to preserving soil health and nutrient cycling in harsh environmental settings.

4 Microbial Adaptations Toward Cold Environments

Microorganisms residing in cold environments experience slower metabolic reactions, higher solvent viscosity, reactive oxygen species (ROS), and increased gas solubility under cold environments. These circumstances have the potential to impair biological processes by interfering with solute transport and diffusion, resulting in problems like ice formation and osmotic stress. These microbes have evolved specific cold tolerance mechanisms to withstand these frigid temperatures and maintain metabolic functions. The next section covers some of these significant modifications.

4.1 Protein Synthesis

Despite the fact that temperature changes can interrupt protein synthesis, these bacteria have developed special traits that allow them to continue operating efficiently. Studies have also demonstrated that cold shock is followed by an increase in cold shock gene production and a decrease in housekeeping gene expression. Researchers has led to know that cold stress causes cells to synthesize proteins at a slower rate, and shifts the protein synthesis of psychrophilic proteins that minimize damage and trigger biological reactions to compensate for stunted growth. For example, ribosomal proteins and RNA chaperones make up around 30% of the proteins that are increased in the Antarctic bacteria *P. haloplanktis* at low temperatures, i.e., 4 °C (Purwar & Srivastava, 2024). Research has indicated that psychrophilic proteins frequently contain greater amounts of proline and isoleucine. Additionally, their polypeptide chains contain significantly higher amounts of hydrophobic, nonpolar, aromatic amino acids like tryptophan and phenylalanine, as well as neutral hydrophilic amino acids like serine and asparagine. These proteins are stabilized during cold-induced denaturation due to their increased hydrophobicity (Panja et al., 2020). It is crucial to remember that heat sensitivity is a common feature of psychrophilic proteins with catalytic multidomains. This occurrence might result from the protein's outer surface having fewer salt bridges, which restricts its structural flexibility. Additionally, several cold shock proteins regulate transcription, which may have an impact on DNA supercoiling. For example, the transcription of other cold shock genes is regulated by CspA, a crucial cold shock protein in *E. coli*. Antarctic psychrotrophic bacteria and *Bacillus subtilis* have been identified to have similar proteins with similar functions (Hébraud & Potier, 1999). Similarly, although not being strictly required for cell growth, a number of other cold shock proteins have been discovered that help to mitigate cold shock. For instance, nucleoid-associated DNA-binding protein of *E. coli*, i.e., H-NS, mitigates the negative effects of low temperatures. Furthermore, the trigger factor (TF) is essential for the cotranslational folding of proteins. As the temperature drops, the TF level gradually rises. *E. coli* cells, which overexpress TF has better survival in low-temperature environments, whereas cells with lower TF levels are more susceptible to cold. By encouraging the refolding of cold-damaged proteins, TF is thought to facilitate continuous protein production and folding at low temperatures and preserve the functionality of pre-existing proteins (Phadtare, 2004).

4.2 Protein Folding

It is a crucial process that frequently restricts psychrophile development and causes the creation of certain proteins known as cold shock proteins. In contrast to mesophilic microbes, psychrophilic organisms constantly overexpress these chaperones in response to low temperatures in order to fight off cold-induced denaturation, which breaks down hydrophobic connections and increases the danger of protein misfolding. For newly generated polypeptides to function completely, they must precisely fold into their tertiary and quaternary structures. It is well known that psychrophilic proteins are more flexible, which improves their stability (Feller, 2003). These proteins frequently have lower proline content, fewer electrostatic interactions, and greater hydrogen bonds and van der Waals interactions. Lipases from *Candida antarctica* exhibit exceptional stability, which is explained by their stiffness, increased van der Waals forces, decreased electrostatic interactions, and stronger hydrogen bonds. Moreover, proteins such as IF2 and Hsc66 have been proposed to help ensure appropriate protein folding at low temperatures (Caldas et al., 2000; Lelivelt & Kawula, 1995).

4.3 Cold Shock Proteins, Antifreeze Protein, and Chaperones

The growth and survival of psychrophilic bacteria depend on cold-induced proteins, such as cold shock proteins (CSPs) and cold acclimation proteins (CAPs), which help stabilize RNA and inhibit the production of secondary structures at low temperatures. For example, cold-shock protein HZB mutants from *Bacillus caldolyticus* (PDB: 1HZB) and CspA from *Escherichia coli* (PDB: 1A04) demonstrate their significance in adapting to abrupt temperature changes and preserving cellular integrity via electrostatic interactions. By attaching to ice surfaces, antifreeze proteins (AFPs), including those from *Hypogastrura harveyi* (PDB: 5ANP) and *Pseudopleuronectes americanus* (PDB: 1AI7), stop ice crystals from forming and shield cells from harm caused by cold (Purwar & Srivastava, 2024). According to molecular research, the RNA-binding cold-shock domain (CSD) of CSPs, which are found in many psychrotrophic bacteria, has highly conserved glycine-rich regions. Although they have low thermostability, CSPs are tiny proteins, usually consisting of 50 amino acids, that show particular activity at low to moderate temperatures (Giaquinto et al., 2007). By attaching to the surfaces of ice crystals, antifreeze proteins (also known as Ice Binding Proteins or IBPs) are crucial for cell protection. These consist of ice recrystallization inhibiting proteins (IRIPs) and antifreeze glycoproteins (AFGPs). Research on *Rhodococcus* sp. JG3 focuses on adaptation strategies to cold and salty conditions documented that bacteria use IBPs or related compounds to control environmental stress and affect freezing characteristics (Garcia-Descalzo et al., 2020). Molecular chaperones, such as caseinolytic proteases (Clps), transcription factor (TF), GroEL, DnaK, and GroES, are crucial for the synthesis of recombinant proteins. In *Escherichia coli*, these chaperones are increased during cold shock, which facilitates effective protein folding even at low temperatures. Nucleic acid secondary structures are stabilized by cold, which can interfere with translation, RNA degradation, and transcription (Purwar & Srivastava, 2024). The main purpose of cold-induced proteins involved in RNA metabolism is to either encourage the breakdown of organized RNA or stop secondary structures from forming. Furthermore, secondary structures can be unwound by the helicase DeaD, which makes it easier for cold shock exoribonucleases like PNPase and RNase R to degrade them. According to Bae et al. (2000), it has been demonstrated that both CspA and CspE contribute to transcription at low temperatures by functioning as transcription antiterminators and inhibiting the development of hairpins that would result in premature termination. Furthermore, cold stress induces the production of NusA, another protein having antiterminator function (Bae et al., 2000).

4.4 Cold Active Enzymes

Structural and functional changes in cold-adapted enzymes under cold environments enable them to continue functioning even at temperatures close to freezing. In general, cold-adapted enzymes have less proline or arginine, different arginine to lysine ratios, more polar residues, fewer hydrophobic residues, and fewer disulfide linkages, which are important factors that contribute to these enzymes' cold tolerance (Raymond-Bouchard et al., 2018). The three temperature-sensitive phases of psychrophile enzyme catalysis are substrate recognition and binding, conformational changes that result in transition states and product synthesis, and product release. Psychrophilic enzymes often have lower activation energies and optimal activity at lower temperatures than their mesophilic counterparts. These enzymes retain conserved active site areas that are necessary for catalytic performance over a range of temperatures, while having a lower overall sequence identity (Sarma et al., 2023).

4.5 Cell Envelope and Cell Membrane

In order to maintain cellular integrity and function in cold conditions, microorganisms rely on modifications to the lipid composition of their membranes and cell envelope, reducing lipid saturation and favoring methyl-branched, cis-isomeric, and/or shorter chain length lipids. Particularly for long-chain polyunsaturated fatty acids (LCPUFAs) with carbon chains of C20 or longer, this aids in maintaining the ideal membrane fluidity. By creating hydrophobic areas between lipid bilayers, shielding cells from reactive oxygen species, and maintaining membrane flexibility, PUFAs like eicosapentaenoic acid (EPA), docosahexaenoic acid (DHA), and arachidonic acid (ARA) promote membrane fluidity (Suzuki et al., 2019). *Planococcus halocryophilus* Or1 and other cold-adapted bacteria can fortify their cell envelope by thickening the outer surface with hydrophobic encrustations made of choline, calcium carbonate, and peptidoglycan (Cho et al., 2019). Higher amounts of short-chain and/or unsaturated fatty acids are commonly found in the lipid A component of lipopolysaccharides in these bacteria, which improves membrane fluidity in cold environments. Furthermore, by acting as cryoprotectants and providing protection against freeze-thaw cycles, the exopolysaccharides (EPS) that envelop bacterial cells are essential for cold adaptation. Cold-adapted bacteria frequently produce EPS, especially those that contain mannose as a key component, underscoring their survival strategies in extremely cold conditions. Laurate levels fall, and palmitoleate starts to show up at low temperatures (Carty et al., 1999). Palmitoleate is an unsaturated fatty acid, in contrast to laurate, and counteracts the effects of the cold by lowering the phase transition temperature and by increasing membrane fluidity.

4.6 Change in Metabolic Activity

Primary metabolic pathways are often downregulated in cold environments, but alternate pathways are triggered to help species that have acclimated to the cold. Oxidative stress results from the increased generation of reactive oxygen species caused by the greater solubility of oxygen in colder conditions. Certain heterotrophic bacteria, like *Psychrobacter arcticus*, have special metabolic characteristics that allow them to use oxidized carbon sources like acetate. These characteristics include the absence of glycolysis genes but the presence of gluconeogenic enzymes (Chattopadhyay, 2006). Acetyl-CoA metabolism is upregulated while energy-producing proteins are downregulated. Similarly, in reaction to cold temperatures, *Nesterenkonia* sp. AN1 and *Exiguobacterium sibiricum* exhibit enhanced expression of enzymes involved in energy metabolism and the glyoxylate cycle, respectively (Rodrigues et al., 2009).

4.7 Cryoprotectant Production

For cold adaptation, some substances are essential, including glycine, betaine, glycerol, trehalose, sucrose, mannitol, and sorbitol. They aid by stabilizing cellular membranes, scavenging free radicals, inhibiting the aggregation of macromolecules, and lowering the freezing point of the cytoplasm. These substances also lessen water loss and cell shrinkage, stabilize proteins and membranes, lower the colloidal glass transition temperature (Tg), and restore osmotic balance. Different cold adaptation mechanisms were discovered by comparing the genomes and global phenotypic traits of two Antarctic *Pseudoalteromonas* strains, *P. haloplanktis* TAC125 and *Pseudoalteromonas* sp.TB41, cultivated at 4 °C and 15 °C. Notably, in *P. haloplanktis* TAC125, protein S-thiolation, which is controlled by glutathione and glutathionylspermidine, has documented as a possible novel mechanism for cold adaptation (Mocali et al., 2017). The bacterium's increased uptake of substances, including spermine, glutathione, and ornithine, which are probably linked to cryotolerance, demonstrated the significance of glutathione metabolism in this investigation. These substances might contribute to strengthening resistance to oxidative stress and tying together compatible solutes. When the bacterium was cultivated at 4 °C, metabolomic analysis of the Arctic isolate *Mesorhizobium* sp. StrainN33 revealed an increase in the accumulation of sarcosine, threonine, and valine, which are probably cryoprotectants (Purwar & Srivastava, 2024).

4.8 Pigment Production

Psychrophilic microbes frequently produce pigment, which helps them survive in hostile surroundings. These pigments have a number of uses, such as supplying energy for photosynthesis, protecting against environmental stressors like desiccation, high temperatures, and UV rays, and enhancing the body's defenses against other bacteria. These pigments are essential for allowing these microorganisms to flourish in cold environments (Ramasamy et al., 2023). Numerous colors, including carotenoids, prodigiosin, melanin, violacein, indigoidine, and scytonemin, are produced by psychrophilic microbes. In ice cores, the percentage of pigmented bacteria tends to rise with depth. The most prevalent colonies are yellow in hue. It is believed that carotenoids and other pigments help cold-adapted organisms by controlling the fluidity of cell membranes and increasing their stiffness. A study by Shen et al. (2018) also documented that the content of zeaxanthin-like pigments increased by 26–65% following freeze-thaw cycles, indicating that these pigments enhance cryoprotection via adjusting membrane fluidity. In a similar vein, *Serratia marcescens* exhibits its role in temperature adaptation by producing red pigments (prodigiosin) at lower temperatures (25 °C) as opposed to higher temperatures (37 °C) (Paillié-Jiménez et al., 2020).

4.9 Polyhydroxyalkanoates Metabolism

Many bacteria native to cold environments are capable of synthesizing polyhydroxyalkanoates, which are essential physiological reserve polymers. As dynamic storage for carbon and reducing equivalents, these polymers build up in situations of uneven development, such as when there is an excess of carbon in comparison to other nutrients like nitrogen or phosphorus. PHAs support cell motility, cellular redox equilibrium, resistance to oxidative stress, and cryoprotection (Grzesiak et al., 2024). They increase the survival of bacteria and give them resistance to environmental stressors. After 24 h at 10 °C, PHA-depolymerase activity noticeably increases in freezing conditions. The main proteins linked to PHA granules, phasins, serve a number of purposes, including enhancing bacterial fitness and providing stress resistance. Genome analysis of *Colwellia psychrerythraea* revealed that it can synthesize and break down fatty acids, which is connected to its ability to synthesize PHA (Tribelli & López, 2018). Moreover, *Pseudomonas extremaustralis* is reported to produce a variety of PHAs, including medium-chain-length PHAs (mclPHAs) and short-chain-length PHAs (sclPHAs), such as polyhydroxybutyrate (PHB). Particularly, the synthesis of PHB (polyhydroxybutyrate) was found to be essential for both survival in freezing temperatures and growth at low temperatures.

5 Industrial and Ecological Significance of Cold-Adaptive Bacteria

The versatility of psychrophilic microorganisms to function in extreme conditions makes them valuable for industrial applications and ecological balance. This section explores the industrial and ecological roles of cold-adaptive bacteria, emphasizing their significance in sustainable development.

5.1 Industrial Application

Cold-adapted bacterial enzymes exhibit distinctive properties that make them highly effective and economically viable for various biotechnological applications. Byproducts of psychrophiles are extensively utilized in industries such as food processing, baking, textile manufacturing, bioremediation, and biotransformation. Given their versatility and eco-friendly advantages, cold-adapted enzymes continue to gain prominence across diverse industrial sectors.

5.1.1 Molecular biology

DNA ligases play a crucial role in DNA replication, recombination, and repair by facilitating the formation of phosphodiester bonds between adjacent 5′-phosphate and 3′-hydroxyl groups in double-stranded DNA. These enzymes are widely utilized in molecular biology for genetic engineering and cloning applications. Notably, the psychrophilic bacterium *Pseudoalteromonas haloplanktis* TAE72 has been reported to produce a DNA ligase that remains active even at temperatures as low as 4 °C (Georlette et al., 2000). In recent years, several cold-adapted enzymes, including alkaline phosphatases, single- and double-stranded nucleases, and uracil-DNA N-glycosylases, have gained commercial significance as molecular biology tools. Various biotechnology companies, such as New England Biolabs, Inc., ArcticZymes, Takara-Clontech, and Affymetrix, Inc., have introduced these enzymes for research and industrial applications due to their efficiency in low-temperature environments (Lanes et al., 2002; Awazu et al., 2011; Muller-Greven et al., 2013).

5.1.2 Detergent Industry

A sustainable alternative to harmful chemicals in the detergent industry is the use of enzymes, which offer an eco-friendly and effective solution. Cold-active enzymes are gaining traction in the commercial sector due to their environmental benefits and sustainability. Various cold-adapted enzymes, including proteases, amylases, and lipases, have demonstrated high efficiency in low-temperature cleaning applications (Barroca et al., 2017). Additionally, psychrozymes incorporated into wipes and specialized cleaning formulations enable the effective cleaning of solid objects that cannot withstand high-temperature washing (Khan et al., 2023b). The use of cold-active enzymes in detergents contributes to significant energy savings by reducing washing temperatures. Studies indicate that lowering the wash temperature by just

10 °C can decrease electricity consumption by approximately 30% (Hamid et al., 2022). Several bacterial strains, such as *Acinetobacter* sp., *Bacillus* sp., *Planococcus* sp., *Pseudomonasaeruginosa*, *Pseudoalteromonasarctica*, and *Serratia marcescens*, produce proteases that serve as effective detergent additives for cold-water washing. In recent years, the detergent industry has introduced commercial cold-adapted proteases designed for low-temperature applications. For example, two such enzymes, Purafect® and Properase®, were developed and released by a company based in Palo Alto, CA, USA (Sarmiento et al., 2015).

5.1.3 Pharmaceuticals and Bioproducts

Pharmaceuticals and Bioproducts: Enzyme stability is a crucial factor in determining its suitability for industrial and pharmaceutical applications. Many compounds used in medicine, cosmetics, fragrances, and pharmaceuticals are highly thermolabile. Due to their structural flexibility and ability to function at low temperatures, psychrophilic enzymes provide an optimal solution for synthesizing these heat-sensitive ingredients (Collins & Margesin, 2019). Furthermore, several bioactive compounds, including antifungal, anticancer, and antitumor agents, have been successfully derived from cold-adapted bacterial strains by using cold-active enzymes (Banerjee et al., 2016). One of the key advantages of cold-adapted enzymes is their thermolability, allowing them to be easily inactivated with moderate heat (Bhatia et al., 2021). This eliminates the need for chemical deactivation, making them ideal for applications such as live vaccine development. Additionally, these enzymes play a significant role in the semisynthesis and enzymatic modification of fine chemicals and pharmaceuticals. For instance, cold-active β-d-galactosidase, known for its strong hydrolytic activity against lactose and efficient glycosyl transfer capability, holds considerable potential in the pharmaceutical industry. It has been employed in the biosynthesis of alkyl glycosides, which serve as essential nonionic surfactants in drug formulations (Makowski et al., 2009). Furthermore, the transglycosylation properties of cold-active enzymes facilitate the modification of antibiotics, ergot alkaloids, and flavonol glycosides such as myricitrin, enhancing their bioavailability and therapeutic effects (Pawlak-Szukalska et al., 2014). Cold-tolerant proteases also play a crucial role in producing bioactive peptides, short amino acid chains with antioxidant and antihypertensive properties that can support the immune system and help combat severe diseases, including cancer. Additional examples of cold-adapted enzymes used in biomedicine include nitroreductase, which acts as a prodrug-activating enzyme in cancer therapy, and α-galactosidase, which has applications in blood transfusion treatments (Çelik & Yetiş, 2012).

5.1.4 Bioremediation of Contaminants

Bioremediation of Contaminants: In addition to their role in the discovery of bioactive compounds, cold-adapted microorganisms have demonstrated significant potential in bioremediation, effectively degrading a wide range of organic pollutants such as mineral oil hydrocarbons, phenolic compounds, polycyclic aromatic hydrocarbons, pesticides, and other persistent contaminants (Yap et al., 2021). These microorganisms play a crucial role in breaking down hydrocarbons, pesticides, and heavy metals in cold environments, where their psychrotolerant nature enables efficient bioremediation in Arctic and alpine ecosystems. Several cold-adapted bacterial species, including *Pseudomonas* sp., *Rhodococcus* sp., *Oleispira antarctica*, and *Sphingomonas* spp., have been identified for their ability to degrade petroleum hydrocarbons (Miri et al., 2019; Yap et al., 2021). Additionally, cold-active enzymes such as lipases, proteases, and xylanases are frequently investigated for their applications in environmental cleanup efforts (Al-Maqtari et al., 2019; Miri et al., 2019). These enzymes contribute to sustainable bioremediation processes, offering an eco-friendly solution for managing pollution in cold regions.

5.1.5 Food industry

The use of mesophilic enzymes in the food and brewing industries has been well-established. However, the introduction of psychrophilic enzymes as alternatives has shown promising outcomes. Cold-adapted microbial enzymes exhibit higher catalytic efficiency at low and moderate temperatures compared to their mesophilic counterparts. This advantage makes cold-active enzymes particularly beneficial in the food industry, as they require fewer biochemical inputs, shorten processing times, reduce energy consumption, and can be easily inactivated with mild heat (Kumar et al., 2021). For instance, a cold-adapted protease derived from *Pseudoalteromonas* sp. released more free amino acids from milk protein at 4 °C than mesophilic papain, indicating its broad substrate specificity and potential application in low-temperature food processing (Wang et al., 2008). Similarly, an alkaline peptidase from *Penicillium chrysogenum* has shown greater suitability for cheese production than commonly used mesophilic and thermostable microbial proteases (Zhu et al., 2009). In the fruit juice industry, cold-active pectinases are highly sought after to preserve the quality and nutritional content of fruit juices while optimizing key processes such as liquefaction, clarification, and juice extraction at lower temperatures. Moreover, any remaining enzyme activity can be efficiently deactivated with mild heat treatment after processing, ensuring better control over the final product (Adapa et al., 2014).

6 Ecological Significance

6.1 Microbial-mediated cold stress management in agricultural crops:

One of the primary consequences of chilling stress is the rigidification of plant cell membranes due to reduced fluidity, along with the accumulation of cryoprotectants and an increased capacity to tolerate oxidative stress. Despite employing several mechanisms for survival in cold conditions, plants often experience a net decrease in growth and yield under such circumstances (Soualiou et al., 2022). PGPRs play a crucial role in enhancing plant cold tolerance by inducing several genes that enable plants to withstand cold stress (Zubair et al., 2019; Khan et al., 2023a). In a study by

Turan et al. (2013), the inoculation of *Azospirillum brasilense*, *Bacillus megaterium*, *B. subtilis*, and *Raoultella terrigena* into wheat and barley was found to aid in the recovery of plants from cold injury. This was attributed to the maintenance of antioxidant enzyme activity and reduced freezing injury.

6.2 Nutrient cycling and plant growth promotion in cold ecosystems:

Cold-adapted bacteria play a crucial role in the cycling of essential nutrients such as carbon, nitrogen, phosphorus, and sulfur in cold environments, helping to maintain ecological balance and sustain both plant and microbial communities (Pandey et al., 2023; Khan et al., 2024). Their symbiotic and associative interactions with plants enhance nutrient absorption, improve stress tolerance, and promote plant growth in harsh conditions (Suyal et al., 2023). A key mechanism through which these plant growth-promoting rhizobacteria (PGPR) enhance plant development is by synthesizing stimulatory phytohormones in the root zone. These hormones promote the formation and elongation of root hairs, improving the plant's ability to absorb water and essential nutrients from the soil (Khan et al., 2019; Chaffai et al., 2024). Among the numerous geochemical cycles influenced by microbes, phosphate solubilization plays a crucial role due to the essential function of phosphorus in plant nutrition. Although soil contains adequate phosphorus to support plant growth, much of it exists in immobilized forms, necessitating microbial-mediated transformation for plant availability (Khan et al., 2023c). Das et al. (2003) were the first to report phosphate solubilization at low temperatures, investigating cold-tolerant *Pseudomonas* mutants for their ability to solubilize phosphate at 10 °C. Their findings indicated that these mutants exhibited greater efficiency in phosphate solubilization at 10 °C compared to their wild-type strains, with higher activity observed at this lower temperature than at 25 °C. Iron is a vital micronutrient for plants, serving as a cofactor for numerous enzymes involved in redox processes (Khan & Singh, 2021). However, a significant portion of iron in the soil exists in the highly insoluble ferric hydroxide form, making it a limiting factor for plant growth even in iron-rich environments (Punetha et al., 2023). To overcome this challenge, microorganisms have developed specialized iron acquisition strategies, including the synthesis of siderophore compounds that chelate iron and facilitate its uptake (Singh et al., 2021). Katiyar and Goel (2004) engineered a cold-tolerant mutant strain of *Pseudomonas fluorescens* with significantly enhanced siderophore production, approximately 17 times higher than the wild-type strain. This mutant also demonstrated improved colonization of the rhizosphere. When tested on *Vigna radiata* plants, it effectively promoted growth at both 25 °C and 10 °C. Moreover, by thriving in extreme environments, these bacteria contribute to microbial diversity, which is essential for ecosystem stability and resilience against climate change. They also play a significant role in improving soil structure, accelerating the decomposition of organic matter, and boosting microbial activity, which aids in restoring degraded soils in cold regions.

7 Conclusion

The chapter has provided a substantial demonstration of the molecular, physiological, and structural aspects of bacterial adaptation to temperature extremes, which allow microorganisms to flourish in extreme conditions. These adaptable characteristics provide important information about how bacteria react to stress and the possibility of using thermophilic and psychrophilic organisms for biotechnology. The topic opens up important research topics, i.e., investigation of metabolic pathways, the function of protectants, and the functional study of unexplored cold or heat-active enzymes that have potential applications in environmental sustainability, bioremediation, and the creation of environmentally friendly biotechnological processes. In addition to providing more thorough insights into how bacteria adapt at the cellular and molecular levels, the application of multiomics techniques, such as metagenomics and machine learning technologies, may result in the discovery of new biomolecules with significant commercial potential. Another fascinating possibility is the ongoing research into cold and heat-active enzymes and their uses in a variety of sectors, including food preservation and detergents. The use of such extremophiles as models for researching environmental stress adaptation will be crucial in the development of sustainable biotechnological solutions as climate change continues to affect ecosystems. As a result, current and upcoming studies in this area should greatly advance scientific understanding and real-world applications across a range of industries.

References

Adapa, V., Ramya, L. N., Pulicherla, K. K., & Rao, K. S. (2014). Cold active pectinases: Advancing the food industry to the next generation. *Applied Biochemistry and Biotechnology, 172*, 2324–2337.

Ajeje, S. B., Hu, Y., Song, G., Peter, S. B., Afful, R. G., Sun, F., Asadollahi, M. A., Amiri, H., Abdulkhani, A., & Sun, H. (2021). Thermostable cellulases/xylanases from thermophilic and hyperthermophilic microorganisms: Current perspective. *Frontiers in Bioengineering and Biotechnology, 9*, 794304. https://doi.org/10.3389/fbioe.2021.794304

Al-Maqtari, Q. A., Waleed, A. A., & Mahdi, A. A. (2019). Cold-active enzymes and their applications in industrial fields-a review. *International Journal of Research in Agricultural Sciences, 6*(4), 2348–3997.

Albers, S. V., Van de Vossenberg, J. L., Driessen, A. J., & Konings, W. N. (2001). Bioenergetics and solute uptake under extreme conditions. *Extremophiles, 5*, 285–294.

Amend, J. P., & Shock, E. L. (2001). Energetics of overall metabolic reactions of thermophilic and hyperthermophilic Archaea and Bacteria. *FEMS Microbiology Reviews, 25*(2), 175–243.

Awazu, N., Shodai, T., Takakura, H., Kitagawa, M., Mukai, H., & Kato, I. (2011). U.S. Patent No. 8,034,597. Washington, DC: U.S. Patent and Trademark Office.

Bae, W., Xia, B., Inouye, M., & Severinov, K. (2000). Escherichia coli CspA-family RNA chaperones are transcription antiterminators. *Proceedings of the National Academy of Sciences, 97*(14), 7784–7789.

Bala, A., & Singh, B. (2019). Development of an environmental-benign process for efficient pretreatment and Saccharification of Saccharum biomasses for bioethanol production. *Renewable Energy, 130*, 12–24. https://doi.org/10.1016/j.renene.2018.06.033

Banerjee, R., Halder, A., & Natta, A. (2016). Psychrophilic microorganisms: Habitats and exploitation potentials. *European Journal of Biotechnology and Bioscience, 4*(3), 16–24.

Barroca, M., Santos, G., Gerday, C., & Collins, T. (2017). Biotechnological aspects of cold-active enzymes. In *Psychrophiles: From biodiversity to biotechnology* (pp. 461–475).

Bhandiwad, A., Guseva, A., & Lynd, L. (2013). Metabolic engineering of Thermoanaerobacterium thermosaccharolyticum for increased n-Butanol production. *Advances in Microbiology, 3*, 46–51.

Bhatia, R. K., Ullah, S., Hoque, M. Z., Ahmad, I., Yang, Y. H., Bhatt, A. K., & Bhatia, S. K. (2021). Psychrophiles: A source of cold-adapted enzymes for energy efficient biotechnological industrial processes. *Journal of Environmental Chemical Engineering, 9*(1), 104607.

Bisht, T. S., Sharma, S. K., Sati, R. C., Rao, V. K., Yadav, V. K., Dixit, A. K., et al. (2015). Improvement of efficiency of oil extraction from wild apricot kernels by using enzymes. *Journal of Food Science and Technology, 52*, 1543–1551. https://doi.org/10.1007/s13197013-1155-z

Boutz, D. R., Cascio, D., Whitelegge, J., Perry, L. J., & Yeates, T. O. (2007). Discovery of a thermophilic protein complex stabilized by topologically interlinked chains. *Journal of Molecular Biology, 368*(5), 1332–1344.

Caldas, T., Laalami, S., & Richarme, G. (2000). Chaperone properties of bacterial elongation factor EF-G and initiation factor IF2. *The Journal of Biological Chemistry, 275*, 855–860.

Candido, R. G., Mori, N. R., & Gonçalves, A. R. (2019). Sugarcane straw as feedstock for 2G ethanol: Evaluation of pretreatments and enzymatic hydrolysis. *Industrial Crops and Products, 142*, 111845. https://doi.org/10.1016/j.indcrop.2019.111845

Cario, A., Grossi, V., Schaeffer, P., & Oger, P. M. (2015). Membrane homeoviscous adaptation in the piezo-hyperthermophilic archaeon Thermococcus barophilus. *Frontiers in Microbiology, 6*, 1152.

Carty, S. M., Sreekumar, K. R., & Raetz, C. R. (1999). Effect of cold shock on lipid A biosynthesis in Escherichia coli: Induction at 12 °C of an acyltransferase specific for palmitoleoyl-acyl carrier protein. *The Journal of Biological Chemistry, 274*(14), 9677–9685.

Çelik, A., & Yetiş, G. (2012). An unusually cold active nitroreductase for prodrug activations. *Bioorganic & Medicinal Chemistry, 20*(11), 3540–3550.

Chaffai, R., Ganesan, M., & Cherif, A. (2024). Plant growth-promoting rhizobacteria (PGPR) and plant growth-promoting fungi (PGPF) for alleviating abiotic stress in plants. In *Plant adaptation to abiotic stress: From signaling pathways and microbiomes to molecular mechanisms* (pp. 457–496). Springer Nature.

Chakraborty, S., Gupta, R., Jain, K. K., Hemansi, S., & Kuhad, R. C. (2016). Chapter 17—Cellulases: Application in wine and brewery industry. In *New and future developments in microbial biotechnology and bioengineering* (pp. 193–200). Elsevier. https://doi.org/10.1016/B978-0444-63507-5.00017-4

Charlier, D., & Droogmans, L. (2005). Microbial life at high temperature, the challenges, the strategies. *Cellular and Molecular Life Sciences, 62*(24), 2974–2984. https://doi.org/10.1007/s00018-005-5251-8. PMID: 16314930; PMCID: PMC11139153.

Chatterjee, S., Bhattacharjee, I., & Chandra, G. (2010). Biosorption of heavy metals from industrial waste water by Geobacillus thermodenitrificans. *Journal of Hazardous Materials, 175*, 117–125.

Chattopadhyay, M. K. (2006). Mechanism of bacterial adaptation to low temperature. *Journal of Biosciences, 31*, 157–165.

Chernyh, N. A., Gavrilov, S. N., Sorokin, V. V., German, K. E., Sergeant, C., Simonoff, M., et al. (2007). Characterization of technetium (vII) reduction by cell suspensions of thermophilic bacteria and archaea. *Applied Microbiology and Biotechnology, 76*, 467–472.

Cho, S. M., Kim, S., Cho, H., Lee, H., Lee, J. H., Lee, H., et al. (2019). Type II ice-binding proteins isolated from an arctic microalga are similar to adhesin-like proteins and increase freezing tolerance in transgenic plants. *Plant & Cell Physiology, 60*(12), 2744–2757.

Collins, T., & Margesin, R. (2019). Psychrophilic lifestyles: Mechanisms of adaptation and biotechnological tools. *Applied Microbiology and Biotechnology, 103*, 2857–2871.

Combie, J., Combie, J., Albert, F. G., Tran, K. V., Cabrera, J., Correira, H. J., et al. (2001). Extremophilic organisms as an un explored source of antifungal compounds. *The Journal of Antibiotics, 54*, 56–65.

Coronel-León, J., de Grau, G., Grau-Campistany, A., Farfan, M., Rabanal, F., Manresa, A., & Marqués, A. M. (2015). Biosurfactant production by AL 1.1, a Bacillus licheniformis strain isolated from Antarctica: Production, chemical characterization and properties. *Annales de Microbiologie, 65*, 2065–2078.

Dahiya, S., & Singh, B. (2019). Microbial xylanases in bread making. In *Reference module in food science* (pp. 140–149). Academic. https://doi.org/10.1016/B978-0-08-100596-5.21644-2

Daryasafar, A., Jamialahmadi, M., Moghaddam, M. B., & Moslemi, B. (2016). Using biosurfactant producing bacteria isolated from an Iranian oil feld for application in microbial enhanced oil recovery. *Petroleum Science and Technology, 34*, 739–746.

Das, K., Katiyar, V., & Goel, R. (2003). 'P' solubilization potential of plant growth promoting Pseudomonas mutants at low temperature. *Microbiological Research, 158*(4), 359–362.

de Miguel Bouzas, T., Barros-Velázquez, J., & Gonzalez Villa, T. (2006). Industrial applications of hyperthermophilic enzymes: A review. *Protein and Peptide Letters, 13*(7), 645–651.

Driessen, A. J., & Albers, S. V. (2007). Membrane adaptations of (hyper) thermophiles to high temperatures. In *Physiology and biochemistry of extremophiles* (pp. 104–116).

Feller, G. (2003). Molecular adaptations to cold in psychrophilic enzymes. *Cellular and Molecular Life Sciences: CMLS, 60*, 648–662.

Garcia-Descalzo, L., Gil-Lozano, C., Munoz-Iglesias, V., Prieto-Ballesteros, O., Azua-Bustos, A., & Fairén, A. G. (2020). Can halophilic and psychrophilic microorganisms modify the freezing/ melting curve of cold salty solutions? Implications for Mars habitability. *Astrobiology, 20*(9), 1067–1075.

Georlette, D., Jonsson, Z. O., Van Petegem, F., Chessa, J. P., Van Beeumen, J., Hübscher, U., & Gerday, C. (2000). A DNA ligase from the psychrophile Pseudoalteromonas haloplanktis gives insights into the adaptation of proteins to low temperatures. *European Journal of Biochemistry, 267*(12), 3502–3512.

Giaquinto, L., Curmi, P. M., Siddiqui, K. S., Poljak, A., DeLong, E., DasSarma, S., & Cavicchioli, R. (2007). Structure and function of cold shock proteins in archaea. *Journal of Bacteriology, 189*(15), 5738–5748.

Gimbi, D. M., & Kitabatake, N. (2002). Changes in alpha-and beta amylase activities during seed germination of African finger millet. *International Journal of Food Sciences and Nutrition, 53*, 481–488.

Gomez, E. J., Delgado, J. A., & Gonzalez, J. M. (2020). Persistence of microbial extracellular enzymes in soils under different temperatures and water availabilities. *Ecology and Evolution, 10*, 10167–10176.

Grzesiak, J., Rogala, M. M., Gawor, J., Kouřilová, X., & Obruča, S. (2024). Polyhydroxyalkanoate involvement in stress-survival of two psychrophilic bacterial strains from the High Arctic. *Applied Microbiology and Biotechnology, 108*(1), 273.

Guerrero-Barajas, C., Ordaz, A., Garibay-Orijel, C., García-Solares, S. M., Bastida-González, F., & Zárate-Segura, P. B. (2014). Enhanced sulfate reduction and trichloroethylene (TCE) biodegradation in a UASB reactor operated with a sludge developed from hydrothermal vents sediments: Process and microbial ecology. *International Biodeterioration & Biodegradation, 94*, 182–191.

Hamid, B., Bashir, Z., Yatoo, A. M., Mohiddin, F., Majeed, N., Bansal, M., et al. (2022). Cold-active enzymes and their potential industrial applications—A review. *Molecules, 27*(18), 5885.

Hawwa, R., Aikens, J., Turner, R. J., Santarsiero, B. D., & Mesecar, A. D. (2009). Structural basis for thermostability revealed through the identification and characterization of a highly thermostable phosphotriesterase-like lactonase from Geobacillus stearothermophilus. *Archives of Biochemistry and Biophysics, 488*(2), 109–120.

Hébraud, M., & Potier, P. (1999). Cold shock response and low temperature adaptation in psychrotrophic bacteria. *Journal of Molecular Microbiology and Biotechnology, 1*(2), 211–219.

Hickey, D. A., & Singer, G. A. (2004). Genomic and proteomic adaptations to growth at high temperature. *Genome Biology, 5*, 1–7.

Hu, C., Yang, J., Qi, Z., Wu, H., Wang, B., Zou, F., et al. (2022). Heat shock proteins: Biological functions, pathological roles, and therapeutic opportunities. *Medicine Communications, 3*(3), e161.

Katiyar, V., & Goel, R. (2004). Improved plant growth from seed bacterization using siderophore overproducing cold resistant mutant of Pseudomonas fluorescens. *Journal of Microbiology and Biotechnology, 14*(4), 653–657.

Khan, A., & Singh, A. V. (2021). Multifarious effect of ACC deaminase and EPS producing Pseudomonas sp. and Serratia marcescens to augment drought stress tolerance and nutrient status of wheat. *World Journal of Microbiology and Biotechnology, 37*(12), 198.

Khan, A., Singh, J., Upadhayay, V. K., Singh, A. V., & Shah, S. (2019). Microbial biofortification: A green technology through plant growth promoting microorganisms. In *Sustainable green technologies for environmental management* (pp. 255–269).

Khan, A., Upadhayay, V. K., Panwar, M., & Singh, A. V. (2020). Soil microbiota: A key bioagent for revitalization of soil health in hilly regions. In *Microbiological advancements for higher altitude agro-ecosystems & sustainability* (pp. 183–200).

Khan, A., Singh, A. V., Pareek, N., Arya, P., Upadhayay, V. K., Kumar Jugran, A., et al. (2023a). Credibility assessment of cold adaptive Pseudomonas jesenni MP1 and P. palleroniana N26 on growth, rhizosphere dynamics, nutrient status, and yield of the kidney bean cultivated in Indian Central Himalaya. *Frontiers in Plant Science, 14*, 1042053.

Khan, A., Punetha, A., Kukreti, B., Sharma, R. S., Panthari, D., Naudiyal, N., et al. (2023b). Microbial Community dynamics of Antarctica: Their ecological potential and industrial importance. In *Microbial bioactive compounds: Industrial and agricultural applications* (pp. 261–289). Springer Nature.

Khan, A., Singh, A. V., Gautam, S. S., Agarwal, A., Punetha, A., Upadhayay, V. K., et al. (2023c). Microbial bioformulation: A microbial assisted biostimulating fertilization technique for sustainable agriculture. *Frontiers in Plant Science, 14*, 1270039.

Khan, A., Singh, A. V., Kukreti, B., Pandey, D. T., Upadhayay, V. K., Kumar, R., & Goel, R. (2024). Deciphering the impact of cold-adapted bioinoculants on rhizosphere dynamics, biofortification, and yield of kidney bean across varied altitudinal zones. *Science of the Total Environment, 927*, 172204.

Kochhar, N., Shrivastava, S., Ghosh, A., Rawat, V. S., Sodhi, K. K., & Kumar, M. (2022). Perspectives on the microorganism of extreme environments and their applications. *Current Research in Microbial Sciences, 3*, 100134.

Kohli, I., Joshi, N. C., Mohapatra, S., & Varma, A. (2020). Extremophile—An adaptive strategy for extreme conditions and applications. *Current Genomics, 21*(2), 96–110.

Kuhad, R. C., Deswal, D., Sharma, S., Bhattacharya, A., Jain, K. K., Kaur, A., et al. (2016). Revisiting cellulase production and redefining current strategies based on major challenges. *Renewable and Sustainable Energy Reviews, 55*, 249–272. https://doi.org/10.1016/j.rser.2015.10.132

Kumar, A., Mukhia, S., & Kumar, R. (2021). Industrial applications of cold-adapted enzymes: Challenges, innovations and future perspective. *3 Biotech, 11*(10), 426.

Laksanalamai, P., Whitehead, T. A., & Robb, F. T. (2004). *Nature Reviews. Microbiology, 2*, 315–324.

Lanes, O., Leiros, I., Smalås, A. O., & Willassen, N. (2002). Identification, cloning, and expression of uracil-DNA glycosylase from Atlantic cod (Gadus morhua): Characterization and homology modeling of the cold-active catalytic domain. *Extremophiles, 6*, 73–86.

Lelivelt, M. J., & Kawula, T. H. (1995). Hsc66, an Hsp70 homolog in *Escherichia coli*, is induced by cold shock but not by heat shock. *Journal of Bacteriology, 177*, 4900–4907.

Makowski, K., Białkowska, A., Olczak, J., Kur, J., & Turkiewicz, M. (2009). Antarctic, cold-adapted β-galactosidase of Pseudoalteromonas sp. 22b as an effective tool for alkyl galactopyranosides synthesis. *Enzyme and Microbial Technology, 44*(2), 59–64.

Mandelli, F., et al. (2017). Thermal adaptation strategies of the extremophile bacterium Thermus filiformis based on multi-omics analysis. *Extremophiles, 21*, 775–788.

Margalef, O., Sardans, J., Maspons, J., Molowny-Horas, R., Fernández-Martínez, M., Janssens, I. A., Richter, A., Ciais, P., Obersteiner, M., & Peñuelas, J. (2021). The effect of global change on soil phosphatase activity. *Global Change Biology, 27*, 5989–6003.

Marzban, G., & Tesei, D. (2025). The extremophiles: Adaptation mechanisms and biotechnological applications. *Biology, 14*(4), 412.

Mesbah, N. M. (2022). Industrial biotechnology based on enzymes from extreme environments. *Frontiers in Bioengineering and Biotechnology, 10*, 870083.

Min, B., Kim, K., Li, V., Cho, S., & Kim, H. (2020). Changes in cell membrane fatty acid composition of Streptococcus thermophilus in response to gradually increasing heat temperature. *Journal of Microbiology and Biotechnology, 30*(5), 739.

Miri, S., Naghdi, M., Rouissi, T., Kaur Brar, S., & Martel, R. (2019). Recent biotechnological advances in petroleum hydrocarbons degradation under cold climate conditions: A review. *Critical Reviews in Environmental Science and Technology, 49*(7), 553–586.

Mnif, S., Sayadi, S., & Chamkha, M. (2014). Biodegradative potential and characterization of a novel aromatic-degrading bacterium isolated from a geothermal oil field under saline and thermophilic conditions. *International Biodeterioration & Biodegradation, 86*, 258–264.

Mocali, S., Chiellini, C., Fabiani, A., Decuzzi, S., de Pascale, D., Parrilli, E., et al. (2017). Ecology of cold environments: New insights of bacterial metabolic adaptation through an integrated genomic-phenomic approach. *Scientific Reports, 7*(1), 839.

Muller-Greven, J. C., Post, M. A., & Kubu, C. J. (2013). U.S. Patent No. 8,486,665. Washington, DC: U.S. Patent and Trademark Office.

Nakashima, H., Fukuchi, S., & Nishikawa, K. (2003). Compositional changes in RNA, DNA and proteins for bacterial adaptation to higher and lower temperatures. *Journal of Biochemistry, 133*, 507–513.

Pailliè-Jiménez, M. E., Stincone, P., & Brandelli, A. (2020). Natural pigments of microbial origin. *Frontiers in Sustainable Food Systems, 4*, 590439.

Pandey, S. C., Pande, V., Sati, D., Khan, A., Singh, A. V., Punetha, A., et al. (2023). Microbe-mediated abiotic stress management for sustainable agriculture. In *Unravelling plant-microbe synergy* (pp. 245–262). Academic.

Panja, A. S., Maiti, S., & Bandyopadhyay, B. (2020). Protein stability governed by its structural plasticity is inferred by physicochemical factors and salt bridges. *Scientific Reports, 10*, 1822.

Parmar, A., Singh, N. K., Pandey, A., Gnansounou, E., & Madamwar, D. (2011). Cyanobacteria and microalgae: A positive prospect for biofuels. *Bioresource Technology, 102*, 10163–10172.

Patel, K., & Dudhagara, P. (2020). Optimization of Xylanase production by Bacillus Tequilensis Strain UD-3 using economical agricultural substrate and its application in rice straw pulp bleaching. *Biocatalysis and Agricultural Biotechnology, 30*, 101846. https://doi.org/10.1016/j.bcab.2020.101846

Pawlak-Szukalska, A., Wanarska, M., Popinigis, A. T., & Kur, J. (2014). A novel cold-active β-d-galactosidase with transglycosylation activity from the Antarctic Arthrobacter sp. 32cB–gene cloning, purification and characterization. *Process Biochemistry, 49*(12), 2122–2133.

Phadtare, S. (2004). Recent developments in bacterial cold-shock response. *Current Issues in Molecular Biology, 6*(2), 125–136.

Punetha, A., Khan, A., Punetha, S., & Pandey, D. T. (2023). Role of hidden microbes in sustainable agriculture. In *Advanced microbial technology for sustainable agriculture and environment* (pp. 103–117). Academic.

Purwar, S., & Srivastava, S. (2024). Adaptations of psychrophilic microorganism to low-temperature environments. In *Applied microbiology: Theory & technology* (pp. 168–188).

Ramasamy, K. P., Mahawar, L., Rajasabapathy, R., Rajeshwari, K., Miceli, C., & Pucciarelli, S. (2023). Comprehensive insights on environmental adaptation strategies in Antarctic bacteria and biotechnological applications of cold adapted molecules. *Frontiers in Microbiology, 14*, 1197797.

Raven, N., Ladwa, N., Cossar, D., & Sharp, R. (1992). Continuous culture of the hyperthermophilic archaeum *Pyrococcus furiosus*. *Applied Microbiology and Biotechnology, 38*, 263–267.

Raymond-Bouchard, I., Goordial, J., Zolotarov, Y., Ronholm, J., Stromvik, M., Bakermans, C., & Whyte, L. G. (2018). Conserved genomic and amino acid traits of cold adaptation in subzero-growing Arctic permafrost bacteria. *FEMS Microbiology Ecology, 94*(4), fiy023.

Rodrigues, D. F., da Jesus, E., Ayala-del-Río, H. L., Pellizari, V. H., Gilichinsky, D., Sepulveda-Torres, L., & Tiedje, J. M. (2009). Biogeography of two cold-adapted genera: Psychrobacter and exiguobacterium. *The ISME Journal, 3*(6), 658–665.

Rosa, A. P., Dias, T., Mouazen, A. M., Cruz, C., & Santana, M. M. (2023). Finding optimal microorganisms to increase crop productivity and sustainability under drought—A structured reflection. *Journal of Plant Interactions, 18*, 2178680.

Ryu, H.-W., Yoo, S.-K., Choi, J. M., Cho, K.-S., & Cha, D. K. (2009). Thermophilic biofiltration of H2S and isolation of a thermophilic and heterotrophic H2S-degrading bacterium, Bacillus sp. TSO3. *Journal of Hazardous Materials, 168*, 501–506.

Sahlan, M., & Yohda, M. (2013). Molecular chaperones in thermophilic eubacteria and archaea. In *Thermophilic microbes in environmental and industrial biotechnology: Biotechnology of thermophiles* (pp. 375–394).

Santana, M. M., Dias, T., Gonzalez, J. M., & Cruz, C. (2021). Transformation of organic and inorganic sulfur-adding perspectives to new players in soil and rhizosphere. *Soil Biology and Biochemistry, 160*, 108306.

Sar, P., Kazy, S. K., Paul, D., & Sarkar, A. (2013). Metal bioremediation by thermophilic microorganisms. In *Thermophilic microbes in environmental and industrial biotechnology* (pp. 171–201). Springer.

Sarma, J., Sengupta, A., Laskar, M. K., Sengupta, S., Tenguria, S., & Kumar, A. (2023). Microbial adaptations in extreme environmental conditions. In *Bacterial survival in the hostile environment* (pp. 193–206). Academic.

Sarmiento, F., Peralta, R., & Blamey, J. M. (2015). Cold and hot extremozymes: Industrial relevance and current trends. *Frontiers in Bioengineering and Biotechnology, 3*, 148.

Schmerling, C., Kouril, T., Snoep, J., Bräsen, C., & Siebers, B. (2022). Enhanced underground metabolism challenges life at high temperature–metabolic thermoadaptation in hyperthermophilic archaea. *Current Opinion in Systems Biology, 30*, 100423.

Sen, S. K., Raut, S., Dora, T. K., & Mohapatra, P. K. D. (2014). Contribution of hot spring bacterial consortium in cadmium and lead bioremediation through quadratic programming model. *Journal of Hazardous Materials, 265*, 47–60.

Sharafi, H., Abdoli, M., Hajfarajollah, H., Samie, N., Alidoust, L., Abbasi, H., et al. (2014). First report of a lipopeptide biosurfactant from thermophilic bacterium Aneurini bacillus thermoaerophilus MK01 newly isolated from municipal landfill site. *Applied Biochemistry and Biotechnology, 173*, 1236–1249.

Shen, L., Liu, Y., Wang, N., Jiao, N., Xu, B., & Liu, X. (2018). Variation with depth of the abundance, diversity and pigmentation of culturable bacteria in a deep ice core from the Yuzhufeng Glacier, Tibetan Plateau. *Extremophiles, 22*, 29–38.

Singh, A. V., Khan, A., & Joshi, M. (2021). Plant–microbe interaction: A sustainable strategy to elevate salinity tolerance in plants. In *Microbes and signaling biomolecules against plant stress: Strategies of plant-microbe relationships for better survival* (pp. 37–54).

Soualiou, S., Duan, F., Li, X., & Zhou, W. (2022). Crop production under cold stress: An understanding of plant responses, acclimation processes, and management strategies. *Plant Physiology and Biochemistry, 190*, 47–61.

Stetter, K. O. (1999). Extremophiles and their adaptation to hot environments. *FEBS Letters, 452*(1–2), 22–25.

Storey, J. M., & Storey, K. B. (2023). Chaperone proteins: Universal roles in surviving environmental stress. *Cell Stress & Chaperones, 28*(5), 455–466.

Sun, M., Zheng, H., Meng, L., Sun, J., Song, H., Bao, Y., et al. (2015). Direct cloning, expression of a thermostable xylanase gene from the metagenomic DNA of cow dung compost and enzymatic production of xylo oligosaccharides from corncob. *Biotechnology Letters, 37*, 1877–1886. https://doi.org/10.1007/s10529-015-1857-6

Suyal, D. C., Khan, A., Singh, A. V., Agarwal, A., Pareek, N., Sah, V. K., & Goel, R. (2023). Impact assessment of cold-adapted Pseudomonas jesenii MP1 and Pseudomonas palleroniana N26 on Phaseolus vulgaris yield and soil health. *Frontiers in Agronomy, 5*, 1121757.

Suzuki, H., Hulatt, C. J., Wijffels, R. H., & Kiron, V. (2019). Growth and LC-PUFA production of the cold-adapted microalga Koliella Antarctica in photobioreactors. *Journal of Applied Phycology, 31*(2), 981–997.

Tekkol, G. E., Sargin, S., Karaçanci, S., Pembeci, C., Mandaci, S., Akgun, I. H., et al. (2017). Production of GH11 xylanase for bakery industry by solid state fermentation. *Journal of Biotechnology, 256*(S55), 1. https://doi.org/10.1016/j.jbiotec.2017.06.985

Trent, J. D., Nimmesgern, E., Wall, J. S., Hartl, F. U., & Horwich, A. L. (1991). *Nature, 354*, 490–493.

Tribelli, P. M., & López, N. I. (2018). Reporting key features in cold-adapted bacteria. *Life, 8*(1), 8.

Turan, M., Güllüce, M., Çakmak, R., & Şahin, F. (2013). Effect of plant growth-promoting rhizobacteria strain on freezing injury and antioxidant enzyme activity of wheat and barley. *Journal of Plant Nutrition, 36*(5), 731–748.

Urbieta, M. S., Donati, E. R., Chan, K. G., Shahar, S., Sin, L. L., & Goh, K. M. (2015). Thermophiles in the genomic era: Biodiversity, science, and applications. *Biotechnology Advances, 33*(6), 633–647.

Uzuner, S., & Cekmecelioglu, D. (2019). Chapter 3—Enzymes in the beverage industry. In *Enzymes in food biotechnology* (pp. 29–43). Academic. https://doi.org/10.1016/B978-0-12-813280-7.00003-7

Wang, Q., Hou, Y., Xu, Z., Miao, J., & Li, G. (2008). Optimization of cold-active protease production by the psychrophilic bacterium Colwellia sp. NJ341 with response surface methodology. *Bioresource Technology, 99*(6), 1926–1931.

Wang, Q., Cen, Z., & Zhao, J. (2015). The survival mechanisms of thermophiles at high temperatures: An angle of omics. *Physiology, 30*(2), 97–106.

Wani, A. K., Akhtar, N., Sher, F., Navarrete, A. A., & Américo-Pinheiro, J. H. P. (2022). Microbial adaptation to different environmental conditions: Molecular perspective of evolved genetic and cellular systems. *Archives of Microbiology, 204*(2), 144.

Wu, H., Cheng, X., Zhu, Y., Zeng, W., Chen, G., & Liang, Z. (2018). Purification and characterization of a cellulase-free, thermostable endo-xylanase from streptomyces griseorubens LH-3 and its use in biobleaching on eucalyptus Kraft Pulp. *Journal of Bioscience and Bioengineering, 125*, 46–51. https://doi.org/10.1016/j.jbiosc.2017.08.006

Yan, S., Zhang, N., Li, J., Wang, Y., Liu, Y., Cao, M., & Yan, Q. (2021). Characterization of humic acids from original coal and its oxidization production. *Scientific Reports, 11*, 15381.

Yang, J., Ma, T., Shang-guan, F., & Han, Z. (2020). Improving the catalytic activity of thermostable xylanase from *Thermotoga maritima* via mutagenesis of non-catalytic residues at glycone subsites. *Enzyme and Microbial Technology, 139*, 109579. https://doi.org/10.1016/j.enzmictec.2020.109579

Yap, H. S., Zakaria, N. N., Zulkharnain, A., Sabri, S., Gomez-Fuentes, C., & Ahmad, S. A. (2021). Bibliometric analysis of hydrocarbon bioremediation in cold regions and a review on enhanced soil bioremediation. *Biology, 10*(5), 354.

Zhu, H. Y., Tian, Y., Hou, Y. H., & Wang, T. H. (2009). Purification and characterization of the cold-active alkaline protease from marine cold-adaptive Penicillium chrysogenum FS010. *Molecular Biology Reports, 36*, 2169–2174.

Zubair, M., Hanif, A., Farzand, A., Sheikh, T. M. M., Khan, A. R., Suleman, M., et al. (2019). Genetic screening and expression analysis of psychrophilic Bacillus spp. reveal their potential to alleviate cold stress and modulate phytohormones in wheat. *Microorganisms, 7*(9), 337.

Microbial Resilience: Adapting to the Challenges of Climate Change

Vijay Krishna, Ayush Om, Kumar Vishal, Prashant Kumar, Asha Kumari, Mahendra Singh Bhinda, and Sanjay Kumar Sanadya

Abstract

Microbial communities are critical drivers of ecosystem functioning, mediating essential processes such as carbon and nitrogen cycling, nutrient recycling, and plant health through mutualistic and pathogenic interactions. Despite extensive research on disturbance impacts, uncertainties remain regarding the factors that govern microbial community stability and the consequences of such changes for ecosystem resilience. Beneficial microbes, i.e., mycorrhiza, Rhizobium, cyanobacteria, and plant growth-promoting rhizobacteria support soil productivity and climate-smart agriculture, adapting via genetic, epigenetic, and physiological mechanisms. A systems-level understanding that integrates microbial networks, environmental gradients, and functional redundancy is essential for linking community composition to the ecosystem. Advances in biotechnological tools, including metabolomics, metagenomics, synthetic biology, and microbiome engineering, enable the development of synthetic microbial consortia (SMC) and extremophile-based bioinoculants that enhance plant productivity, stress tolerance, and soil health. Effective microbial inoculant formulation, quality control, and regulatory frameworks are crucial for translating laboratory findings to field applications. Harnessing microbial diversity through integrative approaches offers a sustainable pathway for improving agricultural resilience,

V. Krishna · A. Om · K. Vishal · P. Kumar · A. Kumari (✉)
ICAR-Indian Agricultural Research Institute, Hazaribag, Jharkhand, India
e-mail: asha.kumari@icar.org.in

M. S. Bhinda
ICAR- Vivekananda Parvatiya Krishi Anusandhan Sansthan, Almora, Uttarakhand, India

S. K. Sanadya
CSK HP Agricultural University, Palampur, Himachal Pradesh, India

G. Bhandari et al. (eds.), *Microbial Resilience*,
https://doi.org/10.1007/978-981-95-8883-1_5

reducing greenhouse gas emissions, and ensuring food security under changing climatic conditions.

Keywords

Microbe · Climate change · Biotechnology · Diversity · Nutrients · Mitigation

1 Introduction

Microbial communities are closely connected to ecosystem functioning due to their essential roles in carbon (C) and nitrogen (N) cycling, as well as their feedback effects on plant communities as both mutualists and pathogens (Van der Heijden et al., 2008). While extensive research has explored the effects of various disturbances on soil microbial communities and their functioning (Griffiths & Philippot, 2013), significant uncertainties persist regarding the factors governing soil microbial community stability and the consequences of disturbance-induced changes for their ability to endure subsequent disturbances. This may be partly because most studies have assessed the stability of broad microbial properties, such as biomass and respiration, rather than community structure (number of different taxa and their relative abundances). Yet, shifts in the abundances or relative contributions of community members may influence microbial community stability—changes that may remain undetected when focusing solely on bulk microbial properties. Furthermore, soils are complex, highly heterogeneous environments, and the factors influencing microbial community stability in soils may differ from those in other ecosystems.

Soil is vital for life, supporting terrestrial biogeochemical and nutrient cycling, serving as a foundation for food production, and acting as a natural filter that helps provide clean groundwater (Fig. 1). It is an incredibly complex system, housing communities across multiple trophic levels, featuring intricate substrate transport mechanisms, and exhibiting extreme temporal and spatial heterogeneity, along with numerous internal and external feedbacks that shape community structure and function. Microorganisms drive many processes that support ecosystem services, such as decomposition, mineralization, nutrient cycling, disease regulation, and pollutant breakdown. Any disturbance to the soil can disrupt microbial activity, alter these processes, and impact overall ecosystem performance. The high complexity and biodiversity of soils enable extensive feedbacks and interactions, most of which are overlooked in traditional "compartmentalized" approaches where specific groups are studied in isolation. A more holistic approach that considers critical biotic and abiotic interactions (e.g., network modeling) can improve understanding of the mechanisms that regulate soil systems, enhancing our ability to predict and manage these systems under stress.

Climate change is projected to lead to more frequent droughts and heavy rainfall, rising temperatures, and increased litter inputs and plant root exudates due to elevated atmospheric CO_2 levels, all of which significantly influence soil microbial

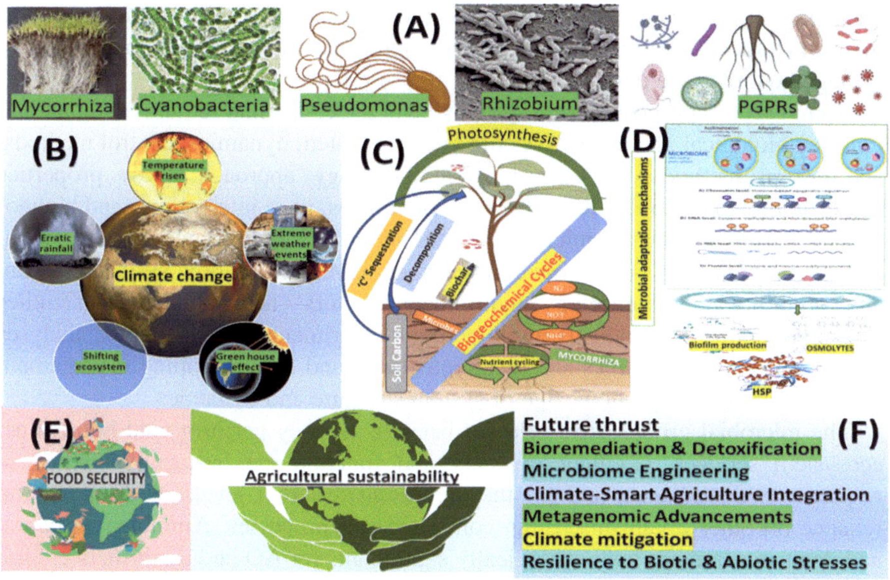

Fig. 1 Microbial resilience and adaptation strategies in climate-smart agriculture. This comprehensive diagram illustrates the multifaceted role of beneficial microbes in climate resilience: (**a**) key beneficial microbes, including mycorrhiza, cyanobacteria, rhizobium, pseudomonas, and PGPRs-support plant health and soil productivity, (**b**) major climate change drivers such as temperature rise, erratic rainfall, extreme weather events, greenhouse effects, and ecosystem shifts that challenge agricultural sustainability, (**c**) plant-microbe-soil interactions driving biogeochemical cycles, including nutrient cycling, biological nitrogen fixation, carbon sequestration, and decomposition, (**d**) microbial adaptation mechanisms including genetic, epigenetic (Leray et al., 2021) and physiological responses such as heat shock proteins (HSPs), osmolyte production (Paul, 2013), and biofilm formation (Sharma et al., 2023), (**e**) the contribution of resilient microbes toward achieving food security and long-term agricultural sustainability, and (**f**) future thrust areas for innovations, including bioremediation, microbiome engineering, metagenomics, climate-smart agriculture integration, and building resilience to biotic and abiotic stresses for climate mitigation

community structure and function (Bardgett et al., 2013). In this context, we focus on pulse disturbances linked to climate change, such as drought, increased rainfall, and higher litter inputs, because these disturbances have clear start and end points, enabling the evaluation of both resistance and resilience in microbial community composition. Because linking microbial community structure to function is challenging, research often focuses on the microbial groups directly involved in delivering specific ecosystem services. This aligns with Ridder (2008), who argued that ecosystem services are performed not by entire ecosystems, but by specific components within those systems, suggesting that understanding the whole system may not be necessary. Such an approach simplifies microbial systems into distinct compartments linked to specific functions, making investigation more manageable. However, given the soil system's complexity with its multiple interacting parts and feedback loops, studying individual components in isolation can only yield a partial understanding. Many other components exert indirect but significant influences on these functions. Thus, compartmentalized approaches may fail to support effective management of these intricate systems.

Kitano (2002) proposed that a systems-level understanding requires insights into four key biological properties: system structure, system dynamics, control methods, and design methods. For a soil systems-biology approach, these properties translate into: (1) mapping the network of organisms, gene interactions, and biogeochemical pathways in soil microbial communities, (2) understanding the temporal and spatial variability in microbial systems, (3) investigating how system structure and variability are influenced by environmental factors—achievable through studies along environmental gradients, multifactorial experiments, and metabolic analyses such as metatranscriptomics and metaproteomics, and (4) developing management strategies informed by knowledge of system structure and dynamics.

The microbial groups were selected because (1) they perform vital but distinct ecosystem services, (2) their diversity and function are frequently studied in response to disturbance, and (3) they represent different levels of diversity and reliance on other organisms within connected soil processes. Ammonia oxidation and denitrification are both ecologically significant, yet AO and denitrifiers exhibit contrasting characteristics. AOs are phylogenetically narrow, with limited functional redundancy, whereas denitrifiers are highly diverse and likely to show greater redundancy. AO and denitrifiers also respond differently to environmental factors: organic matter is crucial for denitrifiers but not for nitrifiers, and moisture—which limits oxygen availability—boosts denitrifier activity while inhibiting nitrifiers (Stark & Firestone, 1995). Consequently, factors like landform or plant type can exert opposing effects on nitrification and denitrification. These same regulators influence symbiotic organisms. Symbionts (both mutualists and pathogens) rely on close host interactions, which often lead to them being studied in isolated, pair-wise associations rather than as part of broader, complex soil systems. However, symbionts also interact with free-living soil biota and with each other—directly and through host-mediated feedbacks (Bever et al., 2013). Understanding these interactions is crucial for a more complete view of soil system dynamics. This chapter highlights the vital role of microbes in maintaining soil health by facilitating the decomposition of SOM. While global understanding of soil microbial dynamics and functions continues to grow, knowledge of the rhizospheric network remains limited, despite its significant influence on regulating soil-plant systems.

2 Microbial Diversity and Ecosystem Function

The concept of community ecology originally emerged from plant and animal ecology. Communities are described as multispecies assemblages where organisms coexist within a shared environment and interact with one another. This field aims to understand how these assemblages are structured, the nature of their functional interactions, and how community composition changes across space and time. Clements (1916) characterized the community as a "supra-organism," suggesting it possesses a distinct level of organization, with tightly connected interactions among its members forming a causal system that leads to emergent properties. According to this view, community boundaries are determined less by physical dimensions and more by the range over which strong interactions occur between populations (Levins & Lewontin, 1985). On the other hand, Gleason's (1926) individualistic concept proposes that species co-occur primarily because they share similar tolerances to environmental conditions, without necessarily interacting. As a result, the way communities are defined may reflect the ecologist's perspective rather than inherent biological characteristics.

Defining microbial communities presents even greater challenges. The idea of a "contiguous environment" and the nature of "interactions" become less straightforward. Microorganisms respond to and influence their microenvironments, which are typically on the scale of microns rather than millimeters (Young et al., 2008), except for multicellular structures like fungal hyphae. However, in water-saturated sediments and stable aquatic columns, the consumption of substrates and release of metabolic products can create chemical gradients extending over meters (Wakeham et al., 2007). This means that microorganisms can metabolically "interact" across large distances. As a result, the strength and scale of these interactions may vary significantly depending on how the researcher defines the microbial community.

To bring more clarity to the term "microbial community," it would be beneficial for microbial ecologists to explicitly define what they mean within each specific study. Microbes that strongly interact within a localized microenvironment can be considered a local community. However, most habitats are patchy in both organism distribution and physicochemical conditions—even seemingly uniform oligotrophic planktonic environments may harbor nutrient-rich microhabitats like marine snow (Azam & Malfatti, 2007). This patchwork of local communities has been referred to as a phenomenological community (Sterelny, 2006), representing a set of macroscale habitats identified by researchers where microbial assemblages persist in spatial association. A more refined approach is to define an indexical community—a group of populations that directly interact with a key population or drive a specific biogeochemical process, alongside other local populations influencing those direct interactions (Sterelny, 2006).

Recent advancements in community ecology increasingly acknowledge that biological assemblages cannot be fully described without considering the surrounding abiotic environment. Recognizing the close relationship between microbes and their microscale physical and chemical environments is crucial for properly delineating microbial communities (O'Donnell et al., 2007). Accordingly, it may be more insightful to define microbial communities not from a broad, macroscale perspective (e.g., aquatic vs. terrestrial environments), but rather from a bottom-up approach—starting with an analysis of the microenvironment's physicochemical properties and scaling up to a spatial domain ("contiguous environment") defined by the area over which significant direct or indirect interactions occur. This approach depends on detailed characterization of the local environment, but technological advancements continue to improve microscale analysis capabilities (Young et al., 2008).

Nowadays, with the ongoing decline in soil quality, maintaining a healthy soil system relies on a complex interplay of physical, chemical, and biological indicators. Healthy soils strike a balance between meeting agricultural demands and supporting community well-being. Soil organic matter (SOM) plays a crucial role in preserving soil health and quality, neutralizing harmful compounds, suppressing pathogens, and promoting environmental sustainability. It involves interactions among both internal and external soil components to support a sustainable food production system. Beneficial soil microbes are key players, driving essential biological transformations and managing different carbon pools along with macro- and micronutrients, which support soil plant-microbe interactions.

2.1 Soil Microbes

Soil microbes are integral to nutrient cycling and plant health. Nitrogen-fixing bacteria (e.g., *Rhizobium*) and mycorrhizal fungi enhance soil fertility by forming symbiotic relationships with plants. However, agricultural practices such as pesticide use and tillage disrupt microbial diversity. Thiele-Bruhn et al. (2012) demonstrated that antibiotics in manure reduce bacterial biomass by 15–30%, while Hartmann et al. (2015) linked glyphosate to declines in *Pseudomonadaceae* populations. Sustainable practices like crop rotation and organic amendments can restore microbial equilibrium.

2.2 Marine Microbes

Marine microbes regulate global carbon cycles, with phytoplankton contributing ~50% of Earth's photosynthesis. The Tara Oceans project revealed that viral lysis releases 145 gigatons of carbon annually, fuelling microbial loops (Sunagawa et al., 2015). Hydrocarbon-degrading bacteria, such as *Alcanivorax borkumensis*, are critical in oil spill remediation (Hazen et al., 2010). Salinity gradients and temperature fluctuations further shape community structure, as seen in estuarine ecosystems.

2.3 Extremophiles

Extremophilic microorganisms are capable of surviving and thriving in environments with extreme conditions. These remarkable organisms include thermophiles (high-temperature adapted), psychrophiles (cold-adapted), halophiles (salt-tolerant), acidophiles and alkaliphiles (pH extremes-adapted), piezophiles (pressure-tolerant), radiophiles (radiation-resistant), xerophiles (desiccation-tolerant), anaerobes (oxygen-independent), and various other specialized forms (Passarini et al., 2022). *Deinococcus radiodurans* repairs DNA damage caused by ionizing radiation (Slade & Radman, 2011), while *Thermus aquaticus* from Yellowstone hot springs revolutionized PCR with its heat-stable enzymes (Brock, 1978). Acidophiles like *Acidithiobacillus ferrooxidans* thrive in pH < 3, enabling bioleaching of metals (Johnson, 2014).

2.4 Industrial and Urban Microbes

Microbes are harnessed for biotechnology, including bioleaching and wastewater treatment. *Saccharomyces cerevisiae* ferments biofuels, reducing reliance on fossil fuels. Urban environments host unique microbiomes; subway systems harbor antibiotic-resistant genes. Dietary shifts rapidly alter human gut microbiota, impacting health (David et al., 2014). Microbial adaptability underpins their success across ecosystems. Understanding their responses to environmental stressors informs conservation, agriculture, and industry. Future research should integrate omics technologies to unravel microbial interactions at scale.

3 Microbial Roles in Major Biogeochemical Cycles

Soil microbes are crucial players in nutrient cycling, which supports life. They significantly contribute to the cycling of carbon, nitrogen (N), and phosphorus—essential elements for creating vital biomolecules like amino acids, proteins, DNA, and RNA, the building blocks of life. Many essential plant nutrients originate from the weathering of minerals. Silicate minerals, including feldspar, mica, and hornblende, release calcium, magnesium, and potassium, while apatite serves as the main phosphorus source. Soil bacteria and fungi are key in facilitating mineral weathering, promoting ion cycling, and enhancing plant nutrition.

Microbes have a critical function in the recycling of nutrients, an activity necessary to balance ecosystems and facilitate life on planet Earth. Recycling of nutrients occurs through decomposition of organic waste and the process of converting inorganic substances to become available to be reused by plants, animals, and other living organisms. This process is significant to the ecosystems' sustainability as it provides continuous availability of elements that are indispensable for life, like carbon, nitrogen, phosphorus, and sulfur. This section delves into the functions of microbes in the recycling of nutrients based on the examples of carbon, nitrogen, phosphorus, and sulfur cycles (Table 1).

3.1 Carbon Cycle

Microorganisms play essential roles in carbon cycling, a key component of all living organisms. Primary producers capture carbon dioxide and convert it into organic matter. In terrestrial ecosystems, plants are the main producers of organic material, though surface-dwelling algae and cyanobacteria, both free-living and symbiotic

Table 1 Role of microbes in biogeochemical cycles

Microorganisms	Role	References
Mycorrhizal fungi	Soil carbon stabilization by transferring plant-derived carbon into soil aggregates; enhance soil fertility and nutrient uptake; play a role in phosphorus cycling; affected by nitrogen addition and drying conditions	Treseder (2004), Gutknecht et al. (2012)
Prochlorococcus	Fix a significant portion (25%) of oceanic CO_2	Jiao et al. (2010)
Bacillus and pseudomonas	Produce extracellular polymeric substances (EPSs) that enhance soil aggregation and carbon retention; *Bacillus megaterium* var. *phosphaticum* and *Bacillus mucilaginosus* solubilize bone phosphorus	Rillig et al. (2015)
Rhizobium, Mesorhizobium, Frankia	Symbiotic biological nitrogen fixation; engineered strains could reduce fertilizer use	Herridge et al. (2008) and Geddes et al. (2019)
Azotobacter, Burkholderia, Clostridium, certain methanogens	Free-living nitrogen fixation	Herridge et al. (2008)
Azotobacter vinelandii	In consortium with mycorrhizal fungi, promotes root development and nutrient uptake in wheat	Dal Cortivo et al. (2018)
Nitrosospira, Nitrosomonas (AOB)	Drive the first step of nitrification (NH_3 to nitrite); AOB prevalent in topsoil with high NH_3	Stark and Firestone (1995), Dal Cortivo et al. (2018)
Nitrososphaera (AOA)	Drive the first step of nitrification (NH_3 to nitrite); AOA dominate in acidic subsoils with lower NH_3	Stark and Firestone (1995), Bell et al. (2014)
Nitrobacter, Nitrospira	Convert nitrite to nitrate during nitrification	–
Proteobacteria, Actinobacteria, Firmicutes, fungi	Capable of denitrification	Bardgett et al. (2013) and Bever et al. (2013)
Pseudomonas stutzeri	Example of a denitrifying bacteria that converts nitrate to N_2 gas, mitigating N_2O emissions	Cayuela et al. (2014)
Alcaligenes, Escherichia	Involved in dissimilatory nitrate reduction to ammonium (DNRA)	–

(continued)

Table 1 (continued)

Microorganisms	Role	References
Anammox bacteria (*Kuenenia*, *Brocadia*)	Perform anaerobic ammonium oxidation (anammox), converting ammonium directly to nitrogen gas	Ling et al. (2015)
Methylococcus, *Methylobacterium*	Oxidize methane (CH_4) to CO_2 in aerobic soils	Conrad (2009) and Ling et al. (2015)
Alcaligenes eutrophus	Contains cation/proton antiporter provide resistance to Cd^{2+}, Zn^{2+}, and Co^{2+}	–
Arthrobacter nicotinovorans HIM	Can degrade the pesticide atrazine	–
Terrabacter sp. *strain DDE-1*	Can metabolize DDE, a byproduct of DDT	–
Ligninolytic fungi (*Phanaerochaete chrysosporium*)	Degrade a wide range of pollutants (e.g., pentachlorophenol, dioxins) through cometabolism	–
Deinococcus radiodurans	Resists radiation and repairs DNA damage; upregulates cold shock proteins	Slade and Radman (2011)
Thermus aquaticus	Produces heat-stable enzymes for PCR	Brock (1978)
Acidithiobacillus ferrooxidans	Thrives in acidic environments and used in bioleaching of metals	Johnson (2014)
PGPMs (*Pseudomonas*, *Azospirillum*, *Gluconacetobacter*, *Bacillus*, fungi)	Improve agronomic efficiency, enhance nutrient availability, promote plant growth, protect against pathogens	Hamilton et al. (2016), Meena and Jha (2018), Singh et al. (2018), Dey et al. (2018), Pouvreau et al. (2018), El Chami et al. (2020), Gorguner and Kavvas (2020), Etesami and Glick (2020), Lyu et al. (2020), De Souza et al. (2020), Hagagy et al. (2021), Khan et al. (2021), Yadav (2021), Mukhopadhyay et al. (2021), Bungau et al. (2021), Song et al. (2021), Lone and Malik (2021), Rizvi et al. (2021), Passarini et al. (2022), Wong (2022), and Trivedi et al. (2022)
P. putida, *Comamonas testosterone*, *C. freundii*, *Enterobacter cloacae*	Enhance phosphate mobilization and improve crop productivity in synthetic consortia	Baas et al. (2016)
Bacteria with ACC deaminase activity	Reduce stress and enhance crop productivity in nutrient-deficient conditions in synthetic consortia	Vitullo et al. (2012) and Tsolakidou et al. (2019)

(like in lichens), also significantly contribute to carbon fixation in certain environments. Additionally, autotrophic microbes within soil can fix CO_2. Organic matter produced through primary production exists in living organisms and the nonliving organic material derived from them. Heterotrophic bacteria and fungi act as nature's recyclers, breaking down nonliving organic material. These soil saprotrophs finalize the carbon cycle by converting organic material from primary producers back into carbon dioxide through respiration.

3.2 Nitrogen Cycle

All organisms need nitrogen because it is a vital component of proteins and nucleic acids. Animals obtain nitrogen from organic sources, while plants rely on inorganic forms like ammonium and nitrate or simpler nitrogen compounds such as single amino acids (e.g., glycine) (Schimel & Bennett, 2004). Most microbes can utilize ammonium or nitrate for growth. Microbes are crucial players in the nitrogen cycle, performing several key processes that other organisms cannot, including nitrogen fixation, dissimilatory nitrate reduction to ammonia (DNRA), nitrification, anammox, and denitrification. Since nitrogen is often the primary limiting nutrient for plant biomass production in terrestrial ecosystems, the rates of these microbial processes frequently control overall ecosystem productivity. Some nitrogen cycle steps, like nitrogen fixation or nitrification, are performed by a limited number of microbial groups and are known as narrow processes, while others, such as DNRA, involve many different microbial groups and are considered broad processes. The release of ammonium from decomposing soil organic matter is called ammonification.

Only bacteria and archaea are capable of biological N-fixation, where atmospheric nitrogen gas is converted into ammonium. This is the sole natural process introducing new nitrogen into the biosphere, making it essential for ecosystem functioning. N-fixation relies on the enzyme nitrogenase, which is highly sensitive to oxygen and requires a low-oxygen environment to function. It's an energy-intensive process, consuming 16 moles of ATP for every mole of nitrogen fixed. The ammonium produced through this process is incorporated into amino acids, which are then assembled into proteins. In nitrogen-limited environments, nitrogen-fixing microbes gain a competitive advantage. Free-living microbes like *Azotobacter, Burkholderia, Clostridium*, and certain methanogens are among those capable of nitrogen fixation. Some of these microbes may associate with the plant rhizosphere, while others form symbiotic relationships with plants, such as *Rhizobium, Mesorhizobium*, and *Frankia*. Plant exudates can provide energy to support nitrogen fixation. In New Zealand's agricultural soils, rhizobia that form root nodules in symbiosis with introduced legumes like clover, lucerne, and lotus contribute significantly to nitrogen input. Similarly, native legumes such as Sophora and Clianthus form symbiotic associations with *Mesorhizobium* or *Rhizobium leguminosarum* (Weir et al., 2004). Interestingly, rhizobia strains found on native legumes differ from those associated with weed legumes like gorse. Nitrogen fixation rates by symbiotic rhizobia are typically two to three orders of magnitude higher than those achieved by free-living soil bacteria.

During nitrification, ammonia or ammonium ions are oxidized first to nitrite and then to nitrate. This aerobic process in soil is primarily carried out by a select group of autotrophic bacteria and Crenarchaea. The conversion of ammonia to nitrite is mainly driven by bacteria such as *Nitrosospira* and *Nitrosomonas*, alongside *Crenarchaeum* and *Nitrososphaera*. Meanwhile, the transformation of nitrite into nitrate is performed by bacteria like *Nitrobacter* and *Nitrospira*. Nitrifying microbes harness the energy released from this process to fix carbon dioxide. Nitrification is particularly important in soil systems because the conversion from positively charged ammonium ions (NH_4^+) to negatively charged nitrate ions (NO_3^-) promotes nitrate leaching, as nitrate is less likely to bind to negatively charged clay particles. To reduce nitrate loss, nitrification inhibitors are commonly applied to New Zealand soils. Studies have shown that ammonia-oxidizing bacteria (AOB) are more prevalent in topsoil, whereas ammonia-oxidizing archaea (AOA) dominate in certain subsoils, with AOB thriving in high-ammonia environments and AOA preferring lower ammonia concentrations.

Denitrification, on the other hand, is a microbial respiratory pathway where nitrogen oxides serve as alternative electron acceptors under oxygen-limited conditions. This sequential reduction converts nitrate (NO_3^-) to nitrite (NO_2^-), then to nitric oxide (NO), and finally to either nitrous oxide (N_2O)—a potent greenhouse gas or nitrogen gas (N_2), which is environmentally harmless. Denitrification predominantly occurs in waterlogged, anaerobic soils and serves as the main biological route by which fixed nitrogen returns to the atmosphere, completing the nitrogen cycle. However, this process also results in significant nitrogen loss from soils, impacting crop productivity. A wide array of phylogenetically diverse bacteria, including members of Proteobacteria, Actinobacteria, and Firmicutes, as well as fungi and other soil eukaryotes, are capable of denitrification. Notably, many denitrifiers lack some of the necessary enzymes, resulting in "incomplete denitrification," which often produces N_2O rather than N_2. For instance, most fungi and roughly a third of sequenced bacterial denitrifiers (Philippot et al., 2011) lack N_2O reductase. This incomplete process is a significant contributor to greenhouse gas emissions from New Zealand's pastoral agriculture.

Another pathway, known as dissimilatory nitrate reduction to ammonium (DNRA), involves facultative anaerobic bacteria like Alcaligenes and Escherichia, which reduce nitrate to nitrite under anaerobic conditions. The resulting nitrite may either be excreted or further reduced to ammonia through hydroxylamine under specific conditions. Unlike denitrifiers, these organisms do not produce gaseous nitrogen products. Anammox (anaerobic ammonium oxidation) bacteria offer an alternative nitrogen removal route, converting ammonium directly to N_2 gas in the presence of both oxidized and reduced nitrogen compounds under anaerobic conditions. Anammox bacteria found in permafrost and agricultural soils belong to a unique, monophyletic group within the Planctomycetes, with genera like "*Kuenenia*" and "*Brocadia*" identified through phylogenetic studies.

3.3 Phosphorus Cycle

Phosphorus (P) is relatively scarce in the environment, and its availability is further limited by its tendency to precipitate in the presence of divalent and trivalent cations under neutral and alkaline pH conditions. Microbes influence phosphorus cycling in two primary ways. First, they mineralize organic phosphorus (mainly found as phosphate esters) into inorganic phosphate through the action of phosphatase enzymes, which are widely produced by bacteria and fungi. Then, convert insoluble, immobilized phosphorus into soluble or more mobile forms, a process typically driven by the production of organic acids. Microbes release enough phosphorus to meet their own needs, while also providing phosphorus for plants and other soil organisms.

Mycorrhizal fungi contribute by producing oxalate, which helps release phosphate from insoluble mineral forms. This phosphorus mobilization by fungal symbionts is a key strategy that supports plants in overcoming phosphorus limitations. For instance, several ectomycorrhizal basidiomycete fungi possess high-affinity phosphate transporters, which are activated in extraradical hyphae when their host plants experience phosphorus deficiency. In New Zealand pasture soils, phosphorus-solubilizing bacteria have been identified among Proteobacteria (*Pseudomonas*) as well as Actinobacteria, Firmicutes, and Bacteroidetes. However, their abundance and diversity are influenced by farm management practices, with the highest populations typically found in phosphorus-deficient soils. Additionally, evidence suggests that long-term application of phosphorus-rich fertilizers can modify the diversity of Actinobacteria and arbuscular mycorrhizal fungi in pasture soils.

3.4 Sulfur Cycle

The sulfur cycle is the conversion of sulfur among its different oxidation states, and microorganisms are central to these conversions. Sulfur is necessary for amino acid and vitamin synthesis, and its cycling is directly associated with the carbon and nitrogen cycles. Sulfur-oxidizing bacteria, including Thiobacillus, oxidize low-valent sulfur compounds (H_2S) to sulfate (SO_4^{2-}), and this can be assimilated by plants and other organisms. On the other hand, sulfate-reducing bacteria (*Desulfovibrio*) reduce sulfate to sulfide under anaerobic conditions, which completes the cycle. In the marine environment, microbial sulfur cycling is especially critical in the deep ocean, where sulfur-oxidizing bacteria engage in symbiosis with invertebrates and supply them with energy via chemosynthesis.

3.5 Detoxification of Heavy Metals

Microbes exposed to heavy metals in the environment have developed various detoxification strategies over time. In soils, common heavy metal contaminants include copper, mercury (Hg), zinc (Zn), lead, cobalt (Co), and cadmium (Cd). In New Zealand's agricultural soils, cadmium buildup has been associated with the application of superphosphate fertilizers. Heavy metals can be harmful to soil microbes due to their strong affinity for thiol groups in biomolecules like proteins. To prevent cellular damage, bacteria have evolved three main mechanisms to tolerate metals. The first mechanism involves sequestering metals by binding them to cellular components, which lowers the concentration of free metal ions in the cytoplasm. Metals can bind to cell membranes, cell walls, and extracellular polymeric substances (EPS). EPS can capture several metals, including copper and lead. The second mechanism involves detoxifying metals by reducing intracellular ions. For example, mercury ions (Hg^{2+}) can be reduced to elemental mercury (Hg^0) by the enzyme mercury reductase, encoded by the merA gene. The resulting Hg^0 then diffuses out of the cell due to its high volatility. The third mechanism involves actively pumping metal ions out of the cell using efflux systems. One example is the CZC cation/proton antiporter, found in *Alcaligenes eutrophus*, which provides resistance to cadmium (Cd^{2+}), zinc (Zn^{2+}), and cobalt (Co^{2+}) by expelling metal ions from the cytoplasm across the cell membrane into the surrounding environment. These microbial processes are now being explored for use in bioremediation efforts to clean up heavy metal-contaminated waste.

3.6 Biodegradation

Soil-dwelling heterotrophic bacteria (*Pseudomonas, Sphingomonas*, and *Mycobacterium*) are frequently associated with oil degradation. *Pseudomonas*, for instance, has been extensively studied, and scientists have identified the genes and enzymes that enable it to break down alkanes, monoaromatic compounds, naphthalene, and phenanthrene as sole carbon sources under aerobic conditions. Enhancing oil degradation in contaminated soil usually involves adding nutrients like nitrogen (N) and phosphorus (P), and sometimes oxygen and water. In most cases, adding hydrocarbon-degrading bacteria is unnecessary because these microbes are naturally widespread in soil. When oil spills occur, their populations rapidly increase. High hydrocarbon concentrations can deplete N and P, which are essential for microbial growth, limiting the activity of these bacteria.

Both bacteria and fungi contribute to pesticide degradation. For example, *Arthrobacter nicotinovorans* HIM, isolated from New Zealand agricultural soil, can use the pesticide atrazine as its sole carbon and nitrogen source. This bacterium also breaks down other triazine compounds like simazine, terbuylazine, propazine, and cyanazine. Some pesticides degrade quickly in soil, reducing their effectiveness against pests. DDT is widely used in New Zealand to control grass grubs, but persists as resistant to degradation. Under aerobic conditions, DDT converts into DDE, which was once thought to be a dead-end product. However, *Terrabacter* sp. strain DDE-1, isolated from Winchmore Research Station soil, was found to metabolize DDE when biphenyl was provided as an alternative carbon source.

Ligninolytic fungi, such as the white rot fungus *Phanaerochaete chrysosporium*, are capable of degrading a wide variety of environmental pollutants, including pentachlorophenol and dioxins, through cometabolism, where alternative carbon sources like sawdust, straw, or corn cobs support their growth. This remarkable ability stems from the enzymes these fungi evolved to break down lignin. New Zealand strains of white rot fungi, along with Zygomycetes, have successfully degraded pentachlorophenol and certain dioxin and furan compounds in soil from an old wood-treatment site in Whakatane. Successful in situ biodegradation relies on the contaminant, microbes, and environment.

Generally, microbes can more easily break down natural organic contaminants, while synthetic compounds such as DDT and aldrin are more persistent. Hydrophobic contaminants, which poorly dissolve in water, often bind to soil organic matter or clay particles. This can lower their toxicity but also reduce their biodegradability. In some cases, high contaminant concentrations hinder degradation because, although these compounds serve as carbon and energy sources for microbes, they can become toxic at elevated levels. For degradation to occur, the right microbes must be present in sufficient numbers, which partly depends on how long they have been exposed to the pollutant. Some bacteria exhibit motility and chemotaxis, enabling them to sense and move toward contaminants, while fungi extend filaments in the direction of the pollutant. Having the right microbes isn't enough—environmental conditions must also support microbial activity. Factors such as pH, temperature, moisture, nutrient levels, and oxygen availability all influence microbial growth and function. Most microbes thrive within specific ranges of these conditions. Hydrocarbons, for example, break down more easily in oxygen-rich environments, whereas dechlorination processes (e.g., trichloroethylene degradation) often require anaerobic conditions. Some compounds, like alkylated benzenes, degrade under both aerobic and anaerobic environments. Nutrients that commonly limit biodegradation include NPK and iron.

4 Climate Change on Microbial Communities

Microbial communities are pivotal to ecosystem stability, driving biogeochemical cycles, nutrient recycling, and symbiotic relationships with flora and fauna. However, climate change—characterized by rising temperatures, shifting precipitation patterns, elevated atmospheric CO_2, and an increase in extreme weather events is reshaping microbial diversity and function. These transformations carry significant implications for ecosystem resilience, agricultural productivity, and human health.

4.1 Temperature

Rising temperatures enhance microbial metabolic rates, accelerating the decomposition of organic matter and increasing emissions of greenhouse gases such as CO_2 and CH_4, which further exacerbate climate change. In terrestrial ecosystems, warmer soils stimulate a surge in nutrient release; however, prolonged warming may disadvantage cold-adapted microbes, reducing overall microbial diversity and biomass. Over time, microbial communities shift toward thermophilic species, potentially disrupting nutrient cycling and carbon sequestration. For instance, warming experiments in grassland soils have demonstrated a shift from diverse bacterial and fungal assemblages to thermophilic-dominated communities, altering nitrogen cycling dynamics and influencing plant productivity.

4.2 Precipitation Patterns

Climate change intensifies both droughts and heavy rainfall events, leading to distinct microbial responses. Droughts induce soil desiccation, diminishing microbial activity and promoting the dominance of drought-resistant taxa like Actinobacteria. This shift impedes decomposition and nutrient turnover, impairing plant growth. Conversely, excessive rainfall and flooding create anoxic conditions, disadvantaging aerobic microbes. Anaerobic specialists, such as methanogenic archaea and sulfate-reducing bacteria, proliferate under these conditions, enhancing CH_4 and H_2S emissions—contributing to atmospheric greenhouse gas concentrations and soil degradation.

4.3 Carbon Dioxide

Elevated atmospheric CO_2 enhances plant growth, resulting in increased carbon-rich root exudates that stimulate microbial activity and carbon cycling. However, this carbon influx may provoke nitrogen immobilization by microbes, potentially limiting nitrogen availability for plants and influencing ecosystem productivity. Additionally, ocean acidification driven by dissolved CO_2 reduces seawater pH, threatening marine microbes, particularly calcifiers like coccolithophores, while favoring acid-tolerant bacterial taxa. This imbalance disrupts marine food webs and compromises oceanic carbon sequestration. Free-Air CO_2 Enrichment (FACE) experiments in forest ecosystems illustrate increased root biomass and microbial activity, yet nitrogen availability declines due to microbial nitrogen immobilization, influencing plant nutrient dynamics.

5 Microbial Mechanisms of Environmental Resilience

Microbial communities exhibit remarkable genetic and physiological plasticity, enabling rapid adaptation to environmental perturbations induced by climate change. This section explores key genetic and physiological mechanisms underpinning microbial resilience, including mutations, horizontal gene transfer (HGT), epigenetic modifications, membrane adjustments, stress protein synthesis, and metabolic shifts.

5.1 Mutations and Natural Selection

Spontaneous mutations within microbial genomes can confer adaptive advantages under novel environmental stresses. Soil microbes exposed to chronic warming, for example, may acquire mutations in metabolic pathways that enhance thermotolerance. These mutations propagate through natural selection, fostering the emergence of climate-resilient microbial strains. This evolutionary adaptability underscores the pivotal role of genetic variation in maintaining microbial function amid dynamic environmental pressures.

5.2 Horizontal Gene Transfer (HGT)

HGT facilitates the rapid acquisition of advantageous traits from neighboring organisms. Marine bacteria subjected to thermal stress, for example, may acquire plasmid-borne genes conferring antibiotic resistance or heat tolerance. This genetic exchange enhances microbial resilience, fostering community stability in fluctuating environments. HGT's role in promoting genetic diversity and environmental adaptation is particularly evident in dynamic marine ecosystems impacted by climate change.

5.3 Epigenetic Modifications

Epigenetic mechanisms, including DNA methylation and histone modifications, enable rapid phenotypic adjustments without altering genetic sequences. In drought-tolerant fungi, alterations in methylation patterns of stress-response genes have been documented, enhancing resilience under water-limited conditions. These reversible modifications fine-tune gene expression, enabling swift physiological responses to environmental stressors, underscoring the importance of epigenetic regulation in microbial adaptation.

5.4 Physiological Adaptation

Microorganisms demonstrate exceptional physiological plasticity, employing diverse biochemical and cellular strategies to maintain homeostasis under climate-induced stress. This section details critical physiological adaptations, including membrane composition modifications, stress protein synthesis, and dormancy strategies.

5.5 Fatty Acid Profile Adjustments

Microbes modulate membrane fluidity by adjusting phospholipid fatty acid saturation levels. Psychrophilic bacteria, for instance, increase unsaturated fatty acid content to maintain membrane flexibility in cold environments. In contrast, thermophiles enhance saturated and branched-chain fatty acid content, stabilizing membranes against heat-induced damage. This lipid plasticity ensures cellular integrity, facilitating survival across extreme thermal gradients.

5.6 Hopanoid Production

Gram-negative bacteria produce hopanoids—steroid-like molecules—to reinforce membrane stability under stress. Acidophilic bacteria elevate hopanoid synthesis in low-pH environments, enhancing membrane rigidity and preserving nutrient uptake and energy production functions. This adaptation supports microbial survival in acidic habitats, demonstrating membrane composition's critical role in climate resilience.

5.7 Heat Shock Proteins (HSPs)

Molecular chaperones such as DnaK and GroEL prevent protein denaturation during heat stress. In *Escherichia coli*, thermal shock triggers a 20-fold upregulation of these chaperones, ensuring protein folding fidelity and preserving cellular function. This rapid stress response exemplifies microbial resilience to abrupt temperature increases.

5.8 Cold Shock Proteins (CSPs)

Cold shock proteins (CSPs) maintain RNA stability and enhance translation under cold stress. *Deinococcus radiodurans* upregulates CSPs to prevent harmful secondary RNA structures from impeding ribosome function. This adaptation supports cellular processes, ensuring survival in cold environments. Microbial genetic and physiological plasticity underscores their capacity to endure climate change-induced environmental fluctuations. Understanding these adaptive strategies is crucial for predicting ecosystem responses and informing biotechnological innovations aimed at mitigating climate change impacts.

5.9 Physiological Plasticity

Microbes adjust enzyme production, membrane composition, and dormancy strategies to cope with stress. For example, actinobacteria increase trehalose synthesis under drought, enhancing desiccation tolerance (Berlemont et al., 2014).

5.10 Community Composition Shifts

Taxonomic turnover, driven by competitive exclusion or priority effects, alters ecosystem function. After rewetting, rare bacterial taxa rapidly dominate, restoring decomposition rates (Aanderud et al., 2015).

5.11 Evolutionary Adaptation

Horizontal gene transfer and selection for stress-tolerant genotypes enable rapid adaptation. Experimental evolution studies demonstrate microbial lineages adapting to salinity or temperature extremes within hundreds of generations (Gonzalez & Bell, 2013). Microbial responses to climate change are multifaceted, shaped by abiotic factors, biotic interactions, and historical contexts. While general trends emerge—such as fungal decline under nitrogen deposition the inherent diversity of microbial systems necessitates context-specific predictions. Integrating genomic, metabolic, and ecological data will enhance models of microbial feedbacks, informing mitigation strategies in a warming world.

6 Microbial Responses to Climate Change Factors

Microbes are key mediators of ecosystem processes, catalyzing biogeochemical cycles and exhibiting rapid responses to environmental perturbations. Climate change affects important abiotic factors (temperature, precipitation, atmospheric CO_2, and nitrogen deposition) that shape microbial communities in various ways. This chapter reviews the current knowledge of the response of microbiomes to these factors, highlighting the physiological, compositional, and evolutionary mechanisms. Understanding these responses is crucial for predicting ecological resilience and feedback to global climate systems.

6.1 Temperature

Increasing temperatures have direct effects on microbial metabolism, most commonly with warmer environments stimulating enzymatic activity and decomposition of organic matter (Schimel et al., 2007). Soil microbial community composition has also been shown to respond to warming, particularly in systems with a carbon limitation (Allison et al., 2010). For instance, in the Jasper Ridge Global Change Experiment (JRGCE; Gutknecht et al., 2012), adding nitrogen to a 0.7 °C soil temperature increase reduced fungal biomarkers and fungal-to-bacterial ratios, indicating interactive effects (Gutknecht et al., 2012). Conversely, microbial communities from historically fluctuating environments exhibit greater resilience to warming due to preadapted physiological plasticity (Hawkes & Keitt, 2015). Microbial taxa have various thermal optima, such that generalists proliferate under variable conditions, whereas specialists become dominant in more stable environments. Populations experiencing gradual warming can undergo evolutionary rescue—rapid genetic adaptation—but there may be limits to how they respond due to local adaptation to their previous conditions (Gonzalez & Bell, 2013). Soil bacteria in temperate grasslands, for instance, exhibit metabolic acclimatization to short-term warming, whereas long-term shifts favor heat-tolerant taxa (DeAngelis et al., 2015).

6.2 Precipitation and Soil Moisture

Alterations in precipitation patterns significantly impact soil moisture, which serves as an essential determinant of microbial activity. Drought conditions lead to a reduction in microbial biomass as a result of restricted substrate diffusion. Conversely, rewetting events initiate surges in respiration and nutrient mineralization. Microbial communities residing in arid environments, such as the grasslands of the Chihuahuan Desert, exhibit a greater capacity to withstand fluctuations in moisture levels. In contrast, soils that historically maintain mesic conditions display functional delays in response (Bell et al., 2014). Furthermore, fungal communities, and specifically arbuscular mycorrhizae, tend to be more adversely affected by drying conditions compared to bacteria, which can be attributed to the vulnerabilities within their hyphal networks (Barnard et al., 2006).

An uptick in rainfall or flooding tends to alter redox conditions, creating a conducive environment for anaerobic organisms such as methanogens and sulfate-reducing bacteria. For instance, the thawing of permafrost releases previously stored organic carbon, which, in turn, stimulates methanogenesis and escalates greenhouse gas emissions (Ling et al., 2015). On the other hand, extended periods of waterlogging may hinder aerobic decomposers, thereby modifying the pathways involved in carbon cycling (Tiemann & Billings, 2011).

6.3 Atmospheric CO_2 elevation

Elevated levels of CO_2 enhance the process of plant photosynthesis, which subsequently increases rhizodeposition and soil carbon contributions. This phenomenon often stimulates microbial biomass, especially mycorrhizal fungi, which thrive on carbon allocated by their host plants (Treseder, 2004). However, within the context of the JRGCE, the effects of elevated CO_2 on microbial activity were minimal when considered in isolation, likely attributable to stable plant productivity (Gutknecht et al., 2012). However, when CO_2 enrichment is combined with nitrogen addition, there can be a pronounced exacerbation of phosphorus limitation, leading to a reconfiguration of microbial enzymatic profiles (Liang et al., 2015). Microbial responses to CO_2 influence soil carbon storage. Increased extracellular enzyme production under high CO_2 may enhance decomposition, offsetting carbon gains from plant inputs (Allison et al., 2010). Conversely, priming effects where fresh carbon stimulates decomposition of older soil organic matter can reduce sequestration (Kuzyakov, 2010).

6.4 Nitrogen Deposition

Nitrogen addition consistently reduces fungal biomass and fungal-to-bacterial ratios, as observed in grasslands and forests (Treseder, 2008; Gutknecht et al., 2012). This reflects fungal sensitivity to nitrogen-induced pH changes and reduced plant carbon allocation to mycorrhizae. Bacteria, especially copiotrophic taxa, often proliferate under nitrogen enrichment due to increased labile carbon availability (Fierer et al., 2012).

Nitrogen deposition alters microbial functional guilds. Ammonia-oxidizing archaea (AOA) and bacteria (AOB) respond differentially; AOA dominate in acidic soils, while AOB thrive in neutral conditions. Elevated nitrogen also stimulates denitrification, though chronic deposition can saturate microbial processing, leading to nitrate leaching and eutrophication (Barnard et al., 2006). Climate factors rarely act in isolation. Warming and nitrogen deposition synergistically reduce fungal biomarkers, while warming combined with CO_2 may mitigate microbial carbon limitation (Gutknecht et al., 2012). Conversely, drought can negate nitrogen effects by limiting substrate availability (Evans & Wallenstein, 2014). Such interactions underscore the complexity of microbial responses in multistressor environments. Prior environmental conditions shape microbial resilience. Communities from fluctuating environments exhibit broader niche breadth, enabling faster acclimatization to change (Hawkes & Keitt, 2015). In contrast, stable environments foster specialists vulnerable to novel stressors, highlighting the role of historical legacies in structuring responses (de Vries & Shade, 2013).

7 Role of Microbes in Climate Change Mitigation

Microorganisms are indispensable agents in global biogeochemical cycles, influencing greenhouse gas (GHG) fluxes, carbon sequestration, and nutrient transformations. Their metabolic versatility enables them to mitigate climate change through processes such as carbon storage, methane oxidation, and nitrogen cycle regulation. This section explores the mechanisms by which microbes contribute to climate change mitigation, emphasizing their roles in terrestrial, aquatic, and engineered systems.

7.1 Carbon Sequestration

Soil microbes mediate the decomposition and stabilization of organic matter, directly impacting carbon sequestration. Mycorrhizal fungi, for instance, transfer plant-derived carbon into soil aggregates, where it is protected from microbial degradation (Treseder, 2004). In grasslands, microbial lipid biomarkers correlate with soil carbon content, suggesting that shifts in microbial community structure influence long-term carbon storage (Gutknecht et al., 2012). *Bacillus* and *Pseudomonas* bacteria produce extracellular polymeric substances that bind soil particles, enhancing aggregate formation and carbon retention (Rillig et al., 2015).

7.2 Oceanic Carbon Pump

Marine microbes drive the biological carbon pump, sequestering atmospheric CO_2 via photosynthesis and transferring it to deep-sea sediments. Cyanobacteria (e.g., *Prochlorococcus*) fix 25% of oceanic CO_2 while heterotrophic bacteria mineralize organic matter, influencing carbon export efficiency (Jiao et al., 2010). Iron-limited regions, such as the Southern Ocean, show increased carbon fixation when microbial communities are supplemented with iron, highlighting their role in regulating oceanic carbon sinks (Boyd et al., 2007).

7.3 Methanotrophy

Methanotrophic bacteria (e.g., *Methylococcus*) oxidize methane (CH_4) in aerobic soils, converting it to CO_2, which has a lower global warming potential. Peatlands and rice paddies host diverse methanotrophs that reduce CH_4 emissions by 20–40% (Conrad, 2009). In Arctic permafrost, thaw-induced drainage creates aerobic zones where methanotrophs mitigate CH_4 release from anaerobic methanogens (Ling et al., 2015).

7.4 Methanogenesis Mitigation

Strategies to suppress methanogens in ruminants, such as feed additives (3-nitrooxypropanol), reduce enteric CH_4 emissions by 30% without altering microbial diversity (Hristov et al., 2015). In wetlands, water table management limits anaerobic conditions, curbing methanogen activity (Hemes et al., 2018).

7.5 Nitrogen Cycle Regulation

Denitrifying bacteria (e.g., *Pseudomonas stutzeri*) convert nitrate (NO_3) to N_2 gas, mitigating nitrous oxide (N_2O) emissions. In agricultural soils, biochar amendments enhance denitrifier activity, reducing N_2O emissions by up to 50% (Cayuela et al., 2014). Microbial consortia in hypoxic sediments also express nitrous oxide reductase (*nosZ*), catalyzing N_2O to N_2.

7.6 Nitrogen-Fixing Symbioses

Legume-rhizobia symbioses reduce synthetic fertilizer demand by fixing 40–60 million metric tons of nitrogen annually (Herridge et al., 2008). Engineered rhizobia strains with enhanced nitrogenase efficiency could further lower agricultural N_2O emissions (Geddes et al., 2019).

7.7 Microbial Biofuel Production

Algae (*Chlorella*) and cyanobacteria convert CO_2 into biodiesel via photosynthetic lipid accumulation (Chisti, 2007). Methanogenic archaea in anaerobic digesters transform organic waste into biogas (60–70% CH_4), displacing fossil fuels.

7.8 Electroactive Microbes

Microbial fuel cells harness electroactive bacteria (e.g., *Geobacter*) to oxidize organic matter, generating electricity while sequestering CO_2 (Logan et al., 2006). This technology mitigates emissions from wastewater treatment, which accounts for 3% of global electricity use (McCarty et al., 2011).

7.9 Engineering Microbial Consortia

Synthetic biology could optimize microbial CO_2 fixation pathways (e.g., CRISPR-edited cyanobacteria) (Gleizer et al., 2019).

7.10 Microbiome Management

Agricultural practices that foster beneficial microbes (e.g., cover cropping) can enhance soil carbon storage (Lehmann et al., 2020).

7.11 Quantifying Microbial Fluxes

Advanced models integrating genomic and geochemical data will improve predictions of microbial climate feedbacks (Wieder et al., 2015). Microbes are pivotal to climate change mitigation, yet their efficacy depends on ecosystem context, microbial diversity, and anthropogenic interventions. Harnessing microbial processes through targeted strategies offers a sustainable pathway to offset GHG emissions but requires interdisciplinary collaboration to address knowledge gaps and ecological complexities.

8 Microbiome-Based Strategies for Climate-Resilient Agriculture

Climate and microorganisms jointly regulate plant growth, influencing crop yield, quality, and sustainability in global food systems (Hagagy et al., 2021). Microbes play central roles in biogeochemical cycles—particularly of carbon, nitrogen, and oxygen (Khan et al., 2021)—yet climate change and excessive agrochemical use increasingly disrupt their ecological functions. Biotechnological advances have facilitated the emergence of climate-resilient agriculture (Meena & Jha, 2018), emphasizing the urgent need for a "next-generation Green Revolution" based on sustainable microbial innovations (El Chami et al., 2020).

Extremophilic microorganisms, capable of surviving under multiple stressors such as high salinity, extreme temperatures, pH imbalance, radiation, or hypoxia (Yadav, 2021), have evolved unique adaptations useful for developing synthetic microbial communities that enhance crop tolerance to abiotic and biotic stresses (De Souza et al., 2020). These microorganisms can function directly as endophytes or rhizospheric partners (Lyu et al., 2020), or indirectly through the production of phytohormones (e.g., gibberellins, auxins), extremozymes (e.g., cellulases, proteases), and other metabolites for bioremediation and biodegradation (Mohanty & Swain, 2018). Their genetic potential can be further exploited using metagenomics and CRISPR-based genome editing to improve plant resilience, nutrient cycling, and stress adaptability (Sethi, 2022).

The strategic use of indigenous microbial communities and coevolved plant-microbe associations has proven effective in mitigating climate-induced agricultural stress (Bungau et al., 2021). The design of constructed or synthetic microbiomes, grounded in ecological and evolutionary principles, offers promising applications for sustainable agriculture (Hamilton et al., 2016; Singh et al., 2018). In arid and saline regions, symbiotic microorganisms isolated from stress-tolerant plants significantly enhance crop survival and productivity (Etesami & Glick, 2020; Mukhopadhyay et al., 2021). Extremophiles, with their specialized enzymes and bioactive compounds, represent untapped biological resources for improving soil fertility, developing stress-resilient crops, and supporting environmental sustainability (Al-Shaibani et al., 2021; Rizvi et al., 2021).

8.1 Synthetic Endophytes

Synthetic endophytes represent more than just experimental models for environmental research. They provide a molecular framework that enhances our understanding of natural phenomena while allowing these processes to be studied under controlled experimental conditions (Lone & Malik, 2021). The knowledge gained from these model systems can subsequently be validated through greenhouse or in vitro experiments. Such reference systems are invaluable for establishing causal relationships between endophytic microorganisms and their plant hosts (Omomowo & Babalola, 2019). As we move forward, it will be essential to deepen our fundamental understanding of synthetic endophyte systems and translate this knowledge into practical applications for the agricultural industry and crop production systems. This comprehensive approach—integrating microbiology, biotechnology, and ecological principles—holds tremendous potential for developing sustainable, climate-resilient agricultural systems capable of meeting the food security challenges of the twenty-first century.

8.2 Metabolomics

Metabolomic approaches have proven highly effective in characterizing and diagnosing crop diseases and their causal agents (Emwas et al., 2013). However, their application in exploring complex tripartite interactions involving plants, pathogenic microbes, and beneficial microbes remains limited. As global agricultural systems seek to reduce foodborne illnesses, improve crop productivity, and stabilize agro-supply chains, a deeper understanding of microbe-derived biopesticides within these interactions is increasingly essential. Advanced metabolomic platforms have been employed to elucidate these dynamics. For instance, GC-TOF-MS and HILIC-Q-TOF-MS analyses revealed that consortia of arbuscular mycorrhizal fungi (AMF) and plant growth-promoting rhizobacteria (PGPR) influence metabolic pathways in Durum wheat (Adeniji & Babalola, 2020). Similarly, HS-SPME-GC-MS and LC-HRMS techniques characterized volatile and nonvolatile metabolite changes during fungal interactions, demonstrating complete suppression of grapevine pathogens *Eutypalata* and *Botryosphaeria obtusa* by 2-nonanone within 1 week (Azzollini et al., 2018). ^{1}H NMR metabolomics has also been pivotal in elucidating metabolic alterations in *Fusarium graminearum* strains, identifying distinct metabolite shifts—upregulation of L-serine, L-phenylalanine, D-glucose, L-glutamate, L-methionine, pyroglutamate, and citrate, and downregulation of D-myoinositol, threonine, malate, and L-sucrose—following fungicide exposure (Sevastos et al., 2018).

Comprehensive GC-MS and LC-MS studies on transgenic crops carrying the *Lr34* resistance gene revealed significant metabolic modifications, including elevated glucose and fructose levels and reduced dehydroascorbate concentrations (Bucher et al., 2017; Sun et al., 2022). Furthermore, Orbitrap MS coupled with GC-MS has illuminated plant defense responses, identifying phytoalexins, coumarins, flavonoids, and 1-methyltryptophan as biomarkers of induced resistance in soybean and tomato (Aliferis et al., 2014; Adeniji & Babalola, 2020). Recent innovations in nontargeted and semitargeted metabolomics, coupled with advanced chemometric and bioinformatics tools, are enhancing data resolution and interpretability. These approaches promise to accelerate biomarker discovery and the development of metabolite-based crop protection strategies, contributing to sustainable agriculture and global food security.

8.3 Advanced Biotechnological Approaches

The fields of metabolic engineering and synthetic biology are poised to revolutionize crop improvement strategies. A defining characteristic of these disciplines is the implementation of iterative design-build-test-learn cycles, where successive rounds of experimentation and modification lead to progressively refined solutions (Pouvreau et al., 2018). The ultimate objective of rational microbiome design is to develop systems that allow precise modification of specific microbial functions and phenotypes directly within their natural environments under optimal operational conditions (Pouvreau et al., 2018). The integration of in situ metagenomic analysis with CRISPR-Cas gene-editing technology represents a particularly promising approach. This combination enables precise manipulation of microbial metabolic networks, facilitating the effective combination of naturally occurring and synthetically engineered microbial communities (Song et al., 2021). Endophytic microorganisms are emerging as particularly important players in developing climate-resilient agricultural systems. One effective strategy for mitigating the impacts of abiotic and biotic stresses associated with climate change involves the application of extremophilic endophytic microorganisms (Dey et al., 2018).

8.4 Microbiome Engineering

8.4.1 Soil Modification Strategies

Soil modification encompasses a range of organic and inorganic amendments as well as agricultural technologies employed to enhance soil productivity and modulate plant-microbe interactions (Samuel et al., 2018). These practices have demonstrated significant potential for improving crop yields while influencing the composition and activity of soil microbial communities.

8.4.2 Inorganic Soil Amendments

Inorganic modifications, including vermiculite, lime, perlite, and sand, serve multiple functions in agricultural systems. These materials are particularly effective in mitigating the negative impacts of soil salinity and acidity on crop growth (Wang et al., 2015; Ganesh et al., 2017). Beyond their primary functions, these inorganic amendments can promote increased plant biomass through enhancement of root exudate production, stimulation of more extensive and biologically active microbial communities, and improved plant performance in nitrogen-deficient soils. Comparative studies have revealed that combined application of inorganic supplements with organic amendments can boost crop production by up to 30%, significantly outperforming the 8% increase typically observed with organic fertilizers alone (Wang et al., 2015).

8.4.3 Organic Soil Amendments

Organic modifications involve the incorporation of various biological materials that serve as nutrient sources for heterotrophic microorganisms, including bacteria, fungi, and beneficial soil invertebrates (Ganesh et al., 2017). These amendments help in the restoration of depleted soil fertility, the reduction of agricultural pollution, and the sustained improvement of microbial activity and diversity. Conventional organic fertilizers, used for decades, represent a well-understood example of such amendments (Saeid & Chojnacka, 2019). These materials are typically rich in nitrogen, phosphorus, and organic matter, influencing both the soil microbiota and physicochemical properties to enhance agricultural productivity (Elouear et al., 2016).

Research has demonstrated several successful applications of organic amendments in improving soil fertility and plant productivity. The application of sheep dung, for instance, has been found to stimulate microbial proliferation in the rhizosphere while simultaneously reducing the concentrations of toxic metals such as lead, cadmium, and zinc in crops like alfalfa. This practice has also been associated with significant improvements in crop yield. Similarly, bone-based amendments have long served as sustainable sources of phosphorus since the nineteenth century. Microbial species such as *Bacillus megaterium* var. *phosphaticum* and *Bacillus mucilaginosus* are known to play pivotal roles in solubilizing bone-derived phosphorus and rendering it bioavailable to plants, achieving these outcomes even in the absence of organic acid production (Elouear et al., 2016). Furthermore, the utilization of crop residues presents an opportunity to valorize organic carbon and nutrient-rich waste, leading to the development of biostimulants through recycling and vermicomposting processes while simultaneously enhancing soil structural stability (Saeid & Chojnacka, 2019). Vermicompost end products, in particular, are recognized for their environmental sustainability, high nutrient density, and efficacy as soil conditioners that support plant growth and development (Usmani et al., 2017).

8.5 Biostimulants

Biostimulants, including algae extracts, amino acids, humic and fulvic acids, and plant-derived hormones, exert their effects through diverse physiological and biochemical pathways that collectively enhance plant performance and soil health. These compounds play a crucial role in facilitating nutrient decomposition and modulating the composition and biomass of the soil microbiome, thereby contributing to improved nutrient availability and plant resilience (Hellequin et al., 2018). Compost, on the other hand, serves as a well-established and sustainable alternative to synthetic fertilizers, offering the dual benefit of nutrient recycling and agricultural waste management. The biocomposting process is primarily driven by microbial oxidative activity, which transforms reusable organic compounds into stable humic materials. This transformation not only enriches the nutrient profile of soils but also supports plant conservation and growth, particularly within organic and ecologically managed agricultural systems.

8.6 Synthetic Microbial Consortia

SMC represents a transformative innovation in agricultural microbiology, grounded in synthetic biology principles aimed at designing functional microbial communities that emulate or enhance natural plant-microbe interactions. These engineered systems overcome major limitations of conventional biofertilizers—such as narrow host specificity, poor competitiveness against native soil microbiota, and low adaptability to fluctuating environments (Hart et al., 2018).

The development of effective SMCs involves a systematic, multistage process. The first phase centers on the strategic selection of microbial strains based on their functional traits, ecological compatibility, and potential synergistic interactions with the target crop. Selected strains are isolated and stabilized under laboratory conditions to ensure viability and functional consistency. Subsequent optimization focuses on balancing microbial relationships to enhance collective metabolic efficiency, resilience, and nutrient cycling capabilities. The final stages involve rigorous validation in greenhouse and field trials to evaluate performance, adaptability, and ecological stability (Kong et al., 2018). Several studies demonstrate the potential of SMCs in improving crop productivity. A consortium comprising *Pseudomonas putida*, *Comamonas testosterone*, *Citrobacter freundii*, and *Enterobacter cloacae* significantly enhanced phosphate solubilization, nearly doubling yield in field trials (Baas et al., 2016). Similarly, synergistic interactions between *Azotobacter vinelandii* and *Rhizophagus irregularis* promoted root development and nutrient uptake in wheat (Dal Cortivo et al., 2018). SMCs enriched with ACC deaminase-producing strains have shown antimicrobial activity against *Fusarium oxysporum* f. sp. *lycopersici*, mitigating disease severity and enhancing tomato productivity (Tsolakidou et al., 2019; Vitullo et al., 2012). Moreover, rhizosphere consortia regulate plant growth through phytohormone synthesis (Stringlis et al., 2018). Despite promising outcomes, field-scale adoption of SMCs faces challenges related to crop specificity, soil variability, performance consistency, cost-effectiveness, and regulatory approval. Future research should integrate SMCs with precision microbiome engineering, sustainable soil management, and advanced breeding strategies to reduce chemical inputs and enhance resilience. Ultimately, SMCs hold immense promise as sustainable biotechnological tools for boosting crop productivity, improving soil health, and supporting global food security.

8.7 Critical Considerations for Microbial Inoculant Formulation

Microbial inoculant formulation requires careful attention to maintaining microbial viability, as the loss of living cells during storage or handling often renders inoculants ineffective before field application. Populations frequently fall below the critical threshold of 10^3 colonies per gram of soil required for agricultural efficacy, leading to poor performance in the field (Chaín et al., 2020; Sharma et al., 2020). To overcome this limitation, the choice of carrier material becomes crucial. Carriers must possess optimal water retention capacity and adequate aeration to sustain microbial metabolism and proliferation while ensuring prolonged shelf life. In addition to these physical and chemical properties, carriers must also be practical, cost-effective, readily available, and easy to apply under diverse agricultural conditions.

Liquid formulations of microbial inoculants are carefully designed mixtures consisting of several essential components in precise proportions. Typically, they include nutrient broth (10–40%) to support microbial growth, dispersants (1–5%) and suspension agents (1–3%) to maintain homogeneity, surfactants (3–8%) to enhance distribution and adherence, and liquid carriers (35–65% by weight), such as water, oil, or a combination of both, to provide stability and ease of use (Kalra et al., 2010). The optimal formulation ensures that microbial populations remain viable and active until field application, thereby improving the overall consistency and efficiency of bioinoculant products.

8.8 Quality Control and Standardization Protocols

Ensuring the quality and consistency of microbial inoculants is essential for their field performance and commercial success. Quality control and standardization protocols form the backbone of reliable inoculant production. Formulation stabilization plays a critical role in maintaining the physical and chemical uniformity of products. The incorporation of polysaccharide gums helps sustain optimal viscosity, while the use of colloidal clays, starch cellulose, or synthetic polymers prevents sedimentation and ensures long-term stability. Equally vital is microbial characterization, which relies on polyphasic identification techniques encompassing morphological, chemotaxonomic, and genotypic analyses. Advanced molecular tools such as 16S rDNA sequencing, ITS analysis, and DNA-DNA hybridization further ensure precise identification and authenticity of both bacterial and fungal strains used in formulations (Young et al., 2021).

Safety and regulatory compliance are integral components of inoculant quality assurance. Adherence to the American Biological Safety Association (ABSA) guidelines ensures a comprehensive evaluation of biosafety parameters. Moreover, conducting field demonstrations and farmer training programs enhances awareness, promotes responsible use, and bridges the gap between laboratory innovation and on-farm application. However, the lack of universal regulatory standards for beneficial microbial products has led to regionally diverse frameworks, often restricted to rhizobial inoculants (Bharti et al., 2017; Sanap et al., 2009). Most regulations specify minimum viable cell counts and permissible contamination levels, while a few countries have established stricter protocols involving national laboratory evaluations and detailed regulatory reviews before commercialization (Bharti et al., 2017; Aguilar et al., 2019).

For effective implementation, microbial inoculant programs must emphasize environmental safety by verifying the sustainability and ecological compatibility of biocontrol agents (Herrmann et al., 2015; Luna et al., 2022). Production standards should mandate the use of pure cultures in sterile carriers, maintenance of at least 10^6

viable cells per seed, and rigorous contamination control. Furthermore, success depends on effective knowledge dissemination through farmer-oriented demonstration projects, integration of social science insights, and strategic use of information and communication technologies (ICT) for awareness, capacity building, and market outreach.

8.9 Policy Framework for Microbiome Storage and Application

The development of comprehensive policies for microbiome storage and application represents a critical challenge in modern agriculture, requiring careful consideration of functionality preservation, safety protocols, and efficient handling procedures (Bhattacharjee et al., 2022). Current global practices reveal significant disparities in microbial inoculant management, particularly across developing nations in Africa, Latin America, and parts of Asia, where standardized protocols for beneficial soil microbiota production are frequently absent. In these regions, microbial formulations are often prepared using nonsterile carriers, compromising their effectiveness (Koskey et al., 2021). While some countries, including Brazil, Rwanda, Zambia, and Argentina, have established regulatory frameworks, these guidelines are either inadequately integrated into production processes or inconsistently implemented (Azevedo & Quecine, 2019; Magnoli et al., 2020).

9 Conclusion and Future Perspectives

Soil microbial communities are central to ecosystem stability, nutrient cycling, and plant productivity. Their diversity, functional redundancy, and interactions determine how soils respond and recover from disturbances, including climate-induced stresses such as drought, flooding, and elevated CO_2 levels. Despite significant progress in agricultural biotechnology, microbial inoculant applications continue to face key challenges that limit their effectiveness. Current strategies often focus on individual microbial strains, which frequently exhibit poor competitiveness against native soil microbiota and limited colonization, reducing their ability to enhance plant growth and stress tolerance. Environmental variability further threatens microbial viability and field performance. Sustainability is another critical concern, necessitating the development of environmentally friendly, safe, and cost-effective formulations that are accessible and practical for farmers. Addressing these challenges requires a multifaceted approach, including the establishment of universal quality standards, improved strain selection, and advanced formulation technologies to enhance microbial stability and efficacy. Equally important is deepening our understanding of microbial community interactions and their role in plant-soil health. Effective knowledge transfer through field demonstrations, farmer training, and digital communication tools is essential to bridge the gap between research and practical application. Harnessing microbial processes, both naturally occurring and engineered, offers promising avenues for sustainable agriculture, environmental management, and mitigation of greenhouse gas emissions. By integrating microbial ecology, biotechnology, and sustainable management, future strategies can strengthen ecosystem resilience, food security, and climate adaptation.

References

Aanderud, Z. T., et al. (2015). Resuscitation of the rare biosphere contributes to pulses of ecosystem activity. *Frontiers in Microbiology, 6*, 24.

Adeniji, A. A., & Babalola, O. O. (2020). Metabolomic applications for understanding complex tripartite plant-microbes interactions: Strategies and perspectives. *Biotechnology Reports, 25*, e00425.

Aguilar, C. N., Hernández-Almanza, A., Sabu, A., Loredo, A., Sudhakaran, S., Davila-Medina, D., Morales-Martínez, T. K., Sepúlveda-Torre, L., Sugathan, S., & Teixeira, J. A. (2019). Microbial technology: Advances and challenges. In *Advances in food bioproducts and bioprocessing technologies* (pp. 3–24). CRC Press.

Al-Shaibani, M. M., Radin Mohamed, R. M. S., Sidik, N. M., Enshasy, H. A. E., Al-Gheethi, A., Noman, E., Al-Mekhlafi, N. A., & Zin, N. M. (2021). Biodiversity of secondary metabolites compounds isolated from phylum actinobacteria and its therapeutic applications. *Molecules, 26*, 4504.

Aliferis, K. A., Faubert, D., & Jabaji, S. (2014). A metabolic profiling strategy for the dissection of plant defense against fungal pathogens. *PLoS ONE, 9*, e111930.

Allison, S. D., Wallenstein, M. D., & Bradford, M. A. (2010). Soil-carbon response to warming dependent on microbial physiology. *Nature Geoscience, 3*, 336–340.

Azam, F., & Malfatti, F. (2007). Microbial structuring of marine ecosystems. *Nature Reviews. Microbiology, 5*(10), 782–791.

Azevedo, J. L., & Quecine, M. C. (2019). Biodiversity and biotechnological applications of microorganisms associated with tropical plants. In *Microbiome in plant health and disease* (pp. 293–313). Springer.

Azzollini, A., Boggia, L., Boccard, J., Sgorbini, B., Lecoultre, N., Allard, P.-M., Rubiolo, P., Rudaz, S., Gindro, K., & Bicchi, C. (2018). Dynamics of metabolite induction in fungal co-cultures by metabolomics at both volatile and non-volatile levels. *Frontiers in Microbiology, 9*, 72.

Baas, P., Bell, C., Mancini, L. M., Lee, M. N., Conant, R. T., & Wallenstein, M. D. (2016). Phosphorus mobilizing consortium Mammoth P™ enhances plant growth. *PeerJ, 4*, e2121.

Bardgett, R. D., et al. (2013). Hierarchical responses of plant–soil interactions to climate change. *Journal of Ecology, 101*(2), 334–343.

Barnard, R., et al. (2006). Several components of global change alter nitrifying and denitrifying activities in an annual grassland. *Functional Ecology, 20*, 557–564.

Bell, C. W., et al. (2014). Soil microbial responses to 7 years of altered precipitation in a desert grassland. *Global Change Biology, 20*, 1657–1673.

Berlemont, R., et al. (2014). Cellulolytic potential under environmental changes in microbial communities from grassland litter. *Frontiers in Microbiology, 5*, 639.

Bever, J. D., et al. (2013). Microbial phylotype composition and diversity predicts plant productivity. *Ecology Letters, 16*(2), 167–174.

Bharti, N., Sharma, S. K., Saini, S., Verma, A., Nimonkar, Y., & Prakash, O. (2017). Microbial plant probiotics: Problems in application and formulation. In *Probiotics and plant health* (pp. 317–335). Springer.

Bhattacharjee, A., Dubey, S., & Sharma, S. (2022). Storage of soil microbiome for application in sustainable agriculture: Prospects and challenges. *Environmental Science and Pollution Research, 29*, 3171–3183.

Boyd, P. W., et al. (2007). Mesoscale iron enrichment experiments 1993–2005. *Science, 315*, 612–617.

Brock, T. D. (1978). *Thermophilic microorganisms and life at high temperatures* (465p.). Springer.

Bucher, R., Veyel, D., Willmitzer, L., Krattinger, S., Keller, B., & Bigler, L. (2017). Combined GC-and UHPLC-HR-MS based metabolomics to analyze durable anti-fungal resistance processes in cereals. *CHIMIA International Journal for Chemistry, 71*, 156–159.

Bungau, S., Behl, T., Aleya, L., Bourgeade, P., Aloui-Sossé, B., Purza, A. L., Abid, A., & Samuel, A. D. (2021). Expatiating the impact of anthropogenic aspects and climatic factors on long-term soil monitoring and management. *Environmental Science and Pollution Research, 28*, 30528–30550.

Cayuela, M. L., et al. (2014). Biochar reduces N_2O emissions from soils. *Plant and Soil, 381*, 1–13.

Chaín, J. M., Tubert, E., Graciano, C., Castagno, L. N., Recchi, M., Pieckenstain, F. L., Estrella, M. J., Gudesblat, G., Amodeo, G., & Baroli, I. (2020). Growth promotion and protection from drought in Eucalyptus grandis seedlings inoculated with beneficial bacteria embedded in a superabsorbent polymer. *Scientific Reports, 10*, 18221.

Chisti, Y. (2007). Biodiesel from microalgae. *Biotechnology Advances, 25*, 294–306.

Clements, F. E. (1916). *Plant succession: An analysis of the development of vegetation* (512p) (Publication no. 242). Carnegie Institution of Washington.

Conrad, R. (2009). The global methane cycle: Recent advances. *Annual Review of Microbiology, 63*, 191–205.

Dal Cortivo, C., Barion, G., Ferrari, M., Visioli, G., Dramis, L., Panozzo, A., & Vamerali, T. (2018). Effects of field inoculation with VAM and bacteria consortia on root growth and nutrients uptake in common wheat. *Sustainability, 10*, 3286.

David, L. A., et al. (2014). Diet rapidly and reproducibly alters the human gut microbiome. *Nature, 505*(7484), 559–563.

De Souza, R. S. C., Armanhi, J. S. L., & Arruda, P. (2020). From microbiome to traits: Designing synthetic microbial communities for improved crop resiliency. *Frontiers in Plant Science, 11*, 1179.

de Vries, F. T., & Shade, A. (2013). Controls on soil microbial community stability under climate change. *Frontiers in Microbiology, 4*, 265.

DeAngelis, K. M., et al. (2015). Long-term forest soil warming alters microbial communities. *Frontiers in Microbiology, 6*, 104.

Dey, R., Pal, K., Thomas, M., Sherathia, D., Mandaliya, V., Bhadania, R., Patel, M., Maida, P., Mehta, D., & Nawade, B. (2018). Endophytic microorganisms: Future tools for climate resilient agriculture. In *Microbes for climate resilient agriculture* (pp. 235–253). Wiley.

El Chami, D., Daccache, A., & El Moujabber, M. (2020). How can sustainable agriculture increase climate resilience? A systematic review. *Sustainability, 12*, 3119.

Elouear, Z., Bouhamed, F., Boujelben, N., & Bouzid, J. (2016). Application of sheep manure and potassium fertilizer to contaminated soil and its effect on zinc, cadmium and lead accumulation by alfalfa plants. *Sustainable Environment Research, 26*, 131–135.

Emwas, A.-H. M., Salek, R. M., Griffin, J. L., & Merzaban, J. (2013). NMR-based metabolomics in human disease diagnosis: Applications, limitations, and recommendations. *Metabolomics, 9*, 1048–1072.

Etesami, H., & Glick, B. R. (2020). Halotolerant plant growth–promoting bacteria: Prospects for alleviating salinity stress in plants. *Environmental and Experimental Botany, 178*, 104124.

Evans, S. E., & Wallenstein, M. D. (2014). Climate change alters ecological strategies of soil bacteria. *Ecology Letters, 17*, 155–164.

Fierer, N., et al. (2012). Cross-biome metagenomic analyses of soil microbes. *Proceedings of the National Academy of Sciences, 109*, 21390–21395.

Ganesh, K. S., Sundaramoorthy, P., Nagarajan, M., & Xavier, R. L. (2017). Role of organic amendments in sustainable agriculture. In *Sustainable agriculture towards food security* (pp. 111–124). Springer.

Geddes, B. A., et al. (2019). Engineering transkingdom signalling in plants. *Nature Microbiology, 4*, 656–662.

Gleason, H. A. (1926). The individualistic concept of the plant association. *Bulletin of the Torrey Botanical Club, 53*(1), 7–26.

Gleizer, S., et al. (2019). Conversion of E. coli to generate biomass from CO_2. *Cell, 179*, 1255–1263.

Gonzalez, A., & Bell, G. (2013). Evolutionary rescue to abrupt environmental change. *Philosophical Transactions of the Royal Society B, 368*, 20120079.

Gorguner, M., & Kavvas, M. L. (2020). Modeling impacts of future climate change on reservoir storages and irrigation water demands in a Mediterranean basin. *Science of the Total Environment, 748*, 141246.

Griffiths, B. S., & Philippot, L. (2013). Resistance and resilience of soil microbes. *FEMS Microbiology Reviews, 37*(2), 112–129.

Gutknecht, J. L. M., et al. (2012). Microbial responses to simulated global change. *Global Change Biology, 18*, 2256–2269.

Hagagy, N. I., Saddiq, A. A., Hamedo, H. A., & Selim, S. A. (2021). Extremophiles inhabiting unique ecosystems in Egypt. In *Extreme environments* (pp. 252–262). CRC Press.

Hamilton, C. E., Bever, J. D., Labbé, J., Yang, X., & Yin, H. (2016). Mitigating climate change through managing constructed-microbial communities in agriculture. *Agriculture, Ecosystems and Environment, 216*, 304–308.

Hart, M. M., Antunes, P. M., Chaudhary, V. B., & Abbott, L. K. (2018). Fungal inoculants in the field: Is the reward greater than the risk? *Functional Ecology, 32*, 126–135.

Hartmann, M., et al. (2015). Soil microbial community dynamics. *Frontiers in Microbiology, 6*, 1543.

Hawkes, C. V., & Keitt, T. H. (2015). Resilience vs. historical contingency in microbial responses. *Ecology Letters, 18*, 612–625.

Hazen, T. C., et al. (2010). Deep-sea oil plume enriches psychrophilic oil-degraders. *Science, 330*(6001), 204–208.

Hellequin, E., Monard, C., Quaiser, A., Henriot, M., Klarzynski, O., & Binet, F. (2018). Specific recruitment of soil bacteria and fungi decomposers following a biostimulant application increased crop residues mineralization. *PLoS ONE, 13*, e0209089.

Hemes, K. S., et al. (2018). Methane cost of carbon sequestration in wetlands. *Geophysical Research Letters, 45*, 6081–6091.

Herridge, D. F., et al. (2008). Global inputs of biological nitrogen fixation. *Plant and Soil, 311*, 1–18.

Herrmann, L., Atieno, M., Brau, L., & Lesueur, D. (2015). Microbial quality of commercial inoculants to increase BNF and nutrient use efficiency. In *Biological nitrogen fixation* (pp. 1031–1040). Wiley.

Hristov, A. N., et al. (2015). Mitigation of methane and nitrous oxide from livestock. *Journal of Animal Science, 93*, 1767–1789.

Jiao, N., et al. (2010). Microbial production of recalcitrant dissolved organic matter. *Nature Reviews. Microbiology, 8*, 593–599.

Johnson, D. B. (2014). Extremophiles and biotechnology. *Nature Reviews. Microbiology, 12*(2), 115–124.

Kalra, A., Chandra, M., Awasthi, A., Singh, A. K., & Khanuja, S. P. S. (2010). Natural compounds enhancing growth and survival of rhizobial inoculants in vermicompost-based formulations. *Biology and Fertility of Soils, 46*, 521–524.

Khan, I., Khan, F., Ahmad, S., Pandey, P., & Khan, M. M. (2021). Microbes and climate: A tangled relation. In *Microbiomes and the global climate change* (pp. 3–15). Springer.

Kitano, H. (2002). Systems biology: A brief overview. *Science, 295*(5560), 1662–1664.

Kong, Z., Hart, M., & Liu, H. (2018). Paving the way from the lab to the field: Using synthetic microbial consortia to produce high-quality crops. *Frontiers in Plant Science, 9*, 1467.

Koskey, G., Mburu, S. W., Awino, R., Njeru, E. M., & Maingi, J. M. (2021). Potential use of beneficial microorganisms for soil amelioration, phytopathogen biocontrol, and sustainable crop production in smallholder agroecosystems. *Frontiers in Sustainable Food Systems, 5*, 130.
Kuzyakov, Y. (2010). Priming effects: Interactions between living and dead organic matter. *Soil Biology and Biochemistry, 42*, 1363–1371.
Lehmann, J., et al. (2020). Persistence of soil organic carbon. *Nature Geoscience, 13*, 529–534.
Leray, M., et al. (2021). Natural experiments and long-term monitoring are critical to understand and predict marine host–microbe ecology and evolution. *PLoS Biology, 19*(8), e3001322.
Levins, R., & Lewontin, R. C. (1985). Dialectics and reduction in ecology. In *The dialectical biologist*. Harvard University Press.
Liang, C., et al. (2015). Microbial lipid and amino sugar responses to global changes. *Frontiers in Microbiology, 6*, 385.
Ling, Y. C., et al. (2015). Iron- and sulfate-reducing bacteria in acid sulfate soils. *Frontiers in Microbiology, 6*, 624.
Logan, B. E., et al. (2006). Microbial fuel cells: Methodology and technology. *Environmental Science & Technology, 40*, 5181–5192.
Lone, S. A., & Malik, A. (2021). *Microbiomes and the global climate change* (374p.). Springer.
Luna, G. M., Quero, G. M., Kokou, F., & Kormas, K. (2022). Time to integrate biotechnological approaches into fish gut microbiome research. *Current Opinion in Biotechnology, 73*, 121.
Lyu, D., Backer, R., Subramanian, S., & Smith, D. L. (2020). Phytomicrobiome coordination signals hold potential for climate change-resilient agriculture. *Frontiers in Plant Science, 11*, 634.
Magnoli, K., Carranza, C. S., Aluffi, M. E., Magnoli, C. E., & Barberis, C. L. (2020). Herbicides based on 2, 4-D: Its behavior in agricultural environments and microbial biodegradation aspects. A review. *Environmental Science and Pollution Research, 27*, 38501–38512.
McCarty, P. L., et al. (2011). Wastewater treatment as a net energy producer. *Environmental Science & Technology, 45*, 7100–7106.
Meena, R. P., & Jha, A. (2018). Conservation agriculture for climate change resilience: A microbiological perspective. In *Microbes for climate resilient agriculture* (pp. 165–190). Wiley.
Mohanty, S., & Swain, C. K. (2018). Role of microbes in climate smart agriculture. In *Microorganisms for green revolution* (pp. 129–140). Springer.
Mukhopadhyay, R., Sarkar, B., Jat, H. S., Sharma, P. C., & Bolan, N. S. (2021). Soil salinity under climate change: Challenges for sustainable agriculture and food security. *Journal of Environmental Management, 280*, 111736.
O'Donnell, A. G., et al. (2007). Visualization and modelling in soil microbiology. *Nature Reviews. Microbiology, 5*(8), 689–699.
Omomowo, O. I., & Babalola, O. O. (2019). Bacterial and fungal endophytes: Tiny giants with immense beneficial potential for plant growth and sustainable agricultural productivity. *Microorganisms, 7*, 481.
Passarini, M. R. Z., Duarte, A. W. F., Rosa, L. H., de Oliveira, V. M., & Ottoni, J. R. (2022). Extremofuels: Production of biofuels by extremophile microbes as an alternative to avoid climate change effects. In *Microbiome under changing climate* (pp. 237–256). Elsevier.
Paul, D. (2013). Osmotic stress adaptations in rhizobacteria. *Journal of Basic Microbiology, 53*, 101–110.
Philippot, L., et al. (2011). Denitrifiers lacking N_2O reductase drive emissions. *Global Change Biology, 17*(3), 1497–1504.
Pouvreau, B., Vanhercke, T., & Singh, S. (2018). From plant metabolic engineering to plant synthetic biology: The evolution of the design/build/test/learn cycle. *Plant Science, 273*, 3–12.
Ridder, B. (2008). Questioning ecosystem services for biodiversity conservation. *Biodiversity and Conservation, 17*(3), 781–790.
Rillig, M. C., et al. (2015). Microbial communities in wetlands. *Nature Reviews. Microbiology, 13*, 121–132.

Rizvi, A., Ahmed, B., Khan, M. S., Umar, S., & Lee, J. (2021). Psychrophilic bacterial phosphate-biofertilizers: A novel extremophile for sustainable crop production under cold environment. *Microorganisms, 9*, 2451.

Saeid, A., & Chojnacka, K. (2019). Fertilizers: Need for new strategies. In *Organic farming* (pp. 91–116). Elsevier.

Samuel, A. D., Bungau, S., Tit, D. M., Melinte, C. E., Purza, L., & Badea, G. E. (2018). Effects of long-term application of organic and mineral fertilizers on soil enzymes. *Revista de Chimie, 69*, 2608–2612.

Sanap, D., Satpute, T., Dudhate, D., Babar, A., & Nagure, D. (2009). Constraints and suggestions in soybean crop under organic and inorganic system. *International Journal of Agricultural Sciences, 5*, 608–610.

Schimel, J. P., & Bennett, J. (2004). Nitrogen mineralization challenges. *Ecology, 85*(3), 591–602.

Schimel, J., et al. (2007). Microbial stress-response physiology and ecosystem function. *Ecology, 88*, 1386–1394.

Sethi, S. (2022). Utilization of beneficial fungal strain/bacterial strains in climate-resilient agriculture. In *Microbiome under changing climate* (pp. 313–331). Elsevier.

Sevastos, A., Kalampokis, I., Panagiotopoulou, A., Pelecanou, M., & Aliferis, K. (2018). Implication of *Fusarium graminearum* primary metabolism in its resistance to benzimidazole fungicides as revealed by 1H NMR metabolomics. *Pesticide Biochemistry and Physiology, 148*, 50–61.

Sharma, V., Salwan, R., & Al-Ani, L. K. T. (2020). *Molecular aspects of plant beneficial microbes in agriculture*. Academic.

Sharma, S., Mohler, J., Mahajan, S. D., Schwartz, S. A., Bruggemann, L., & Aalinkeel, R. (2023). Microbial biofilm: A review on formation, infection, antibiotic resistance, control measures, and innovative treatment. *Microorganisms, 11*(6), 1614.

Singh, B. K., Trivedi, P., Singh, S., Macdonald, C. A., & Verma, J. P. (2018). Emerging microbiome technologies for sustainable increase in farm productivity and environmental security. *Microbiology Australia, 39*, 17–23.

Slade, D., & Radman, M. (2011). Oxidative stress resistance in *Deinococcus radiodurans*. *Microbiology and Molecular Biology Reviews, 75*(1), 133–191.

Song, C., Jin, K., & Raaijmakers, J. M. (2021). Designing a home for beneficial plant microbiomes. *Current Opinion in Plant Biology, 62*, 102025.

Stark, J. M., & Firestone, M. K. (1995). Soil moisture effects on nitrifying bacteria. *Applied and Environmental Microbiology, 61*(1), 218–221.

Sterelny, K. (2006). Local ecological communities. *Philosophy in Science, 73*(2), 215–231.

Stringlis, I. A., Yu, K., Feussner, K., De Jonge, R., Van Bentum, S., Van Verk, M. C., Berendsen, R. L., Bakker, P. A., Feussner, I., & Pieterse, C. M. (2018). MYB72-dependent coumarin exudation shapes root microbiome assembly to promote plant health. *Proc Natl Acad Sci USA, 115*, E5213–E5222.

Sun, J., Duan, Z., Zhang, Y., Cao, S., Tang, Z., & Abozeid, A. (2022). Metabolite profiles provide insights into underlying mechanism in *Bupleurum* (Apiaceae) in response to three levels of phosphorus fertilization. *Plants, 11*, 752.

Sunagawa, S., et al. (2015). Structure and function of the global ocean microbiome. *Science, 348*(6237), 1261359.

Thiele-Bruhn, S., et al. (2012). Antibiotics in soil. *Environmental Science & Technology, 46*(15), 8641–8648.

Tiemann, L. K., & Billings, S. A. (2011). Soil nitrogen cycling under extreme moisture. *Soil Biology and Biochemistry, 43*, 1837–1847.

Treseder, K. K. (2004). Mycorrhizal responses to N, P, and CO_2. *Ecology, 85*, 1172–1179.

Treseder, K. K. (2008). Nitrogen additions and microbial biomass. *Ecology Letters, 11*, 1111–1120.

Trivedi, P., Batista, B. D., Bazany, K. E., & Singh, B. K. (2022). Plant–microbiome interactions under a changing world: Responses, consequences and perspectives. *The New Phytologist, 234*(4), 1951–1959.

Tsolakidou, M.-D., Stringlis, I. A., Fanega-Sleziak, N., Papageorgiou, S., Tsalakou, A., & Pantelides, I. S. (2019). Rhizosphere-enriched microbes as a pool to design synthetic communities for reproducible beneficial outputs. *FEMS Microbiology Ecology, 95*, fiz138.

Usmani, Z., Kumar, V., & Mritunjay, S. K. (2017). Vermicomposting of coal fly ash using epigeic and epi-endogeic earthworm species: Nutrient dynamics and metal remediation. *RSC Advances, 7*, 4876–4890.

Van der Heijden, M. G. A., et al. (2008). Soil microbes as drivers of plant diversity. *Ecology Letters, 11*(3), 296–310.

Vitullo, D., Di Pietro, A., Romano, A., Lanzotti, V., & Lima, G. (2012). Role of new bacterial surfactins in the antifungal interaction between *Bacillus amyloliquefaciens* and *Fusarium oxysporum*. *Plant Pathology, 61*, 689–699.

Wakeham, S. G., et al. (2007). Microbial ecology of the Black Sea water column. *Organic Geochemistry, 38*(12), 2070–2097.

Wang, S., Tan, Y., Fan, H., Ruan, H., & Zheng, A. (2015). Responses of soil microarthropods to inorganic and organic fertilizers in a poplar plantation in a coastal area of eastern China. *Applied Soil Ecology, 89*, 69–75.

Weir, B. S., et al. (2004). Diverse Mesorhizobium strains nodulate New Zealand legumes. *Applied and Environmental Microbiology, 70*(10), 5980–5987.

Wieder, W. R., et al. (2015). Representing microbial processes in soil models. *Biogeochemistry, 128*, 7–14.

Wong, C. K. F. (2022). Application of microbes in climate-resilient crops. In *Application of microbes in environmental and microbial biotechnology* (pp. 93–112). Springer.

Yadav, A. N. (2021). Biodiversity and bioprospecting of extremophilic microbiomes for agro-environmental sustainability. *Journal of Applied Biology, 9*, 1–6.

Young, I. M., et al. (2008). Microbial distribution in soils: Physics and scaling. *Advances in Agronomy, 100*, 81–121.

Young, C.-C., Shen, F.-T., & Singh, S. (2021). Strategies for the exploration and development of biofertilizer. In *Bacteria in agrobiology: Plant probiotics* (pp. 127–139). Springer.

Effect of Soil Moisture on Microbial Communities

Ayisha Shibin, Kamali Raja, Thottiyan Saeed,
Yama Santhoshi Lavanya, and Minakshi Grover

Abstract

Soil moisture is a critical component that influences various biological, chemical, and physical processes in the soil ecosystem. Soil moisture has a profound impact on microbial communities in terms of microbial diversity, community composition, and functions. This chapter discusses various climatic and anthropogenic factors that influence soil moisture and the subsequent impacts on soil microbial communities and related functions. Climatic factors, including precipitation, temperature, and evaporation rates, significantly alter the soil moisture dynamics, resulting in deficit or excess moisture conditions. Anthropogenic activities, including land use changes, irrigation practices, deforestation, etc., also impact soil moisture levels. Optimal moisture levels enhance microbial metabolic activities, promoting nutrient cycling and soil fertility. Conversely, extreme moisture conditions—such as drought or flooding—disrupt microbial processes, leading to altered microbial diversity and community structures. Microorganisms use various strategies to adapt to or cope with the changing moisture regime, such as the accumulation of osmoprotectants, production of exopolymeric substances, entering a dormant state, etc. The soil texture and structure also influence moisture availability and microbial habitat. Changes in the soil moisture content influence the availability of substrates and nutrient cycling in the soil, thus impacting soil biological health and its agricultural productivity. Effective mitigation strategies for managing soil moisture are essential for maintaining microbial activity and ensuring agricultural sustainability. Strategies including irrigation management, soil amendments, and conservation practices are

A. Shibin · K. Raja · T. Saeed · M. Grover (✉)
ICAR-Indian Agricultural Research Institute, New Delhi, India

Y. S. Lavanya
Professor Jayashankar Telangana State Agricultural University Press, Hyderabad, Telangana, India

G. Bhandari et al. (eds.), *Microbial Resilience*,
https://doi.org/10.1007/978-981-95-8883-1_6

highlighted as key approaches to optimize soil moisture levels. In conclusion, understanding the factors influencing soil moisture and their impact on microbial communities is vital for sustainable resource management and agricultural productivity. This study underscores the need for integrated approaches to manage soil moisture, considering both climatic and anthropogenic influences, to support microbial health and ecosystem functions.

Keywords

Adaptation and Mitigation · Exopolysaccharides · Microbiome · Plant-microbe interactions · Soil Enzymes · Synthetic communities

1 Introduction

Soil moisture is a primary limiting factor in agriculture because plants require water for crucial growth and development functions; technically, it is defined as the volume of water retained within the unsaturated zone of the soil, representing the water held in soil pores, though its precise definition varies depending on whether it is expressed in relative, absolute, or indirect terms, or based on the reference storage utilized. This soil water acts as a vital source for the atmosphere, primarily escaping through evapotranspiration (which encompasses both plant transpiration and evaporation from bare soil), a process crucial to the continental water cycle as it contributes to the return of up to 60% of total land precipitation back to the atmosphere (Oki & Kanae, 2006); however, in practical applications, only a portion of the total soil moisture is typically significant or quantifiable, making it essential to evaluate soil moisture in relation to a specific volume of soil.

Volumetric soil moisture (θ) is the ratio of the volume of water to the soil volume (V). In some case soil moisture definitions, V depends on space and time (rooting depth or water table depth). The maximum soil moisture content is the water-holding capacity (θSAT) of the soil that varies between 0 and θSAT, whereas saturation ratio ($\theta s = \theta/\theta SAT$) varies between 0 and 1. Another important term used to define soil moisture is field capacity (FC), which is the water content retained in the soil after complete drainage postsaturation. The maximum water in volume available to plants is the field capacity (θFC) minus the permanent wilting point (θWILT). Water cannot be held above FC against gravitational force, and below the wilting point, the water is held too strongly by the soil matrix, making it inaccessible to plants. Soil properties such as texture modulate the field capacity and the wilting and thus vary geographically. Based on field capacity and the wilting point, "soil moisture index" ($SMI = \theta - \theta wilt/\theta FC - \theta wilt$) can be expressed. The SMI is thus a measure of soil moisture content as a ratio of the total storage available to plants (between 0 and 1). Finally, the soil moisture can also be expressed as "soil moisture potential," characterizing the binding of soil moisture to the soil matrix. The saturation water content θSAT, the field capacity θFC, and the permanent wilting point θWILT are typically defined as corresponding to soil moisture potentials of approximately

−1 hPa, ca. −100 to −300 hPa, and ca. −15,000 hPa (1.5 MPa), respectively (Seneviratne et al., 2010).

When soil moisture deviates from the optimal range due to either dropping below field capacity to the permanent wilting point (moisture deficit) or rising toward saturation it induces moisture stress in the soil, which profoundly affects all aspects of plant growth, including water relations, photosynthesis, respiration, metabolic and hormonal reactions, nutrition, development, and yield, thereby impacting terrestrial carbon uptake and food production (van der Molen et al., 2011; Vicente-Serrano et al., 2010). Term moisture stress is commonly used for moisture deficit, it applies equally to excess moisture conditions; furthermore, plant moisture stress becomes severe when the soil water potential is low and environmental factors hinder water absorption, triggering various changes within the plants due to altered translocation, water absorption, and transpiration, which in turn influences the land-atmosphere interface by modulating surface temperature (T), evapotranspiration (ET), precipitation (P) and the energy balance (Teuling et al., 2013; Jung et al., 2010).

Similar to plants, microbial life and activity also depend heavily on soil moisture potential as they need a steady environment to grow. The bacteria, fungi, and other microorganisms that live in soil are extremely sensitive to moisture variations in the soil. The effects of soil moisture on microbial communities are complex and influenced by several environmental conditions, such as temperature, soil type, depth, and land management practices. Understanding these interactions is essential for predicting and managing the response of soil microbial communities to changing environmental conditions.

This chapter highlights the effects of soil moisture on microbial diversity and function, adaptation mechanisms, and strategies to mitigate the effects of stress in agriculture.

2 Climatic Change and Soil Moisture

Climate change is expected to significantly impact the distribution and intensity of precipitation, leading to increased variability in rainfall. These erratic precipitation patterns disrupt soil moisture levels with drier soils in the regions receiving low precipitation and waterlogged conditions in regions prone to heavy and intense precipitation, resulting in cascading impact on soil physicochemical and biological properties, plant-microbe interactions, and overall productivity and sustainability of the agroecosystems. Rising temperatures also trigger soil moisture loss due to increased evaporation and transpiration losses, which further increase the irrigation demand for the crops. Climate models predict an increase in global-mean precipitation in response to increased radiative forcing, with increased precipitation over large parts of the land surface, but significant decreases over subtropical regions, including Southern Africa, Central America, and the Mediterranean Basin, indicating increased meteorological drought in these regions. Furthermore, model projections also indicate decreases in moisture levels of surface (top 10 cm) soil at the global level, with regional declines of up to 10%. Furthermore, none of the regions, including those with increased precipitation, show significant increases in surface soil moisture, indicating a spatially extensive increase in drought conditions under warmer climatic conditions, mainly driven by increased evaporation (Berg & Sheffield 2018). However, Dai (2011) indicated that an increase in surface drought might be alleviated under elevated CO_2 levels, due to higher water use efficiency of the plants resulting in reduced evaporation demand. Salinity is another important abiotic stress caused due to several reasons, including low precipitation and dry climatic conditions, high rate of evaporation, seawater inundations, waterlogging, indiscriminate use of chemical inputs, inappropriate irrigation practices, etc. (Tarolli et al., 2024). The increased salt levels decrease the free water availability by increasing the osmotic stress in the soil. Thus, different abiotic stresses are interrelated, and one type of stress may add to the severity of the other stresses (Prajapati et al., 2024).

3 Impact of Soil Moisture on Microbial Communities

3.1 Impact of Soil Moisture on Community Composition

The impact of soil moisture on microbial community composition is an important aspect of soil ecology. Soil moisture significantly influences the diversity, abundance, structure, and function of microbial communities, which in turn affects various ecosystem processes (Naylor & Coleman-Derr, 2018). Soil moisture level plays an important role in regulating biogeochemical nutrient cycling and oxygen levels in the soil (Gleeson et al., 2008). Low moisture conditions improve aeration, favoring aerobic microorganisms; however, it causes soil desiccation, which leads to reduced microbial activity and favors fungi and spore-forming bacteria (Drenovsky et al., 2004). Higher soil moisture generally improves microbial growth, species richness, and diversity and increases microbial biomass (Chen et al., 2020). Flooding of water reduces soil oxygen levels and favors facultative and obligate anaerobic microorganisms. For instance, in an alpine steppe ecosystem, high soil moisture facilitated the growth of Gram-negative bacteria, while Gram-positive bacteria and fungi were more abundant in lower moisture conditions (Chen et al., 2020). Soil moisture also interacts with various environmental factors to shape microbial communities. For example, in wetland ecosystems, characterized by high moisture levels, soil microorganisms were found to be more sensitive to temperature (Zhao et al., 2024). In a warming experiment, drought conditions led to a 50–80% reduction in microbial population size compared to the control (Sheik et al., 2011). This significant decrease in microbial abundance can potentially affect various soil processes, including nutrient cycling and organic matter decomposition. A study by Hao et al. (2021) investigated the influence of moisture availability on the microbial community structure in the depth profiles of field soils under corn and soybean cultivation. The abundance of the phylum Proteobacteria increased below a depth of 90 cm in cultivated soils, whereas other phyla showed reduced abundance in deeper soils. Land-management practices influenced the response of microbial communities to soil moisture. For example, plastic film mulch, which affects soil moisture retention, was found to play a greater role in regulating fungal community structure compared to bacterial communities (Liu et al., 2021).

Various studies reported the effects of moisture on soil microbial composition, one such study was conducted by Hartmann et al. (2017). A decade-long irrigation experiment was conducted at a water-limited pine forest in the central European Alps. As the forest soils transitioned from arid to moisture-rich conditions, a striking shift in microbial diversity emerged. Phyla such as Proteobacteria, Zygomycota, and Planctomycetes flourished in the irrigated plots. Enhanced microbial activity accelerated the mineralization of soil organic matter, leading to increased carbon release. Conversely, the drier plots harbored an abundance of microbial groups adapted to resource scarcity, including Gemmatimonadetes, Actinobacteria, Armatimonadetes, and Acidobacteria. These shifts reflected fundamental survival strategies such as metabolic versatility for degrading complex organic compounds, increased carbon-use efficiency, and cellular modifications (e.g., accumulation of osmoprotectants, formation of dormant structures, etc.) against desiccation and nutrient limitations.

In another interesting study, Banerjee et al. (2016) investigated the impact of soil moisture on microbial community structure and nitrous oxide (N_2O) emissions in agricultural soils. They examined soils under wetting, drying, and static moisture conditions to assess how current and previous moisture levels affected bacterial community composition, gene transcription, and N_2O production. The study indicated that microbial community composition was dependent on previous moisture conditions. As soils dried, the bacterial rRNA represented fewer and more evenly distributed genera. The authors hypothesized that this was linked to the evenness of resource distribution, which was controlled by differences in substrate diffusion in wetting versus drying conditions. This finding highlights the importance of considering soil moisture history when studying microbial communities and their functions. Interestingly, the study revealed that N_2O flux depended on both current and previous soil moisture status.

3.2 Effects of Soil Moisture on Microbial Metabolic Activities

3.2.1 Effect of Moisture Conditions on Microbial Respiration

Soil moisture plays a crucial role in regulating microbial metabolic activities and respiration rates. Microbial activity generally increases with rising soil moisture levels until it reaches an optimal threshold, beyond which it may decline due to oxygen limitation. Respiration rates are often highest immediately upon rewetting, decreasing exponentially over time (Meisner et al., 2015). In air-dried soils, bacterial growth rates remain low but increase rapidly as moisture content rises, facilitating metabolic activities and respiration (Iovieno & Baath, 2008). As soil moisture increases, the diffusion of substrates and enzymes improves, enhancing microbial respiration, which indicates its metabolism (Baldrian et al., 2010). However, the relationship between soil moisture and microbial respiration is not strictly linear and depends on factors such as soil type, microbial community composition, organic carbon bioavailability, oxygen diffusion rates, soil pore structure, clay content, and microbial drought resistance (Yan et al., 2016).

3.2.2 Effect of Different Soil Moisture Levels on Soil Microbial Enzyme Activity

Soil enzymes play a crucial role in biochemical processes within soil ecosystems, influencing nutrient cycling and organic matter decomposition (Burns et al., 2013). Among the key environmental factors affecting soil enzymatic activity, soil moisture is particularly significant. It regulates microbial biomass and enzyme activities, with higher moisture content generally promoting microbial functions in forest soils (Baldrian et al., 2010). Soil dehydrogenase, an intracellular oxidoreductase enzyme, is used as an indicator of overall microbial activity. A study conducted under laboratory conditions (Sagar et al., 2014) indicated maximum soil dehydrogenase activity at FC and a decrease in enzyme activity above or below the FC. Soil moisture also influences the activity of extracellular enzymes such as laccase, Mn-peroxidase, β-glucosidase, β-xylosidase, chitinase, and acid phosphatase, which are involved in organic matter decomposition. However, the other factors, such as temperature, also influence the microbial response to moisture. For example, microbial extracellular enzyme activity at mesophilic temperatures (~20 °C) that favor mesophilic microflora was highest under high water availability (aw > 0.95), while at thermophilic temperatures (~60 °C) favoring thermophilic microflora, enzyme activity peaked under dry conditions (aw 0.65) (Gomez et al., 2020). This contradicts the general assumption that wetter conditions always enhance microbial activity.

The sensitivity of enzyme production to soil moisture fluctuations also depends on the historical moisture regime. In drier environments, enzyme activity becomes more responsive to changes in moisture levels (Averill et al., 2016). Additionally, precipitation can modify soil enzyme activities by altering substrate diffusion rates and the availability of inhibitory compounds (Xu et al., 2015). Studies have shown a positive correlation between soil moisture and enzyme activity (Luo et al., 2020). Seasonal variations also play a role; for instance, microbial growth is more favorable in spring, while invertase and catalase activity peak in the autumn but decline in winter (Noe et al., 2012). Ultimately, the extracellular enzyme pool is shaped by both microbial community strategies and soil microclimatic conditions (Allison & Treseder, 2008; Steinweg et al., 2013).

3.2.3 Effect of Different Soil Moisture Levels on Overall Microbial Productivity

Soil moisture content plays a crucial role in shaping microbial productivity by influencing the carbon and nitrogen cycles. Microbial carbon use efficiency (CUE) declines under moisture stress in soil (Tiemann & Billings 2011). When CUE is low, a larger portion of the available carbon is respired (released as CO_2) instead of being used for microbial growth, resulting in reduced microbial productivity. Similarly, a soil with 60% water holding capacity (WHC) was identified as an optimal threshold for changes in the proportion of labile organic carbon and native soil organic matter utilized by microbes. When soil moisture content is below 60% WHC, the ratio between labile organic carbon and native soil organic matter increases with increasing moisture levels, but gradually decreases when soil moisture content exceeds 60% WHC (Bian et al., 2022). These findings highlight the importance of maintaining appropriate soil moisture levels for optimal microbial activity and productivity in the soil.

3.2.4 Effect of Different Soil Moisture Levels on Nutrient Cycling

Soil moisture content plays a crucial role in nutrient cycling, influencing microbial activity, organic matter decomposition, and nutrient availability. In well-hydrated soils, microbes thrive, decomposing organic material and making nutrients more accessible to plants. On the other hand, dry conditions can hinder microbial activity, slowing down decomposition and limiting nutrient supply. This strong link between soil moisture and microbial function highlights the complex interactions that regulate nutrient cycling in the soil. Low water availability reduces decomposition and nutrient cycling rate (Guntiñas et al., 2013) because the thin water films around soil particles reduce the diffusion of enzymes and nutrients, which are the substrate needed by the microbes for mineralization and immobilization (Papendick & Campbell, 1981). Biological nitrogen fixation is a process by which certain microorganisms convert atmospheric nitrogen into a form that plants can use (Sun et al., 2021). Water availability plays a crucial role in this process, with soil moisture content significantly influencing plant nitrogen assimilation (Aranjuelo et al., 2007). Drought stress has been reported to alter nitrogen fixation products, affecting nodule nitrogen-fixation enzyme activity through a feedback mechanism (Sanko-Sawczenko et al., 2019). Similarly, rhizobial symbionts in legumes exhibit strong sensitivity to water stress, which can impair their symbiotic efficiency (Devries et al., 1989). Nitrogenase activity in alfalfa was found to be higher under moderate drought conditions compared to well-watered conditions, suggesting that mild water stress enhances enzyme activity. Excessive moisture severely inhibits root respiration and rhizobial activity, thereby reducing biological nitrogen fixation capacity (Wang et al., 2024). Beyond plant-associated nitrogen fixation, soil biological nitrogen fixation also exhibits moisture sensitivity, with temperature-dependent variations. Li et al. (2018) reported that at 15 °C, changes in moisture had minimal impact on BNF rates. However, at 25 °C, increased moisture enhanced nitrogen fixation, whereas at 35 °C, intermediate moisture levels led to peak BNF activity, while both low and high moisture levels suppressed fixation. This indicates that optimal soil moisture for BNF is temperature-dependent, and excessive water at elevated temperatures may inhibit the process.

4 Strategies Used by Microbes to Cope with Moisture Changes

Microorganisms use diverse mechanisms to cope with moisture stress, as presented in Fig. 1 and elaborated in the following subsections.

4.1 Production of Osmoprotectants and Related Compounds

Microorganisms exhibit a remarkable capacity to adapt to changing environmental conditions by deploying underlying defense mechanisms, which primarily involve the intracellular accumulation of compatible solutes that shield macromolecules and cell membranes from damage. These molecules, which do not inhibit overall cellular functions, have been labeled as "compatible solutes" (also called osmolytes), and they include various types of organic molecules used for osmotic balance, such as sugars, amino acids and their derivatives, polyols and derivatives, betaines, and ectoines and peptides. Compatible solutes can be synthesized de novo or, if available in the medium, they can be taken up by the organisms; in addition to their well-studied function as osmoprotectants, they possess protein-stabilizing properties that

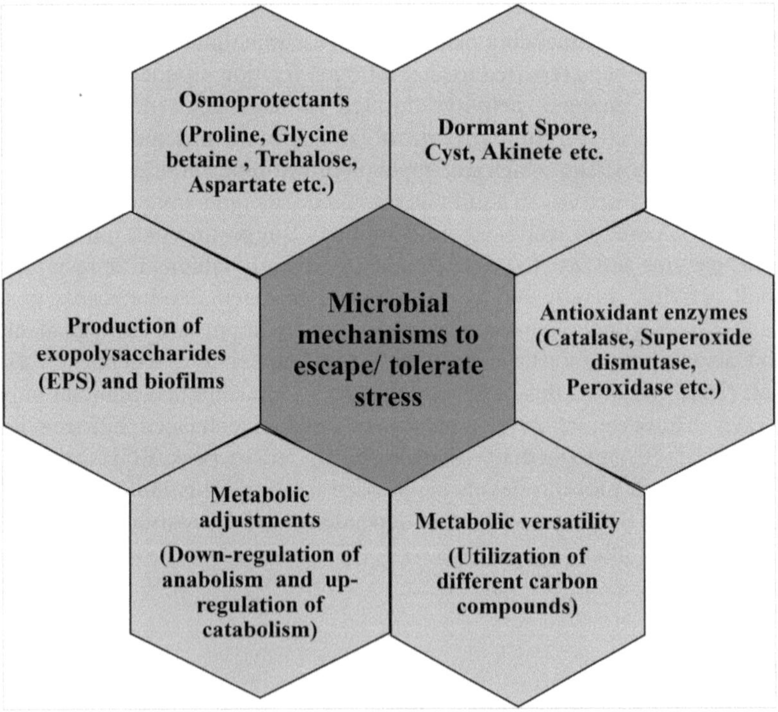

Fig. 1 Mechanisms used by microorganisms to escape/tolerate abiotic stress

support the correct folding of polypeptides under denaturing conditions both in vitro and in vivo, and these stabilizing effects on macromolecules and biosynthetic processes likely contribute to their overall physiological role as stress protectants (Paul, 2013). For example, *Escherichia coli* utilizes a variety of compounds for stress adaptation, including proline, trehalose, betaine, and dimethyl sulfoniopropionate, although the effectiveness of glutamate and proline is limited by the allosteric control of proline biosynthesis in *E. coli* and the negative charge of glutamate, respectively (Purvis et al., 2005).

Glycine betaine, carnitine, and proline are common compatible solutes for bacterial osmotic protection, and glycine betaine is most preferred over them. Gram-positive bacteria accumulate proline by de novo synthesis, while Gram-negative bacteria enhance transport. *Pseudomonas fluorescence* MSP-393, accumulation of aspartic acid and glutamic acid was about five-folds in cells during osmotic stress, whereas serine increased about seven-folds (Paul & Nair, 2008).

Many reports suggest that rhizobacteria use the osmoprotectants exuded from plants to different extents depending on the type of compatible solutes. The nitrogen-fixing *Azospirillum* living in the intercellular spaces of plant-roots directly receives osmoprotectants accumulated in the root. Glycine betaine is synthesized by many rhizobacteria in response to osmotic stress caused by salinity and drought, and enhances stress resilience in bacteria by stabilizing proteins and membranes. It forms hydrogen bonds and electrostatic interactions, preserving the structural integrity of macromolecules under stressed conditions. Some bacteria can catabolize glycine betaine, using it as a sole source of carbon and nitrogen. This catabolic pathway involves the conversion of choline to glycine betaine and then to glycine. The inoculation with osmolyte-producing rhizobacteria has been reported to increase osmolyte production in the host plant, resulting in better stress adaptation (Ansary et al., 2012).

4.2 Role of Dormancy and Spore Formation

A common response of bacteria toward different abiotic stresses is to enter into a dormant stage of reduced or an inert state of metabolic activity and resume normal metabolism during favorable conditions. Many bacteria, during stressful situations, enter a dormant state by cyst formation. The encystment involves thickening of the bacterial cell wall that protects and helps the bacteria survive desiccation and other stressful conditions by slowing down their metabolism and other activities. Once the favorable conditions return, the cyst wall dissolves (excystation), allowing the cell to resume its normal metabolic activities (Dos Reis et al., 2016; Somers et al., 2004). In *Bacillus* and *Clostridium*, dormancy involves the formation of highly resistant spores, which can survive a broad range of biotic and abiotic stress for long periods of time. The desiccation tolerance of spores is largely attributed to the accumulation of dipicolinic acid (DPA) and alpha/beta-type small, acid-soluble spore proteins (SASPs) in the intracellular spore space, protecting DNA from oxidative damage (Setlow, 2014).

Some cyanobacterial members found in dry environments, specifically Nostocales and Stigonematales, can form akinetes under conditions of nutrient starvation and desiccation. The low cytoplasmic water content of akinetes, along with the accumulation of cyanophycin and glycogen, helps in xeric tolerance for longer periods (Kaplan-Levy et al., 2010).

Metabolic dormancy is another type of dormancy that represents a key way of desiccation tolerance in nonsporulating prokaryotes. This shift to a viable but nonculturable (VBNC) state has been reported for bacteria such as *Sinorhizobium meliloti*.

4.3 Metabolic Adaptations

Bacteria differentially up- and down-regulate different metabolic pathways to deal with changes in metabolic flux and to compensate for changing energy demands, which ultimately leads to a shift from anabolic to catabolic metabolism (Kocharunchitt et al., 2012). Photosynthetic machinery is down-regulated in response to desiccation in *Anabaena* sp. PCC712037, which prevents ROS and thus protects the cell from oxidative stress.

Under desiccating conditions, glucose is diverted to trehalose synthesis, and thus fatty acid oxidation becomes a highly efficient energy source. The *Pontibacter* sp., which was isolated from desert soil, encodes proteins for the utilization of a wide array of carbon substrates, as well as key enzymes for CO_2 fixation that are absent from the same species isolated from nonxeric environments. Many energy-consuming processes, such as motility and chemotaxis, are also down-regulated in cells under stressed conditions.

4.4 Production of Antioxidant Enzymes

The accumulation of ROS as a product of cellular metabolism is high under stress conditions. ROS have a tendency to oxidize nucleic acids, lipids, proteins, and sugars.

The first-line defense antioxidants include glutathione peroxidase, catalase, and superoxide dismutase. The antioxidant enzyme activities have been reported in PGP-inoculated stressed plants, thereby alleviating the effects of stress (Chakraborty et al., 2013). Gusain et al. (2015) reported the increase in superoxide dismutase, peroxidase, catalase, and ascorbate peroxidase activities in rice cultivars under different levels of drought stress with inoculation of *Arthrobacter nitroguajacolicus*, *Pseudomonas fluorescens*, *Bacillus cereus*, *Pseudomonas jessenii,* and *Pseudomonas synxantha*.

4.5 Production of Exopolysaccharides (EPS)

EPS are high-molecular weight carbohydrate which helps bacteria to develop biofilms and adhere to plant roots and soil particles. EPS producing bacteria are actively found in the soil organic matter and provide protection to cells against water stress through facilitating the interaction of plants with microbes (Tahir et al., 2019). It has been observed that an increased production of EPS by *Bacillus endophyticus* and *Pseudomonas aeruginosa* under osmotic stress helps in the mitigation of subsiding stress. The inoculation of *Arabidopsis thaliana* seedlings with EPS producing strains was reported to ameliorate the adverse effects of osmotic stress (Ghosh et al., 2019).

5 Soil Moisture and Plant-Microbe Interactions

5.1 Soil Moisture and Rhizosphere Effect

Root exudation is an important mechanism that helps plants to interact with the soil in the rhizosphere region. The availability of nutrients from root exudates makes the rhizosphere a hotspot for microbial activity. Sugars, amino acids, and organic acids are among the primary exudates produced by plant roots and discharged into the soil. The composition of root exudates varies with plant type and environmental factors (Lin et al., 2023). The majority of root exudates are passively removed and secreted by the root and consumed by bacteria in the rhizosphere. Soil moisture has a substantial influence on root exudation patterns. Under moisture deficit conditions, plant roots undergo osmotic adjustment, and a greater proportion of photosynthetic C is dedicated to exudates rather than biomass production. The types of root exudates vary in response to drought (Williams & de Vries, 2020), which in turn influences the microbial diversity, community structure, and activity in the rhizosphere accordingly (Lin et al., 2023). When soil moisture levels are low, increased exudation of organic acids and amino acids can improve nutrient availability and microbial interactions in the rhizosphere. Conversely, high soil moisture can lessen the demand for such exudates by increasing nutrient availability. This dynamic adjustment helps plants maximize nutrient intake and preserve root health under varied environmental conditions (Singh et al., 2023).

Root exudates account for primary metabolites such as sugar, amino acids, and organic acids, as well as secondary metabolites such as flavonoids, strigolactones, and terpenoids. Strigolactones work with the drought stress hormone abscisic acid to help plants adapt to drought stress (Korek & Marzec, 2023). Thus, root exudates can serve as a direct communication channel between plants and rhizosphere microorganisms, potentially influencing plant tolerance and survival during periods of acute abiotic stress.

5.2 Interactions Between Soil Microbes and Plants Under Different Moisture Conditions

Plants are associated with dynamic microbial populations that influence plant growth, physiology, and development. Beneficial microbes mostly include nitrogen-fixing bacteria, plant growth-promoting bacteria, and mycorrhizal fungi (Singh et al., 2023). Environmental conditions, including moisture stress, influence these plant-microbe interactions directly and indirectly. Figure 2 presents the beneficial plant-microbe interactions under moisture-deficit stress conditions in general.

5.2.1 Symbiotic Interactions and Moisture Stress

Mycorrhizal connections and bacteria-mediated nitrogen fixation are examples of symbiotic relationships between plants and microorganisms. Both interactions are important for plant health and growth because they allow plants to absorb nutrients (such as phosphorus and nitrogen) more efficiently during stressed conditions.

5.2.1.1 Mycorrhizal Fungi and Moisture Stress

Arbuscular Mycorrhizal (AM) symbiosis is the most common symbiotic relationship that occurs between terrestrial flowering plant roots and members of the phylum Glomeromycota that are obligate biotrophs. AM regulates plant growth and

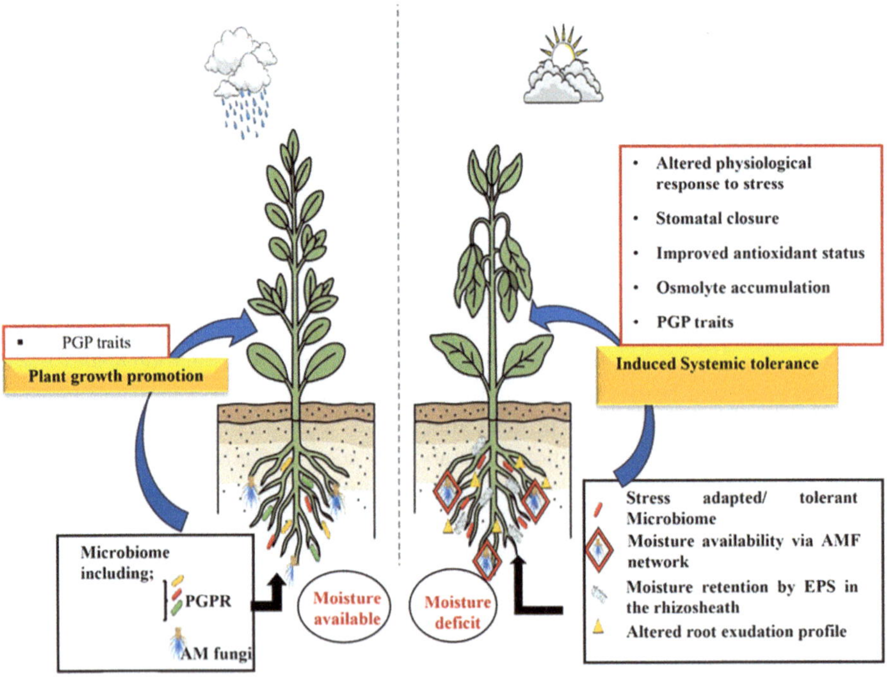

Fig. 2 Plant-microbe beneficial interactions under normal and moisture-deficient conditions

development by enhancing water and mineral nutrient absorption from the soil and consuming carbohydrates and fats from plants in exchange. Soil moisture levels are critical to mycorrhizal connections. Arbuscular mycorrhizal fungi are less moisture-dependent but are nonetheless affected, whereas ectomycorrhizal fungi are extremely sensitive to soil moisture levels. Furthermore, moisture affects fungal spore germination, growth, and virulence (Mishra et al., 2024). For example, the ectomycorrhizal fungus *Pisolithus tinctorius* needs considerable soil moisture to thrive in association with pines. *Rhizophagus irregularis*, an arbuscular mycorrhizal fungus, thrives best in moderate moisture conditions, but can adapt to a range of other settings (Wahab et al., 2023).

AM associations have been found to alleviate drought stress in diverse crops in a number of ways. Maize plants benefit from arbuscular mycorrhizal fungus during drought stress, which manages water loss by reducing aquaporin-related gene expression in the root (Quiroga et al., 2017). Furthermore, mycorrhizal root colonization in *Citrus trifoliata* (trifoliate orange) stimulates H_2O_2 efflux, which helps to reduce oxidative damage during drought stress (Huang et al., 2017). Similar interactions have been described for a variety of different plant-mycorrhizal associations (including rose, sorghum, chickpea, rice, and many tree species). For example, lowland rice requires flooded field conditions for the majority of the growing season. However, the majority of rice cultivation areas are experiencing drought stress due to erratic precipitation and increased frequency of intermittent dry spells. In this regard, AM symbiosis has been found helpful in increasing plant yield through improvement in photosynthetic efficiency, buildup of antioxidant molecule glutathione, and reduced H_2O_2 production in rice during the recovery phase following a 15-day drought spell (Ruiz-Sánchez et al., 2010). Arbuscular mycorrhizal (AM) symbiosis helps plants adjust to drought conditions by regulating the concentration of free proline and soluble carbohydrates, thereby adjusting the soil water potential, as demonstrated in sweet potato, where AMF inoculation enhanced water deficit tolerance, improved plant growth parameters, and increased photosynthetic efficiency (Yooyongwech et al., 2016). Similarly, while strawberry plants are naturally vulnerable to drought stress due to their shallow root systems and high leaf area, the combined application of silicon (Si) and AMF had a synergistic impact on plant biomass: Si accumulation alone mitigates the effects of protracted drought by activating photosynthetic enzymes, enhancing water use efficiency, improving nutrient uptake, promoting root development, and increasing hydraulic conductivity, and when combined with AMF, this effect was amplified, leading to increased photosynthetic rate, higher water content, and enhanced zinc absorption (Moradtalab et al., 2019). Furthermore, a separate study confirmed the impact of AMF on the overall performance of maize under drought stress, showing that the mediation occurred through the enhanced efficiency of photosystem II, increased chlorophyll concentration, and higher levels of soluble sugar and unsaturated fatty acids (Hu et al., 2020).

5.2.1.2 Legume-Rhizobium Symbiosis and Moisture Stress

Drought severely impacts the legume-rhizobium symbiosis because an insufficient water supply prevents the necessary interchange of signaling molecules between the legume and the rhizobia, leading to inefficient or ineffective nodulation and impaired biological nitrogen fixation (BNF). Lack of adequate moisture is also detrimental to the density of nitrogen-fixing bacterial populations and can disrupt the process of root hair infection by inhibiting the production of infectious threads. The major physiological responses to water stress in this system include a decrease in the intensity of carbon metabolism in the nodules, reduced nitrogenase activity, feedback regulation caused by the accumulation of nitrogen fixation products, and the modification of rhizobia cells (Yeremko et al., 2025). Collectively, these effects result in lower nitrogen availability for the plant, consequently restricting overall growth and diminishing agricultural yields in water-limited scenarios (Dollete et al., 2024).

5.2.2 Nonsymbiotic Plant-Microbe Interactions and Moisture Stress

Plant Growth Promoting Rhizobacteria (PGPR) are known to help crops tolerate water deficiencies by releasing phytohormones like ABA, ROS-scavenging enzymes, and other drought stress-inducible responses (Batool et al., 2020), while plant cells respond to drought by increasing compatible solutes to regulate root water content without decreasing water potential, thus weakening the osmotic impact of drought (Karlowsky et al., 2018). Furthermore, the root exudates composition is critical as it determines the selection and communication signal for beneficial rhizospheric bacteria, influencing plant drought tolerance; for instance, the higher amount of malic acid in maize root exudates compared to other organic acids (like succinic and fumaric, etc.) makes it an active chemoattractant for various PGPR, including *Azospirillum* spp. and *Bacillus subtilis*, and the malic acid released by roots of Arabidopsis and Glycine max helps differentiate between beneficial and detrimental soil microorganisms (Rudrappa et al., 2008). PGPR such as *Bacillus subtilis* protect timothy plants by increasing osmolyte production, and other rhizobacteria, including *Pseudomonas putida* and *Bacillus amyloliquefaciens*, also improve the drought tolerance of *Cicer arietinum*; however, moisture stress modifies both the quality and quantity of root exudation, causing a shift in the rhizosphere community composition based on the available resources (Lin et al., 2023). In environments with low soil moisture, which have lower nutrient quality and oxygen accessibility compared to water-rich soil, soil bacteria maximize the breakdown and utilization of polysaccharides rather than oligosaccharides as carbon sources (Kurm et al., 2017; Martiny et al., 2017). Gram-positive bacteria are generally more effective at utilizing carbon sources than Gram-negative types, as exemplified by Gram-positive bacteria using inorganic nitrogen in droughted soil to enzymatically destroy complex organic molecules, and Actinobacteria thriving in moisture-limited soils due to their ability to exploit refractory carbon resources. In contrast, Gram-negative bacteria such as Proteobacteria and Acidobacteria rely solely on readily available carbon and nitrogen sources found in root exudates; therefore, the unequal availability of carbon and nitrogen sources in both the soil and root exudates causes significant shifts in the bacterial community (Martiny et al., 2017). Several reviews have been published on microbial-induced adaptation and mitigation of abiotic stress (Grover et al., 2011, 2021), and Table 1 summarizes recent research on the role of microorganisms in moisture stress management.

5.2.3 Pathogenic Interactions and Moisture Stress

Moisture deficit stress has a wide-ranging impact on pathogen infection rates and illness severity. Diseases such as pea root rot caused by *Aphanomyces euteiches*, wheat take-all caused by *Gaeumannomyces graminis* var. *tritici*, onion white rot caused by *Sclerotium cepivorum*, and grapevine black foot caused by *Ilyonectria/ Dactylonectria* spp. become more severe due to drought stress. Drought, however, reduced the severity of kiwifruit sclerotinia rot (*Sclerotinia sclerotiorum*) and radiata pine red needle cast (*Phytophthora pluvialis*) (Wakelin et al., 2018). In general, necrotrophic pathogens favor drought-induced death by exhausting plant resources through repair and compartmentalization, whereas biotrophic pathogens that induce diseases are expected to be less severe under drought due to the strong relationship between pathogen performance and tree nutritional status. However, once established in the stressed plants, biotrophs are likely to enhance the severity of drought-related consequences on plants as they can deplete carbohydrate reserves critical for plants' drought resistance (Oliva et al., 2014).

The pathogen *Phytophthora infestans*, causing potato blight, lives in moist environments, promoting spore germination and infection (Ivanov et al., 2021). *Fusarium oxysporum,* a pathogen causing wilt, requires adequate soil moisture for its spores to develop and propagate (Purohit et al., 2023). Prolonged drought induces water stress in forest trees, which increases susceptibility to infection by pathogens causing dieback disease from the genus *Phytophthora*, hence enabling the formation of possibly new illnesses (Hossain et al., 2019). Drought can also cause the formation of novel diseases that can resist harsh environmental conditions and exploit changes in plant physiology in response to stress. For example, dryness favors the infection of the chickpea plant by the dry root rot fungal pathogen *Macrophomina phaseolina* (Rai et al., 2022). Drought-induced suppression of plant basal immune responses exacerbated potato yellow vein virus infection and disease symptoms. These alterations impact the host-virus-vector (greenhouse whitefly) relationship, resulting in increased horizontal virus transmission (Vasquez et al., 2022).

6 Mitigation Strategies for Managing Soil Moisture for Optimal Microbial Activity

Maintaining soil health and microbial activity, which in turn affects plant development, nutrient cycling, and general ecosystem processes, depends critically on soil moisture. Microbial activity can be inhibited by both dryness and excessive rainfall, which can lower soil fertility and increase plant stress. For sustainable agricultural

Table 1 Effect of microbial inoculants on plant growth and drought stress tolerance

Plants	Microbes	Experimental setup	Mechanism/key findings	References
Maize	*Bacillus* spp. (BEB1, BEB2, BEB3, BEB4)	Pot experiment	Enhanced plant growth, nutrient uptake, and photosynthetic pigments. Reduced oxidative stress and lipid peroxidation. Modulated antioxidant enzymes: ascorbate peroxidase, peroxidase, catalase, superoxide dismutase, leading to drought stress amelioration	Azeem et al. (2022)
Wheat	*Bacillus paramycoides* (DT-85), *Bacillus paranthracis* (DT-97)	Pot experiment	Enhanced drought tolerance *via* PGP traits (phosphate, exopolysaccharides, siderophores, IAA, gibberellic acid, ammonia) and increased antioxidant enzymes (SOD, catalase, peroxidase, proline)	Yadav et al. (2022)
Soybean	*Ambispora leptoticha* (AMF) and *Bradyrhizobium liaoningense*	Pot experiment	Dual inoculation significantly improved plant growth (biovolume index, root volume, total dry biomass, seed yield), especially under severe drought (50% FC). biomass increased by 58% and mean seed weight by 88% compared to control	Ashwin et al. (2022)
Wheat	*Acinetobacter* sp. (LEW9), *Chryseobacterium* sp. (LEW3), *Klebsiella* sp. (LEW16)	Pot experiment	PGPR enhanced drought stress tolerance in wheat through EPS production. Strain LEW16 significantly improved root morphology, seedling growth, and root colonization under PEG-induced drought stress. These strains also exhibited P and Zn solubilization and HCN production	Latif et al. (2022)

(continued)

Table 1 (continued)

Plants	Microbes	Experimental setup	Mechanism/key findings	References
Maize	*Azospirillum argentinense strains* Az19 and Az39	Field study	Strain Az19 outperformed Az39 in mitigating drought effects. Az19 maintained maize growth under water deficit, improved fertile plant % and ear weight, and enhanced drought tolerance across different genotypes in field trials—resulting in better grain yield	Garcia et al. (2023)
Pearl millet	*Azospirillum formosense, Bacillus* spp.	Seed bioassay	Coinoculation enhanced PGP traits (nitrogenase, phosphate solubilization, siderophore, IAA), seed germination, root traits, and seedling vigor under drought stress. Synergistic interaction improved stress adaptation and yield potential	Yaadesh et al. (2023)
Maize	*Azospirillum brasilense* (Az1 and Az2)	Field study	Inoculation improved chlorophyll content and grain yield (~10%) across water and nitrogen conditions. Under water stress, inoculation reduced environmental stress effects by ~11%. Az2 with 120 kg N/ha under irrigation gave the highest yield	Marques et al. (2023)
Rice	*Trichoderma asperelloides, Trichoderma brevicompactum*	Pot experiment	Enhanced root length, proline, phenolic content; reduced MDA, H_2O_2. Upregulated DHN-4, AQU-2, DREB genes, improving drought tolerance in tolerant (BRRI dhan56) and susceptible (IR64) varieties	Quazi et al. (2024)

(continued)

Table 1 (continued)

Plants	Microbes	Experimental setup	Mechanism/key findings	References
Soybean	*Bradyrhizobium* sp. (DTB4, DTR30), *AMF (Glomus clarum, Funneliformis mosseae, Gigaspora margarita)*	Field study	Bioinoculants improved growth traits (shoot/root length, dry weight, leaf area), nutrient uptake (N, P, K), chlorophyll, nodulation, and yield under drought stress. Reduced proline content, increased bacterial count, mycorrhizal colonization, and soil enzyme activities (dehydrogenase, phosphatase)	Nader et al. (2024)
Tea	Arbuscular mycorrhizal fungus (*Claroideoglomus etunicatum*)	Pot experiment	AMF inoculation enhanced plant height, leaf/root biomass, root morphology (lateral roots, root hair density, length, diameter), and nutrient uptake (P, K, Ca, Mg, Fe, Mn, Zn). Increased abscisic acid, brassinosteroids, gibberellins, and IAA in roots, improving drought adaptability	Liu et al. (2024)
Rice	*Streptomyces pactum* Act12	Field study	Inoculation improved chlorophyll content, photosynthetic efficiency, antioxidant enzyme activity (SOD, POD), and osmolyte content (proline, soluble proteins)	Yang et al. (2024)
Soybean	*Bradyrhizobium japonicum*	Field study	Rhizobial inoculation with fungicide enhanced antioxidant enzyme activity (SOD by 32.7%, catalase by 90.6%) under drought, regulated lipoperoxidation, and improved nitrogen fixation efficiency by 25.6%	Nyzhnyk et al. (2024)

(continued)

Table 1 (continued)

Plants	Microbes	Experimental setup	Mechanism/key findings	References
Tomato	*Bacillus cereus* AN14, *Bacillus thuringiensis* AN10, *Bacillus* sp. AN22	Pot experiment	Selected *Bacillus* strains improved drought tolerance, delayed wilting, and enhanced fruit weight under water stress. Microbiota-treated plants showed the highest yield	Taiwo and Akintokun (2025)
Maize	*Bacillus subtilis*, *Bacillus* sp., Arbuscular mycorrhizal fungi (AMF)	Pot experiment	Biostimulants (microbial consortia + organic amendment) improved biomass, root mycorrhization, osmolytes, stomatal conductance, (total sugars and proline), and antioxidant enzyme polyphenol oxidase. Reduced stress markers (malondialdehyde, hydrogen peroxide)	Ech-Chatir et al. (2025)
Wheat	*Streptomyces pactum* Act12, *Streptomyces rochei D74*	Field study	Combined inoculation improved chlorophyll content (32.60%), PSII efficiency (43.07%), and antioxidant enzyme activities (SOD 19.32%, POD 75.44%). Reduced MDA levels (61.61%) and enhanced proline (90.44%) and soluble protein (60.17%) under drought. Yield increased by 26.19% due to more effective spikes and grains per spike	Shan et al. (2025)

systems, controlling soil moisture is therefore essential to promoting the best possible microbial performance. With an emphasis on microbial health, we examine several methods in this section to improve soil moisture retention, lower evaporation, and lessen the consequences of extreme weather events like drought and flooding.

6.1 Approaches to Enhance Soil Water Retention and Reduce Evaporation

A variety of strategies can be used to improve soil water retention and lower evaporation, guaranteeing a consistent flow of hydration for plants and microbial communities.

6.1.1 Use of Soil Amendments

Soil amendments are organic or inorganic materials added to the soil to improve its structure, water-holding capacity, and overall fertility. Table 2 presents the amendment strategies to enhance soil moisture and microbial activities.

Table 2 Amendments/practices for managing soil moisture and their effects on microbial activity

Strategy/practice	Effect on soil moisture	Microbial benefits	References
Biological soil amendments	Bio-based fertilizers and compost teas to enhance microbial balance	Improved root growth, microbial activity, and soil moisture under salinity/drought	Sani and Yong (2021)
Cover cropping	Plants reduce evaporation and increase infiltration while protecting the soil surface	Increased microbial biomass, fungal abundance, and enzyme activity	Quintarelli et al. (2022)
Hydrogels	Water-absorbing polymers improve retention and reduce irrigation frequency	Increased microbial population and soil respiration under water-deficit stress	Patra et al. (2022)
Superabsorbent polymer (SAP) and crop-residue mulch (CRM)	Water-absorbing polymers and crop residue mulches help in moisture conservation	Increased activity of soil dehydrogenase, alkaline phosphatase, acid phosphatase, and urease activities	Bana et al. (2023)
Clay-based additives	Enhance capillary moisture retention and stabilize microstructures	Improved microbial wettability, pore distribution, and structure stability	Khan et al. (2023)
Microbial gel carriers	Encapsulation of microbes in gel matrices for slow-release in rhizosphere	Prolonged microbial activity and hydration around roots during dry periods	Pereira et al. (2023)
Green manure	Growing and incorporating leguminous or cover crops into soil before main crop season. Enhance soil moisture and fertility	Boosted microbial nitrogen fixation, soil organic matter	Khan et al. (2024)
Organic matter (compost, manure, biochar)	Improve soil structure, porosity, nutrient retention, and water infiltration	Preserved microbial function, improved enzyme activity, and bacterial diversity	Verma et al. (2024)
Conservation tillage	Minimize soil disturbance and retain residue cover for moisture conservation	Enhanced microbial activity, enzyme function, and moisture retention	Srivastava (2025)

6.1.1.1 Organic Matter (Compost, Manure, and Biochar)

By increasing the porosity of the soil, organic additions such as compost, well-rotted manure, and biochar aid in improving the structure of the soil. By enabling soil particles to bond and form aggregates, this improves water infiltration and retention. Additionally, organic matter aids in improving nutrient retention by raising the cation exchange capacity. Application of SWP 550 biochar preserved superior microbial function under dry conditions. Biochar had an impact on soil moisture and microbial activity, with low moisture decreasing bacterial richness and enzyme activity. Because of its large surface area that promotes microbial colonization, SWP 700 biochar improved soil enzyme activity, bacterial diversity, and evenness (Dissanayake et al., 2022).

6.1.1.2 Hydrogels

The addition of hydrogels, which are water-absorbing polymers, to soil can improve its ability to retain water and lessen the frequency of irrigation. The application of hydrogel-nanonatural char composite (0.3%) at 85% water holding capacity (WHC) increased tomato growth and nutrient uptake by improving soil moisture retention and reducing the adverse impacts of water-deficit stress. Similarly, with 0.6% nanocomposite under 75% WHC, the maximum soil respiration rate (61% increase) and microbial population size (89% increase) were recorded, indicating that the composite also markedly increased soil microbial activity (Nassaj-Bokharaei et al., 2021). Particularly helpful in arid regions and sandy soils that naturally drain rapidly are hydrogels.

6.1.1.3 Clay and Clay-Based Additives

Clay can assist decrease evaporation-induced water loss in high-sandy soils by improving the soil's ability to retain water. Near the plant root zone, clay particles aid in the capillary effect that retains moisture. Microbial activity was greatly impacted by soil moisture dynamics, which resulted in improved soil microstructural stability through the binding effects of microbial exudates that encouraged particle cementation and pore space remodeling. Microbial wettability, pore water distribution, and microbial activity all showed positive relationships, highlighting the function of moisture-driven microbial processes in enhancing soil stability and fortifying soil structure under fluctuating moisture circumstances (Vu et al., 2022).

6.1.2 Cover Crops

Legumes, grasses, and cereals are examples of cover crops that can be very important for enhancing soil moisture retention. These crops lessen the amount of direct sunlight that reaches the soil's surface by creating ground cover, which lowers evaporation. Their root systems also enhance water infiltration and soil structure, which makes the environment more hospitable for microorganisms. Oats and various cover crop combinations considerably increased total microbial biomass, fungal abundance, and enzyme activities linked to carbon and nitrogen cycling. Cover cropping in a crop-fallow cycle also improved soil microbial activity and community structure. When compared to fallow, these enhancements, particularly under oat and diverse-mix treatments, improved soil moisture retention and increased the abundance of arbuscular mycorrhizal fungi by 84%, improving soil health in semi-arid environments (Thapa et al., 2021). Cover crops also help to prevent soil erosion and reduce the risk of moisture loss during periods of intense rainfall or wind.

6.1.3 Conservation Tillage

Conservation tillage is the term for farming methods that reduce soil disturbance. Soil aggregates can be broken down by conventional tillage techniques, which lowers the soil's ability to retain water and promotes evaporation. In comparison to conventional tillage, long-term conservation tillage increased β-glucosidase and dehydrogenase activities in the upper soil layers and enhanced soil microbial activity and moisture retention by minimizing soil disturbance and preserving crop residue cover (Nugroho et al., 2023). Sustainable soil health and microbial function were promoted by conservation tillage, which improved organic matter and balanced soil moisture, which facilitated gradual nutrient release and increased soil nutrient-supplying capacity. Soil structure, organic matter retention, and soil moisture content are all enhanced by conservation tillage methods like reduced-till or no-till farming. Conservation tillage additionally reduces moisture loss and offers a stable environment for soil bacteria by decreasing the surface area exposed to the sun and wind.

6.2 Approaches to Deal with Drought and Flooding Conditions

Drought and flooding are two of the most challenging conditions affecting soil moisture and microbial health. Both extremes can disrupt microbial communities and reduce soil fertility. However, there are strategies that farmers can adopt to mitigate the impacts of these conditions.

6.2.1 Use of Soil Amendments and Moisture Conservation

Compost, manure, and biochar are examples of organic additions that can increase soil resistance to flooding and drought. These additions lessen waterlogging during periods of high rainfall and aid in the soil's ability to retain moisture during dry spells. By binding water in the soil, lowering evaporation, and giving plants and bacteria a steady source of water, organic additions assist the soil in retaining moisture during drought conditions. These changes enhance drainage during floods, facilitating the more effective removal of surplus water and avoiding waterlogging. Organic amendments improve soil structure and promote soil aeration, which lessens the negative effects of too much water.

6.2.2 Drainage Systems

An effective drainage system is essential for controlling flooding. An overabundance of water in the soil can damage the microbial ecology and drown plant roots. Tile drains, French drains, and surface drains are a few examples of proper drainage systems that may efficiently handle excess water and avoid waterlogging. These devices allow oxygen to reach plant roots and soil bacteria by diverting water away from the root zone.

6.2.3 Drought-Tolerant Crops and Water-Saving Techniques

Growing drought-tolerant crops can reduce the stress on the soil by ensuring that crops are more resilient to low-moisture conditions. Mulching, drip irrigation, and rainwater collection are some methods that can assist in preserving soil moisture and lessening the need for outside water sources. In contrast to the high-yielding cultivar Co51, the drought-tolerant rice landrace Norungan maintained higher soil moisture, microbial biomass carbon, and enzyme activity, indicating superior rhizosphere resistance under moisture stress. According to Nunna and Balachandar (2024), Norungan increased soil microbial activity and biochemical functions by attracting particular microbial populations, which helped the soil become more drought-tolerant. Using mulching techniques, especially with wool and straw mulch, increased microbial activity and soil moisture retention, which increased crop yields in a variety of soil types and irrigation schedules. According to the study, mulching increases the amount of active carbon and enzyme activity; the effects of mulching vary depending on the degradability of the mulch material, the frequency of irrigation, and the soil's ability to retain water (Papdi et al., 2022).

6.3 Microbial Resources in Mitigating Soil Moisture Stress Effects

Microorganisms play an essential role in mitigating the effects of soil moisture stress. The presence of healthy and diverse microbial communities can help maintain soil fertility and ensure that plants have access to nutrients and water, even under challenging moisture conditions.

6.3.1 Microbial Inoculants

Microbial inoculants, such as beneficial bacteria and fungi, can be introduced to the soil to enhance moisture retention and improve nutrient availability. Arbuscular mycorrhizal fungi, for instance, develop symbiotic interactions with plant roots to enhance nutrient and water absorption. Similarly, by increasing their ability to withstand stress, drought-resistant bacteria can aid plants in surviving in arid environments. When maize plants were inoculated with Pseudomonas putida after being exposed to drought stress, plant biomass, RWC, and leaf water potential all increased (Sandhya et al., 2010). Bacterial inoculation of wheat plants boosted water intake during drought circumstances by increasing root growth and lateral root formation (Dar et al., 2018). PGPR isolated from xerophytic rhizospheres on rice variety CO51 under varying moisture stress levels was assessed in this work. Because of their higher proline accumulation and antioxidant activity, seeds treated with *Bacillus velezensis* VKSB5 and *Bacillus altitudinis* MLSB2 demonstrated enhanced germination, growth, and drought resistance. According to Rokins et al. (2022), these results demonstrate the potential of microbial inoculants to alleviate drought stress and promote early rice growth. Through the production of ACC deaminase and exopolysaccharides, drought-tolerant phyllosphere bacterial isolates enhanced cotton growth under moisture stress. *Acinetobacter* sp. exhibited improved root length, fresh weight, and boll production, as well as the highest seed germination (76.67%) and seedling vigor (45.81%) (Sharath et al., 2021). These findings demonstrate how microbial inoculants may enhance soil nitrogen status and promote crop resilience in drought-prone environments.

6.3.2 Biological Soil Amendments

Beneficial microorganisms found in biological soil amendments, like compost teas or bio-based fertilizers, can improve microbial diversity and soil moisture retention. These products aid in reestablishing the soil's microbial equilibrium, which is especially beneficial following dry spells or floods. In order to assist sweet sorghum withstand drought and salinity stress, soil amendments that combined organic acids, bio-based materials, and beneficial bacteria enhanced soil moisture retention and microbial activity. The treatments showed promise for maintaining crop growth in saline-alkaline conditions by improving root vigor, photosynthesis, and nutrient availability (Xue et al., 2024).

6.3.3 Microbial Biofilms

Clusters of microorganisms known as microbial biofilms stick to surfaces like soil particles or plant roots. These biofilms aid in retaining moisture in the soil and creating a barrier that stops water loss from happening too quickly. By enhancing root characteristics and antioxidant defense systems, biofilm-forming PGPR improved soil moisture retention and wheat resistance under water deficit. These drought-tolerant bacteria showed their function in maintaining microbial activity and plant growth in arid circumstances by increasing relative water content and harvest index while decreasing oxidative stress (Karimi et al., 2022). Through efficient root colonization and stress modulation, the biofilm-forming PGPR strains *Bacillus subtilis* FAB1 and *Pseudomonas azotoformans* FAP3 improved soil moisture retention and microbial activity, hence increasing wheat drought tolerance. The ability of biofilm-engineered rhizobacteria to maintain crop growth under water stress was demonstrated by their consortium's considerable increase in soil enzyme activities and physiological features (Ansari et al., 2023). Additionally, some bacteria release compounds that enhance soil aggregation and water retention.

6.4 Innovative Practices in Soil Moisture and Microbial Management

As the challenges of climate change intensify, new research and technological advancements offer hope for managing soil moisture and enhancing microbial health in increasingly unpredictable environments. Some of these cutting-edge practices and technological advancements include:

6.4.1 Soil Moisture Sensors

Technological advances in soil moisture sensors allow farmers to monitor soil moisture levels in real time, enabling precise irrigation practices. These sensors help avoid over-watering or under-watering, ensuring that the soil maintains optimal moisture levels for microbial activity.

6.4.2 Precision Agriculture

The use of satellite imagery, drones, and artificial intelligence (AI) is revolutionizing soil moisture management. These technologies can help identify areas of the farm that are at risk of drought or flooding, allowing for targeted interventions.

6.4.3 Microbial Sensors

The existence and activity of advantageous bacteria in the soil-plant system can be inferred from microbial sensors, such as biosensors that identify particular microbial metabolites or genetic markers. Microbial inoculants may be applied promptly and precisely thanks to this real-time monitoring, which guarantees that they are used to reduce drought stress when and where it is most needed (Haddi et al., 2020; Akanmu et al., 2021).

6.4.4 Synthetic Microbial Communities

The creation of artificial microbial communities with distinct interactions and functional redundancy is made easier by synthetic biology. Researchers can improve the overall stability and robustness of microbial therapies by creating consortia with complementary characteristics. For example, a synthetic consortium may consist of drought-tolerant PGPR, phosphate-solubilizing fungi, and nitrogen-fixing bacteria, each of which performs unique tasks that together improve plant resilience (Tripathi et al., 2020; Vocciante et al., 2022). But there are drawbacks to using synthetic biology in agriculture as well, such as potential unforeseen ecological effects, public acceptance issues, and regulatory barriers. To make sure that the created bacteria do not harm natural ecosystems or endanger human health, ethical considerations and thorough safety evaluations are crucial. Notwithstanding these difficulties, synthetic biology holds enormous promise for creating tailored microbial solutions for drought tolerance, providing a route to more robust and sustainable agricultural systems.

6.4.5 Genomics and Microbiome Analysis

Advancements in genomics and microbiome research have transformed the study and manipulation of microbial communities. High-throughput sequencing technologies, such as next-generation and third-generation sequencing, enable in-depth profiling of soil- and plant-associated microbiomes. Metagenomics, in particular, facilitates the analysis of microbial genetic diversity without the need for cultivation, offering crucial insights into microbial functions and ecological roles in drought tolerance (El-Shazoly et al., 2024; Abbasi & Mohammadi, 2023; Sharma et al., 2022).

6.4.6 Bioinformatics

Bioinformatics and machine learning are crucial for analyzing genomics and metagenomics data from high-throughput sequencing, aiding in the identification of beneficial microbial biomarkers, functional predictions, and microbe-plant-environment interactions. These insights drive the development of microbial inoculants and optimized consortia for drought resilience, advancing sustainable agriculture through cutting-edge genomics and microbiome research (Basu et al., 2021; Riaz et al., 2021).

Case Studies and Success Stories

As climate change intensifies drought stress in agriculture, innovative microbial-based strategies are emerging as effective solutions to enhance crop resilience and sustain yields. The integration of beneficial microbes, such as mycorrhizal fungi, nitrogen-fixing bacteria, and PGPR, has demonstrated significant improvements in soil moisture retention, nutrient uptake, and drought tolerance across diverse cropping systems. In India's rice paddies, the use of a microbial consortium consisting of mycorrhizal fungi, nitrogen-fixing bacteria, and drought-tolerant PGPR has greatly increased rice yields in water-stressed environments. Higher grain yields and more consistency in rice production have resulted from the consortium's facilitation of improved nutrient uptake, soil moisture retention, and plant drought resistance. In areas where traditional irrigation methods are limited by water constraints, this integrated microbial strategy has proven especially advantageous, offering a long-term way to sustain rice yield (Alsharif et al., 2020; Das et al., 2022; Poppeliers et al., 2023).

A prominent illustration is found in the semiarid areas of sub-Saharan Africa, where smallholder farmers have begun using PGPR inoculants to increase the yields of sorghum and maize during dry spells. It has been demonstrated that applying PGPR strains such as *Bacillus subtilis* and *Azospirillum brasilense* in these areas improves water use efficiency, boosts nutrient uptake, and promotes root growth. Farmers have reported notable gains in crop yields, less dependency on irrigation, and improved soil fertility, all of which have improved food security and economic stability in areas vulnerable to drought (Afridi et al., 2022; Lu et al., 2023; Muter, 2023).

A thorough analysis highlighting the function of several microorganisms in reducing drought stress in plants was carried out by Shaffique et al. (2022). The meta-analysis showed that in drought conditions, microbial inoculation could result in notable improvements in plant growth characteristics. In comparison to noninoculated controls, inoculated plants exhibited an average increase in shoot biomass of 30%, root biomass of 25%, and water use efficiency of 20%. Osmotic adjustment, phytohormonal regulation, and increased antioxidant enzyme activity are some of the mechanisms at play. Farmers who have implemented innovative practices like precision irrigation, microbial inoculants, and conservation tillage have reported better soil moisture management and increased resilience to extreme weather events. These success stories highlight the potential of integrating advanced technologies and traditional practices to improve soil health and microbial activity.

7 Summary and Future Directions

Soil moisture is essential for plant growth and development as it represents the water present in soil pores. Soil moisture also plays a crucial role in shaping microbial community composition, diversity, and abundance. Factors influencing soil moisture include climatic factors and anthropogenic factors. Well-hydrated soils promote microbial activity and decomposition, making nutrients more accessible to plants. Dry conditions can hinder microbial activity, slowing down decomposition and limiting nutrient supply. Soil moisture impacts microbial communities by influencing community composition and metabolic activities, thus affecting nutrient cycling, overall productivity, and ecosystem functioning. Soil moisture influences the quantity and composition of root exudates, which in turn affects microbial diversity and activity in the rhizosphere, impacting plant-microbe interactions. Mitigation strategies for managing soil moisture include using soil amendments for enhanced water retention and reduced evaporation, the use of cover crops, tools to enhance water use efficiency, stress-adapted/tolerant plant beneficial microorganisms, and designer communities to impart tolerance to crops and boost productivity under stressful conditions.

References

Abbasi, S., & Mohammadi, P. (2023). Improving drought tolerance and glycyrrhizin content of licorice plant by EPS-producing cyanobacteria. *Industrial Crops and Products, 197*, 116578.

Afridi, M. S., Ali, S., Salam, A., César Terra, W., Hafeez, A., Sumaira, et al. (2022). Plant microbiome engineering: Hopes or hypes. *Biology, 11*(12), 1782.

Akanmu, A. O., Babalola, O. O., Venturi, V., Ayilara, M. S., Adeleke, B. S., Amoo, A. E., et al. (2021). Plant disease management: Leveraging on the plant-microbe-soil interface in the biorational use of organic amendments. *Frontiers in Plant Science, 12*, 700507.

Allison, S. D., & Treseder, K. K. (2008). Warming and drying suppress microbial activity and carbon cycling in boreal forest soils. *Global Change Biology, 14*(12), 2898–2909.

Alsharif, W., Saad, M. M., & Hirt, H. (2020). Desert microbes for boosting sustainable agriculture in extreme environments. *Frontiers in Microbiology, 11*, 1666.

Ansari, F. A., Ahmad, I., & Pichtel, J. (2023). Synergistic effects of biofilm-producing PGPR strains on wheat plant colonization, growth and soil resilience under drought stress. *Saudi Journal of Biological Sciences, 30*(6), 103664.

Ansary, M. H., Rahmani, H. A., Ardakani, M. R., Paknejad, F., Habibi, D., & Mafakheri, S. (2012). Effect of *Pseudomonas fluorescence* on proline and phytohormonal status of maize (*Zea mays* L.) under water deficit stress. *Annals of Biological Research, 3*(2), 1054–1062.

Aranjuelo, I., Irigoyen, J. J., & Sánchez-Díaz, M. (2007). Effect of elevated temperature and water availability on CO_2 exchange and nitrogen fixation of nodulated alfalfa plants. *Environmental and Experimental Botany, 59*(2), 99–108.

Ashwin, R., Bagyaraj, D. J., & Raju, B. M. (2022). Dual inoculation with rhizobia and arbuscular mycorrhizal fungus improves water stress tolerance and productivity in soybean. *Plant Stress, 4*, 100084.

Averill, C., Waring, B. G., & Hawkes, C. V. (2016). Historical precipitation predictably alters the shape and magnitude of microbial functional response to soil moisture. *Global Change Biology, 22*(5), 1957–1964.

Azeem, M., Haider, M. Z., Javed, S., Saleem, M. H., & Alatawi, A. (2022). Drought stress amelioration in maize (*Zea mays* L.) by inoculation of *Bacillus* spp. strains under sterile soil conditions. *Agriculture, 12*(1), 50.

Baldrian, P., Merhautová, V., Petránková, M., Cajthaml, T., & Šnajdr, J. (2010). Distribution of microbial biomass and activity of extracellular enzymes in a hardwood forest soil reflect soil moisture content. *Applied Soil Ecology, 46*(2), 177–182.

Bana, R. S., Grover, M., Singh, D., Bamboriya, S. D., Godara, S., Kumar, M., Kumar, A., Sharma, S., Shekhawat, P. S., Lomte, D., Bhuva, H. M., Babar, S. R., Suryawanshi, R. T., Vasuki, V., Singh, N., Khandelwal, V., Kumar, A., Singh, A., & Tara Satyavathi, C. (2023). Enhanced pearl millet yield stability, water use efficiency and soil microbial activity using superabsorbent polymers and crop residue recycling across diverse ecologies. *European Journal of Agronomy, 148*, 126876. ISSN 1161-0301.

Banerjee, S., Helgason, B., Wang, L., Winsley, T., Ferrari, B. C., & Siciliano, S. D. (2016). Legacy effects of soil moisture on microbial community structure and N_2O emissions. *Soil Biology and Biochemistry, 95*, 40–50.

Basu, A., Prasad, P., Das, S. N., Kalam, S., Sayyed, R. Z., Reddy, M. S., & El Enshasy, H. (2021). Plant growth promoting rhizobacteria (PGPR) as green bioinoculants: Recent developments, constraints, and prospects. *Sustainability, 13*(3), 1140.

Batool, T., Ali, S., Seleiman, M. F., Naveed, N. H., Ali, A., Ahmed, K., et al. (2020). Plant growth promoting rhizobacteria alleviates drought stress in potato in response to suppressive oxidative stress and antioxidant enzymes activities. *Scientific Reports, 10*(1), 16975.

Berg, A., & Sheffield, J. (2018). Climate change and drought: The soil moisture perspective. *Current Climate Change Reports, 4*, 180–191. https://doi.org/10.1007/s40641-018-0095-0

Bian, H., Li, C., Zhu, J., Xu, L., Li, M., Zheng, S., & He, N. (2022). Soil moisture affects the rapid response of microbes to labile organic C addition. *Frontiers in Ecology and Evolution, 10*, 857185.

Burns, R. G., DeForest, J. L., Marxsen, J., Sinsabaugh, R. L., Stromberger, M. E., Wallenstein, M. D., et al. (2013). Soil enzymes in a changing environment: Current knowledge and future directions. *Soil Biology and Biochemistry, 58*, 216–234.

Chakraborty, U., Chakraborty, B. N., Chakraborty, A. P., & Dey, P. L. (2013). Water stress amelioration and plant growth promotion in wheat plants by osmotic stress tolerant bacteria. *World Journal of Microbiology and Biotechnology, 29*, 789–803.

Chen, Y., Ma, S., Jiang, H., Hu, Y., & Lu, X. (2020). Influences of litter diversity and soil moisture on soil microbial communities in decomposing mixed litter of alpine steppe species. *Geoderma, 377*, 114577.

Dai, A. (2011). Drought under global warming: A review. *Wiley Interdisciplinary Reviews: Climate Change, 2*(1), 45–65. https://doi.org/10.1002/wcc.81

Dar, Z. M., Masood, A., Mughal, A. H., Asif, M., & Malik, M. A. (2018). Review on drought tolerance in plants induced by plant growth promoting rhizobacteria. *International Journal of Current Microbiology and Applied Sciences, 7*(5).

Das, P. P., Singh, K. R., Nagpure, G., Mansoori, A., Singh, R. P., Ghazi, I. A., et al. (2022). Plant-soil-microbes: A tripartite interaction for nutrient acquisition and better plant growth for sustainable agricultural practices. *Environmental Research, 214*, 113821.

Devries, J. D., Bennett, J. M., Albrecht, S. L., & Boote, K. J. (1989). Water relations, nitrogenase activity and root development of three grain legumes in response to soil water deficits. *Field Crop Res, 21*(3–4), 215–226.

Dissanayake, P. D., Palansooriya, K. N., Sang, M. K., Oh, D. X., Park, J., Hwang, S. Y., et al. (2022). Combined effect of biochar and soil moisture on soil chemical properties and microbial community composition in microplastic-contaminated agricultural soil. *Soil Use and Management, 38*(3), 1446–1458.

Dollete, D., Lumactud, R. A., Carlyle, C. N., Szczyglowski, K., Hill, B., & Thilakarathna, M. S. (2024). Effect of drought stress on symbiotic nitrogen fixation, soil nitrogen availability and soil microbial diversity in forage legumes. *Plant and Soil, 495*(1), 445–467.

Dos Reis, S. P., Marques, D. N., Lima, A. M., & De Souza, C. R. B. (2016). Plant molecular adaptations and strategies under drought stress. In *Drought stress tolerance in plants* (Vol. 2, pp. 91–122). Springer Science and Business Media LLC.

Drenovsky, R. E., Vo, D., Graham, K. J., & Scow, K. M. (2004). Soil water content and organic carbon availability are major determinants of soil microbial community composition. *Microbial Ecology, 48*, 424–430.

Ech-Chatir, L., Ouhaddou, R., Ikan, C., Boutasknit, A., Hashem, A., Avila-Quezada, G. D., et al. (2025). Co-inoculation of compost with arbuscular mycorrhizal fungi and endophytic bacteria to alleviate the adverse impact of drought stress in maize. *Pakistan Journal of Botany, 57*, 4.

El-Shazoly, R. M., Aloufi, A. S., & Fawzy, M. A. (2024). The potential use of arthrospira (Spirulina platensis) as a biostimulant for drought tolerance in wheat (Triticum aestivum L.) for sustainable agriculture. *Journal of Plant Growth Regulation*, 1–18.

García, J. E., Ruiz, M., Maroniche, G. A., Creus, C., Puente, M., Zawoznik, M. S., & Groppa, M. D. (2023). Inoculation with Azospirillum argentinense Az19 improves the yield of maize subjected to water deficit at key stages of plant development. *Revista Argentina de Microbiología, 55*(3), 255–261.

Ghosh, D., Gupta, A., & Mohapatra, S. (2019). A comparative analysis of exopolysaccharide and phytohormone secretions by four drought-tolerant rhizobacterial strains and their impact on osmotic-stress mitigation in Arabidopsis thaliana. *World Journal of Microbiology and Biotechnology, 35*, 1–15.

Gleeson, D. B., Herrmann, A. M., Livesley, S. J., & Murphy, D. V. (2008). Influence of water potential on nitrification and structure of nitrifying bacterial communities in semiarid soils. *Applied Soil Ecology, 40*(1), 189–194.

Gomez, E. J., Delgado, J. A., & Gonzalez, J. M. (2020). Environmental factors affect the response of microbial extracellular enzyme activity in soils when determined as a function of water availability and temperature. *Ecology and Evolution, 10*(18), 10105–10115.

Grover, M., Ali, S. Z., Sandhya, V., Rasul, A., & Venkateswarlu, B. (2011). Role of microorganisms in adaptation of agriculture crops to abiotic stresses. *World Journal of Microbiology and Biotechnology, 27*(5), 1231–1240. https://doi.org/10.1007/s11274-010-0572-7

Grover, M., Bodhankar, S., Sharma, A., Sharma, P., Singh, J., & Nain, L. (2021). PGPR mediated alterations in root traits: Way toward sustainable crop production. *Frontiers in Sustainable Food Systems, 4*, 618230. https://doi.org/10.3389/fsufs.2020.618230

Guntiñas, M. E., Gil-Sotres, F., Leirós, M. C., & Trasar-Cepeda, C. (2013). Sensitivity of soil respiration to moisture and temperature. *Journal of Soil Science and Plant Nutrition, 13*(2), 445–461.

Gusain, Y. S., Singh, U. S., & Sharma, A. K. (2015). Bacterial mediated amelioration of drought stress in drought tolerant and susceptible cultivars of rice (Oryza sativa L.). *African Journal of Biotechnology, 14*(9), 764–773.

Haddi, K., Turchen, L. M., Viteri Jumbo, L. O., Guedes, R. N., Pereira, E. J., Aguiar, R. W., & Oliveira, E. E. (2020). Rethinking biorational insecticides for pest management: Unintended effects and consequences. *Pest Management Science, 76*(7), 2286–2293.

Han, L., Qin, H., Wang, J., Yao, D., Zhang, L., Guo, J., & Zhu, B. (2023). Immediate response of paddy soil microbial community and structure to moisture changes and nitrogen fertilizer application. *Frontiers in Microbiology, 14*, 1130298.

Hao, J., Chai, Y. N., Lopes, L. D., Ordóñez, R. A., Wright, E. E., Archontoulis, S., & Schachtman, D. P. (2021). The effects of soil depth on the structure of microbial communities in agricultural soils in Iowa (United States). *Applied and Environmental Microbiology, 87*(4), e02673–e02620.

Hartmann, M., Brunner, I., Hagedorn, F., Bardgett, R. D., Stierli, B., Herzog, C., et al. (2017). A decade of irrigation transforms the soil microbiome of a semi-arid pine forest. *Molecular Ecology, 26*(4), 1190–1206.

Hossain, M., Veneklaas, E. J., Hardy, G. E. S. J., & Poot, P. (2019). Tree host–pathogen interactions as influenced by drought timing: Linking physiological performance, biochemical defence and disease severity. *Tree Physiology, 39*(1), 6–18.

Hu, Y., Xie, W., & Chen, B. (2020). Arbuscular mycorrhiza improved drought tolerance of maize seedlings by altering photosystem II efficiency and the levels of key metabolites. *Chemical and Biological Technologies in Agriculture, 7*, 1–14.

Huang, Y. M., Zou, Y. N., & Wu, Q. S. (2017). Alleviation of drought stress by mycorrhizas is related to increased root H_2O_2 efflux in trifoliate orange. *Scientific Reports, 7*(1), 42335.

Iovieno, P., & Baath, E. (2008). Effect of drying and rewetting on bacterial growth rates in soil. *FEMS Microbiology Ecology, 65*(3), 400–407.

Ivanov, A. A., Ukladov, E. O., & Golubeva, T. S. (2021). *Phytophthora infestans*: An overview of methods and attempts to combat late blight. *Journal of Fungi, 7*(12), 1071.

Jung, M., et al. (2010). Recent decline in the global land evapotranspiration trend due to limited moisture supply. *Nature, 467*, 951–954. https://doi.org/10.1038/nature09396

Kaplan-Levy, R. N., Hadas, O., Summers, M. L., Rücker, J., & Sukenik, A. (2010). Akinetes: Dormant cells of cyanobacteria. In E. Lubzens, J. Cerda, & M. Clark (Eds.), Dormancy and resistance in harsh environments (pp. 5–27). Springer. https://doi.org/10.1007/978-3-642-12422-8_2

Karimi, E., Aliasgharzad, N., Esfandiari, E., Hassanpouraghdam, M. B., Neu, T. R., Buscot, F., et al. (2022). Biofilm forming rhizobacteria affect the physiological and biochemical responses of wheat to drought. *AMB Express, 12*(1), 93.

Karlowsky, S., Augusti, A., Ingrisch, J., Akanda, M. K. U., Bahn, M., & Gleixner, G. (2018). Drought-induced accumulation of root exudates supports post-drought recovery of microbes in mountain grassland. *Frontiers in Plant Science, 9*, 1593.

Khan, S., Ajmal, S., Hussain, T., & Rahman, M. U. (2023). Clay-based materials for enhanced water treatment: Adsorption mechanisms, challenges, and future directions. *Journal of Umm Al-Qura University for Applied Sciences*, 1–16.

Khan, M. T., Aleinikovienė, J., & Butkevičienė, L. M. (2024). Innovative organic fertilizers and cover crops: Perspectives for sustainable agriculture in the era of climate change and organic agriculture. *Agronomy, 14*(12), 2871.

Kocharunchitt, C., King, T., Gobius, K., Bowman, J. P., & Ross, T. (2012). Integrated transcriptomic and proteomic analysis of the physiological response of *Escherichia coli* O157: H7 Sakai to steady-state conditions of cold and water activity stress. *Molecular & Cellular Proteomics, 11*(1).

Korek, M., & Marzec, M. (2023). Strigolactones and abscisic acid interactions affect plant development and response to abiotic stresses. *BMC Plant Biology, 23*(1), 314.

Kurm, V., Van Der Putten, W. H., De Boer, W., Naus-Wiezer, S., & Hol, W. G. (2017). Low abundant soil bacteria can be metabolically versatile and fast growing. *Ecology, 98*(2), 555–564.

Latif, M., Bukhari, S. A. H., Alrajhi, A. A., Alotaibi, F. S., Ahmad, M., Shahzad, A. N., et al. (2022). Inducing drought tolerance in wheat through exopolysaccharide-producing rhizobacteria. *Agronomy, 12*(5), 1140.

Li, D., Zhang, Q., Xiao, K., Wang, Z., & Wang, K. (2018). Divergent responses of biological nitrogen fixation in soil, litter and moss to temperature and moisture in a karst forest, southwest China. *Soil Biology and Biochemistry, 118*, 1–7.

Lin, H. A., Coker, H. R., Howe, J. A., Tfaily, M. M., Nagy, E. M., Antony-Babu, S., et al. (2023). Progressive drought alters the root exudate metabolome and differentially activates metabolic pathways in cotton (Gossypium hirsutum). *Frontiers in Plant Science, 14*, 1244591.

Liu, J., Li, S., Yue, S., Tian, J., Chen, H., Jiang, H., et al. (2021). Soil microbial community and network changes after long-term use of plastic mulch and nitrogen fertilization on semiarid farmland. *Geoderma, 396*, 115086.

Liu, C. Y., Hao, Y., Wu, X. L., Dai, F. J., Abd-Allah, E. F., Wu, Q. S., & Liu, S. R. (2024). Arbuscular mycorrhizal fungi improve drought tolerance of tea plants via modulating root architecture and hormones. *Plant Growth Regulation, 102*(1), 13–22.

Lu, X. L., Ding, K., Dong, X. X., Li, G., & Ma, J. (2023). A bioaugmentation strategy for promoting the humification process during composting by microbial inoculants: A review. *Organic Fertilizers-New Advances and Applications.*

Luo, S. P., He, B. H., Zeng, Q. P., Li, N. J., & Yang, L. (2020). Effects of seasonal variation on soil microbial community structure and enzyme activity in a Masson pine forest in Southwest China. *Journal of Mountain Science, 17*(6), 1398–1409.

Marques, D. M., Magalhães, P. C., Marriel, I. E., Gomes Júnior, C. C., Da Silva, A. B., Da Silva, N. T. A., & De Souza, T. C. (2023). Azospirillum brasilense reduces the effects of water stress and increases maize yield in irrigated areas with high soil nitrogen doses. *Journal of Plant Growth Regulation, 42*(7), 4263–4274.

Martiny, J. B., Martiny, A. C., Weihe, C., Lu, Y., Berlemont, R., Brodie, E. L., et al. (2017). Microbial legacies alter decomposition in response to simulated global change. *The ISME Journal, 11*(2), 490–499.

Meisner, A., Rousk, J., & Baath, E. (2015). Prolonged drought changes the bacterial growth response to rewetting. *Soil Biology and Biochemistry, 88*, 314–322.

Mishra, S., Srivastava, A., Singh, A., Pandey, G. C., & Srivastava, G. (2024). An overview of symbiotic and pathogenic interactions at the fungi-plant interface under environmental constraints. *Frontiers in Fungal Biology, 5*, 1363460.

van der Molen, M. K., et al. (2011). Drought and ecosystem carbon cycling. *Agricultural and Forest Meteorology, 151*, 765–773. https://doi.org/10.1016/j.agrformet.2011.01.018

Moradtalab, N., Hajiboland, R., Aliasgharzad, N., Hartmann, T. E., & Neumann, G. (2019). Silicon and the association with an arbuscular-mycorrhizal fungus (Rhizophagus clarus) mitigate the adverse effects of drought stress on strawberry. *Agronomy, 9*(1), 41.

Muter, O. (2023). Current trends in bioaugmentation tools for bioremediation: A critical review of advances and knowledge gaps. *Microorganisms, 11*(3), 710.

Nader, A. A., Hauka, F. I., Afify, A. H., & El-Sawah, A. M. (2024). Drought-tolerant bacteria and arbuscular mycorrhizal fungi mitigate the detrimental effects of drought stress induced by withholding irrigation at critical growth stages of soybean (Glycine max, L.). *Microorganisms, 12*(6), 1123.

Nassaj-Bokharaei, S., Motesharezedeh, B., Etesami, H., & Motamedi, E. (2021). Effect of hydrogel composite reinforced with natural char nanoparticles on improvement of soil biological properties and the growth of water deficit-stressed tomato plant. *Ecotoxicology and Environmental Safety, 223*, 112576.

Naylor, D., & Coleman-Derr, D. (2018). Drought stress and root-associated bacterial communities. *Frontiers in Plant Science, 8*, 2223.

Noe, L., Ascher, J., Ceccherini, M. T., Abril, A., & Pietramellara, G. (2012). Molecular discrimination of bacteria (organic versus mineral soil layers) of dry woodlands of Argentina. *Journal of Arid Environments, 85*, 18–26.

Nugroho, P. A., Juhos, K., Prettl, N., Madarász, B., & Kotroczo, Z. (2023). Long-term conservation tillage results in a more balanced soil microbiological activity and higher nutrient supply capacity. *International Soil and Water Conservation Research, 11*(3), 528–537.

Nunna, S. A. D., & Balachandar, D. (2024). Rhizobacterial community structure differs between landrace and cultivar of rice under drought conditions. *Current Microbiology, 81*(10), 334.

Nyzhnyk, T., Kots, S., & Pukhtaievych, P. (2024). Rhizobium inoculant and seed-applied fungicide effects improve the drought tolerance of soybean plants as an effective agroecological solution under climate change conditions. *Frontiers in Bioscience-Elite, 16*(3), 23.

Oki, T., & Kanae, S. (2006). Global hydrological cycles and world water resources. *Science, 313*, 1068–1072. https://doi.org/10.1126/science.1128845

Oliva, J., Stenlid, J., & Martinez-Vilalta, J. (2014). The effect of fungal pathogens on the water and carbon economy of trees: Implications for drought-induced mortality. *The New Phytologist, 203*(4), 1028–1035.

Papdi, E., Veres, A., Kovács, F., & Juhos, K. (2022). How different mulch materials regulate soil moisture and microbiological activity? *Journal of Central European Green Innovation, 10*(Suppl 3), 26–38.

Papendick, R. I., & Campbell, G. S. (1981). Theory and measurement of water potential. *Water Potential Relations in Soil Microbiology, 9*, 1–22.

Patra, S. K., Poddar, R., Brestic, M., Acharjee, P. U., Bhattacharya, P., Sengupta, S., et al. (2022). Prospects of hydrogels in agriculture for enhancing crop and water productivity under water deficit condition. *International Journal of Polymeric Science, 2022*(1), 4914836.

Paul, D. (2013). Osmotic stress adaptations in rhizobacteria. *Journal of Basic Microbiology, 53*(2), 101–110.

Paul, D., & Nair, S. (2008). Stress adaptations in a plant growth promoting rhizobacterium (PGPR) with increasing salinity in the coastal agricultural soils. *Journal of Basic Microbiology, 48*(5), 378–384.

Pereira, J. F., Oliveira, A. L. M., Sartori, D., Yamashita, F., & Mali, S. (2023). Perspectives on the use of biopolymeric matrices as carriers for plant-growth promoting bacteria in agricultural systems. *Microorganisms, 11*(2), 467.

Poppeliers, S. W., Sánchez-Gil, J. J., & de Jonge, R. (2023). Microbes to support plant health: Understanding bioinoculant success in complex conditions. *Current Opinion in Microbiology, 73*, 102286.

Prajapati, N. K., Meena, R., Kumar, P., Chand, K., Kumar, P., & Kumar, S. (2024). Cabbage breeding tools for biotic and abiotic resistance. *Romanian Journal of Horticulture, 5*, 23–32. https://doi.org/10.51258/RJH.2024.03

Purohit, A., Ghosh, S., Chaudhuri, R. K., & Chakraborti, D. (2023). Biological control of Fusarium wilt in legumes. In *Plant stress mitigators* (pp. 435–454). Academic.

Purvis, J. E., Yomano, L. P., & Ingram, L. O. (2005). Enhanced trehalose production improves growth of *Escherichia coli* under osmotic stress. *Applied and Environmental Microbiology, 71*(7), 3761–3769.

Quazi, S. A. J., Shozib, H. B., Khaton, A., & Zaidi, N. W. (2024). Role of *Trichoderma asperelloides* and *Trichoderma brevicompactum* in improving drought tolerance in rice. *Plant Stress, 12*, 100457.

Quintarelli, V., Radicetti, E., Allevato, E., Stazi, S. R., Haider, G., Abideen, Z., et al. (2022). Cover crops for sustainable cropping systems: A review. *Agriculture, 12*(12), 2076.

Quiroga, G., Erice, G., Aroca, R., Chaumont, F., & Ruiz-Lozano, J. M. (2017). Enhanced drought stress tolerance by the arbuscular mycorrhizal symbiosis in a drought-sensitive maize cultivar is related to a broader and differential regulation of host plant aquaporins than in a drought-tolerant cultivar. *Frontiers in Plant Science, 8*, 1056.

Rai, A., Irulappan, V., & Senthil-Kumar, M. (2022). Dry root rot of chickpea: A disease favored by drought. *Plant Disease, 106*(2), 346–356.

Riaz, U., Murtaza, G., Anum, W., Samreen, T., Sarfraz, M., & Nazir, M. Z. (2021). Plant growth-promoting rhizobacteria (PGPR) as biofertilizers and biopesticides. In *Microbiota and biofertilizers: A sustainable continuum for plant and soil health* (pp. 181–196).

Rokins, P. G., Gomathi, V., & Mary, J. K. (2022). Plant growth-promoting rhizobacteria mediated moisture stress alleviation in the early stages of Rice (*Oryza sativa* L.) variety CO 51. *Journal of Applied & Natural Science, 14*(4).

Rudrappa, T., Czymmek, K. J., Paré, P. W., & Bais, H. P. (2008). Root-secreted malic acid recruits beneficial soil bacteria. *Plant Physiology, 148*(3), 1547–1556.

Ruiz-Sánchez, M., Aroca, R., Muñoz, Y., Polón, R., & Ruiz-Lozano, J. M. (2010). The arbuscular mycorrhizal symbiosis enhances the photosynthetic efficiency and the antioxidative response of rice plants subjected to drought stress. *Journal of Plant Physiology, 167*(11), 862–869.

Sagar, R., Hagavane, R. D., & Chaudhari Zanje, H. B. (2014). Influence of soil moisture regimes and sources of organic manures on dehydrogenase activities in inceptisol. *World Journal of Agricultural Sciences, 10*(3), 81–84.

Sandhya, V. S. K. Z., Ali, S. Z., Grover, M., Reddy, G., & Venkateswarlu, B. (2010). Effect of plant growth promoting Pseudomonas spp. on compatible solutes, antioxidant status and plant growth of maize under drought stress. *Plant Growth Regulation, 62*, 21–30.

Sani, M. N. H., & Yong, J. W. (2021). Harnessing synergistic biostimulatory processes: A plausible approach for enhanced crop growth and resilience in organic farming. *Biology, 11*(1), 41.

Sanko-Sawczenko, I., Łotocka, B., Mielecki, J., Rekosz-Burlaga, H., & Czarnocka, W. (2019). Transcriptomic changes in *Medicago truncatula* and Lotus japonicus root nodules during drought stress. *International Journal of Molecular Sciences, 20*(5), 1204.

Seneviratne, S. I., Corti, T., Davin, E. L., Hirschi, M., Jaeger, E. B., Lehner, I., Orlowsky, B., & Teuling, A. J. (2010). Investigating soil moisture–climate interactions in a changing climate: A review. *Earth-Science Reviews, 99*(3–4), 125–161. https://doi.org/10.1016/j.earscirev.2010.02.004

Setlow, P. (2014). Germination of spores of Bacillus species: What we know and do not know. *Journal of Bacteriology, 196*(7), 1297–1305.

Shaffique, S., Khan, M. A., Imran, M., Kang, S. M., Park, Y. S., Wani, S. H., & Lee, I. J. (2022). Research progress in the field of microbial mitigation of drought stress in plants. *Frontiers in Plant Science, 13*, 870626.

Shan, H., Wen, H., Zhang, J., Wang, Y., Lu, L., Liu, Y., et al. (2025). Synergistic role of streptomyces composite inoculants in mitigating wheat drought stress under field conditions. *Plants, 14*(3), 366.

Sharath, S., Triveni, S., Nagaraju, Y., Latha, P. C., & Vidyasagar, B. (2021). The role of phyllosphere bacteria in improving cotton growth and yield under drought conditions. *Frontiers in Agronomy, 3*, 680466.

Sharma, A., Usha, Gupta, S., Chaubey, K. K., & Singh, S. V. (2022). An overview of microbial-mediated alleviation of abiotic stress response in plant. In *Plant stress mitigators: Action and application* (pp. 581–596).

Sheik, C. S., Beasley, W. H., Elshahed, M. S., Zhou, X., Luo, Y., & Krumholz, L. R. (2011). Effect of warming and drought on grassland microbial communities. *The ISME Journal, 5*(10), 1692–1700.

Singh, N., Chattopadhyay, D., & Gupta, S. K. (2023). Updating the impact of drought on root exudation: A strigolactones perspective. *Journal of Plant Growth Regulation, 42*(8), 5131–5151.

Somers, E., Vanderleyden, J., & Srinivasan, M. (2004). Rhizosphere bacterial signalling: A love parade beneath our feet. *Critical Reviews in Microbiology, 30*(4), 205–240.

Srivastava, R. K. (2025). Conservation tillage practices on GHG emissions, soil health and overall agricultural sustainability. *Soil Use and Management, 41*(2), e70096.

Steinweg, J. M., Dukes, J. S., Paul, E. A., & Wallenstein, M. D. (2013). Microbial responses to multi-factor climate change: Effects on soil enzymes. *Frontiers in Microbiology, 4*, 146.

Sun, W. E. N. L. I., Shahrajabian, M. H., & Cheng, Q. I. (2021). Nitrogen fixation and diazotrophs—A review. *Romanian Biotechnology Letters, 26*(4), 2834–2845.

Tahir, M., Khalid, U., Khan, M. B., Shahid, M., Ahmad, I., Akram, M., et al. (2019). Auxin and 1-aminocyclopropane-1-carboxylate deaminase activity exhibiting rhizobacteria improved maize quality and productivity under drought conditions

Taiwo, M. O., & Akintokun, A. K. (2025). Plant growth-promoting bacteria delayed wilting and improved tomato yield when grown under water stress condition. *Journal of Plant Nutrition*, 1–19.

Tarolli, P., Luo, J., Park, E., Barcaccia, G., & Masin, R. (2024). Soil salinization in agriculture: Mitigation and adaptation strategies combining nature-based solutions and bioengineering. *iScience, 27*(2), 108830. https://doi.org/10.1016/j.isci.2024.108830

Teuling, A. J., et al. (2013). Evapotranspiration amplifies European summer drought. *Geophysical Research Letters, 40*, 2071–2075. https://doi.org/10.1002/grl.50495

Thapa, V. R., Ghimire, R., Acosta-Martínez, V., Marsalis, M. A., & Schipanski, M. E. (2021). Cover crop biomass and species composition affect soil microbial community structure and enzyme activities in semiarid cropping systems. *Applied Soil Ecology, 157*, 103735.

Tiemann, L. K., & Billings, S. A. (2011). Changes in variability of soil moisture alter microbial community C and N resource use. *Soil Biology and Biochemistry, 43*(9), 1837–1847.

Tripathi, A. N., Meena, B. R., Pandey, K. K., & Singh, J. (2020). Microbial bioagents in agriculture: Current status and prospects. In *New frontiers in stress management for durable agriculture* (pp. 331–368).

Vasquez, D. F., Hernandez, A., Torres, D., Borrero-Echeverry, F., Zuluaga, P., & Rincon, D. F. (2022). Drought as a modulator of plant–virus–vector interactions: Effects on symptom expression, plant immunity and vector behaviour. *Plant Pathology, 71*(6), 1282–1292.

Verma, S., Pradhan, S. S., Singh, A., & Kushuwaha, M. (2024). Effect of organic manure on different soil properties: A review. *International Journal of Plant & Soil Science, 36*(5), 182–187.

Vocciante, M., Grifoni, M., Fusini, D., Petruzzelli, G., & Franchi, E. (2022). The role of plant growth-promoting rhizobacteria (PGPR) in mitigating plant's environmental stresses. *Applied Sciences, 12*(3), 1231.

Vu, E., Schaumann, G. E., & Buchmann, C. (2022). The contribution of microbial activity to soil–water interactions and soil microstructural stability of a silty loam soil under moisture dynamics. *Geoderma, 417*, 115822.

Wahab, A., Muhammad, M., Munir, A., Abdi, G., Zaman, W., Ayaz, A., et al. (2023). Role of arbuscular mycorrhizal fungi in regulating growth, enhancing productivity, and potentially influencing ecosystems under abiotic and biotic stresses. *Plants, 12*(17), 3102.

Wakelin, S. A., Gomez-Gallego, M., Jones, E., Smaill, S., Lear, G., & Lambie, S. (2018). Climate change induced drought impacts on plant diseases in New Zealand. *Australasian Plant Pathology, 47*, 101–114.

Wang, Y., He, C., Meng, P., Sun, S., Gao, J., Zhang, J., & Cai, J. (2024). Effects of different moisture conditions on biological nitrogen fixation capacity and nutritive value of alfalfa (*Medicago sativa* L.) seedlings. *Journal of Soil Science and Plant Nutrition*, 1–13.

Williams, A., & de Vries, F. T. (2020). Plant root exudation under drought: Implications for ecosystem functioning. *The New Phytologist, 225*(5), 1899–1905.

Xu, Z., Yu, G., Zhang, X., Ge, J., He, N., Wang, Q., & Wang, D. (2015). The variations in soil microbial communities, enzyme activities and their relationships with soil organic matter decomposition along the northern slope of Changbai Mountain. *Applied Soil Ecology, 86*, 19–29.

Xue, W., Yang, S., Liu, X., Qian, M., Wang, H., Yang, H., et al. (2024). Enhanced sweet sorghum growth and soil quality in coastal saline–alkali soils through organic acid-containing bio-based materials and microbial synergy. *Agronomy, 15*(1), 56.

Yaadesh, S., Tomar, G. S., Kaushik, R., Prasanna, R., & Grover, M. (2023). A zospirillum–Bacillus associations: Synergistic effects on in vitro PGP traits and growth of pearl millet at early seedling stage under limited moisture conditions. *3 Biotech, 13*(3), 90.

Yadav, V. K., Yadav, R. C., Choudhary, P., Sharma, S. K., & Bhagat, N. (2022). Mitigation of drought stress in wheat (Triticum aestivum L.) by inoculation of drought tolerant Bacillus paramycoides DT-85 and Bacillus paranthracis DT-97. *Journal Of Applied Biology & Biotechnology, 10*(1), 59–69.

Yan, Z., Liu, C., Todd-Brown, K. E., Liu, Y., Bond-Lamberty, B., & Bailey, V. L. (2016). Pore-scale investigation on the response of heterotrophic respiration to moisture conditions in heterogeneous soils. *Biogeochemistry, 131*, 121–134.

Yang, B., Wen, H., Wang, S., Zhang, J., Wang, Y., Zhang, T., et al. (2024). Enhancing drought resistance and yield of wheat through inoculation with Streptomyces pactum Act12 in drought field environments. *Agronomy, 14*(4), 692.

Yeremko, L., Czopek, K., Staniak, M., Marenych, M., & Hanhur, V. (2025). Role of environmental factors in legume-rhizobium symbiosis: A review. *Biomolecules, 15*(1), 118.

Vicente-Serrano, S. M., Beguería, S., & López-Moreno, J. I. (2010). A multiscalar drought index sensitive to global warming: The standardized precipitation evapotranspiration index. *Journal of Climate, 23*(7), 1696–1718. https://doi.org/10.1175/2009JCLI2909.1

Yooyongwech, S., Samphumphuang, T., Tisarum, R., Theerawitaya, C., & Cha-Um, S. (2016). Arbuscular mycorrhizal fungi (AMF) improved water deficit tolerance in two different sweet potato genotypes involves osmotic adjustments via soluble sugar and free proline. *Scientia Horticulturae, 198*, 107–117.

Zhao, J., Xie, X., Jiang, Y., Li, J., Fu, Q., Qiu, Y., et al. (2024). Effects of simulated warming on soil microbial community diversity and composition across diverse ecosystems. *Science of the Total Environment, 911*, 168793.

Microbial Response to Shift in Nutrient Availability

Pragati Srivastava, Samiksha Joshi, Neetika Naudiyal, and Bharti Kukreti

Abstract

For ecosystems to remain stable, functional, and resilient, microbial reactions to changes in nutrient supply are crucial. A mix of ecological, evolutionary, and physiological variables influences these reactions. It is essential to comprehend these dynamics in order to manage ecosystems, particularly when considering pollution, land-use patterns, and environmental change. The proper propagation of microorganisms in the lab and their optimal growth in their natural habitats depend primarily on nutrients, which are essential for microbial growth. Nutrients that provide energy, carbon, and other essential components are needed. Depending on the cellular and metabolic processes of the organism, different nutrients are used to promote growth. All living organisms need the same nutrients, which comprise carbon, nitrogen, sulfur, phosphorus, potassium, magnesium, calcium, oxygen, iron, and other trace elements. Essential nutrients are those that an organism needs in order to function. The availability of specific nutrients dictates organismal growth by controlling and limiting activation of cellular and metabolic pathways necessary for progress. In industrial

P. Srivastava (✉)
Department of Molecular biology and Biotechnology, ICAR-National Institute for Plant Biotechnology, IARI, PUSA, New Delhi, India

S. Joshi
School of Biosciences, Swami Rama Himalayan University, Dehradun, Uttarakhand, India

N. Naudiyal
Department of Biochemistry, CBSH, G. B. Pant University of Agriculture and Technology, Pantnagar, Uttarakhand, India

B. Kukreti
Department of Microbiology, College of Basic Sciences and Humanities, Govind Ballabh Pant University of Agriculture and Technology, Pantnagar, Uttarakhand, India

G. Bhandari et al. (eds.), *Microbial Resilience*,
https://doi.org/10.1007/978-981-95-8883-1_7

microbiology, this concept is critical, as microbial growth and production are dictated by proper cellular growth and metabolism. The production of vital components is typically restricted by the availability and concentration of a limiting nutrient. Hence, it is critical to identify the required nutrients and ensure these are given in the culturing of microorganisms. This chapter covers all the aspects referring to nutrient limitation and resource competition, adjustment in microbial metabolic pathways, physiological stress response, microbial community dynamics, microbial adaptation and evolution, and the impact of human activities and environmental changes, referring to pollution, eutrophication, and other agricultural practices.

Keywords

Microorganisms · Nutrient · Ecosystem · Stress response · Microbial growth

1 Introduction

Microbial communities adapt through physiological, metabolic, and genetic processes in response to changes in the availability of nutrients (Nguyen et al., 2021). Microbial interactions, biogeochemical cycles, and ecosystem functioning can all be impacted by these reactions. Microbes use vital resources from their surroundings for growth and reproduction (Pradhan et al., 2025). For instance, testable predictions of metabolic activity at the whole genome scale are provided by genome-scale metabolic models that employ flux balance analysis. The planet's biogeochemistry, ecosystem function, and macro-organism health are all influenced by microbial populations (Zhu et al., 2023).

Recent advances underscore the complex and dynamic adaptations of microbial communities to fluctuations in nutrient availability, integrating physiological, metabolic, genetic, and ecological perspectives. Microbial communities employ diverse adaptive strategies as nutrient gradients in their environments shift, influencing not only their interactions and structure but also broader ecosystem processes and plant productivity. Emerging evidence points to the crucial roles played by root-microbe and microbe-mineral interactions in modulating plant nutrient acquisition, soil fertility, and resilience to environmental changes. These adaptations frequently involve a suite of microbial processes—ranging from the production of organic acids and siderophores to the formation of symbiotic networks such as those seen in rhizobia-legume or mycorrhizal associations—that regulate nutrient cycling and bioavailability Thus, one of the main challenges facing the domains of ecology, evolution, public health, and biotechnology is comprehending and forecasting the dynamics of microbial populations. Numerous predictions about competition, metabolic interactions, collaboration, and diversification within microbial communities, as well as product creation and its optimization for use in biotechnology, are based on mathematical models of microbial growth and metabolism (Perez-Garcia et al., 2016). Environmental conditions, the makeup of the microbial community, and the type of nutrient change all affect how microorganisms react to changes in nutrient availability. These elements influence how microorganisms interact, compete, and adapt in their environments. These are the main elements influencing microbial reactions:

1. *Nutrient Type*: Different nutrients (carbon, nitrogen, phosphorus, sulfur, etc.) trigger distinct microbial responses. For example, carbon limitation may slow microbial metabolism, while nitrogen limitation may increase nitrogen fixation in some bacteria. Microbial carbon use efficiency (CUE) and nitrogen use efficiency (NUE) are increasingly recognized as pivotal parameters for assessing the fate of soil organic carbon, soil health, and ecosystem sustainability. Recent studies demonstrate that both CUE and NUE are sensitive to shifts in soil nutrient status, moisture, and anthropogenic nutrient enrichment, ultimately affecting greenhouse gas emissions, plant growth, and biogeochemical cycling. For instance, increased nutrient inputs can trigger compositional rearrangements within soil microbial communities, enhance carbon sequestration, or facilitate rapid microbial growth that alters the balance between respiration and biomass production. This underlines the importance of context-specific modeling and experimental validation to inform management practices promoting nutrient cycling efficiency and soil carbon retention, particularly under the stressors of climate change and intensive agriculture.

Carbon Use Efficiency A sudden influx of nutrients (eutrophication) can lead to microbial blooms, while depletion may favor slow-growing, stress-resistant species. The buildup and loss of soil carbon due to microbial growth and respiration is measured by a comprehensive metric called microbial carbon use efficiency (CUE), which is taken into consideration when determining the fate of soil organic carbon (SOC) (Adingo et al., 2021). Because microbial CUE is sensitive to changes in soil nutrients, including phosphorus and nitrogen (N), it is important to evaluate how microbial CUE responds to nutrient inputs brought on by human activity and climate change, as well as how it contributes to the accumulation of SOC. In carbon-limited conditions, microbes rely on alternative metabolic pathways (e.g., fermentation, chemolithotrophy) (Rajpurohit & Eiteman, 2022). Some may degrade recalcitrant carbon sources like lignin, whereas in carbon-rich conditions, Heterotrophic bacteria (e.g., *Pseudomonas* spp.) rapidly increase, leading to CO_2 release and potential hypoxic conditions due to oxygen consumption (Joye et al., 2022).

Nitrogen Use Efficiency

A key component of soil and ecosystem health is nitrogen use efficiency (NUE), which is influenced by microbial dynamics and fluctuating nutrient availability. It describes how well microbes absorb, change, and hold onto nitrogen when exposed to different nutrient sources (Govindasamy et al., 2023). Microbial communities have an impact on soil fertility, greenhouse gas emissions, and plant productivity through their important involvement in the nitrogen cycle (You et al., 2022).

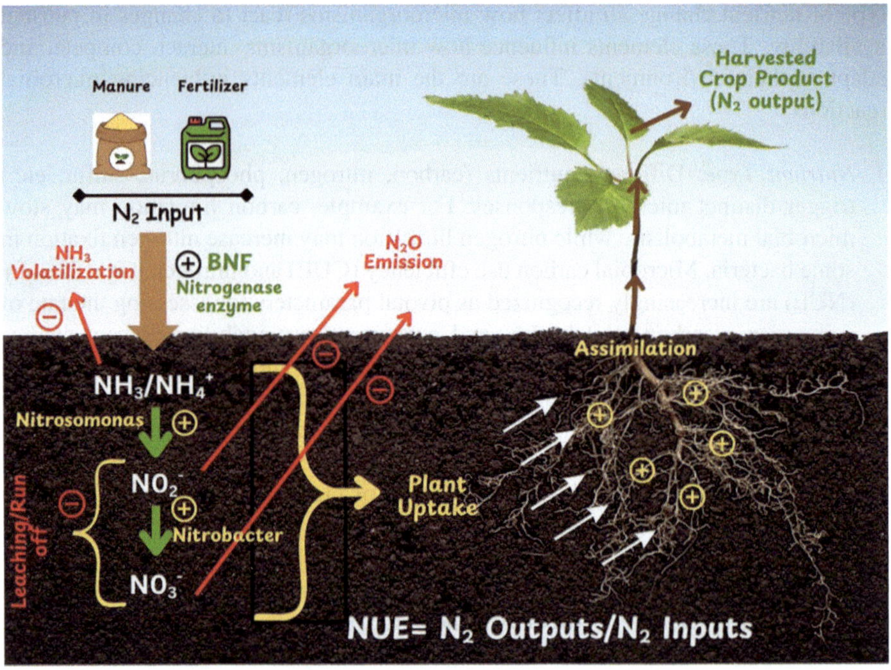

Fig. 1 Visualizing microbial nitrogen use efficiency (NUE) in soil: uptake, retention, and transformation. + and − signs indicate gain and loss of N_2

1. *Microbial Nitrogen Uptake and Retention*

Microbes compete with plants for nitrogen (N), particularly under low N availability. High NUE in microbial communities means they efficiently incorporate available nitrogen into their biomass, reducing nitrogen losses through leaching or gaseous emissions (Fig. 1) (Yadav et al., 2017).

2. *Shifts in Microbial Community Composition*

Under high N availability: Fast-growing, r-strategist microbes (such as certain bacteria) dominate, utilizing excess nitrogen rapidly. Under low N availability, K-strategist microbes (like fungi) become dominant, as they can efficiently recycle nitrogen from organic matter (Preena et al., 2021) (Table 1).

3. *Changes in Microbial Enzyme Activity*

Nitrogen-limited environments lead to increased production of nitrogen-acquiring enzymes (e.g., proteases, chitinases) by microbes to break down organic matter. In nitrogen-rich environments, microbial production of carbon-acquiring enzymes may increase, altering carbon-nitrogen balance (Kunito et al., 2024).

Table 1 Impact of nutrient enrichment on microbial community composition

S. No	Nutrient enrichment type	Example system/ study	Impact on microbial community composition	Citations
1.	Nitrogen (N) addition	Temperate steppe grassland	Promotes copiotrophic bacteria (Proteobacteria, Firmicutes, Bacteroidota); inhibits oligotrophic bacteria (Acidobacteria, Chloroflexi, Verrucomicrobiota)	Zhang et al. (2024)
2.	Phosphorus (P) addition	Temperate steppe grassland	Increases abundance of Acidobacteria and Chloroflexi; decreases Actinobacteria and Proteobacteria; enhances bacterial alpha diversity	Zhang et al. (2024)
3.	Combined N + P addition	Temperate steppe grassland	Alters community structure, increases DOC, and shifts functional groups (e.g., chemoheterotrophy, denitrification)	Zhang et al. (2024)
4.	Nutrient-rich medium	Sugarcane bagasse compost	Reduces microbial diversity; favors generalist taxa; suppresses lignocellulolytic specialists	Mello et al. (2016)
5.	Nutrient-limited medium	Sugarcane bagasse compost	Increases microbial diversity; enriches lignocellulolytic specialists (e.g., Cellulomonas, Paenibacillus)	Mello et al. (2016)
6.	N and C enrichment	Marine lakes	Drives convergence of microbial communities toward heterotrophic networks (e.g., Vibrio, Alteromonas); increases heterotrophy	Wilson et al. (2024)
7.	Long-term fertilization	Oligotroph-dominated soils	Drives convergence of microbial communities toward heterotrophic networks (e.g., Vibrio, Alteromonas); increases heterotrophy	Bledsoe et al. (2020)
8.	Integrated nutrient management	Agricultural soils	Improves soil fertility and microbial biomass; sustains microbial diversity with dominance of Proteobacteria	Datta et al. (2025)

4. *Interactions with Other Biogeochemical Cycles*

Since microorganisms require a balanced carbon-to-nitrogen ratio (C:N) in order to flourish, carbon availability affects microbial NUE (Gaudel et al., 2024). Because microorganisms need phosphorus for ATP synthesis and nucleic acid creation, phosphorus availability also has an impact on NUE (Sun et al., 2023). Microbes cause nitrogen loss by nitrification when there is an excess of nitrogen available, which results in nitrate leaching. When nitrate is converted to gaseous nitrogen by denitrifiers, a powerful greenhouse gas called N_2O is released (Thakur & Medhi, 2019). Certain microorganisms store nitrogen in storage compounds such as polyphosphate granules. Others modify their metabolic processes to make use of different nitrogen sources, including organic nitrogen molecules or amino acids (Racki et al., 2017).

2 Implications for Agriculture and Ecosystem Management

Fertilizer optimization, utilizing and comprehending microbial nitrogen use efficiency (NUE), can help customize nitrogen fertilization to minimize waste and environmental damage (Meng et al., 2022). Additionally, promoting a varied microbial community and improving soil microbial health can increase NUE and resilience to nutrient variations (Fierer & Jackson, 2006; Philippot et al., 2013). Pollution can be decreased by controlling soil properties such as pH and organic matter content, which significantly influence microbial nitrogen absorption and transformations (Kuzyakov & Xu, 2013). Furthermore, microbes have varying capacities to process complex organic molecules compared to simpler inorganic nutrients like ammonium or phosphate (Drinkwater et al., 1998).

3 Microbial Community Composition

Diverse communities may be more resilient to nutrient shifts, as different microbes can exploit various resources (Philippot et al., 2021). If multiple microbes perform similar functions, the ecosystem can maintain stability despite changes in community composition (Allison & Martiny, 2008). Some microbes may rapidly outcompete others when conditions change, leading to shifts in community structure. Environmental condition influences microbial metabolic rates and enzyme activity, impacting how microbes utilize nutrients, and pH affects nutrient solubility and microbial enzyme function, determining which microbes can thrive (Gale, 1943; Nannipieri et al., 2002; Wani et al., 2022). Fenchel and Finlay (2008) stated that moisture and oxygen availability, i.e., aerobic vs. anaerobic conditions, alter microbial metabolic pathways (e.g., fermentation vs. respiration). In aquatic environments, stratification may create nutrient-rich and nutrient-poor layers, affecting microbial distribution (Jørgensen et al., 2022; Reineke & Schlömann, 2023).

4 Microbial Adaptation Mechanisms

Some microbes have the ability to switch metabolic pathways (e.g., facultative anaerobes switching between respiration and fermentation) (Madigan et al., 2021). Gene regulation and horizontal gene transfer (HGT) enable microbes to acquire new traits for nutrient utilization (Frost et al., 2005). In response to nutrient scarcity, some microbes enter dormancy or form spores to survive until conditions improve (Lennon & Jones, 2011). Fast-growing microbes may outcompete others for nutrients, while some produce antimicrobial compounds to suppress competitors (Hibbing et al., 2010). Mycorrhizal fungi enhance plant nutrient uptake, benefiting both partners (Smith & Read, 2008). Nitrogen-fixing bacteria supply plants with nitrogen in exchange for carbon (Oldroyd et al., 2011). Grazing by protozoa or viral infections can regulate microbial populations and influence community structure (Suttle, 2007). Communities previously exposed to nutrient fluctuations may

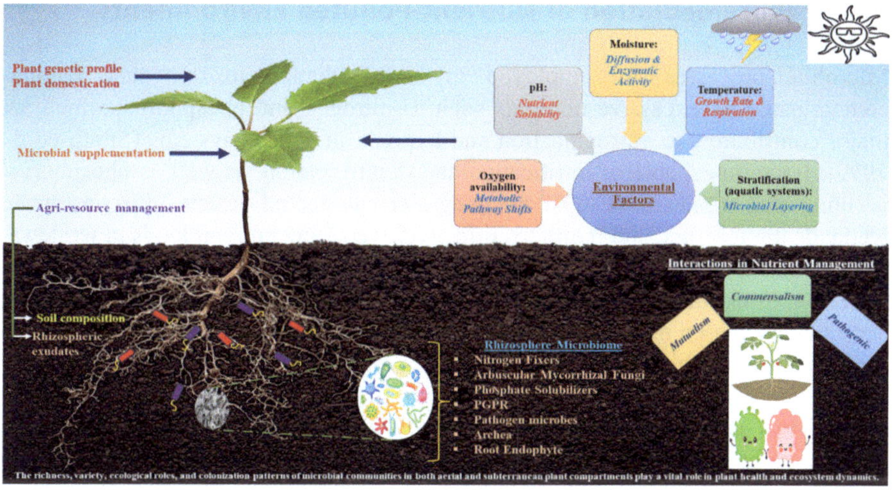

Fig. 2 Multifactorial interactions governing plant-microbiome dynamics and rhizospheric nutrient management (the figure depicts how plant traits, microbial communities, soil composition, and environmental factors interact to regulate nutrient cycling and plant health in the rhizosphere)

develop resilience, allowing them to respond more efficiently to future shifts, exhibiting microbial memory of previous exposure and adaptation (Shade et al., 2012). Environments with frequent nutrient disturbances may harbor microbes adapted to rapid shifts (Allison & Martiny, 2008) (Fig. 2).

5 Technological Applications and Implications

Many practical uses in biotechnology, agriculture, and environmental management depend on how microorganisms react to the availability of nutrients. Microorganisms' physiological, metabolic, and genetic adaptations to shifting nutritional conditions have been used to provide effective and long-lasting solutions for pollution reduction, nutrient recycling, and ecosystem restoration (Kuypers et al., 2018; Bünemann et al., 2018). For example, microbial responses to nutrient gradients have been harnessed in bioremediation strategies, agricultural biofertilizers, and wastewater treatment technologies (Timmusk et al., 2017; Uschner & Klipp, 2019). These microbial traits—such as resource competition, dormancy, metabolic switching, and horizontal gene transfer—are critical for maintaining ecosystem functions under fluctuating environmental conditions (Wu et al., 2019). This section examines important technical fields where vital operations are supported by microbial dynamics.

6 Bioremediation of Nutrient-Polluted Environments

Microbial processes are employed in bioremediation to remediate habitats contaminated by excessive nutrients such as nitrogen and phosphorus, which are major contributors to eutrophication and hypoxia in aquatic systems (Smith et al., 1999). Bacteria capable of nitrification and denitrification, as well as phosphorus-accumulating organisms (PAOs), are integral to engineered systems like constructed wetlands, bioreactors, and biofilters, effectively reducing nutrient loads in wastewater and agricultural run-off. In order to purify and repair habitats affected by nutrient pollution, especially excess nitrogen and phosphorus, bioremediation uses microbial activities. Increased nutrient loads in aquatic environments can cause hypoxia, algal blooms, and eutrophication (Conley et al., 2009). The nitrogen cycle depends on microorganisms like nitrifying bacteria (*Nitrosomonas, Nitrobacter*) and denitrifying bacteria (*Pseudomonas, Paracoccus*), which reduce nitrogen buildup by converting ammonia to nitrate and then nitrogen gas (Kuypers et al., 2018; Daims et al., 2016). Furthermore, in engineered systems, phosphorus-accumulating organisms (PAOs) are used to sequester surplus phosphorus in their biomass, lowering the danger of eutrophication (Uschner & Klipp, 2019). The design and functioning of artificial wetlands, bioreactors, and biofilters that reduce the nutrient load in industrial effluents, municipal wastewater, and agricultural runoff depend heavily on these microbial capacities (García et al., 2010; Vymazal, 2011).

7 Microbial Engineering in Agriculture

A promising tactic to improve crop productivity, soil fertility, and ecological sustainability is microbial engineering. Critical soil functions are supported by the creation and use of microbial inoculants, which are made up of certain strains of helpful bacteria and fungi (Bashan et al., 2014). Legumes, for example, develop symbiotic partnerships with nitrogen-fixing bacteria like *Rhizobium* spp., which transform atmospheric nitrogen into a form that is useful to the organism (Oldroyd & Dixon, 2014). The release of organic acids by phosphate-solubilizing microorganisms increases the bioavailability of insoluble phosphorus compounds (Sharma et al., 2013). Arbuscular mycorrhizal fungi also improve the intake of water, micronutrients, and phosphorus by forming mutualistic relationships with plant roots (Smith & Smith, 2011). In addition to encouraging nutrient uptake, these inoculants strengthen plants' defenses against biotic and abiotic stressors (Timmusk et al., 2017). Therefore, microbial engineering promotes agroecological intensification and lessens dependency on artificial fertilizers (Bender et al., 2016).In agriculture, microbial biotechnology—including the development of biofertilizers and microbial inoculants—improves soil fertility, plant growth, and crop yield while reducing reliance on chemical fertilizers. Biofertilizers comprising nitrogen-fixing bacteria (like Rhizobium) and phosphate-solubilizing microbes increase the bioavailability of key nutrients, promoting sustainable farming practices. Arbuscular mycorrhizal fungi and plant growth-promoting rhizobacteria (PGPR) further enhance nutrient uptake and bolster resistance to biotic and abiotic stresses. Microbial dynamics inform sustainable agricultural approaches such as crop rotation, organic amendments, and precision nutrient management—methods that enhance soil health, minimize chemical input, and foster resilient agroecosystems. Recent advances integrate microbial product application with digital tools, enabling AI-driven, site-specific adjustments to maximize efficiency and sustainability in nutrient management systems.

The creation of sustainable farming methods that include biological concepts in land management has been aided by knowledge of microbial ecology and nutrient cycling (Bünemann et al., 2018). For instance, crop rotation improves nutrient retention in the soil, promotes microbial diversity, and breaks the cycles of pests and diseases (Larkin, 2015). Carbon substrates from the use of organic fertilizers, such as compost, manure, and green manure, promote microbial activity and enhance soil structure (Elfstrand et al., 2007). Additionally, by supporting colonies of nutrient-mineralizing bacteria and decomposers, these amendments help release nutrients in response to plant needs (Schmidt et al., 2018; Abdul et al., 2025). By optimizing fertilizer application time and dose based on microbiological evaluations, nutrient management systems can minimize losses from leaching and volatilization (Bargaz et al., 2018; Elazzazy et al., 2025). When combined, these strategies support the ideas of soil health preservation and regenerative agriculture (LaCanne & Lundgren, 2018).

8 Biotechnology in Wastewater Treatment

Microbial consortia are foundational to advanced wastewater treatment, where they facilitate processes like nitrification, denitrification, and enhanced biological phosphorus removal (EBPR) (Davis et al., 2025). Novel monitoring tools, such as metagenomics and real-time biosensors, now allow for precise assessment and dynamic optimization of microbial community structure and function under varied operational conditions, improving treatment outcomes and system resilience.

9 Implications and Future Trends

The integration of genomics, synthetic biology, and digital agriculture has accelerated the rational design of custom microbial solutions tailored to specific soil, crop, or waste management scenarios. These advances enable precision resource use, reduce environmental pollution, enhance crop performance, and support the shift toward regenerative, circular bioeconomies in agriculture. As microbial technologies become mainstream, they offer scalable, cost-effective approaches for promoting soil health, food security, and environmental sustainability.

Microbial technologies have rapidly advanced, driving innovation in biotechnology, agriculture, and environmental management by leveraging the physiological and metabolic flexibility of microorganisms to address global sustainability challenges. Recent research emphasizes several cutting-edge developments and applications.

10 Bioremediation and Environmental Management

Modern bioremediation exploits microbial consortia and genetic engineering to transform polluted environments by degrading persistent pollutants and recycling nutrients. Recent approaches include immobilized bacterial consortia and microalgae-based systems, which exhibit enhanced efficiency for nutrient and contaminant removal from wastewater. Microbial fuel cells and biofilm reactors now contribute to cost-effective and scalable wastewater treatment, with omics and CRISPR tools enabling optimization of microbial consortia for diverse pollutants. Recent studies also underscore microalgae's capacity to assimilate nitrogen and phosphorus from industrial effluents while producing value-added biomass, further supporting circular bioeconomy models.

11 Microbial Engineering and Sustainable Agriculture

The development and application of microbial inoculants—beneficial bacteria and fungi formulated as biofertilizers—have become transformative for sustainable agriculture. Engineered and native strains are being tailored to promote nutrient uptake, strengthen disease resistance, and boost stress tolerance in crops, reducing dependence on synthetic fertilizers and chemicals. Inoculants not only improve crop yields but also enhance soil structure and fertility; for example, by producing acid radicals and breaking down recalcitrant soil compounds, they make nutrients more bioavailable and restore degraded soils. The use of microbial inoculants is also expanding in livestock nutrition, where they improve animal health and output. The market for agricultural microbial inoculants is expected to grow at a Compound Annual Growth Rate (CAGR) of 7.8% between 2024 and 2029, reflecting increasing adoption and policy support for sustainable practices.

12 Microbial Production of Nutraceuticals and Bioproducts

Microbial biotechnology also powers the industrial-scale production of nutraceuticals and bioactive molecules through fermentation and synthetic biology. Microbial cell factories efficiently convert simple carbon and nitrogen sources into polyunsaturated fatty acids, vitamins, peptides, and antioxidants, offering sustainable alternatives to traditional plant- and animal-derived products. Precision genome editing tools like CRISPR/Cas9 are now widely used for strain development, increasing yield, and diversifying the range of commercially valuable biocompounds.

13 Future Directions and Societal Implications

Integration of multiomics, adaptive evolution, and machine learning is further revolutionizing the precision and scalability of these technologies. International regulations—for example, on microbial inoculant registration—are adapting to promote safe, effective, and globally harmonized deployment of microbial solutions. These advancements collectively support the transition toward regenerative agriculture, eco-efficient waste management, and circular bioeconomies, while contributing to food security and resilience in the face of climate change. The systematic management of microbial consortia that carry out crucial nutrient removal tasks is the foundation of wastewater treatment biotechnology (Rittmann & McCarty, 2012). Autotrophic nitrifiers convert ammonia to nitrate in activated sludge systems, whereas heterotrophic bacteria break down organic materials (Daims et al., 2006). Anaerobic ammonium oxidation (anammox) bacteria and denitrifiers work together to remove nitrogen in anoxic environments (Kartal et al., 2010). Utilizing the metabolic capacities of phosphorus-accumulating organisms (PAOs), which intracellularly absorb and store phosphorus under cyclic aerobic and anaerobic circumstances, enhanced biological phosphorus removal (EBPR) systems work (Izadi et al., 2020). Process optimization requires a comprehensive understanding of microbial nutrient responses due to the temporal variability of influent composition (Chen et al., 2017). Assessing the structure, function, and resilience of microbial communities in dynamic operational situations is becoming more and more common with the use of advanced tools like metagenomics, meta-transcriptomics, and real-time monitoring (Ju & Zhang, 2015; Sun et al., 2019).

References

Abdul, M. A. B., Rahman, M. U., & Ali, M. (2025). Microbial biotechnology for soil health and plant nutrition: Mechanisms and future prospects. *Applied Agriculture Sciences, 3*(1), 1–15.

Adingo, S., Yu, J. R., Xuelu, L., Li, X., Jing, S., & Xiaong, Z. (2021). Variation of soil microbial carbon use efficiency (CUE) and its Influence mechanism in the context of global environmental change: A review. *PeerJ, 9*, e12131.

Allison, S. D., & Martiny, J. B. (2008). Resistance, resilience, and redundancy in microbial communities. *Proceedings of the National Academy of Sciences, 105*(supplement_1), 11512–11519.

Bargaz, A., Lyamlouli, K., Chtouki, M., Zeroual, Y., & Dhiba, D. (2018). Soil microbial resources for improving fertilizers efficiency in an integrated plant nutrient management system. *Frontiers in Microbiology, 9*, 1606.

Bashan, Y., de Bashan, L. E., Prabhu, S. R., & Hernandez, J. P. (2014). Advances in plant growth-promoting bacterial inoculant technology: Formulations and practical perspectives (1998–2013). *Plant and Soil, 378*, 1–33.

Bender, S. F., Wagg, C., & van der Heijden, M. G. A. (2016). An underground revolution: Biodiversity and soil ecological engineering for agricultural sustainability. *Trends in Ecology & Evolution, 31*(6), 440–452.

Bledsoe, R. B., Goodwillie, C., & Peralta, A. L. (2020). Long-term nutrient enrichment of an oligotroph-dominated wetland increases bacterial diversity in bulk soils and plant rhizospheres. *Msphere, 5*(3), 10–1128.

Bünemann, E. K., Bongiorno, G., Bai, Z., Creamer, R. E., De Deyn, G., de Goede, R., et al. (2018). Soil quality—A critical review. *Soil Biology and Biochemistry, 120*, 105–125.

Chen, Y., Lan, S., Wang, L., Dong, S., Zhou, H., Tan, Z., & Li, X. (2017). A review: Driving factors and regulation strategies of microbial community structure and dynamics in wastewater treatment systems. *Chemosphere, 174*, 173–182.

Conley, D. J., Paerl, H. W., Howarth, R. W., Boesch, D. F., Seitzinger, S. P., Havens, K. E., et al. (2009). Controlling eutrophication: Nitrogen and phosphorus. *Science, 323*(5917), 1014–1015.

Daims, H., Taylor, M. W., & Wagner, M. (2006). Wastewater treatment: A model system for microbial ecology. *Trends in Biotechnology, 24*(11), 483–489. https://doi.org/10.1016/j.tibtech.2006.09.002

Daims, H., Lebedeva, E. V., Pjevac, P., Han, P., Herbold, C., Albertsen, M., et al. (2016). Complete nitrification by Nitrospira bacteria. *Nature, 528*(7583), 504–509.

Datta, S., Mazumdar, S. P., Majumdar, B., Alam, N. M., Chattopadhyay, L., Ghosh, S., et al. (2025). Impact of integrated nutrient management on soil microbiome diversity and health in rice based cropping system: Insights from long-term agricultural practices. *Rhizosphere, 33*, 101048.

Davis, B. C., Vikesland, P. J., & Pruden, A. (2025). Evaluating quantitative metagenomics for environmental monitoring of antibiotic resistance and establishing detection limits. *Environmental Science & Technology, 59*(12), 6192–6202.

Drinkwater, L. E., Wagoner, P., & Sarrantonio, M. (1998). Legume-based cropping systems have reduced carbon and nitrogen losses. *Nature, 396*(6708), 262–265.

Elazzazy, A. M., Baeshen, M. N., Alasmi, K. M., Alqurashi, S. I., Desouky, S. E., & Khattab, S. M. (2025). Where biology meets engineering: Scaling up microbial nutraceuticals to bridge nutrition, therapeutics, and global impact. *Microorganisms, 13*(3), 566.

Elfstrand, S., Båth, B., & Mårtensson, A. (2007). Influence of various forms of green manure amendment on soil microbial community composition, enzyme activity and nutrient levels in leek. *Applied Soil Ecology, 36*(1), 70–82.

Fenchel, T., & Finlay, B. (2008). Oxygen and the spatial structure of microbial communities. *Biological Reviews, 83*(4), 553–569.

Fierer, N., & Jackson, R. B. (2006). The diversity and biogeography of soil bacterial communities. *Proceedings of the National Academy of Sciences, 103*(3), 626–631.

Frost, L. S., Leplae, R., Summers, A. O., & Toussaint, A. (2005). Mobile genetic elements: The agents of open-source evolution. *Nature Reviews. Microbiology, 3*(9), 722–732.

Gale, E. F. (1943). Factors influencing the enzymic activities of bacteria. *Bacteriological Reviews, 7*(3), 139–173.

García, J., Rousseau, D. P., Morató, J., Lesage, E., Matamoros, V., & Bayona, J. M. (2010). Contaminant removal processes in subsurface-flow constructed wetlands: A review. *Critical Reviews in Environmental Science and Technology, 40*(7), 561–661.

Gaudel, G., Xing, L., Raseduzzaman, M., Poudel, M., Dong, W., & Hu, C. (2024). Soil microbes, carbon, nitrogen, and the carbon to nitrogen ratio indicate priming effects across terrestrial ecosystems. *Journal of Soils and Sediments, 24*(1), 307–322.

Govindasamy, P., Muthusamy, S. K., Bagavathiannan, M., Mowrer, J., Jagannadham, P. T. K., Maity, A., et al. (2023). Nitrogen use efficiency—A key to enhance crop productivity under a changing climate. *Frontiers in Plant Science, 14*, 1121073.

Hibbing, M. E., Fuqua, C., Parsek, M. R., & Peterson, S. B. (2010). Bacterial competition: Surviving and thriving in the microbial jungle. *Nature Reviews. Microbiology, 8*(1), 15–25.

Izadi, P., Izadi, P., & Eldyasti, A. (2020). Design, operation and technology configurations for enhanced biological phosphorus removal (EBPR) process: A review. *Reviews in Environmental Science and Biotechnology, 19*, 561–593.

Jørgensen, B. B., Wenzhöfer, F., Egger, M., & Glud, R. N. (2022). Sediment oxygen consumption: Role in the global marine carbon cycle. *Earth-Science Reviews, 228*, 103987.

Joye, S. B., Bowles, M. W., & Ziervogel, K. (2022). Marine biogeochemical cycles. *The Marine Microbiome*, 623–671.

Ju, F., & Zhang, T. (2015). Bacterial assembly and temporal dynamics in activated sludge of a full-scale municipal wastewater treatment plant. *The ISME Journal, 9*(3), 683–695. https://doi.org/10.1038/ismej.2014.162

Kartal, B., Kuenen, J. G., & van Loosdrecht, M. C. M. (2010). Sewage treatment with anammox. *Science, 328*(5979), 702–703. https://doi.org/10.1126/science.1185941

Kunito, T., Moro, H., Mise, K., Sawada, K., Otsuka, S., Nagaoka, K., & Fujita, K. (2024). Ecoenzymatic stoichiometry as a temporally integrated indicator of nutrient availability in soils. *Soil Science & Plant Nutrition, 70*(4), 246–269.

Kuypers, M. M. M., Marchant, H. K., & Kartal, B. (2018). The microbial nitrogen-cycling network. *Nature Reviews. Microbiology, 16*(5), 263–276.

Kuzyakov, Y., & Xu, X. (2013). Competition between roots and microorganisms for nitrogen: Mechanisms and ecological relevance. *The New Phytologist, 198*(3), 656–669.

LaCanne, C. E., & Lundgren, J. G. (2018). Regenerative agriculture: Merging farming and natural resource conservation profitably. *PeerJ, 6*, e4428.

Larkin, R. P. (2015). Soil health paradigms and implications for disease management. *Annual Review of Phytopathology, 53*, 199–221.

Lennon, J. T., & Jones, S. E. (2011). Microbial seed banks: The ecological and evolutionary implications of dormancy. *Nature Reviews. Microbiology, 9*(2), 119–130.

Madigan, M. T., Bender, K. S., Buckley, D. H., Sattley, W. M., & Stahl, D. A. (2021). *Brock biology of microorganisms* (16th ed.). Pearson.

Mello, B., Alessi, A., McQueen-Mason, S., et al. (2016). Nutrient availability shapes the microbial community structure in sugarcane bagasse compost-derived consortia. *Scientific Reports, 6*, 38781. https://doi.org/10.1038/srep38781

Meng, C., Liu, H., Li, Y., Shen, J., Li, X., Deng, Y., et al. (2022). Influence path identification of topography, soil, hydrology and landscape on phosphorus buffering capacity in typical agricultural catchments in central subtropical China. *Journal of Environmental Management, 315*, 115164.

Nannipieri, P., Kandeler, E., & Ruggiero, P. (2002). Enzyme activities and microbiological and biochemical processes in soil. *Enzymes in the Environment, 30*, 1–33.

Nguyen, J., Lara-Gutiérrez, J., & Stocker, R. (2021). Environmental fluctuations and their effects on microbial communities, populations and individuals. *FEMS Microbiology Reviews, 45*(4), fuaa068.

Oldroyd, G. E. D., & Dixon, R. (2014). Biotechnological solutions to the nitrogen problem. *Current Opinion in Biotechnology, 26*, 19–24.

Oldroyd, G. E. D., Murray, J. D., Poole, P. S., & Downie, J. A. (2011). The rules of engagement in the legume-rhizobial symbiosis. *Annual Review of Genetics, 45*, 119–144.

Perez-Garcia, O., Lear, G., & Singhal, N. (2016). Metabolic network modeling of microbial interactions in natural and engineered environmental systems. *Frontiers in Microbiology, 7*, 673.

Philippot, L., Raaijmakers, J. M., Lemanceau, P., & van der Putten, W. H. (2013). Going back to the roots: The microbial ecology of the rhizosphere. *Nature Reviews. Microbiology, 11*(11), 789–799.

Philippot, L., Griffiths, B. S., & Langenheder, S. (2021). Microbial community resilience across ecosystems and multiple disturbances. *Microbiology and Molecular Biology Reviews, 85*(2), 10–1128.

Pradhan, N., Singh, S., Saxena, G., Pradhan, N., Koul, M., Kharkwal, A. C., & Sayyed, R. (2025). A review on microbe-mineral transformations and their impact on plant growth. *Frontiers in Microbiology, 16*, 1549022. https://doi.org/10.3389/fmicb.2025.1549022. PMID: 40822401; PMCID: PMC12350396.

Preena, P. G., Rejish Kumar, V. J., & Singh, I. S. B. (2021). Nitrification and denitrification in recirculating aquaculture systems: The processes and players. *Reviews in Aquaculture, 13*(4), 2053–2075.

Racki, L. R., Tocheva, E. I., Dieterle, M. G., Sullivan, M. C., Jensen, G. J., & Newman, D. K. (2017). Polyphosphate granule biogenesis is temporally and functionally tied to cell cycle exit during starvation in *Pseudomonas aeruginosa*. *Proceedings of the National Academy of Sciences, 114*(12), E2440–E2449.

Rajpurohit, H., & Eiteman, M. A. (2022). Nutrient-limited operational strategies for the microbial production of biochemicals. *Microorganisms, 10*(11), 2226.

Reineke, W., & Schlömann, M. (2023). Microorganisms at different sites: Living conditions and adaptation strategies. In *Environmental microbiology* (pp. 349–396). Springer.

Rittmann, B. E., & McCarty, P. L. (2012). *Environmental biotechnology: Principles and applications* (2nd ed.). McGraw-Hill Education.

Schmidt, R., Gravuer, K., Bossange, A. V., Mitchell, J., & Scow, K. (2018). Long-term use of cover crops and no-till shift soil microbial community life strategies in agricultural soil. *PLoS ONE, 13*(2), e0192953.

Shade, A., Peter, H., Allison, S. D., Baho, D. L., Berga, M., Bürgmann, H., et al. (2012). Fundamentals of microbial community resistance and resilience. *Frontiers in Microbiology, 3*, 417.

Sharma, S. B., Sayyed, R. Z., Trivedi, M. H., & Gobi, T. A. (2013). Phosphate solubilizing microbes: Sustainable approach for managing phosphorus deficiency in agricultural soils. *Springerplus, 2*, 587.

Smith, S. E., & Read, D. J. (2008). *Mycorrhizal symbiosis* (3rd ed.). Academic.

Smith, S. E., & Smith, F. A. (2011). Roles of arbuscular mycorrhizas in plant nutrition and growth: New paradigms from cellular to ecosystem scales. *Annual Review of Plant Biology, 62*, 227–250.

Smith, V. H., Tilman, G. D., & Nekola, J. C. (1999). Eutrophication: Impacts of excess nutrient inputs on freshwater, marine, and terrestrial ecosystems. *Environmental Pollution, 100*(1–3), 179–196.

Sun, H., Narihiro, T., Ma, X., Zhang, X. X., Ren, H., & Ye, L. (2019). Diverse aromatic-degrading bacteria present in a highly enriched autotrophic nitrifying sludge. *Science of the Total Environment, 666*, 245–251.

Sun, L., Li, J., Qu, L., Wang, X., Sang, C., Wang, J., et al. (2023). Phosphorus limitation reduces microbial nitrogen use efficiency by increasing extracellular enzyme investments. *Geoderma, 432*, 116416.

Suttle, C. A. (2007). Marine viruses—Major players in the global ecosystem. *Nature Reviews. Microbiology, 5*(10), 801–812.

Thakur, I. S., & Medhi, K. (2019). Nitrification and denitrification processes for mitigation of nitrous oxide from waste water treatment plants for biovalorization: Challenges and opportunities. *Bioresource Technology, 282*, 502–513.

Timmusk, S., Behers, L., Muthoni, J., Muraya, A., & Aronsson, A. C. (2017). Perspectives and challenges of microbial application for crop improvement. *Frontiers in Plant Science, 8*, 49.

Uschner, F., & Klipp, E. (2019). Signaling pathways in context. *Current Opinion in Biotechnology, 58*, 155–160.

Vymazal, J. (2011). Constructed wetlands for wastewater treatment: Five decades of experience. *Environmental Science & Technology, 45*(1), 61–69.

Wani, A. K., Akhtar, N., Sher, F., Navarrete, A. A., & Américo-Pinheiro, J. H. P. (2022). Microbial adaptation to different environmental conditions: Molecular perspective of evolved genetic and cellular systems. *Archives of Microbiology, 204*(2), 144.

Wilson, J. M., Abboud, S. S., & Beman, J. M. (2024). Effects of experimental nutrient enrichment and eutrophication on microbial community structure and function in "marine lakes". *Elementa: Science of the Anthropocene, 12*(1), 00007.

Wu, L., Ning, D., Zhang, B., Li, Y., Zhang, P., Shan, X., et al. (2019). Global diversity and biogeography of bacterial communities in wastewater treatment plants. *Nature Microbiology, 4*(7), 1183–1195.

Yadav, M. R., Kumar, R., Parihar, C. M., Yadav, R. K., Jat, S. L., Ram, H., et al. (2017). Strategies for improving nitrogen use efficiency: A review. *Agricultural Reviews, 38*(1), 29–40.
You, X., Wang, S., Du, L., Wu, H., & Wei, Y. (2022). Effects of organic fertilization on functional microbial communities associated with greenhouse gas emissions in paddy soils. *Environmental Research, 213*, 113706.
Zhang, H., Jiang, N., Zhang, S., Zhu, X., Wang, H., Xiu, W., et al. (2024). Soil bacterial community composition is altered more by soil nutrient availability than pH following long-term nutrient addition in a temperate steppe. *Frontiers in Microbiology, 15*, 1455891.
Zhu, Y. G., Zhu, D., Rillig, M. C., Yang, Y., Chu, H., Chen, Q. L., et al. (2023). Ecosystem microbiome science. *mLife, 2*(1), 2–10.

Symbiotic Relationships: Microbes and Plants in a Changing Climate

Hem Ch. Joshi, Prabha Pant, Himani Karki, Babita Patni, and Dhanushka Udayanga

Abstract

Symbiotic relationships between microbes and plants are integral to ecosystem functionalities, especially given the effects of climate change. This chapter attempts to investigate this very mystique between the plants and their microbial partners, mycorrhizal fungi, nitrogen-fixing bacteria, and endophytic microbes, under certain varying climatic conditions. These symbiotic relations assist in improving the stress response of the plants due to drought, salinity, temperature, and other environmental perturbations while minimizing carbon losses and maximizing resource uptake. This chapter further examines the phases of climate change effects on microbial adaptation and the developmental kinetics of plant–microbe relationships. Exploiting these kinds of relationships seems quite useful in enhancing agriculture, restoring an ecosystem, and ameliorating climate change. Incorporating the new technologies brought by genomics and biotechnology into these strategies makes it possible to increase the output of plants and ameliorate the environment due to the microbial interactions, as this chapter will illustrate.

H. Ch. Joshi (✉) · P. Pant · H. Karki
Department of Botany, D.S.B. Campus, Kumaun University, Nainital, Uttarakhand, India

B. Patni
High Altitude Plant Physiology Research Centre, Hemvati Nandan Bahuguna Garhwal University, Srinagar, Uttarakhand, India

D. Udayanga
Department of Applied Sciences, Faculty of Humanities and Sciences, Sri Lanka Institute of Information Technology, Malabe, Sri Lanka

G. Bhandari et al. (eds.), *Microbial Resilience*,
https://doi.org/10.1007/978-981-95-8883-1_8

Keywords

Symbiosis · Plant–microbe interaction · Climate change · Mycorrhizal fungi · Nitrogen fixation · Endophytic microbes · Environmental stress resilience · Carbon sequestration · Sustainable agriculture · Ecosystem restoration

1 Introduction

Plants coexist with a diverse array of microorganisms—including bacteria, fungi, archaea, and viruses—distributed across the rhizosphere, endosphere, phyllosphere, and seeds. These microbial partners influence plant growth, nutrient acquisition, development, stress responses, and immunity. Mutualistic symbioses, such as those involving rhizobia in legumes and arbuscular mycorrhizal fungi (AMF) across most plant families, are critical for nitrogen fixation and phosphorus mobilization, respectively (Chen et al., 2024; Kumar et al., 2023). Endophytic bacteria and fungi further contribute to plant resilience by modulating hormonal pathways, producing antimicrobial compounds, and enhancing tolerance to abiotic stress (Ali et al., 2024). Increasingly, plants and their associated microbiomes are being conceptualized as a functional unit—the plant holobiont—where emergent properties like nutrient cycling and stress buffering cannot be explained by host or microbes alone (Ravelo-Ortega et al., 2023).

Rapid climate change, characterized by elevated temperatures, increased frequency of droughts and floods, CO_2 enrichment, and altered precipitation patterns, poses substantial challenges to plant productivity and ecosystem stability (Zhang et al., 2024). These environmental changes directly impact plant–microbe interactions by modifying soil physicochemical conditions, microbial community structures, and plant signaling networks (Singh et al., 2025). Recent studies demonstrate that beneficial microbes mitigate the negative impacts of climate stress by enhancing water-use efficiency, antioxidant responses, root development, and nutrient uptake (Ghosh et al., 2024; Ali et al., 2024). For example, a synthetic microbial consortium developed from drought-tolerant rhizobacteria improved resilience in both Arabidopsis and tomato by sustaining root architecture and increasing soil microbial diversity under water-deficit conditions (Lundberg et al., 2025). Similarly, oak seedlings exposed to combined warming and drought stresses exhibited altered rhizosphere and phyllosphere microbiome assembly, confirming the sensitivity and adaptive potential of plant–microbe networks under climate extremes (Miller et al., 2024). These findings highlight not only the ecological importance of symbiosis but also its potential as a nature-based solution for climate-adaptive agriculture and forestry (Ravelo-Ortega et al., 2023; Kumar et al., 2023).

Alongside this growing ecological appreciation, several emerging research directions are reshaping how symbioses are investigated and harnessed. Synthetic microbial communities (SynComs), for instance, represent a transition from the use of single-strain inoculants to deliberately constructed, functionally complementary microbial consortia designed to enhance nutrient use efficiency, disease suppression, and abiotic stress tolerance (Chen et al., 2024; Ali et al., 2024). Advances in multiomics approaches—including metagenomics, transcriptomics, metabolomics, and high-throughput phenotyping—are enabling detailed mapping of microbiome assembly and function under diverse environmental scenarios, while computational modeling and artificial intelligence are beginning to predict beneficial interactions with increasing accuracy (Zhang et al., 2024). In parallel, plant breeding programs are beginning to consider traits that influence microbiome recruitment, offering the possibility of microbiome-assisted breeding for climate resilience (Singh et al., 2025). These trends collectively signal a shift toward a more mechanistic and application-oriented understanding of plant–microbe partnerships.

Despite these advances, several critical research gaps remain. The context dependency of symbiotic outcomes is a major limitation, as microbial benefits vary depending on soil characteristics, plant genotype, local microbial communities, and the timing and severity of climatic stressors (Miller et al., 2024). Moreover, the stability and persistence of introduced SynComs under field conditions remain unpredictable, raising concerns about reproducibility and scalability (Kumar et al., 2023). Translating laboratory successes into agricultural and ecological systems that are spatially and temporally heterogeneous requires robust, site-specific strategies supported by ecological modeling and long-term field trials (Ravelo-Ortega et al., 2023). Finally, policy frameworks, biosafety assessments, and socioeconomic considerations surrounding the deployment of engineered or selected microbial consortia remain underdeveloped, yet are crucial for responsible and equitable implementation (Zhang et al., 2024). Taken together, current evidence underscores that plant–microbe symbioses are integral to the adaptation of both natural ecosystems and agricultural systems to a changing climate. They represent one of the most promising nature-based solutions for enhancing plant resilience, sustaining productivity, and maintaining ecosystem functions under increasing environmental stress. This chapter synthesizes recent advances in understanding the molecular, physiological, and ecological dimensions of these symbioses, identifies the challenges that must be addressed to translate scientific knowledge into practice, and outlines pathways by which microbial partnerships may contribute to sustainable, climate-resilient ecosystems in the decades ahead.

2 Types of Plant–Microbe Symbiotic Relationships

Plant–microbe symbioses encompass a wide range of associations that differ in specificity, intimacy, and ecological function (Fig. 1). Among the most extensively studied are the rhizobia–legume interactions, mycorrhizal fungi, endophytic microbes, and the partnerships involving cyanobacteria and lichens. Together, these associations form the ecological and evolutionary foundation of terrestrial ecosystems and are central to the sustainability of agricultural productivity under changing environmental conditions.

One of the most iconic examples of mutualism is the rhizobia–legume symbiosis. This interaction involves nitrogen-fixing bacteria, primarily from the genera

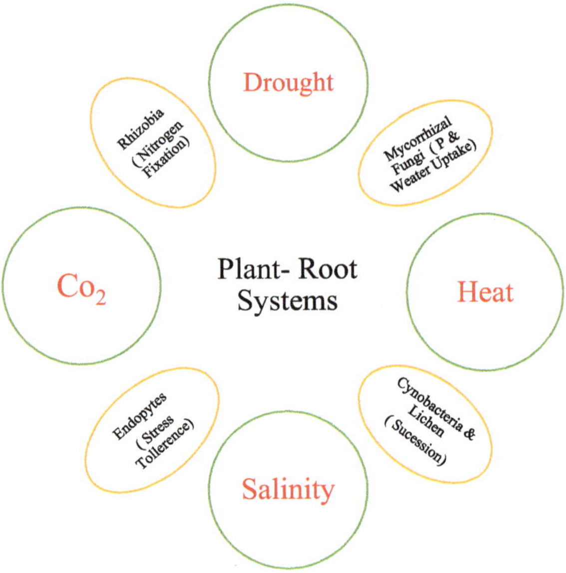

Fig. 1 Plant–microbe symbiosis under climate change

Rhizobium, *Bradyrhizobium*, *Sinorhizobium*, and related taxa, which colonize legume root nodules to convert atmospheric nitrogen (N_2) into plant-available ammonia (NH_3). This biological nitrogen fixation (BNF) reduces dependence on synthetic fertilizers and plays a critical role in global nitrogen cycling (Wang et al., 2024). The symbiosis is tightly regulated by molecular signaling cascades involving flavonoids secreted by the host plant and Nod factors produced by the bacteria, ensuring host specificity and optimal nodule development (Kumar et al., 2023). Beyond nitrogen enrichment, rhizobia also influence soil structure, microbial diversity, and plant stress resilience, making this partnership pivotal for sustainable agriculture and ecosystem functioning (Li et al. 2024a, b).

Another key group of symbiotic partners is mycorrhizal fungi, which form associations with more than 80% of terrestrial plant species. Arbuscular mycorrhizal (AM) fungi, belonging to the phylum Glomeromycota, penetrate root cortical cells to form branched arbuscules, facilitating the exchange of mineral nutrients, particularly phosphorus (P), and micronutrients such as zinc (Zn) and copper (Cu) (Smith & Read, 2023). In contrast, ectomycorrhizal (ECM) fungi form a sheath around the root surface and a Hartig net between epidermal and cortical cells, a configuration particularly common among forest trees, including members of the Pinaceae, Fagaceae, and Betulaceae families (Tedersoo et al., 2023). Both AM and ECM associations improve plant nutrient acquisition, drought and salinity tolerance, and pathogen resistance, while also influencing carbon sequestration and soil structure (Zhang et al., 2024). In the context of climate change, mycorrhizal associations are increasingly recognized for their capacity to mediate plant responses to elevated CO_2 levels and fluctuating precipitation, highlighting their ecological importance in both natural and managed ecosystems (Singh et al., 2025).

Endophytic bacteria and fungi represent another category of hidden allies residing within plant tissues without causing apparent harm. These microbes occupy intercellular spaces or even intracellular niches and have been shown to produce phytohormones, enhance nutrient solubilization, suppress pathogens, and activate plant defense pathways (Ali et al., 2024). Importantly, endophytes often provide protection against abiotic stressors such as heat, drought, salinity, and heavy metal toxicity, mechanisms increasingly valuable under the pressure of global climate change (Miller et al., 2024). For example, *Burkholderia* and *Bacillus* species isolated from desert plants have been shown to enhance drought tolerance in cereals by modulating stomatal conductance and antioxidant activity (Lundberg et al., 2025). Fungal endophytes, including those from the genera *Neotyphodium*, *Trichoderma*, and *Piriformospora*, similarly improve host resilience while influencing root morphology and nutrient uptake (Chen et al., 2024).

Symbiotic relationships also extend beyond the root–soil interface, as seen in partnerships involving cyanobacteria and lichens. Cyanobacteria form associations with a variety of plants, including bryophytes, cycads, ferns (*Azolla*), and angiosperms (*Gunnera*), providing fixed nitrogen and, in some cases, contributing to carbon metabolism (Ravelo-Ortega et al., 2023). Lichens, which represent symbiotic consortia of fungi (mycobionts) and photosynthetic partners (photobionts, typically green algae or cyanobacteria), are among the first colonizers of barren substrates. They play vital roles in primary succession, soil formation, and nutrient cycling in extreme environments (Singh et al., 2025). Moreover, lichens and cyanobacteria contribute to ecosystem stability under stressful conditions, including temperature extremes and desiccation, making them critical components of resilience in fragile ecosystems (Zhang et al., 2024). Collectively, these diverse symbiotic systems illustrate the evolutionary and ecological breadth of plant–microbe partnerships. They are foundational to ecosystem processes such as nutrient cycling, soil stabilization, and plant adaptation, and they represent a largely untapped resource for developing climate-resilient agricultural and forestry practices. Understanding their mechanisms, variability, and potential applications is therefore central to leveraging natural biological systems for sustainable productivity in a changing world.

3 Mechanisms of Symbiotic Adaptation Under Climate Stress

The adaptive potential of plant–microbe symbioses under climate-induced stress is rooted in a network of molecular, physiological, and ecological processes that dynamically respond to environmental cues (Fig. 2). These mechanisms operate at multiple levels, from hormonal signaling in the host plant to biochemical modulation by microbial partners, ensuring that the plant holobiont maintains growth and reproduction despite drought, heat, salinity, and nutrient imbalances.

One of the central layers of this adaptive response is hormonal crosstalk. Symbiotic interactions often reprogram plant hormone pathways to optimize growth and

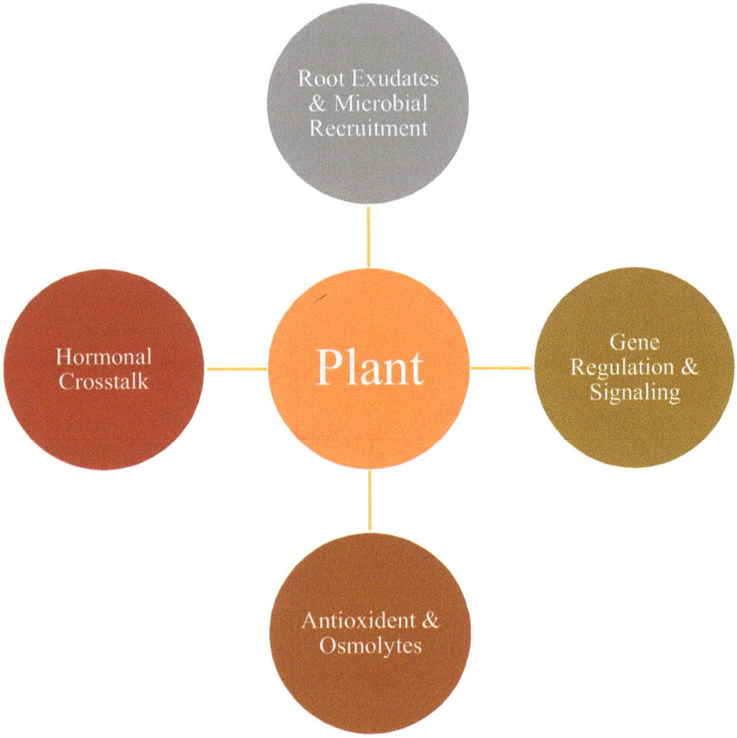

Fig. 2 Microbial mechanism for plant stress tolerance

stress mitigation simultaneously. Ethylene, auxin, jasmonic acid (JA), and salicylic acid (SA) are particularly important in regulating plant immune responses, root architecture, and nodulation efficiency (Ghosh et al., 2024). For example, low ethylene levels promote nodulation, while elevated ethylene in response to stress can inhibit symbiotic signaling unless counterbalanced by microbial production of 1-aminocyclopropane-1-carboxylate (ACC) deaminase, which lowers ethylene accumulation in plant tissues (Kumar et al., 2023). Auxin redistribution mediated by symbiotic microbes enhances lateral root formation, increasing water and nutrient absorption during drought (Ali et al., 2024). Moreover, JA and SA balance immune activation with mutualistic colonization, a process that ensures beneficial microbes are tolerated while pathogens are resisted (Singh et al., 2025).

Root exudates serve as another critical interface for symbiotic adaptation, acting both as signals and nutrient sources that recruit and sustain beneficial microbiota. Under drought and heat stress, plants alter the quantity and composition of root exudates, often increasing the release of osmolytes, amino acids, organic acids, and secondary metabolites that selectively attract drought- or heat-tolerant microbes (Zhang et al., 2024). These shifts in exudation patterns can rapidly restructure the rhizosphere microbiome, favoring taxa capable of producing biofilms, osmoprotectants, or stress-related phytohormones that help maintain root function under extreme conditions (Lundberg et al., 2025). For instance, plants experiencing water deficit have been shown to enrich for Actinobacteria and Firmicutes, which in turn enhance nutrient solubilization and suppress opportunistic pathogens during stress recovery (Miller et al., 2024).

Microbial partners also modulate plant antioxidant and osmolyte systems, which are essential for mitigating oxidative damage induced by abiotic stress. Beneficial microbes often trigger systemic accumulation of antioxidant enzymes such as superoxide dismutase (SOD), catalase (CAT), and peroxidases, reducing reactive oxygen species (ROS) and preserving cellular integrity under high temperature or drought (Ali et al., 2024). Similarly, microbial symbionts stimulate the synthesis of compatible solutes, including proline, glycine betaine, and trehalose, which help maintain osmotic balance and protect proteins and membranes during dehydration and thermal stress (Chen et al., 2024). Such biochemical priming provides a form of "prophylactic" stress tolerance, enabling plants to better endure subsequent episodes of environmental fluctuation.

At the genetic level, symbiotic adaptation involves fine-tuned regulation of host and microbial genes responsible for recognition, colonization, and metabolic exchange. Signal molecules, including Nod factors, mycorrhizal lipochito-oligosaccharides, and volatile organic compounds, activate symbiosis-specific signaling cascades that intersect with abiotic stress pathways (Ravelo-Ortega et al., 2023). Regulatory proteins such as SUPPRESSOR OF MAX2 1 (SMAX1) modulate the delicate balance between strigolactone signaling, symbiosis initiation, and plant growth under nutrient stress (Wang et al., 2024). Furthermore, stress-induced transcription factors and microRNAs can either promote or inhibit symbiotic gene expression, thereby influencing the persistence and effectiveness of mutualistic associations under challenging environmental conditions (Singh et al., 2025). Understanding these regulatory networks is essential for designing strategies to enhance climate resilience through targeted manipulation of plant–microbe interactions. Together, these mechanisms illustrate that symbiotic adaptation to climate stress is not a singular process but a highly integrated and dynamic system. Hormonal signaling, metabolite exchange, biochemical priming, and gene regulation converge to form a flexible framework that enables plants and their microbial partners to survive and thrive in rapidly changing environments. Harnessing these processes through synthetic microbial communities, microbiome-informed breeding, and precision agriculture holds substantial promise for sustaining global food security and ecosystem stability under the pressures of climate change.

4 Climate Change Impacts on Symbiosis

Global climate change exerts multifaceted pressures on plant–microbe symbioses, with cascading consequences for plant health, ecosystem functioning, and agricultural productivity (Table 1). Shifts in temperature regimes, altered precipitation patterns, elevated atmospheric CO_2 concentrations, and progressive changes in soil

Table 1 Types of plant–microbe symbiotic relationships and their roles under climate change

Symbiosis type	Key microbes/ partners	Ecological/ physiological role	Climate-change relevance	References
Rhizobia–legume	*Rhizobium*, *Bradyrhizobium*	Nitrogen fixation, soil fertility improvement	Maintains nitrogen supply in degraded or drought-prone soils	Li et al. (2024b) and Wang et al. (2024)
Mycorrhizal Fungi	AMF (*Glomeromycota*), ECM (*Basidiomycota*, *Ascomycota*)	Enhances P and N uptake, improves water-use efficiency	Increases drought, heat, and salinity tolerance	Tedersoo et al. (2023) and Zhang et al. (2024)
Endophytes	Fungal & bacterial endophytes	Produces secondary metabolites, enhances stress tolerance	Improves heat, drought, and pathogen resistance	Ali et al. (2024) and Singh et al. (2025)
Cyanobacteria & Lichens	*Nostoc*, *Anabaena*, lichenized fungi	Soil stabilization, N and C cycling, pioneer colonization	Supports succession in degraded habitats and extreme climates	Nuñez et al. (2024) and Ravelo-Ortega et al. (2023)

chemistry collectively reshape the composition, diversity, and functionality of microbial communities associated with plants. These changes can either disrupt established mutualisms or favor novel associations, influencing both plant adaptation and ecosystem resilience.

Temperature and precipitation shifts are among the most immediate climate drivers affecting microbial communities. Both warming and altered rainfall regimes influence soil moisture, oxygen availability, and organic matter turnover, parameters that strongly control microbial survival and symbiotic efficacy (Zhang et al., 2024). Elevated temperatures may accelerate microbial metabolic rates but simultaneously reduce diversity by favoring heat-tolerant taxa, sometimes at the expense of beneficial symbionts such as arbuscular mycorrhizal fungi (AMF) or nitrogen-fixing bacteria (Singh et al., 2025). In drought-prone regions, reduced soil water content decreases microbial mobility and symbiotic signaling efficiency, weakening nodulation and mycorrhizal colonization (Kumar et al., 2023). Conversely, flooding events caused by extreme precipitation can lead to hypoxic or anoxic soils, selecting for anaerobic microbial assemblages that are often less effective at supporting plant growth (Miller et al., 2024).

Elevated atmospheric CO_2 modifies plant carbon allocation, indirectly altering plant–microbe interactions. Under higher CO_2 levels, plants often increase photosynthetic activity and root exudation, providing more carbon substrates to the rhizosphere (Ravelo-Ortega et al., 2023). This enhanced carbon flow can stimulate microbial growth and potentially strengthen beneficial interactions, particularly with

AMF and rhizobia, which depend on host-derived carbohydrates (Wang et al., 2024). However, CO_2 enrichment may also shift microbial community composition toward fast-growing, carbon-opportunistic taxa, reducing the relative abundance of slow-growing but ecologically critical symbionts (Ali et al., 2024). The balance between these outcomes is strongly context-dependent, influenced by plant genotype, nutrient availability, and concurrent environmental stressors.

Progressive changes in soil chemistry, driven by altered weathering rates, deposition of atmospheric nitrogen, and management practices responding to climate pressures, further influence symbiotic relationships. Acidification, salinization, and shifts in nutrient stoichiometry can disrupt microbial signaling networks, inhibit nodule formation, or reduce mycorrhizal colonization efficiency (Chen et al., 2024). For example, long-term nitrogen deposition tends to suppress symbiotic nitrogen fixation by reducing the plant's dependence on microbial N acquisition, leading to a decline in rhizobial populations and symbiotic diversity (Li et al. 2024a, b). Similarly, phosphorus imbalances may either stimulate or inhibit mycorrhizal colonization depending on the extent and timing of nutrient enrichment (Tedersoo et al., 2023).

Evidence from long-term ecological studies and herbarium data provides critical insights into these dynamics. Decades of observational data from grasslands, forests, and agricultural systems have documented shifts in symbiotic composition correlating with climate variability (Lundberg et al., 2025). Herbarium-based microbial DNA analyses have revealed historical baselines for plant-associated microbiota, allowing researchers to trace how microbial communities have changed alongside anthropogenic climate trends (Ghosh et al., 2024). For instance, studies comparing century-old specimens with modern counterparts show reductions in nitrogen-fixing bacterial taxa in areas experiencing chronic nitrogen deposition and warming (Miller et al., 2024). Such retrospective perspectives confirm that climate change is not merely altering the abundance of symbionts but is fundamentally reshaping the evolutionary trajectories of plant–microbe partnerships. Taken together, these findings underscore the complexity of climate impacts on plant–microbe symbioses. While certain changes may temporarily enhance mutualistic interactions, others destabilize coevolved networks, increasing the vulnerability of plants and ecosystems to stress. Understanding the environmental thresholds, genetic determinants, and ecological feedbacks that mediate these responses is critical for predicting and managing the future of symbiotic relationships under accelerating global change.

5 Ecological and Evolutionary Perspectives

Plant–microbe symbioses are not only physiological partnerships; they are powerful ecological and evolutionary forces shaping plant communities, ecosystem stability, and adaptive trajectories. As climate change modifies environmental conditions, these symbiotic relationships are increasingly recognized as central drivers of resilience, species distribution, and evolutionary innovation.

Plant–soil feedbacks (PSFs) are a critical ecological mechanism by which plants influence the composition and functionality of soil microbial communities, which in turn affect plant growth and community assembly. In altered ecosystems, PSFs may shift from beneficial to neutral or even negative interactions, depending on disturbance regimes, nutrient cycling, and biotic invasion pressures (Liu et al., 2024). For example, mycorrhizal associations typically enhance nutrient acquisition and growth, yet under high nutrient deposition or soil acidification, their benefits may decline, altering competitive hierarchies and vegetation structure (Bennett et al., 2023). Such changes influence successional trajectories, potentially favoring species with more flexible or opportunistic microbial associations.

Symbiosis also plays a fundamental role in ecological succession and species distribution. Mutualistic microbes often facilitate colonization of stressful or disturbed habitats, aiding primary succession in nutrient-poor soils or postglacial landscapes (Nuñez et al., 2024). Cyanobacteria, lichens, and mycorrhizal fungi contribute to soil development, nutrient availability, and early plant establishment, effectively "engineering" the microhabitats in which subsequent plant and microbial communities evolve. These interactions can influence biogeographic patterns by determining which plant lineages are capable of establishing and persisting under changing climate regimes (Peay et al., 2023).

Evolutionary shifts in symbiotic prevalence provide evidence of how plant–microbe partnerships adapt over time. Grass–endophyte associations, such as those involving *Epichloë* species, illustrate how mutualisms may increase or decrease in frequency depending on environmental pressures, herbivory, and nutrient availability (Bastias et al., 2024). In some cases, formerly rare symbioses become advantageous under new stress regimes, driving rapid changes in host population genetics and microbiome composition. Such evolutionary dynamics highlight the plasticity and coevolutionary potential of plant–microbe partnerships as ecosystems respond to anthropogenic and climatic changes.

At the population and community levels, symbiotic relationships contribute to resilience and adaptation through functional redundancy, diversity buffering, and context-dependent plasticity. Populations with diverse microbial partners may exhibit higher tolerance to drought, heat, or nutrient imbalances, thus stabilizing productivity under variable climates (Kiers & West, 2023). Communities enriched in mutualistic networks are often more resistant to invasion and disease, providing ecosystem-level insurance against environmental perturbation (Liu et al., 2024). However, such buffering capacity is not unlimited; extreme or persistent stressors may exceed the adaptive potential of symbiotic systems, leading to shifts in species dominance and reductions in ecosystem services. Collectively, ecological and evolutionary perspectives underscore that symbiotic relationships are dynamic, context-dependent, and tightly interwoven with broader ecosystem processes. Future research must integrate genomic, ecological, and paleo-ecological approaches to predict how these interactions will evolve, identifying strategies to preserve or restore beneficial symbioses essential for biodiversity and ecosystem functioning in a rapidly changing climate.

6 Harnessing Plant–Microbe Partnerships for Climate-Smart Agriculture

As the agricultural sector faces unprecedented challenges from climate change, harnessing beneficial plant–microbe interactions has emerged as a promising strategy to sustain productivity, soil health, and food security. Microbial symbionts can be leveraged to improve nutrient acquisition, enhance stress tolerance, and reduce dependency on chemical inputs, aligning with global goals for sustainable and climate-resilient farming systems (Table 2).

Microbiome engineering and synthetic microbial communities (SynComs) represent a frontier in agricultural biotechnology. Traditional approaches relied on single-strain inoculants, often with inconsistent performance under field conditions. In contrast, modern strategies use synthetic consortia of beneficial bacteria and fungi designed to perform complementary functions—nutrient solubilization, hormone modulation, pathogen suppression, and stress mitigation—while ensuring stability in complex soil environments (Bai et al., 2024). These SynComs can be tailored to specific crops, soil types, and climatic conditions, offering precision tools for resilience enhancement (Lundberg et al., 2025). Advanced delivery platforms, including seed coatings, encapsulated biofilms, and soil amendments, further improve colonization efficiency and long-term persistence.

Biofertilizers and biostimulants provide another scalable solution for mitigating climate-induced stresses. Nitrogen-fixing bacteria, phosphate-solubilizing microbes, and potassium-mobilizing consortia can reduce reliance on synthetic fertilizers, thereby lowering greenhouse gas emissions associated with chemical production and application (Ali et al., 2024). Biostimulants, such as microbial metabolites, exopolysaccharides, or volatile organic compounds, can prime plant antioxidant

Table 2 Applications of plant–microbe partnerships in climate-smart agriculture

Strategy	Example microbial partners	Target stress	Reported benefits	References
Microbiome engineering (SynComs)	Synthetic consortia of PGPR, AMF, and endophytes	Drought, salinity, heat	Increased yield stability and soil health	Bai et al. (2024) and Lundberg et al. (2025)
Biofertilizers & Biostimulants	Rhizobia, phosphate-solubilizers, potassium-mobilizers	Nutrient deficiency, drought	Reduced fertilizer dependence, improved nutrient cycling	Chen et al. (2024) and Ali et al. (2024)
Microbiome-informed breeding	Host genotypes recruiting beneficial AMF and rhizobia	Heat, water stress	Development of climate-resilient cultivars	Ghosh et al. (2024) and Singh et al. (2025)
Forestry and Restoration	Native ECM fungi, drought-tolerant endophytes	Postfire recovery, reforestation	Improved seedling survival, enhanced carbon sequestration	Peay et al. (2023) and Liu et al. (2024)

systems, osmolyte production, and stress-response pathways, thereby enhancing tolerance to drought, salinity, and heat (Singh et al., 2025). Importantly, combining microbial products with regenerative soil management practices—reduced tillage, organic amendments, and cover cropping—can create synergistic benefits that improve both productivity and ecological resilience.

Integrating microbiome traits into plant breeding programs represents a paradigm shift in crop improvement. Traditional breeding has largely overlooked the role of plant-associated microbiomes, despite their substantial contributions to plant health and productivity. Modern genomic and metagenomic approaches now enable the identification of host genetic loci that influence microbial recruitment and stability, such as genes regulating root exudates and immune signaling (Ghosh et al., 2024). Breeding or engineering plants with enhanced capacities to attract and sustain beneficial microbes could yield cultivars inherently more resilient to climate stresses, reducing input requirements and improving sustainability.

Case studies demonstrate the feasibility and impact of microbial interventions under real-world conditions. In wheat and maize, inoculation with arbuscular mycorrhizal fungi (AMF) and plant growth-promoting rhizobacteria (PGPR) has improved drought and salinity tolerance, leading to yield increases of up to 20–30% under stress conditions (Kumar et al., 2023). Legume–rhizobia inoculants optimized for heat tolerance have sustained nitrogen fixation in soybean during high-temperature events, maintaining protein content and yield stability (Wang et al., 2024). Similarly, heat- and salt-adapted endophytic fungi in rice and barley have been shown to increase photosynthetic efficiency, root development, and grain quality under combined abiotic stresses (Bastias et al., 2024). Collectively, these studies illustrate how tailored microbial partnerships can function as a biological safety net for crops confronting increasingly volatile climates. Future advances will require multidisciplinary integration—combining microbial ecology, crop genetics, soil science, and agronomic innovation—to fully realize the potential of plant–microbe partnerships in climate-smart agriculture. Addressing regulatory frameworks, scalability, and farmer adoption will also be essential to translating laboratory breakthroughs into field-scale resilience.

7 Implications for Forestry and Natural Ecosystems

Plant–microbe symbioses extend beyond agriculture, playing a critical role in sustaining forests and natural ecosystems under accelerating climate change. Trees, shrubs, and herbaceous vegetation rely on complex microbial consortia—mycorrhizal fungi, endophytic bacteria, nitrogen-fixing actinobacteria, and others—that influence nutrient acquisition, water relations, pathogen resistance, and stress tolerance. Leveraging these relationships offers opportunities to enhance reforestation, ecosystem recovery, and climate mitigation strategies.

Microbial-assisted reforestation and postfire recovery have emerged as effective interventions for restoring degraded landscapes. Wildfires, drought, and pest outbreaks have increased in frequency and severity globally, often leaving behind soils depleted of microbial diversity and function (Peay et al., 2023). Inoculating tree seedlings with native mycorrhizal fungi or rhizobacteria accelerates root establishment, nutrient uptake, and competitive ability in stressed environments (Nuñez et al., 2024). Similarly, microbial consortia tailored to specific ecological zones can facilitate the reintroduction of keystone plant species, stabilize soils, and reduce erosion, thereby enhancing ecosystem regeneration trajectories (Kiers & West, 2023). These bio-based strategies complement conventional silvicultural practices by increasing survival rates and ecosystem functional resilience.

Enhancing tree seedling survival under heat, cold, and drought through symbiotic partnerships is a growing focus in forest management. Climate-driven shifts in temperature and precipitation alter seedling physiology and community assembly processes, often reducing establishment success in both managed and wild systems (Bastias et al., 2024). Mycorrhizal fungi—particularly ectomycorrhizal species associated with oaks, pines, and birches—have been shown to improve water-use efficiency, enhance nutrient uptake under stress, and regulate hormonal signaling pathways that mitigate abiotic stress impacts (Liu et al., 2024). Endophytic fungi and bacteria also confer tolerance by modulating antioxidant systems, stabilizing membranes, and facilitating osmolyte accumulation, enabling seedlings to survive otherwise lethal environmental extremes (Bai et al., 2024).

Carbon sequestration and microbial roles in terrestrial carbon stocks are central to climate-change mitigation efforts. Forest ecosystems are among the largest terrestrial carbon sinks, yet their capacity to sequester carbon is influenced by belowground microbial processes. Mycorrhizal fungi mediate carbon allocation from plants to soil, influencing organic matter stabilization and nutrient cycling (Tedersoo et al., 2023). Symbiotic nitrogen fixers contribute to biomass accumulation and ecosystem productivity, indirectly enhancing carbon storage potential (Wang et al., 2024). Moreover, microbial community composition and diversity influence soil respiration rates, determining whether soils act as net carbon sinks or sources under warming scenarios (Zhang et al., 2024). Managing microbial assemblages alongside vegetation composition thus represents a powerful, yet underutilized, lever for increasing forest carbon sequestration and resilience to disturbance. Incorporating microbial ecology into forestry and conservation frameworks offers a pathway toward climate-resilient ecosystem management. By integrating microbiome science with restoration ecology, genetics, and landscape planning, practitioners can design interventions that not only recover biodiversity and productivity but also reinforce long-term ecosystem stability in the face of global change.

8 Conclusion

Plant–microbe symbioses are fundamental to the health, productivity, and resilience of terrestrial ecosystems. In the face of accelerating climate change, these interactions are increasingly recognized not only as biological curiosities but also as practical levers for sustaining biodiversity, agricultural output, and carbon balance. Across scales—from individual plant stress physiology to ecosystem-level carbon and nutrient cycling—microbial partners mediate processes that determine how plants respond to temperature fluctuations, altered precipitation regimes, nutrient imbalances, and biotic pressures.

The synthesis of available evidence underscores that symbiotic relationships serve as buffers against climate stress, facilitating adaptation by enhancing nutrient acquisition, water-use efficiency, and stress signaling modulation. In agriculture, microbiome engineering, biofertilizers, and microbiome-informed breeding programs provide a framework for climate-smart crop systems capable of sustaining yields under increasingly erratic environmental conditions. In forestry and natural ecosystems, microbial-assisted reforestation and targeted inoculation strategies hold promise for accelerating postdisturbance recovery, enhancing seedling survival, and increasing long-term carbon sequestration.

Looking forward, the path toward sustainable, climate-adaptive ecosystems will depend on integrating plant–microbe partnerships into land management, breeding, and conservation practices. This requires multidisciplinary collaboration—linking microbial ecology, genomics, agronomy, forestry, and socio-economic policy—to translate fundamental discoveries into field-ready solutions. Moreover, preserving microbial diversity and understanding the evolutionary trajectories of symbioses are essential for maintaining ecosystem resilience under uncertain climatic futures. Ultimately, harnessing the power of symbiosis offers a biologically grounded, scalable, and environmentally responsible strategy for navigating the profound ecological challenges of the twenty-first century. By strategically aligning natural partnerships with human innovation, it is possible to build resilient agricultural landscapes, restore degraded ecosystems, and reinforce the stability of global carbon and nutrient cycles, ensuring food and ecosystem security for generations to come.

References

Ali, R., Khan, N., & Qureshi, A. (2024). Endophytic microbes as modulators of plant stress tolerance. *Applied Microbiology and Biotechnology, 108*(5), 2257–2273.

Bai, Y., Wang, D., & Wang, Y. (2024). Harnessing synthetic microbial communities for crop resilience. *Nature Plants, 10*, 341–353.

Bastias, D. A., Gundel, P. E., & Martínez-Ghersa, M. A. (2024). Grass–endophyte symbioses: Evolutionary perspectives under climate change. *Trends in Ecology & Evolution, 39*(2), 112–124.

Bennett, J. A., Maherali, H., & Klironomos, J. N. (2023). Plant–soil feedbacks and global change: Current perspectives. *Ecology Letters, 26*(5), 755–774.

Chen, L., Patel, R., Zhao, Y., & Wang, X. (2024). Engineered microbial communities for sustainable agriculture. *Frontiers in Plant Science, 15*, 123456.

Ghosh, A., Patel, P., & Banerjee, S. (2024). Crosstalk between phytohormones and rhizosphere microbiota under abiotic stress. *Plant Cell Reports, 43*(2), 301–315.

Kiers, E. T., & West, S. A. (2023). Mutualisms in a changing world: The rise of context dependency. *Annual Review of Ecology, Evolution, and Systematics, 54*, 145–169.

Kumar, S., et al. (2023). Advances in plant–microbe symbiosis under climate change. *Journal of Integrative Plant Biology, 65*(10), 2035–2050.

Li, R., Khan, N., & Qureshi, A. (2024a). Endophytic microbes as modulators of plant stress tolerance. *Applied Microbiology and Biotechnology, 108*(5), 2257–2273.

Li, X., Sun, J., & He, J. (2024b). Rhizobia–legume interactions: Signaling, efficiency, and agricultural implications. *Plant and Soil, 495*(1–2), 45–66.

Liu, Y., Chen, J., & Ma, K. (2024). Climate-driven shifts in mycorrhizal strategies and forest dynamics. *The New Phytologist, 243*(3), 1215–1231.

Lundberg, D. S., Ramirez, K., & Dangl, J. (2025). Synthetic microbial consortia enhance drought tolerance in crops. *Environmental Microbiome, 30*(1), 55–68.

Miller, J. R., Thompson, P., & Ochoa, J. (2024). Combined warming and drought alter oak seedling microbiomes. *Global Change Biology, 30*(7), e14523.

Nuñez, M. A., Horton, T. R., & Simberloff, D. (2024). Fungal symbioses in primary succession: Implications for restoration. *Ecology, 105*(2), e3902.

Peay, K. G., Fukami, T., & Wardle, D. A. (2023). Fungal networks, plant diversity, and ecosystem resilience. *Annual Review of Ecology, Evolution, and Systematics, 54*, 233–256.

Ravelo-Ortega, J., Martín-Sánchez, L., & Barea, J. M. (2023). The plant holobiont: A key player in sustainable agriculture. *Annual Review of Plant Biology, 74*, 615–641.

Singh, P., Verma, R., & Kaur, A. (2025). Microbiome-assisted breeding for climate resilience. *Trends in Plant Science, 30*(3), 245–258.

Smith, S. E., & Read, D. J. (2023). *Mycorrhizal symbiosis* (4th ed.). Academic.

Tedersoo, L., Bahram, M., & Zobel, M. (2023). Mycorrhizal ecology and global change: Current perspectives and future directions. *Fungal Diversity, 118*, 1–23.

Wang, H., Zhou, Y., & Chen, W. (2024). Biological nitrogen fixation in legumes: Advances and challenges. *Plant Biotechnology Journal, 22*(2), 211–227.

Zhang, Y., Huang, X., & Chen, G. (2024). Climate change and soil microbial networks: Implications for agriculture. *Nature Reviews. Microbiology, 22*(9), 568–584.

Microbial Ecology of Extreme Weather Events

Himani Karki, Prabha Pant, and Hem Ch. Joshi

Abstract

Microbial communities that have the potential to survive in extreme environments are commonly known as the extremophiles. These microbial communities play a crucial role in sustaining the ecosystem. The Extreme Weather Event (EWE) includes geophysical events such as earthquakes and volcanic eruptions; climatological events such as drought; meteorological events such as floods; and hydrological events such as coastal floods, pluvial floods, and sea storms. The frequency of EWE is increasing constantly all around the globe. Extreme weather events can alter the composition and activities of soil microbial communities by altering the physico-chemical properties of soil. The further increase of such EWE not only adversely affects the structure of an ecosystem but also leaves a long-lasting detrimental effect on the ecosystem functioning too. These EWEs are human-induced and impact the whole biotic communities more intensely, which is difficult to quantify. These events are also responsible for the evolutionary adaptations among the microbes to sustain under the extreme environments over the decades. Microbes associated with the extreme environmental conditions also have the capabilities to produce various chemicals and by-products, which are beneficial for a wide range of human well-being. Further investigations on such microbes having the immense potential to survive under the thrilling adverse conditions could be helpful to understand the sustainability or genesis of life on Earth.

Keywords

Soil health · Decomposition · Nutrient cycling · Physico-chemical properties · Ecosystem functioning · Sustainability

H. Karki (✉) · P. Pant · H. Ch. Joshi
Department of Botany, D.S.B. Campus, Kumaun University, Nainital, Uttarakhand, India

G. Bhandari et al. (eds.), *Microbial Resilience*,
https://doi.org/10.1007/978-981-95-8883-1_9

1 Introduction

Certain adaptations have established by extremophiles to endure in unfriendly or extreme environmental conditions, such as “water cages” in halophiles and condensed pore sizes in acidophiles (Coker, 2019). Extremophiles that can subsist in several types of extreme situations are called polyextremophiles (Gupta et al., 2014). Small organic particles, found in extremophilic bacteria that accumulate inside the cells, are known as extremolytes. Extremozymes, isolated from extremophiles, are highly stable at extreme temperatures and pH and are resistant to denaturing agents, detergents, organic solvents, and chaotropic agents (Gupta et al., 2014). Extremophiles produce carotenoids, natural hydrophobic pigments capable of producing colors like bright yellow, red, etc. and can be used in pharmaceutical, cosmetics, food, and textiles industries (Rodrigo-Baños et al., 2015). Extreme weather events are defined as the occurrence of a weather or climate variable 10% above or below a threshold value near the upper or lower end of the range of historically observed records for a particular geographic area by the Intergovernmental Panel on Climate Change (IPCC) (Ummenhofer & Meehl, 2017). Extreme weather events are severe meteorological, hydrological, geophysical, or climatological phenomena that occur worldwide. Geophysical extreme events include earthquakes and tsunamis, as well as volcanic eruptions. Climatological extreme events include extreme temperatures, where high temperatures can lead to drought formation, which can, in turn, result in wildfires. Meteorological extreme events include precipitation extremes, where low precipitation can also lead to drought, and heavy precipitation can result in flooding. Certain extreme events, such as volcanic eruptions, earthquakes, tornadoes, and hurricanes, are only able to occur in certain regions. In contrast, flooding and drought have the ability to strike anywhere in the world. Although both flooding and drought are extreme weather events, they are hydrological opposites, where flooding is defined as complete saturation of the soil by water, which results in standing water, and drought is defined as a lack of water combined with an absence of rainfall for over 15 days. Previous work has shown that flooding and drought/heat events may become more severe or are more likely to occur compared to other extreme weather events due to human-induced climate change (Pidock et al., 2019). This highlights the need to understand the impacts of flooding and drought above other extreme weather events. The IPCC found that both winter and summer flooding may increase in Europe due to significant increases in consecutive rainfall by 2100 (Sillmann et al., 2013). At the same time, Europe may also experience a sevenfold increase in droughts and consecutive droughts by the end of the twenty-first century (Hari et al., 2020). The same is true for flood and drought predictions in Western North America. Drought is predicted to increase by the last half of the twenty-first century, which will result in the driest century on record (Cook et al., 2015).

Soil acts as a unique interface between the lithosphere, the hydrosphere, the atmosphere, and the biosphere and is an irreplaceable natural resource that presides over a number of functions necessary for human life (Bünemann et al., 2018; Lehmann et al., 2020). Soil is also a limited resource, as an inch of topsoil can take 100 years to form due to the slow chemical alteration of weathered rock, coupled with the gradual formation of organic matter, which means that soil is nonrenewable (Amundson, 2021). Approximately 33% of all soils are moderately to highly degraded, and 52% of all agricultural soils have been moderately to severely impacted by soil degradation (ELD Initiative, 2015). The microbial biomass can easily react to changes in nutrients, moisture, temperature, and the type and amount of soil organic matter within a turnover time of less than a year. Thus, soil microbial biomass acts as an indicator of changes in nutrient status, vegetative composition, and climatic conditions (Padaliya et al., 2022; Manral et al., 2023).

Soil organisms include an array of both macrofauna, such as arthropods, earthworms, and nematodes, and microfauna, such as protists, fungi, archaea, and bacteria (Fierer & Jackson, 2006; Wall et al., 2015). It has been estimated that soil biodiversity accounts for approximately 25% of global biodiversity (George et al., 2019).

There are a number of microorganisms that perform specific functions that aid plant growth, and these have been termed plant growth-promoting microbes (PGPMs) (Mishra et al., 2017). PGPMs are mainly comprised of bacteria and fungi, which live within the rhizosphere. The rhizosphere is the fine layer of soil that is directly connected to the roots of the plants and is directly influenced by root exudates (Lugtenberg & Kamilova, 2009). Some of the common examples of hyperthermophilic organisms are *Thermoproteus uzoniensis, Staphylothermus marinus, Pyrodictium abyssi, Pyrococcus furiosus, Hypothermus butylicus, Pyrodictium woesei, Pyrodictium brockii, Pyrodictium occultum*, etc. Extreme weather events, such as hurricanes, floods, droughts, and heatwaves, significantly impact microbial communities in various ecosystems. These events can alter microbial composition, diversity, and functionality, leading to cascading effects on ecosystem processes and biogeochemical cycles (Fig. 1).

The impacts of microbial ecology in extreme weather events are described below.

2 Community Shifts

Rapid change in environmental conditions causes shifts in microbial community structure by favoring certain taxa over others through resource availability, physiological stress, and species interactions. As some microbial populations expand while less competitive ones decline, the overall community composition and function can be dramatically altered.

Changes in temperature: Global warming is a major driver of microbial community shifts, with both short- and long-term effects.

In soil: Warmer temperatures can initially accelerate the decomposition of organic matter by microbes, but this effect is often temporary as the food source becomes depleted. Over the long term, warming favors heat-tolerant microbes over cold-adapted ones, altering soil processes like carbon and nitrogen cycling.

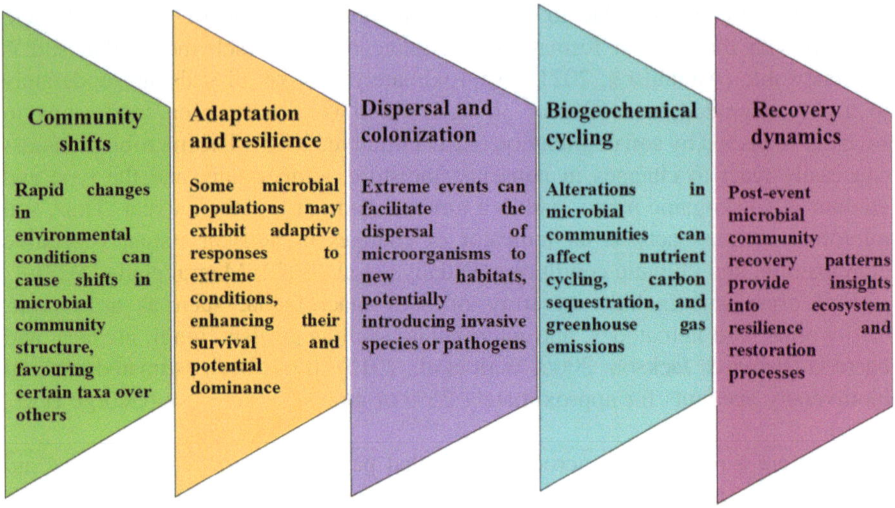

Fig. 1 Impacts of microbial ecology in extreme weather events

In oceans: Rising ocean temperatures have caused heat-loving *Vibrio* bacteria to expand into new geographical regions, increasing the risk of infections. At the same time, warming can increase stratification in the oceans, which limits the flow of nutrients and alters phytoplankton communities.

Change in moisture and rainfall: Altered precipitation patterns, including intensified droughts and floods, strongly influence microbial communities.

Prolonged drought creates unfavorable conditions that reduce the size and diversity of microbial populations. When drought-stricken soils are rewetted, the microbial community composition can shift dramatically toward populations that can better withstand the stress.

In aquatic environments: Increased rainfall and subsequent flooding can overwhelm wastewater systems, introducing excess nutrients, pollutants, and pathogens into rivers and lakes. This causes an influx of waterborne disease agents and also promotes harmful algal blooms.

Shifts in salinity: Rapid changes in salt concentration act as a strong environmental filter, selecting for salt-tolerant microbes.

In saline soils: As salinity increases, the microbial community can shift from being dominated by salt-sensitive bacteria to being dominated by halophilic archaea.

Introduction of pollutants and nutrients: Sudden inputs of nutrients or toxic compounds can disrupt the ecological balance.

Hydrocarbon pollution: In oil-contaminated soil, initial microbial communities are quickly replaced by hydrocarbon-degrading microbes such as *Acinetobacter* and *Gordonia*.

Compost addition: The addition of compost introduces new organisms and organic matter, altering the existing soil microbial network and changing the dominant species.

2.1 Mechanism Driving Community Shifts

Environmental filtering: Abiotic factors, such as temperature, pH, and oxygen availability, directly select for microbes whose traits are best suited to the new conditions. Other taxa whose physiological tolerance limits are exceeded will be outcompeted or become extinct.

Resource availability and competition: Rapid environmental changes alter the quantity and type of resources available to microbes. Organisms that can quickly utilize new, abundant resources (copiotrophs) may outgrow and outcompete those that are adapted to oligotrophic conditions.

3 Adaptation and Resilience

An extreme environment is a habitat characterized by harsh environmental conditions beyond the optimal range for the development of humans or other living organisms. Extreme environments are characterized by various unfavorable conditions, including high or low temperatures, high or low pressures, and acidic or basic pH.

For an area to be considered extreme, certain conditions or aspects of the environment must be very hard for different forms of life to survive. Examples of some extreme environments include the polar region, deserts, volcanic regions, deep ocean trenches, outer space, and every other planet of the Solar System except the Earth. Some of the common extreme environments include areas that are alkaline, acidic, extremely hot or cold, high salt concentration, or without water or oxygen.

3.1 Extreme Environments Are Classified into the Following Groups Based on the Extreme Physicochemical Conditions

Extreme temperature: Two types of extreme environments can be described: cold and hot.

Extremely cold environments are those with environmental temperatures below 5 °C. These can be found in deep ocean niches, at the peaks of high mountains, or in the Polar Regions.

Extremely hot environments are characterized by environmental temperatures higher than 45 °C. These environments are influenced by geothermal activity, such as geysers and fumaroles of continental volcanic areas or deep-sea vents.

Extreme pH: Extreme environments can also be classified as acidic or alkaline according to their pH.

Extreme acidic environments are natural habitats in which the pH is below 5. Extreme alkaline environments are those with a pH above 9.

Extreme ionic strength: Hypersaline environments are environments with an ionic concentration higher than that of seawater (greater than 3.5%).

Extreme pressure environments are those environments under extreme hydrostatic or litho pressure, such as aquatic habitats at depths of 2000 m or more, or deep-subsurface ecosystems.

High-radiation environments are those areas that are exposed to abnormally high radiation doses, including ultraviolet or gamma radiation, such as deserts and the top of high mountains.

Xeric environments are arid habitats with limited water activity. Cold and hot deserts are some examples of these extreme environments.

4 Dispersal and Colonization

Extreme events facilitate the dispersal and colonization of microorganisms to new habitats, potentially introducing invasive species or pathogens through a variety of mechanisms. Climate change intensifies the frequency and severity of many of these extreme events, amplifying the associated risks.

4.1 Floods and Extreme Precipitation

Floods and heavy rainfall are highly effective at moving microbes from one location to another and enabling them to colonize new areas.

Contamination and mixing: Severe storms can overwhelm wastewater treatment systems, causing untreated or partially treated sewage to mix with floodwaters. This disperses human and animal pathogens, as well as antibiotic-resistant genes, into waterways, drinking water supplies, and coastal recreational areas.

After Hurricane Katrina in 2005, standing water provided a breeding ground for mold and other pathogens, leading to an increase in respiratory illness. In Hawaii, outbreaks of leptospirosis, a disease spread by animal urine, have increased following floods caused by typhoons. Droughts and resulting dust storms can mobilize and transport microbes and their genes over vast distances.

Extreme heat can cause microbes to evolve increased thermotolerance, potentially allowing environmental fungi and bacteria to survive and cause disease in humans. Warming ocean temperatures enable cold-intolerant marine Pathogens like *Vibrio* species to expand their geographic range to higher latitudes.

Forced migration: In response to drought, rodent populations may move toward human settlements, increasing the risk of zoonotic disease transmission. Human migrants displaced by disasters may also be exposed to unfamiliar microbes or live in crowded, unsanitary conditions that promote disease spread.

Ecosystem vulnerability: By reducing the resilience of native species, extreme weather events can create opportunities for more robust non-native and invasive species to colonize and take over.

5 Biogeochemical Cycling

Alterations in microbial communities significantly impact biogeochemical cycles, affecting essential processes like nutrient cycling, carbon sequestration, and the emission of greenhouse gases. Microbes are vital for decomposing organic matter, transforming nutrients, and regulating carbon storage in ecosystems. Changes in their composition and activity can either intensify or mitigate climate change feedback loops by altering the balance of carbon and nutrient cycles.

6 Host-Microbe Interactions

Extreme weather events disrupt critical symbiotic relationships between hosts and their microbial partners, impairing essential functions that underpin ecosystem health and productivity. Climate-driven stressors like heat waves, floods, and droughts can alter the balance of these complex relationships, often shifting mutualistic associations toward parasitic ones and causing widespread ecological cascades.

How extreme weather disrupts symbiotic relationships:

Physiological stress: Extreme temperature pushes hosts and symbionts beyond their thermal tolerance, triggering a breakdown of the symbiotic partnership. For example, heat waves can disrupt the delicate balance between corals and the photosynthetic algae (zooxanthellae) living in their tissues, causing the coral to expel the algae in a process called bleaching. The study of Khati et al. (2021) also focuses on the application of nanocompounds to mitigate drought stress, which helps to improve the water use efficiency by plants and microbes.

Altered microenvironments: Changes in precipitation, like heavy rains or severe droughts, fundamentally alter the microenvironments where symbionts live, such as in soil or water. Flooding can create anaerobic conditions that vanish aerobic soil microbes and reduce nutrient uptake in plants. Conversely, drought changes soil chemistry and reduces microbial diversity, stressing plant hosts.

Nutrient and metabolic imbalances: Extreme conditions alter nutrient availability and the metabolic exchange between host and microbe. In insects, elevated temperatures can cause the collapse of heat-susceptible gut bacteria, impacting the host's ability to extract nutrients and sometimes leading to death.

7 Recovery Dynamics

Microbial community recovery dynamics reveal the pathways and mechanisms that drive ecosystem resilience and restoration following a disturbance. Research shows that recovery patterns are highly dependent on the type of disturbance and the ecosystem involved but are often driven by biological factors such as adaptive growth, dormancy, and functional redundancy.

Bioindicators: By tracking the recovery trajectory of microbial communities and their activities, researchers can develop reliable bioindicators to assess the effectiveness of restoration efforts.

Bioremediation: Harnessing the natural ability of microbes to detoxify pollutants is a key restoration technique. Strategies like biostimulation and bioaugmentation can accelerate the recovery of polluted sites.

Regenerative agriculture: Practices like no-till farming and adding compost support soil microbial communities, which in turn enhance soil health and nutrient cycling.

Microbiome rescue: For ecosystems facing severe climate change pressure, "microbiome rescue" techniques can be used to reintroduce lost functional members. This includes leveraging microbial dormancy or dispersal to intentionally direct a community toward a desired resilience trajectory.

Extreme weather zones refer to environments where temperature, pH, salinity, pressure, or radiation levels are beyond the tolerable limits of most life forms (Table 1).

The microbial biomass exhibits a remarkable capacity to respond to fluctuations in nutrient availability, moisture levels, temperature variations, and the nature and quantity of soil organic matter within a turnover period of less than 1 year. Consequently, soil microbial biomass serves as a reliable indicator of alterations in nutrient dynamics, vegetative composition, and climatic conditions (Padaliya et al., 2022; Manral et al., 2023).

Table 1 Microbial diversity across extreme weather zones

Extreme environment	Representative microbes	Adaptations
Hot deserts	*Deinococcus radiodurans*, *Thermus aquaticus*	UV-resistant DNA repair enzymes, thermostable proteins (e.g., Taq polymerase)
Polar regions	*Psychrobacter* spp., *Colwellia psychrerythraea*	Cold-active enzymes, cryoprotectants, unsaturated lipid membranes
Deep-sea vents	*Thermococcus* spp., *Pyrolobus fumarii*	Pressure-resistant membranes, chemosynthesis via sulfur and methane oxidation
Acidic hot springs	*Sulfolobus acidocaldarius*, *Acidithiobacillus ferrooxidans*	Acid-stable proteins, sulfur metabolism
Hypersaline lakes	*Halobacterium salinarum*, *Haloferax volcanii*, *Dunaliella salina*	Salt-in strategy, carotenoid pigment production for UV protection
High-radiation zones	*Deinococcus radiodurans*, *Rubrobacter radiotolerans*	DNA repair systems (RecA, PprA), antioxidant production, manganese-based ROS detoxification

7.1 Extreme Weather Events and Soil Microbial Communities

Wildfires significantly impact soil microbial communities, altering their composition, diversity, and function. These changes are influenced by factors such as fire intensity, soil depth, and postfire management practices. The effects of wildfires on soil microbes are complex and can have long-lasting implications for ecosystem recovery and nutrient cycling. The following sections detail the specific impacts of wildfires on soil microbial communities.

Soil moisture is anticipated to be the primary outcome of the soil microbial community across diverse ecosystems, such as grassland or forest systems (Evans et al., 2022; Furtak & Wolinska, 2023). Furthermore, arid regions are projected to expand in numerous global locales, and a significant proportion of semiarid and arid territories will likely confront desertification in the foreseeable future (Singh & Chudasama, 2021). The establishment of Arbuscular Mycorrhizal Fungi (AMF) is influenced by hydric regimes, which tend to be comparatively elevated in wetlands, and under conditions where oxygen availability is significantly diminished, thus rendering AMF incapable of thriving and colonizing plant roots with ease (Wu et al., 2021). Additionally, the substrate of wetlands is deficient in the requisite oxygen that fungi necessitate for survival. A longitudinal field investigation has disclosed that soil fungi exhibit a reduced sensitivity to drought compared to soil bacteria in grasslands, which demonstrate greater drought resilience and sustain the cycling of carbon and nitrogen amid water-deficient circumstances (Xu et al., 2024). The structural framework, composition, and biogeochemical functions of microbial communities within periodically inundated agroecosystems (e.g., rice paddy soils) have been extensively documented (Shrestha et al., 2008). The historical context and duration of flooding, the availability of organic matter, fertilization practices, and management strategies during both unflooded and flooded intervals all exert influences on community dynamics and their respective functions (Schutz et al., 1989; Rui et al., 2009; Watanabe et al., 2009). Moreover, management practices associated with rice cultivation, encompassing fertilization and organic amendments, may enhance the responses to flooding, which transpire at a more gradual pace in unmanaged soils (Bargali et al., 2018, 2019).

Drought significantly diminished the diversity of bacterial and archaeal taxa, which are represented as amplicon sequence variants (ASVs). Numerous investigations have examined the implications of drought on microbial life (Manzoni et al., 2012, 2014; Evans et al., 2014; Barnard et al., 2013, 2015; Schaeffer et al., 2017; Bouskill et al., 2013). As soils undergo desiccation, the water potential diminishes, thereby subjecting microorganisms to a variety of stressors, including disconnection and, at localized scales, scarcity of resources (Schimel, 2018). This observation emphasizes the resilience of Actinobacteria in arid soils, as previously indicated by their elevated relative abundances (Canarini et al., 2021; Barnard et al., 2013; Ochoa-Hueso et al., 2018; Naylor et al., 2017). Drought exerts direct effects on microbial functions not only by serving as a stressor but also by constraining substrate availability due to diffusional limitations that hinder flow and transport as soil water films become disjointed (Manzoni et al., 2012; Schimel, 2018).

Diminished moisture levels impact microbial communities by limiting substrate availability through the disruption of previously interconnected hydrological pathways that supported diffusion and mass transfer, as well as through the reduction of substrate dissolution and desorption from mineral surfaces, and by impairing motility and, consequently, dispersal (Schimel, 2018). Extended drought conditions have a profound effect on the abundance, structure, and functionality of the soil microbiome (Siebielec et al., 2020). Deceased microbial cells release various essential compounds such as nitrogen, phosphates, amino acids, and polyols, which can be advantageous to other microbial communities and plants. Naylor et al. (2017) documented the variations in the abundance of root-associated microbial communities at the endosphere and rhizosphere levels during both monsoon and dry seasons. Soil microorganisms predominantly synthesize extracellular enzymes, including β-glucosidase, hydrolases, urease, phosphatase, glycosylating enzymes, cellulase, amylase, cyclomaltodextrinase, chitinase, and numerous others (Shukla & Varma, 2011; Hemkemeyer et al., 2021; Mulvaney & Bremner, 1978).

Consequently, soil enzymes assume an essential function in the biodegradation of organic compounds within the soil matrix and serve as the most sensitive indicators of fluctuations in microbial activities that transpire in the soil ecosystem in reaction to various stimuli, including drought conditions (Hu et al., 2011; Hueso et al., 2012). The investigations conducted by Bargali et al. (2018, 2019) demonstrated that elevated levels of microbial biomass carbon, nitrogen, and phosphorus in the hill base forest ecosystem are attributable to the greater structural diversity of the forest, increased soil moisture content, superior litter quality, and reduced sunlight exposure.

Warming influences microbial metabolism, growth rates, and community composition. Warmer soils tend to harbor more fast-growing copiotrophic microbes, reducing carbon sequestration potential due to enhanced decomposition (Crowther et al., 2015). Marine microbes, especially phytoplankton and nitrogen-fixing cyanobacteria, are highly sensitive to pH and salinity shifts. Changes in ocean chemistry can alter microbial-driven nutrient cycling, potentially disrupting marine food webs (Hutchins & Fu, 2017). Droughts and floods significantly influence microbial biomass and activity in both terrestrial and aquatic ecosystems. Flooding may mobilize dormant microbial seed banks, while drought may reduce microbial diversity and nutrient availability (Evans et al., 2014).

8 Conclusion

Extreme weather conditions significantly influence soil microbial communities by altering their composition, diversity, and functional characteristics. Such modifications can result in alterations in ecosystem functionality, particularly concerning nutrient cycling and carbon metabolism. Extended periods of drought diminish soil moisture levels, resulting in a decline in microbial activity, shifts in community structure, and modifications in functional diversity. Drought-resistant taxa, including Actinobacteria and Alpha-proteobacteria, frequently predominate in arid conditions, whereas sensitive groups experience a reduction in prevalence.

Wildfires modify soil characteristics and establish conditions that preferentially support certain microbial taxa, such as those involved in hydrogen and nitrogen recycling, while simultaneously decreasing the abundance of essential fungal functional groups, including mycorrhizal fungi and saprotrophs. Flooding imposes oxygen limitations and shifts microbial communities toward anaerobic taxa, exemplified by Firmicutes, while concurrently diminishing the presence of aerobic groups. Flooding profoundly affects soil microbial communities, influencing diversity, composition, and function. These changes have implications for ecosystem services under climate change. Future research should explore long-term impacts and management strategies to mitigate adverse effects. Drought generally has adverse effects on soil microbial communities; some ecosystems exhibit resilience through adaptive strategies. Understanding these dynamics is crucial for developing sustainable agricultural practices and enhancing ecosystem resilience in the face of climate change.

References

Amundson, R. (2021). Factors of soil formation in the 21st century. *Geoderma, 391*, 114960. https://doi.org/10.1016/j.geoderma.2021.114960

Bargali, K., Manral, V., Padalia, K., Bargali, S. S., & Upadhyay, V. P. (2018). Effect of vegetation type and season on microbial biomass carbon in Central Himalaya forest soils, India. *Catena, 171*, 125–135.

Bargali, S. S., Padalia, K., & Bargali, K. (2019). Effects of tree fostering on soil health and microbial biomass under different land use systems in central Himalaya. *Land Degradation and Development, 30*, 1984–1998.

Barnard, R. L., Osborne, C. A., & Firestone, M. K. (2013). Responses of soil bacterial and fungal communities to extreme desiccation and rewetting. *The ISME Journal, 7*, 2229–2241.

Barnard, R. L., Osborne, C. A., & Firestone, M. K. (2015). Changing precipitation pattern alters soil microbial community response to wet up under a Mediterranean-type climate. *The ISME Journal, 9*, 946–957.

Bouskill, N. J., et al. (2013). Pre-exposure to drought increases the resistance of tropical forest soil bacterial communities to extended drought. *The ISME Journal, 7*, 384–394.

Bünemann, E. K., Bongiorno, G., Bai, Z., Creamer, R. E., De Deyn, G., de Goede, R., Fleskens, L., Geissen, V., Kuyper, T. W., Mäder, P., Pulleman, M., Sukkel, W., van Groenigen, J. W., & Brussaard, L. (2018). Soil quality—A critical review. *Soil Biology and Biochemistry, 120*, 105–125. https://doi.org/10.1016/j.soilbio.2018.01.030

Canarini, A., et al. (2021). Ecologicalmemory of recurrent drought modifies soil processes via changes in soil microbial community. *Nature Communications, 12*, 1–14.

Coker, J. A. (2019). Recent advances in understanding extremophiles. *F1000Res, 8*, F1000 Faculty Rev-1917. https://doi.org/10.12688/f1000research.20765.1

Cook, B. I., Ault, T. R., & Smerdon, J. E. (2015). Unprecedented 21st century drought risk in the American Southwest and Central Plains. *Science Advances, 1*(1), e1400082.

Crowther, T. W., Thomas, S. M., Maynard, D. S., Baldrian, P., Covey, K., Frey, S. D., van Diepen, L. T. A., & Bradford, M. A. (2015). Biotic interactions mediate soil microbial feedbacks to climate change. *Proceedings of the National Academy of Sciences, 112*(22), 7033–7038.

ELD. (2015). *Report for policy and decision makers: Reaping economic and environmental benefits from sustainable land management* (p. 4). Economics of Land Degradation (ELD) Initiative.

Evans, S. E., Wallenstein, M. D., & Burke, I. C. (2014). Is bacterial moisture niche a good predictor of shifts in community composition under long-term drought? *Ecology, 95*, 110–122.

Evans, S. E., Allison, S. D., & Hawkes, C. V. (2022). Microbes, memory and moisture: Predicting microbial moisture responses and their impact on carbon cycling. *Functional Ecology, 36*, 1430–1441.

Fierer, N., & Jackson, R. B. (2006). The diversity and biogeography of soil bacterial communities. *Proceedings of the National Academy of Sciences of the United States of America, 103*(3), 626–631.

Furtak, K., & Wolinska, A. (2023). The impact of extreme weather events as a consequence of climate change on the soil moisture and on the quality of the soil environment and agriculture—A review. *Catena, 231*, 107378.

George, P. B. L., Lallias, D., Creer, S., Seaton, F. M., Kenny, J. G., Eccles, R. M., Griffiths, R. I., Lebron, I., Emmett, B. A., Robinson, D. A., & Jones, D. L. (2019). Divergent national-scale trends of microbial and animal biodiversity revealed across diverse temperate soil ecosystems. *Nature Communications, 10*(1), 1–11.

Gupta, G. N., Srivastava, S., Khare, S. K., & Prakash, V. (2014). Extremophiles: An overview of microorganism from extreme environment. *International Journal of Agriculture Environment and Biotechnology, 7*(2), 371–380.

Hari, V., Rakovec, O., Markonis, Y., Hanel, M., & Kumar, R. (2020). Increased future occurrences of the exceptional 2018–2019 Central European drought under global warming. *Scientific Reports, 10*(1), 1–10.

Hemkemeyer, M., Schwalb, S. A., Heinze, S., Joergensen, R. G., & Wichern, F. (2021). Functions of elements in soil microorganisms. *Microbiological Research, 252*, 126832.

Hu, J., Lin, X., Wang, J., Dai, J., Chen, R., Zhang, J., & Wong, M. H. (2011). Microbial functional diversity, metabolic quotient, and invertase activity of a sandy loam soil as affected by long-term application of organic amendment and mineral fertilizer. *Journal of Soils and Sediments, 11*(2), 271–280.

Hueso, S., García, C., & Hernández, T. (2012). Severe drought conditions modify the microbial community structure, size and activity in amended and unamended soils. *Soil Biology and Biochemistry, 50*, 167–173.

Hutchins, D. A., & Fu, F. (2017). Microorganisms and ocean global change. *Nature Microbiology, 2*(6), 1–11.

Khati, P., Mishra, P. K., Parihar, M., Singh, A. K., Bisht, J. K., & Pattanayak, A. (2021). Drought tress tolerance: An insight to resistance mechanism and adaptation in plants. In A. Sharma (Ed.), *Microbes and signaling biomolecules against plant stress, rhizosphere biology*. https://doi.org/10.1007/978-981-15-7094-0_10

Lehmann, J., Bossio, D. A., Kögel-Knabner, I., & Rillig, M. C. (2020). The concept and future prospects of soil health. *Nature Reviews Earth and Environment, 1*(10), 544–553. https://doi.org/10.1038/s43017-020-0080-8

Lugtenberg, B., & Kamilova, F. (2009). Plant-growth-promoting rhizobacteria. *Annual Review of Microbiology, 63*(1), 541–556.

Manral, V., Bargali, K., Bargali, S. S., Karki, H., & Chaturvedi, R. K. (2023). Seasonal dynamics of soil microbial biomass C, N and P along an altitudinal gradient in Central Himalaya, India. *Sustainability, 15*, 1651. https://doi.org/10.3390/su15021651

Manzoni, S., Schimel, J. P., & Porporato, A. (2012). Responses of soil microbial communities to water stress: Results from a meta-analysis. *Ecology, 93*(4), 930–938. https://doi.org/10.1890/11-0026.1

Manzoni, S., Schaeffer, S. M., Katul, G., Porporato, A., & Schimel, J. P. (2014). A theoretical analysis of microbial eco-physiological and diffusion limitations to carbon cycling in drying soils. *Soil Biology and Biochemistry, 73*, 69–83.

Mishra, J., Singh, R., & Arora, N. K. (2017). Plant growth-promoting microbes: Diverse roles in agriculture and environmental sustainability. In *Probiotics and plant health* (pp. 71–111). Springer Singapore. Available from: https://link.springer.com/chapter/10.1007/978-981-10-3473-24

Mulvaney, R. L., & Bremner, J. M. (1978). Use of p-benzoquinone and hydroquinone for retardation of urea hydrolysis in soils. *Soil Biology and Biochemistry, 10*, 297–302.

Naylor, D., DeGraaf, S., Purdom, E., & Coleman-Derr, D. (2017). Drought and host selection influence bacterial community dynamics in the grass root microbiome. *The ISME Journal, 11*, 2691–2704.

Ochoa-Hueso, R., et al. (2018). Drought consistently alters the composition of soil fungal and bacterial communities in grasslands from two continents. *Global Change Biology, 24*, 2818–2827.

Padaliya, K., Bargali, S. S., Bargali, K., & Manral, V. (2022). Soil microbial biomass phosphorus under different land use systems of Central Himalaya. *Tropical Ecology, 63*, 30–48. https://doi.org/10.1007/s42965-021-00184-z

Pidock, R., Pearce, R., & McSweeney, R. (2019). Mapped: How climate change affects extreme weather around the world. In *Carbon brief: Clear on climate—Attribution* (pp. 1–17). Carbon Brief Ltd.

Rodrigo-Baños, M., Garbayo, I., Vílchez, C., Bonete, M. J., & Martínez-Espinosa, R. M. (2015). Carotenoids from Haloarchaea and their potential in biotechnology. *Marine Drugs, 13*(9), 5508–5532.

Rui, J., Peng, J., & Lu, Y. (2009). Succession of bacterial populations during plant residue decomposition in rice field soil. *Applied and Environmental Microbiology, 75*, 4879–4886. https://doi.org/10.1128/AEM.00702-09

Schaeffer, S. M., Homyak, P. M., Boot, C. M., Roux-Michollet, D., & Schimel, J. P. (2017). Soil carbon and nitrogen dynamics throughout the summer drought in a California annual grassland. *Soil Biology and Biochemistry, 115*, 54–62.

Schimel, J. P. (2018). Life in dry soils: Effects of drought on soil microbial communities and processes. *Annual Review of Ecology, Evolution, and Systematics, 49*(1), 409–432. https://doi.org/10.1146/annur-evecolsys-110617-062614

Schutz, H., Holzapfel-Pschorn, A., Conrad, R., Rennenberg, H., & Seiler, W. A. (1989). 3-year continuous record on the influence of daytime, season, and fertilizer treatment on methane emission rates from an Italian rice paddy. *Journal of Geophysical Research-Atmospheres, 94*, 16405–16416. https://doi.org/10.1029/JD094iD13p16405

Shrestha, M., Abraham, W. R., Shrestha, P. M., Noll, M., & Conrad, R. (2008). Activity and composition of methanotrophic bacterial communities in planted rice soil studied by flux measurements, analyses of pmoA gene and stable isotope probing of phospholipid fatty acids. *Environmental Microbiology, 10*, 400–412. https://doi.org/10.1111/j.1462-2920.2007.01462.x

Shukla, G., & Varma, A. (Eds.). (2011). *Soil enzymology* (pp. 1–392). Springer.

Siebielec, S., Siebielec, G., Klimkowicz-Pawlas, A., Gałazka, A., Grzadziel, J., & Stuczynski, T. (2020). Impact of water stress on microbial community and activity in sandy and loamy soils. *Agronomy, 10*, 1429.

Sillmann, J., Kharin, V. V., Zwiers, F. W., Zhang, X., & Bronaugh, D. (2013). Climate extremes indices in the CMIP5 multimodel ensemble: Part 2. Future climate projections. *Journal of Geophysical Research – Atmospheres, 118*(6), 2473–2493.

Singh, P. K., & Chudasama, H. (2021). Pathways for climate change adaptations in arid and semi-arid regions. *Journal of Cleaner Production, 284*, 124744.

Ummenhofer, C. C., & Meehl, G. A. (2017). Extreme weather and climate events with ecological relevance: A review. *Philosophical Transactions of the Royal Society of London. Series B, Biological Sciences, 372*(1723), 20160135.

Wall, D. H., Nielsen, U. N., & Six, J. (2015). Soil biodiversity and human health. *Nature, 528*(7580), 69–76.

Watanabe, T., Kimura, M., & Asakawa, S. (2009). Distinct members of a stable methanogenic archaeal community transcribe mcrA genes under flooded and drained conditions in Japanese paddy field soil. *Soil Biology and Biochemistry, 41*, 276–285. https://doi.org/10.1016/j.soilbio.2008.10.025

Wu, Y., Chen, C., Li, J., & Wang, G. (2021). Effects of arbuscular mycorrhizal fungi on maize nitrogen uptake strategy under different soil water conditions. *Plant and Soil, 464*, 441–452.

Xu, T., Johnson, D., & Bardgett, R. D. (2024). Defoliation modifies the response of arbuscular mycorrhizal fungi to drought in temperate grassland. *Soil Biology and Biochemistry, 192*, 109386.

Microbial Diversity of a Warming World

Anuj Chaudhary, Parul Chaudhary, and Durgesh Kumar Jaiswal

Abstract

Crop production is a key component of global food security, but anthropogenic climate change threatens ecosystem function and agricultural productivity. The rising warming of the Earth's temperature greatly impacts ecosystems globally, with important consequences for microbial diversity. The health of terrestrial ecosystems depends on soil biodiversity, which highly depends on climatic conditions. Microorganisms, which include bacteria, archaea, fungi, and viruses, play critical roles in ecological processes such as the nutrient cycle and decomposition. As global temperatures rise, microbial populations are projected to vary in composition, distribution, and activity. These shifts are influenced by temperature sensitivity, with some species thriving in warmer conditions while others may face decline. Changes in microbial diversity could destabilize ecosystems that rely on specific microbial functions. Soil health, which is directly impacted by microbial activity, could decline if key microbes are lost or altered due to warming. This chapter discusses current knowledge of warming on soil microbial taxonomic and functional diversity, as well as microbial diversity's contributions to ecosystem processes.

Keywords

Microbial diversity · Fungi · Bacteria · AMF · Soil · Climate change

A. Chaudhary
Department of Biotechnology, Graphic Era Deemed to be University, Dehradun, Uttarakhand, India

P. Chaudhary
School of Agriculture, Graphic Era Hill University, Dehradun, Uttarakhand, India

D. K. Jaiswal (✉)
Department of Biosciences, Graphic Era (Deemed to be University), Dehradun, Uttarakhand, India

G. Bhandari et al. (eds.), *Microbial Resilience*,
https://doi.org/10.1007/978-981-95-8883-1_10

1 Introduction

Climate change is one of the most severe challenges to civilization, with an annual cost of around 1.2 trillion US dollars and nearly 0.4 million agricultural resource deaths. The impact on agriculture is significant, as the average global temperature has increased by 0.74 °C in the last century (EPA, 2023). One of the main causes of biodiversity loss at both local and global levels is human-caused global change, including warming, changes in precipitation, the wetting-drying cycle, drought, and increased atmospheric nitrogen deposition (Bolan et al., 2024). Human societies can better predict these changes and adjust as needed when they better understand the direction and scale of ecological reactions. Our food security is at risk due to these changes, which substantially impact the development and yield of many crops worldwide. For instance, rice production is threatened by increased temperatures and changes in precipitation, while the spread of various plant diseases is altering the growth and yield of staple crops like wheat and maize (Moullec et al., 2019). Planning for managing natural resources and assessing adaptation measures requires regular evaluations of the impact of climate change on ecosystems, both now and in the future.

The warming globe is complex, and rising temperatures can directly alter microbial activity and diversity (Yang et al., 2021). Temperature is a critical environmental factor that influences microorganisms' growth, metabolism, and community organization. While certain microbial species, particularly those acclimated to colder climes, may decrease or shift their geographical range, others may thrive in warmer conditions (Ibáñez et al., 2023). Bacteria, fungi, archaea, and viruses play vital roles in biological progressions such as nutrient cycling, decomposition, and symbiotic connections with plants and animals (Iqbal et al., 2023). As the planet heats, its diversity and interactions may change, potentially affecting ecosystem health and stability. This shift could lead to significant disruptions in the planet's biogeochemical cycles, with far-reaching implications for ecosystem health, human health, and agricultural productivity (Zhao et al., 2022).

Research has shown that warming significantly alters soil microbial diversity, favouring pathogenic fungi and perhaps influencing ecosystem processes in dryland soils, especially when combined with decreased rainfall. Numerous plant pathogenic fungi, such as *Arcremonium*, *Phaeoacremonium*, and *Stagonosporopsis*, showed a considerable rise in sequencing abundance under warming situations (Cuartero et al., 2024). Additionally, warming alters the composition and organization of soil microbial communities; some studies suggest that warming may increase the variety of active bacteria such as *Proteobacteria*, *Xanthomonas*, *Bradyrhizobium*, and *Acidobacteria* (Metze et al., 2024). The planctomycetes population becomes more prevalent in various settings as temperatures rise (Dedysh et al., 2021). It was observed that soil fungal diversity was more sensitive to climatic factors and negatively impacted soil diversity (Yang et al., 2021). Warming has been shown to have varying effects on biodiversity, accessibility of nutrients, and soil carbon storage depending on environmental conditions and ecosystem type. During warming, plant diversity grew in mountain peaks but dropped in desert grassland,

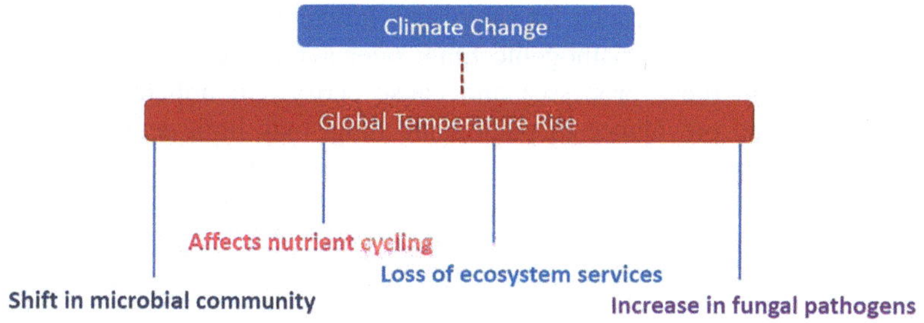

Fig. 1 Climate change and its consequences

and microbial diversity augmented in alpine regions while reduced in temperate grassland (Steinbauer et al., 2018; Wu et al., 2020). It was observed that warming increased the length of arbuscular mycorrhizal fungi (AMF) hyphae, which are the thread-like structures of these beneficial fungi that extend into the soil and facilitate nutrient uptake by plant roots, with root colonization and decreased soil aggregation (Rillig et al., 2002). Grassland degradation and warming combine to change the complexity of microbial networks. Despite its lesser richness, the network is dominated by favourable intra-fungal interactions (Wang et al., 2024). Temperature changes enhanced the *Bemisia tabaci* pest, which causes disease in tomatoes (Sujatha et al., 2025). Assessing the possible effects of global warming on ecosystem services and human well-being requires understanding microbial diversity in the context of climate change (Fig. 1).

2 Warming's Effects on Microbial Diversity

2.1 A Rise in Pathogenic Fungi

Temperature, a critical environmental factor, significantly influences the life cycle, dispersal, and virulence of plant diseases, as well as plant growth and physiological functions (Hunjan & Lore, 2020). It plays a pivotal role in regulating key aspects of pathogen contagion, sporulation, and inoculum dispersal (Kweku et al., 2018). Notably, a rise in temperature often leads to the proliferation and spread of plant diseases, particularly fungi and bacteria, while weakening the host's defensive mechanisms (Velásquez et al., 2018). The warming changes may also lead to an increase in the overwintering and persistence of diseases, thereby extending insect action periods and expanding their geographical ranges. Furthermore, climate changes in host plant phenology can render plants more susceptible or resistant to certain diseases (Pathak et al., 2018). The escalating temperatures could exacerbate the severity of numerous crop diseases such as wheat rust (*Puccinia triticina*), citrus canker (*Xanthomonas* spp.), and potato late blight (*P. infestans*) (Rivero et al., 2022; Singh et al., 2023). However, there are potential solutions and strategies that can be implemented to mitigate these effects, offering hope for the future.

Fungal biodiversity is crucial for ecosystem functions like soil health and nitrogen cycling. By favouring pathogenic fungi over symbiotic ones, climate change may interfere with these processes and cause ecosystem imbalances (Mridu & Kumar, 2024). The prevalence of pathogenic fungi may limit ecosystems' ability to break down organic matter, impacting plant health and soil quality (Yiallouris et al., 2024). Plant pathogenic fungi are demonstrated to grow and multiply more readily in dryland soils under warmer and drier conditions, as predicted by climate change models (Tedersoo et al., 2022; Duan et al., 2025). Warmer weather can accelerate the development of some crop fungal infections, such as *Puccinia triticina*, which causes wheat rust, resulting in higher crop losses (Ebi et al., 2016; Singh et al., 2023). It was observed that *Triphragmium ulmarie* (rust pathogen) was becoming extinct under summer temperatures (Zhan et al., 2018). According to studies, certain plant pathogenic fungi may become more active at rising temperatures, which might impact agricultural yields and jeopardize food security. The capacity of these fungi to create toxins, including mycotoxins, which can have detrimental effects on the health of plants and animals, may also be influenced by warmer temperatures (Hossain et al., 2024). Additionally, plant diseases like *Fusarium*, *Botrytis*, and *Alternaria* may become more common and spread to warmer areas, resulting in higher agricultural losses in products including cereals, fruits, and vegetables. Climate warming favours plant pathogen abundance by shortening the incubation time, resulting in higher relative abundance in soil microbial communities (Singh et al., 2023). This potential for food security risks underlines the threat to the global food supply posed by climate change.

3 Change in Soil Microbial Community Structure

Warming, with its potential to alter the network structure and composition of soil microbial communities, underscores the importance of your work in understanding these effects. These changes could impact the communities' ability to support plant growth and ecosystem services (Zhou et al., 2016). This can affect the health of plants and animals that rely on these microbial communities, as well as ecological processes like nutrient cycling (Guo et al., 2018). The diversity, richness, and enzyme activities of microorganisms were positively impacted by temperature (Kim et al., 2022). Warmer environments may promote thermophilic (heat-loving) microorganisms that flourish at higher temperatures, whereas species suited to cooler climes may decline or migrate to higher latitudes or elevations. Warming considerably boosted aerobic chemoheterotrophy and ureolysis, implying that it improved the capability of bacteria to break down organic matter and greenhouse gas emissions and augmented the soil carbon cycle (Yu et al., 2021). Long-standing warming boosted labile carbon deprivation and nitrogen fixation while decreasing carbon deprivation and denitrification (Guo et al., 2020). The relative abundance of heat-tolerant microorganisms, like *Firmicutes* and *Actinobacteria*, which are known to tolerate high temperatures, may rise in response to rising temperatures. On the other hand, heat-sensitive taxa like *Acidobacteria* and *Proteobacteria* may become less prevalent (Zhao et al., 2024). Growing seasons had different effects of warming on soil microbial biomass. The activity of microbial C- and P-acquiring enzymes was enhanced by warming as well as microbial decomposition of soil labile and recalcitrant C (Ma et al., 2023). By altering soil characteristics, warming impacts microbial biomass, which was observed in numerous alpine habitats (Sun et al., 2016). The beneficial impacts of warming on microbial biomass and stoichiometry may be mitigated by warming-induced decreases in soil moisture aside from available nutrients (Gong et al., 2019; Ma et al., 2023). When warming lasts 3–6 years, soil bacterial alpha diversity dramatically drops, while when warming lasts more than 6 years, bacterial beta diversity rises (Wang et al., 2025). Seasons also impacted how warming altered bacterial diversity, with summer warming linked to less bacterial evenness in the organic horizon. It was observed that neither heat nor warming duration impacted the link between bacterial diversity and ecosystem performance, even though bacterial evenness was reduced in warmed plots (Shinfuku et al., 2024). Additionally, higher temperatures change how taxa interact in microbial communities, which may help explain why declines in diversity have a disproportionate consequence on ecosystem functions (García et al., 2018). Many of the bacterial OTUs that were enhanced by warming had been modified with lignin. Even though carbon is less easily available, these taxa might help improve soil respiration (Pold et al., 2015). Warming drastically reduced ecological multifunctionality in grassland, and various microbial taxa had distinct negative responses (Chen et al., 2023). It was observed that a 2 °C rise in warming temperature showed the highest percentage of positive linkages in bacterial and fungal networks, suggesting that mild warming can encourage symbiotic interactions and positive correlations seen in microbial communities (Liu et al., 2024). Warming boosted soil microbial activity, microbial biomass, and organic carbon in grasslands (Xu et al., 2023). Warming and grazing have significant interactions on soil bacterial diversity and composition (Li et al., 2016). Six years of warming had a substantial impact on the structure of prokaryotic communities (Che et al., 2018). *Actinobacteria* and *Proteobacteria* showed an increasing tendency, while *Chloroflexi* and *Acidobacteria* showed a reduction in relative abundance with warming (Dong et al., 2021). The findings showed that short-term warming changed the beta diversity of soil bacteria in the frozen soil of the forest by increasing the abundance of Ascomycetes and lowering that of *Actinobacteria* (Jiang et al., 2024).

4 Improved Active Bacterial Diversity

Warmer temperatures can alter the diversity and activity of active bacteria in varied environments, with some species benefiting from the conditions while others experience constraints. Climate change has various effects on bacterial diversity, and these changes can have both beneficial and negative consequences for ecosystems, nutrient cycle, and biogeochemistry (Hartmann et al., 2023) (Table 1). This research is crucial in understanding these effects and their implications. Increased carbon release may result from warmer soils having a wider variety of active microbes,

Table 1 Effects of warming on soil microorganisms

S. no	Effects of warming on microbes	References
1.	Reduction in microbial biomass and enzymes	Allison et al. (2010)
2.	Influenced bacterial and fungal diversity	Hayden et al. (2012)
3.	Improved carbohydrate oxidation and carbon utilization	Zhang et al. (2013)
4.	Microbial biomass respiration improved	Schindlbacher et al. (2015)
5.	Decreased soil respiration	Romero-Olivares et al. (2017)
6.	Improved soil microbial abundance	Chen et al. (2015)
7.	Accelerated microbial activity and carbon loss	Walker et al. (2018)
8.	Increased microbial activity	Nottingham et al. (2019)
9.	Altered fungal respiration and enzyme parameters	Allison et al. (2018)
10.	Altered microbial respiration	Bradford et al. (2019)
11.	Shift in microbial community of bacteria rather than fungi	Luláková et al. (2019)
12.	Accelerated the bacterial ability to decompose organic matter	Yu et al. (2021)
13.	Reduced microbial diversity	Zhao et al. (2024)
14.	Improved soil beneficial bacteria	Metze et al. (2024)
15.	Temperature and precipitation affected litter decomposition	Li et al. (2024)

especially bacteria. Warmer temperatures may increase the diversity of bacterial functional groups engaged in the nitrogen cycle, carbon cycling, and sulphur reduction (Ruan et al., 2023). Nitrobacter and Nitrosomonas, nitrifying bacteria, may become more active in warmer soils, resulting in quicker nutrient cycling (Dai et al., 2020). This increased activity can alter soil fertility, plant development, and soil carbon release. Bacteria with symbiotic connections with plants, such as nitrogen-fixing Rhizobium species, may change their activity and variety as temperatures rise (Zhang et al., 2024). These bacteria benefit from higher temperatures and can increase nitrogen availability to plants. As a result, some crops may exhibit enhanced growth in warmer temperatures. However, changes in bacterial populations can also alter the balance of plant-microbe interactions, thereby impacting plant health and production. Increasing root exudates CO_2 enrichment in conjunction with heat can enhance bacterial growth in rhizospheres, improve enzyme activity, and trigger nutrient mobilization (Song et al., 2024). Active soil prokaryotes were more susceptible to warming than total prokaryotes (Che et al., 2018).

5 Effect on Arbuscular Mycorrhizal Fungi (AMF)

Warming can impact AMF's capacity to colonize plant roots and create efficient symbiotic relationships (Duarte & Maherali, 2022). Several factors influence the outcome, including the fungi's temperature tolerance and their host plants, soil

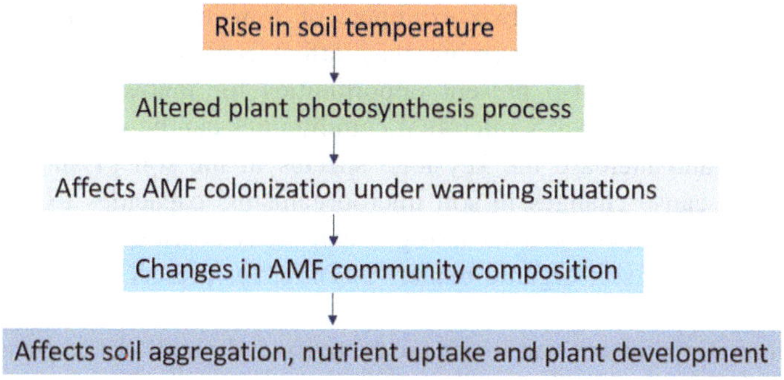

Fig. 2 Impact of warming on arbuscular mycorrhizal fungi

moisture, nutrition availability, and other environmental variables (Fig. 2). Warming may increase the relative abundance of AMF, which benefits plant nutrition and stress tolerance (Cuartero et al., 2024). AMF play an important role in the uptake of critical nutrients such as phosphorus and nitrogen. Warmer temperatures may boost the metabolic rate of fungi, increasing their food absorption capability to a certain extent. However, this process may become less efficient above acceptable temperature limits, as high temperatures can destroy fungal hyphae and reduce total nutrient exchange efficiency between fungi and plants (Alguacil et al., 2021). AMF efficiency variations may impact nitrogen cycling, changing plant productivity and soil fertility. The stability of the ecosystem may be impacted if AMFs are less effective in rising climates because nutrient cycling may slow down and result in less plant growth and output, a potential crisis that requires urgent attention. Increased saprotrophic fungal abundance, improved SOM breakdown, and N mineralization were associated with heat and dryness, significantly decreasing ectomycorrhizal fungi abundance (Querejeta et al., 2021). It was observed that warming and CO_2 increased the relative abundance of pathotrophic fungi while decreasing symbiotrophic such as Ascomycota and Basidiomycota fungi in soil (Gao et al., 2023). Warming and drought affected the abundance and composition of the AMF population (Xie et al., 2024). Warming boosts biomass allocation to roots due to the association of mycorrhizal fungi with plants (Zhou et al., 2022). Warming and precipitation reduced the alpha diversity of the AMF (Zheng et al., 2022).

6 Ecosystem Services

Warming negatively influences ecosystem services, resulting in shifts in species distribution, decreased water availability, increased extreme weather events, and biodiversity concerns. These impacts on human well-being and ecosystem services are deeply interconnected. Climate change can cause agricultural droughts, which in turn impact crop productivity and water availability (Weiskopf et al., 2020). The complexity and importance of these interconnected issues cannot be overstated.

Plant and soil health are two examples of how modifications in microbial populations brought on by warming may affect ecosystem services. However, increased temperatures also present opportunities for innovative solutions. For instance, changes in the soil's microbial communities can be managed to enhance wheat growth and increase the keystone species in the soil (Tian et al., 2022). Warming may cause changes in soil microorganisms' capacities to maintain soil fertility, particularly those that fix nitrogen and break down organic materials. For example, the loss of soil organic matter, essential for preserving soil fertility and structure, may result from the acceleration of decomposition. Increased crop stress from warmer temperatures can lower productivity and yields (León-Sánchez et al., 2020). In addition to directly impacting plant growth, extreme heat, droughts, and changing precipitation patterns can also raise the danger of pests and diseases that prefer warmer climates (Delgado-Baquerizo et al., 2020). This is extremely troubling for food security, especially in areas that depend on commodities like rice, corn, and wheat that are vulnerable to climate change. Warming reduces staple crop yields (e.g. wheat, maize) in tropical and temperate regions due to heat stress and water scarcity. In contrast, high-latitude areas may have temporary advantages.

7 Mechanisms of Microbial Adaptation to a Warming Environment

High temperatures, the preferred habitat of thermophiles, a group of extremophiles primarily comprising Archaea and bacteria, showcase the remarkable adaptability of these organisms. Categorized based on their ideal growth temperatures, they fall into three groups: hyperthermophiles (above 80 °C), extreme thermophiles (up to 80 °C), and moderate thermophiles (45–65 °C) (Somayaji et al., 2022). Despite sharing the same fundamental macromolecular constituents as mesophiles—RNA, DNA, proteins, and lipids—thermophiles employ a variety of ingenious strategies to maintain the stability of these molecules at high temperatures.

Thermophiles have evolved unique mechanisms to ensure the stability of DNA at high temperatures. They utilize DNA-binding proteins to compress and condense DNA from nucleosomes, and proteins that repair DNA to eliminate any products that are cross-linked, alkylated, deaminated, or primary sites. These DNA-binding proteins play a crucial role in the stability of DNA in thermophiles, a fascinating aspect of their adaptation to extreme conditions. Additionally, thermophiles are thought to have smaller genomes and shorter protein sequences than mesophiles, which reduces cells' energy requirements during division (Sabath et al., 2013). Because G-C pairings have a stronger bond, thermophiles' rRNA and tRNA have a larger G-C to A-T ratio, which helps to stabilize these molecules (Wang et al., 2015). Thermophile DNA also includes reverse DNA gyrase, which raises the melting point by creating positive supercoils in the DNA. The cell's polyamines and osmolytes aid in protecting proteins and stabilizing DNA by restoring their equilibrium to its original condition (Neifar et al., 2015). In terms of metabolism, thermophiles adjust by upregulating proteins linked to glycolysis and downregulating proteins linked to respiration, like NADH dehydrogenase and electron transport proteins (Shih & Pan, 2011; Wang et al., 2012). Increases in long-chain and saturated fatty acids are among the membrane modifications that improve membrane stiffness (Siliakus et al., 2017).

In addition to the strategies mentioned earlier, thermophilic actinomycetes, bacteria, and archaea employ a fascinating array of tactics to withstand high temperatures. They enhance their proteins' disulphide, electrostatic, and hydrophobic interactions and produce specialized proteins known as "chaperones." These chaperones play a crucial role in facilitating protein folding, maintaining stable intermediate conformational states, preventing protein aggregation or misfolding in cells, and restoring the functionality of the proteins (Liao et al., 2024; Hertle et al., 2024). Comparative genomic analyses between mesophiles, psychrophiles, and thermophiles have revealed a series of coordinated genetic alterations that are linked to an organism's optimal growth temperature (OGT) and its survival in harsh environments. These alterations include changes in tRNAs and overall G + C composition, folding patterns, and secondary structure frequency and placement. Previous research on the sequencing and structural information of many bacterial species has established a connection between the microbe's OGT and the DNA's dinucleotide makeup (Kohli et al., 2020).

8 Future Directions

Microbial diversity research in the context of climate change is essential for forecasting ecosystem resilience and guiding mitigation plans. In order to understand microbial responses to warmer climates at the genetic and functional levels, future research must prioritize multi-omics tools (metagenomics and transcriptomics). While sophisticated modeling frameworks can incorporate microbial activities into climate projections, bridging micro-scale interactions with global implications, long-term field trials across various biomes will clarify temporal dynamics. It is crucial to use interdisciplinary methods that combine data science, climatology, and microbiology. At the same time, studying extremophiles in analogue settings (such as hot springs) provides information about how microbes will adapt to changing climates. Utilizing microbial potential for climate solutions should be the focus of applied research. Microbes could be engineered using synthetic biology to improve soil carbon sequestration, lower agricultural methane emissions, or break down contaminants. However, navigating unexpected ecological repercussions requires thorough risk evaluations and ethical frameworks, stressing the importance of responsible research and decision-making. Lastly, it is crucial to turn research into policy. Microbial diversity must be incorporated into international climate accords since it affects disease patterns, food security, and carbon budgets. In order to promote stewardship, public engagement programs must emphasize the significance that microbes play in sustainability.

9 Conclusion

In a warming planet, microbial diversity is a dynamic and complicated problem that requires collective efforts. Although some microorganisms might flourish in warmer climates, the overall effects on ecosystems and human health are uncertain and need careful scientific observation. The stability of ecosystems and their ability to continue providing essential services to life on Earth will depend heavily on collaborative efforts to minimize climate change and comprehend microbial responses. Predicting future climatic changes and developing methods to lessen adverse effects on biodiversity and ecosystem stability requires understanding how microbial communities respond to climate change. More investigation is required to understand the processes causing these alterations and forecast the long-term effects of warming on microbial diversity.

References

Alguacil, M. D., Schlaeppi, K., López-García, Á., van der Heijden, M. G., & Querejeta, J. I. (2021). Contrasting responses of arbuscular mycorrhizal fungal families to simulated climate warming and drying in a semiarid shrubland. *Microbial Ecology, 24*, 1–4. https://doi.org/10.1007/s00248-021-01886-6

Allison, S. D., Wallenstein, M. D., & Bradford, M. A. (2010). Soil-carbon response to warming dependent on microbial physiology. *Nature Geoscience, 3*(5), 336–340. https://doi.org/10.1038/ngeo846

Allison, S. D., Romero-Olivares, A. L., Lu, L., Taylor, J. W., & Treseder, K. K. (2018). Temperature acclimation and adaptation of enzyme physiology in *Neurospora discreta*. *Fungal Ecology, 35*, 78–86. https://doi.org/10.1016/j.funeco.2018.07.005

Bolan, S., Padhye, L. P., Jasemizad, T., Govarthanan, M., Karmegam, N., Wijesekara, H., Amarasiri, D., Hou, D., Zhou, P., Biswal, B. K., & Balasubramanian, R. (2024). Impacts of climate change on the fate of contaminants through extreme weather events. *Science of the Total Environment, 909*, 168388.

Bradford, M. A., McCulley, R. L., Crowther, T. W., Oldfield, E. E., Wood, S. A., & Fierer, N. (2019). Cross-biome patterns in soil microbial respiration predictable from evolutionary theory on thermal adaptation. *Nature Ecology & Evolution, 3*(2), 223–231. https://doi.org/10.1038/s41559-018-0771-4

Che, R., Deng, Y., Wang, W., Rui, Y., Zhang, J., Tahmasbian, I., Tang, L., Wang, S., Wang, Y., Xu, Z., & Cui, X. (2018). Long-term warming rather than grazing significantly changed total and active soil procaryotic community structures. *Geoderma, 316*, 10. https://doi.org/10.1016/j.geoderma.2017.12.005

Chen, J., Luo, Y., Xia, J., Jiang, L., Zhou, X., Lu, M., Liang, J., Shi, Z., Shelton, S., & Cao, J. (2015). Stronger warming effects on microbial abundances in colder regions. *Scientific Reports, 5*(1), 18032. https://doi.org/10.1038/srep18032

Chen, W., Zhou, H., Wu, Y., Wang, J., Zhao, Z., Li, Y., Qiao, L., Chen, K., Liu, G., Ritsema, C., & Geissen, V. (2023). Long-term warming impacts grassland ecosystem function: Role of diversity loss in conditionally rare bacterial taxa. *Science of the Total Environment, 892*, 164722. https://doi.org/10.1016/j.scitotenv.2023.164722

Cuartero, J., Querejeta, J. I., Prieto, I., Frey, B., & Alguacil, M. M. (2024). Warming and rainfall reduction alter soil microbial diversity and co-occurrence networks and enhance pathogenic fungi in dryland soils. *Science of the Total Environment, 949*, 175006. https://doi.org/10.1016/j.scitotenv.2024.175006

Dai, Z., Yu, M., Chen, H., Zhao, H., Huang, Y., Su, W., Xia, F., Chang, S. X., Brookes, P. C., Dahlgren, R. A., & Xu, J. (2020). Elevated temperature shifts soil N cycling from microbial immobilization to enhanced mineralization, nitrification and denitrification across global terrestrial ecosystems. *Global Change Biology, 26*(9), 5267–5276. https://doi.org/10.1111/gcb.15211

Dedysh, S. N., Beletsky, A. V., Ivanova, A. A., Kulichevskaya, I. S., Suzina, N. E., Philippov, D. A., Rakitin, A. L., Mardanov, A. V., & Ravin, N. V. (2021). Wide distribution of Phycisphaera-like planctomycetes from WD2101 soil group in peatlands and genome analysis of the first cultivated representative. *Environmental Microbiology, 23*(3), 1510–1526. https://doi.org/10.1111/1462-2920.15360

Delgado-Baquerizo, M., Guerra, C. A., Cano-Díaz, C., Egidi, E., Wang, J. T., Eisenhauer, N., Singh, B. K., & Maestre, F. T. (2020). The proportion of soil-borne pathogens increases with warming at the global scale. *Nature Climate Change, 10*(6), 550–554. https://doi.org/10.1038/s41558-020-0759-3

Dong, X., Liu, C., Ma, D., Wu, Y., Man, H., Wu, X., Li, M., & Zang, S. (2021). Organic carbon mineralization and bacterial community of active layer soils response to short-term warming in the Great Hing'an Mountains of Northeast China. *Frontiers in Microbiology, 12*, 802213. https://doi.org/10.3389/fmicb.2021.802213

Duan, Y., Siegenthaler, A., Skidmore, A. K., Abdullah, H., Chariton, A. A., Laros, I., Rousseau, M., & de Groot, G. A. (2025). Tree vitality predicts plant-pathogenic fungal communities in beech forest canopies. *Forest Ecology and Management, 585*, 122588. https://doi.org/10.1016/j.foreco.2025.122588

Duarte, A. G., & Maherali, H. (2022). A meta-analysis of the effects of climate change on the mutualism between plants and arbuscular mycorrhizal fungi. *Ecology and Evolution, 12*(1), e8518. https://doi.org/10.1002/ece3.8518

Ebi, K. L., Ziska, L. H., & Yohe, G. W. (2016). The shape of impacts to come: Lessons and opportunities for adaptation from uneven increases in global and regional temperatures. *Climatic Change, 139*, 341–349. https://doi.org/10.1007/s10584-016-1816-9

EPA U. (2023). *Climate change impacts on agriculture and food supply*. https://www.epa.gov/climateimpacts/climate-change-impacts-agriculture-and-food-supply

Gao, K., Li, W., Gan, E., Li, J., Jiang, L., & Liu, Y. (2023). Impacts of 10 years of elevated CO2 and warming on soil fungal diversity and network complexity in a Chinese paddy field. *Microbial Ecology, 86*(4), 2386–2399. https://doi.org/10.1007/s00248-023-02248-0

García, F. C., Bestion, E., Warfield, R., & Yvon-Durocher, G. (2018). Changes in temperature alter the relationship between biodiversity and ecosystem functioning. *Proceedings of the National Academy of Sciences of the United States of America., 115*(43), 10989–10994. pmid:30297403.

Gong, S., Zhang, T., & Guo, J. (2019). Warming and nitrogen addition change the soil and soil microbial biomass C: N: P stoichiometry of a meadow steppe. *International Journal of Environmental Research and Public Health, 16*(15), 2705. https://doi.org/10.3390/ijerph16152705

Guo, X., Feng, J., Shi, Z., Zhou, X., Yuan, M., Tao, X., Hale, L., Yuan, T., Wang, J., Qin, Y., & Zhou, A. (2018). Climate warming leads to divergent succession of grassland microbial communities. *Nature Climate Change, 8*(9), 813 818.

Guo, X., Gao, Q., Yuan, M., Wang, G., Zhou, X., Feng, J., Shi, Z., Hale, L., Wu, L., Zhou, A., & Tian, R. (2020). Gene-informed decomposition model predicts lower soil carbon loss due to persistent microbial adaptation to warming. *Nature Communications, 11*(1), 4897. https://doi.org/10.1038/s41467-020-18706-z

Hartmann, M., Herzog, C., Brunner, I., Stierli, B., Meyer, F., Buchmann, N., & Frey, B. (2023). Long-term mitigation of drought changes the functional potential and life-strategies of the forest soil microbiome involved in organic matter decomposition. *Frontiers in Microbiology, 14*, 1267270. https://doi.org/10.3389/fmicb.2023.1267270

Hayden, H. L., Mele, P. M., Bougoure, D. S., Allan, C. Y., Norng, S., Piceno, Y. M., Brodie, E. L., DeSantis, T. Z., Andersen, G. L., Williams, A. L., & Hovenden, M. J. (2012). Changes in the microbial community structure of bacteria, archaea and fungi in response to elevated CO2 and warming in an Australian native grassland soil. *Environmental Microbiology, 14*(12), 3081–3096. https://doi.org/10.1111/j.1462-2920.2012.02855.x

Hertle, E., Ursinus, A., & Martin, J. (2024). Low-temperature features of the psychrophilic chaperonin from *Pseudoalteromonas haloplanktis*. *Archives of Microbiology, 206*(7), 299. https://doi.org/10.1007/s00203-024-04019-y

Hossain, M. M., Sultana, F., Mostafa, M., Ferdus, H., Rahman, M., Rana, J. A., Islam, S. S., Adhikary, S., Sannal, A., Al Emran Hosen, M., & Nayeema, J. (2024). Plant disease dynamics in a changing climate: Impacts, molecular mechanisms, and climate-informed strategies for sustainable management. *Discover Agriculture, 2*(1), 1–35. https://doi.org/10.1007/s44279-024-00144-w

Hunjan, M. S., & Lore, J. S. (2020). Climate change: Impact on plant pathogens, diseases, and their management. In *Crop protection under changing climate* (pp. 85–100). https://doi.org/10.1007/978-3-030-46111-9_4

Ibáñez, A., Garrido-Chamorro, S., & Barreiro, C. (2023). Microorganisms and climate change: A not so invisible effect. *Microbiological Research, 14*(3), 918–947. https://doi.org/10.3390/microbiolres14030064

Iqbal, B., Li, G., Alabbosh, K. F., Hussain, H., Khan, I., Tariq, M., Javed, Q., Naeem, M., & Ahmad, N. (2023). Advancing environmental sustainability through microbial reprogramming in growth improvement, stress alleviation, and phytoremediation. *Plant Stress, 10*, 100283. https://doi.org/10.1016/j.stress.2023.100283

Jiang, Y., Wu, S., Yang, L., Liu, Y., Gao, M., & Ni, H. (2024). Short-term simulated warming changes the beta diversity of bacteria in taiga forests' permafrost by altering the composition of dominant bacterial phyla. *Forests, 15*(4), 693. https://doi.org/10.3390/f15040693

Kim, S. Y., Zhou, X., Freeman, C., & Kang, H. (2022). Changing thermal sensitivity of bacterial communities and soil enzymes in a bog peat in spring, summer and autumn. *Applied Soil Ecology, 173*, 104382. https://doi.org/10.1016/j.apsoil.2021.104382

Kohli, I., Joshi, N. C., Mohapatra, S., & Varma, A. (2020). Extremophile—An adaptive strategy for extreme conditions and applications. *Current Genomics, 21*(2), 96–110. https://doi.org/10.2174/1389202921666200401105908

Kweku, D. W., Bismark, O., Maxwell, A., Desmond, K. A., Danso, K. B., Oti-Mensah, E. A., Quachie, A. T., & Adormaa, B. B. (2018). Greenhouse effect: Greenhouse gases and their impact on global warming. *Journal of Scientific Research and Reports, 17*(6), 1–9. http://www.sciencedomain.org/review-history/23195

León-Sánchez, L., Nicolás, E., Prieto, I., Nortes, P., Maestre, F. T., & Querejeta, J. I. (2020). Altered leaf elemental composition with climate change is linked to reductions in photosynthesis, growth and survival in a semi-arid shrubland. *Journal of Ecology, 108*(1), 47–60. https://doi.org/10.1111/1365-2745.13259

Li, Y., Lin, Q., Wang, S., Li, X., Liu, W., Luo, C., Zhang, Z., Zhu, X., Jiang, L., & Li, X. (2016). Soil bacterial community responses to warming and grazing in a Tibetan alpine meadow. *FEMS Microbiology Ecology, 92*(1), fiv152. https://doi.org/10.1093/femsec/fiv152

Li, K., Song, L., Ran, Q., Yuan, F., Deng, C., & Liu, H. (2024). Global meta-analysis reveals differential effects of climate and litter quality on soil fauna-mediated litter decomposition across size classes. *Geoderma, 450*, 117042. https://doi.org/10.1016/j.geoderma.2024.117042

Liao, Z., Gopalasingam, C. C., Kameya, M., Gerle, C., Shigematsu, H., Ishii, M., Arakawa, T., & Fushinobu, S. (2024). Structural insights into thermophilic chaperonin complexes. *Structure, 32*(6), 679–689. https://doi.org/10.1016/j.str.2024.02.012

Liu, G., Sun, J., Xie, P., Guo, C., Zhu, K., & Tian, K. (2024). Climate warming enhances microbial network complexity by increasing bacterial diversity and fungal interaction strength in litter decomposition. *Science of the Total Environment, 908*, 168444. https://doi.org/10.1016/j.scitotenv.2023.168444

Luláková, P., Perez-Mon, C., Šantrůčková, H., Ruethi, J., & Frey, B. (2019). High-alpine permafrost and active-layer soil microbiomes differ in their response to elevated temperatures. *Frontiers in Microbiology, 10*, 668. https://doi.org/10.3389/fmicb.2019.00668

Ma, Z., Chen, Y., Xu, W., & Liu, M. (2023). Effects of warming on the stoichiometry of soil microbial biomass and extracellular enzymes in an alpine shrubland. *Geoderma, 430*, 116329. https://doi.org/10.1016/j.geoderma.2023.116329

Metze, D., Schnecker, J., De Carlan, C. L. N., Bhattarai, B., et al. (2024). Soil warming increases the number of growing bacterial taxa but not their growth rates. *Science Advances*. https://doi.org/10.1126/sciadv.adk6295

Moullec, F., Barrier, N., Drira, S., Guilhaumon, F., Marsaleix, P., Somot, S., Ulses, C., Velez, L., & Shin, Y. J. (2019). An end-to-end model reveals losers and winners in a warming Mediterranean Sea. *Frontiers in Marine Science, 6*, 345. https://doi.org/10.3389/fmars.2019.00345

Mridu, N. S., & Kumar, J. (2024). Impact of climatic variations on fungi, mycorrhiza and lichens—An appraisal. *Current Research in Environmental & Applied Mycology, 14*(1), 89–107.

Neifar, M., Maktouf, S., Ghorbel, R. E., Jaouani, A., & Cherif, A. (2015). Extremophiles as source of novel bioactive compounds with industrial potential. In *Biotechnology of bioactive compounds: Sources and applications* (pp. 245–267). https://doi.org/10.1002/9781118733103.ch10

Nottingham, A. T., Bååth, E., Reischke, S., Salinas, N., & Meir, P. (2019). Adaptation of soil microbial growth to temperature: Using a tropical elevation gradient to predict future changes. *Global Change Biology, 25*(3), 827–838. https://doi.org/10.1111/gcb.14502

Pathak, R., Singh, S. K., Tak, A., & Gehlot, P. (2018). Impact of climate change on host, pathogen and plant disease adaptation regime: a review. *Biosciences Biotechnology Research Asia, 15*(3), 529–540.

Pold, G., Melillo, J. M., & DeAngelis, K. M. (2015). Two decades of warming increases diversity of a potentially lignolytic bacterial community. *Frontiers in Microbiology, 6*, 480. https://doi.org/10.3389/fmicb.2015.00480

Querejeta, J. I., Schlaeppi, K., López-García, Á., Ondoño, S., Prieto, I., van Der Heijden, M. G., & del Mar Alguacil, M. (2021). Lower relative abundance of ectomycorrhizal fungi under a warmer and drier climate is linked to enhanced soil organic matter decomposition. *The New Phytologist, 232*(3), 1399–1413. https://doi.org/10.1111/nph.17661

Rillig, M. C., Wright, S. F., Shaw, M. R., & Field, C. B. (2002). Artificial climate warming positively affects arbuscular mycorrhizae but decreases soil aggregate water stability in an annual grassland. *Oikos, 97*(1), 52–58. https://doi.org/10.1034/j.1600-0706.2002.970105.x

Rivero, R. M., Mittler, R., Blumwald, E., & Zandalinas, S. I. (2022). Developing climate-resilient crops: Improving plant tolerance to stress combination. *The Plant Journal, 109*(2), 373–389. https://doi.org/10.1111/tpj.15483

Romero-Olivares, A. L., Allison, S. D., & Treseder, K. K. (2017). Soil microbes and their response to experimental warming over time: A meta-analysis of field studies. *Soil Biology and Biochemistry, 107*, 32–40. https://doi.org/10.1016/j.soilbio.2016.12.026

Ruan, Y., Kuzyakov, Y., Liu, X., Zhang, X., Xu, Q., Guo, J., Guo, S., Shen, Q., Yang, Y., & Ling, N. (2023). Elevated temperature and CO2 strongly affect the growth strategies of soil bacteria. *Nature Communications, 14*(1), 391. https://doi.org/10.1038/s41467-023-36086-y

Sabath, N., Ferrada, E., Barve, A., & Wagner, A. (2013). Growth temperature and genome size in bacteria are negatively correlated, suggesting genomic streamlining during thermal adaptation. *Genome Biology and Evolution, 5*(5), 966–977. https://doi.org/10.1093/gbe/evt050

Schindlbacher, A., Schnecker, J., Takriti, M., Borken, W., & Wanek, W. (2015). Microbial physiology and soil CO2 efflux after 9 years of soil warming in a temperate forest–no indications for thermal adaptations. *Global Change Biology, 21*(11), 4265–4277. https://doi.org/10.1111/gcb.12996

Shih, T. W., & Pan, T. M. (2011). Stress responses of thermophilic *Geobacillus* sp. NTU 03 caused by heat and heat-induced stress. *Microbiological Research, 166*(5), 346–359. https://doi.org/10.1016/j.micres.2010.08.001

Shinfuku, M. S., Domeignoz-Horta, L. A., Choudoir, M. J., Frey, S. D., Mitchell, M. F., Ranjan, R., & DeAngelis, K. M. (2024). Seasonal effects of long-term warming on ecosystem function and bacterial diversity. *PLoS One, 19*(10), e0311364. https://doi.org/10.1371/journal.pone.0311364

Siliakus, M. F., van der Oost, J., & Kengen, S. W. (2017). Adaptations of archaeal and bacterial membranes to variations in temperature, pH and pressure. *Extremophiles*, 651–670. https://doi.org/10.1007/s00792-017-0939-x

Singh, B. K., Delgado-Baquerizo, M., Egidi, E., Guirado, E., Leach, J. E., Liu, H., & Trivedi, P. (2023). Climate change impacts on plant pathogens, food security and paths forward. *Nature Reviews. Microbiology, 21*(10), 640–656. https://doi.org/10.1038/s41579-023-00900-7

Somayaji, A., Dhanjal, C. R., Lingamsetty, R., Vinayagam, R., Selvaraj, R., Varadavenkatesan, T., & Govarthanan, M. (2022). An insight into the mechanisms of homeostasis in extremophiles. *Microbiological Research, 263*, 127115. https://doi.org/10.1016/j.micres.2022.127115

Song, B., Li, Y., Yu, Z., Jin, J., Liu, Z., Yang, R., Adams, J. M., & Razavi, B. S. (2024). Changes in enzyme activity, structure and growth strategies of the rhizosphere microbiome influenced by elevated temperature and CO2. *Science of the Total Environment, 954*, 176522. https://doi.org/10.1016/j.scitotenv.2024.176522

Steinbauer, M. J., Grytnes, J. A., Jurasinski, G., Kulonen, A., Lenoir, J., Pauli, H., Rixen, C., Winkler, M., Bardy-Durchhalter, M., Barni, E., & Bjorkman, A. D. (2018). Accelerated increase in plant species richness on mountain summits is linked to warming. *Nature, 556*(7700), 231–234. https://doi.org/10.1038/s41586-018-0005-6

Sujatha, S., Sindhura, K. A. V., Koti, P. S., Hiremath, S., Muttappagol, M., Vinay Kumar, H. D., Shankarappa, K. S., Venkataravanappa, V., Reddy, K. M. S., & Reddy, C. N. L. (2025). Influence of weather and seasonal factors on whitefly dynamics, associated endosymbiotic microbiomes, and *Begomovirus* transmission causing tomato leaf curl disease: Insights from a metagenomic perspective. *Frontiers in Microbiology, 16*, 1555058. https://doi.org/10.3389/fmicb.2025.1555058

Sun, D. D., Li, Y. J., Zhao, W. Q., Zhang, Z. L., Li, D. D., Zhao, C. Z., & Liu, Q. (2016). Effects of experimental warming on soil microbial communities in two contrasting subalpine forest ecosystems, Eastern Tibetan Plateau, China. *Journal of Mountain Science, 13*, 1442–1452. https://doi.org/10.1007/s11629-015-3733-1

Tedersoo, L., Mikryukov, V., Zizka, A., Bahram, M., Hagh-Doust, N., Anslan, S., Prylutskyi, O., Delgado-Baquerizo, M., Maestre, F. T., Pärn, J., & Öpik, M. (2022). Global patterns in endemicity and vulnerability of soil fungi. *Global Change Biology, 28*(22), 6696–6710. https://doi.org/10.1111/gcb.16398

Tian, B., Zhu, M., Pei, Y., Ran, G., Shi, Y., & Ding, J. (2022). Climate warming alters the soil microbial association network and role of keystone taxa in determining wheat quality in the field. *Agriculture, Ecosystems and Environment, 326*, 107817. https://doi.org/10.1016/j.agee.2021.107817

Velásquez, A. C., Castroverde, C. D., & He, S. Y. (2018). Plant–pathogen warfare under changing climate conditions. *Current Biology, 28*(10), R619–R634. https://doi.org/10.1016/j.cub.2018.03.054

Walker, T. W., Kaiser, C., Strasser, F., Herbold, C. W., Leblans, N. I., Woebken, D., Janssens, I. A., Sigurdsson, B. D., & Richter, A. (2018). Microbial temperature sensitivity and biomass change explain soil carbon loss with warming. *Nature Climate Change, 8*(10), 885–889. https://doi.org/10.1038/s41558-018-0259-x

Wang, Z., Tong, W., Wang, Q., Bai, X., Chen, Z., Zhao, J., Xu, N., & Liu, S. (2012). The temperature dependent proteomic analysis of *Thermotoga maritima*. *PLoS One, 7*(10), e46463. https://doi.org/10.1371/journal.pone.0046463

Wang, Q., Cen, Z., & Zhao, J. (2015). The survival mechanisms of thermophiles at high temperatures: An angle of omics. *Physiology, 30*(2), 97–106. https://doi.org/10.1152/physiol.00066.2013

Wang, K., Xue, K., Liu, W., Zhang, B., Wu, W., Zhao, R., Cui, L., Wang, Z., Zhou, S., Tang, L., & Dong, J. (2024). Warming decouples associations between microbial network complexity and ecosystem multifunctionality in alpine grasslands. *Agriculture, Ecosystems and Environment, 374*, 109189. https://doi.org/10.1016/j.agee.2024.109189

Wang, X., Wu, W., Ao, G., Han, M., Liu, M., Yin, R., Feng, J., & Zhu, B. (2025). Minor effects of warming on soil microbial diversity, richness and community structure. *Global Change Biology, 31*(3), e70104. https://doi.org/10.1111/gcb.70104

Weiskopf, S. R., Rubenstein, M. A., Crozier, L. G., Gaichas, S., Griffis, R., Halofsky, J. E., Hyde, K. J., Morelli, T. L., Morisette, J. T., Muñoz, R. C., & Pershing, A. J. (2020). Climate change effects on biodiversity, ecosystems, ecosystem services, and natural resource management in the United States. *Science of the Total Environment, 733*, 137782. https://doi.org/10.1016/j.scitotenv.2020.137782

Wu, S., Gao, X., Lei, J., Zhou, N., & Wang, Y. (2020). Spatial and temporal changes in the normalized difference vegetation index and their driving factors in the desert/grassland biome transition zone of the Sahel Region of Africa. *Remote Sens, 12*(24), 4119. https://doi.org/10.3390/rs12244119

Xie, T., Lin, Y., & Li, X. (2024). Responses of the arbuscular mycorrhizal fungi community to warming coupled with increased drought in an arid desert region. *Geoderma, 441*, 116744. https://doi.org/10.1016/j.geoderma.2023.116744

Xu, H., Huang, L., Chen, J., Zhou, H., Wan, Y., Qu, Q., Wang, M., & Xue, S. (2023). Changes in soil microbial activity and their linkages with soil carbon under global warming. *Catena, 232*, 107419. https://doi.org/10.1016/j.catena.2023.107419

Yang, Y., Li, T., Wang, Y., Cheng, H., Chang, S. X., Liang, C., & An, S. (2021). Negative effects of multiple global change factors on soil microbial diversity. *Soil Biology and Biochemistry, 156*, 108229. https://doi.org/10.1016/j.soilbio.2021.108229

Yiallouris, A., Pana, Z. D., Marangos, G., Tzyrka, I., Karanasios, S., Georgiou, I., Kontopyrgia, K., Triantafyllou, E., Seidel, D., Cornely, O. A., & Johnson, E. O. (2024). Fungal diversity in the soil Mycobiome: Implications for ONE health. *One Health, 16*, 100720. https://doi.org/10.1016/j.onehlt.2024.100720

Yu, Y., Liu, L., Wang, J., Zhang, Y., & Xiao, C. (2021). Effects of warming on the bacterial community and its function in a temperate steppe. *Science of the Total Environment, 792*, 148409. https://doi.org/10.1016/j.scitotenv.2021.148409

Zhan, J., Ericson, L., & Burdon, J. J. (2018). Climate change accelerates local disease extinction rates in a long-term wild host–pathogen association. *Global Change Biology, 24*(8), 3526–3536. https://doi.org/10.1111/gcb.14111

Zhang, N., Liu, W., Yang, H., Yu, X., Gutknecht, J. L., Zhang, Z., Wan, S., & Ma, K. (2013). Soil microbial responses to warming and increased precipitation and their implications for ecosystem C cycling. *Oecologia, 173*, 1125–1142. https://doi.org/10.1007/s00442-013-2685-9

Zhang, Y., Ku, Y. S., Cheung, T. Y., Cheng, S. S., Xin, D., Gombeau, K., Cai, Y., Lam, H. M., & Chan, T. F. (2024). Challenges to rhizobial adaptability in a changing climate: Genetic engineering solutions for stress tolerance. *Microbiological Research, 31*, 127886. https://doi.org/10.1016/j.micres.2024.127886

Zhao, Q., Yu, P., Mahendran, R., Huang, W., Gao, Y., Yang, Z., Ye, T., Wen, B., Wu, Y., Li, S., & Guo, Y. (2022). Global climate change and human health: Pathways and possible solutions. *Eco-Environment & Health, 1*(2), 53–62. https://doi.org/10.1016/j.eehl.2022.04.004

Zhao, J., Xie, X., Jiang, Y., Li, J., Fu, Q., Qiu, Y., Fu, X., Yao, Z., Dai, Z., Qiu, Y., & Chen, H. (2024). Effects of simulated warming on soil microbial community diversity and composition across diverse ecosystems. *Science of the Total Environment, 911*, 168793. https://doi.org/10.1016/j.scitotenv.2023.168793

Zheng, J., Cui, M., Wang, C., Wang, J., Wang, S., Sun, Z., Ren, F., Wan, S., & Han, S. (2022). Elevated CO2, warming, N addition, and increased precipitation affect different aspects of the arbuscular mycorrhizal fungal community. *Science of the Total Environment., 806*, 150522. https://doi.org/10.1016/j.scitotenv.2021.150522

Zhou, J., Deng, Y. E., Shen, L., Wen, C., Yan, Q., Ning, D., Qin, Y., Xue, K., Wu, L., He, Z., & Voordeckers, J. W. (2016). Temperature mediates continental-scale diversity of microbes in forest soils. *Nature Communications, 7*(1), 12083. https://doi.org/10.1038/ncomms12083

Zhou, L., Zhou, X., He, Y., Fu, Y., Du, Z., Lu, M., Sun, X., Li, C., Lu, C., Liu, R., & Zhou, G. (2022). Global systematic review with meta-analysis shows that warming effects on terrestrial plant biomass allocation are influenced by precipitation and mycorrhizal association. *Nature Communications, 13*(1), 4914. https://doi.org/10.1038/s41467-022-32671-9

Pathogenic Microbes and Climate Change: Emerging Risks

Abhiruchi Uniyal, Swati Lohani, Priyanka Khati, Vijaya Rani, and S. Santhiya

Abstract

Pathogenic microbes have taken advantage of climate change to evolve virulence and spread plant diseases. Climate change significantly alters the virulence and dissemination of pathogenic microbes by rising temperatures, changing precipitation patterns, and increasing atmospheric carbon dioxide concentrations, consequently enhancing their survival and reproduction rates. Microbial pathogens may evolve to become more virulent in response to environmental changes, and virulent strains resistant to plant defenses may also evolve. Climate change may also alter plant-microbe interactions with plant helpful microbes that assist plant defenses, take up nutrients, and perform other functions like nitrogen fixation. Increased humidity, altered precipitation patterns, and other climate-related environmental stressors (such as drought or nutrient shortage) may encourage the persistence and spread of pathogens, including their reproductive and dispersal stages, and lower plant resistance and immunity. A multifaceted approach to managing climate change's negative effects on plant health is crucial. To ensure sustainable agricultural production systems and food security in a changing climate, research on plant resilience and the interactions of pathogens with climate change will be essential.

A. Uniyal (✉)
Wildlife Institute of India, Dehradun, Uttarakhand, India

S. Lohani
ICAR VPKAS, Almora, Uttarakhand, India

P. Khati · S. Santhiya
Crop Production Division, ICAR VPKAS, Almora, Uttarakhand, India

V. Rani
ICAR-IARI, Division of Agricultural Microbiology, New Delhi, India

G. Bhandari et al. (eds.), *Microbial Resilience*,
https://doi.org/10.1007/978-981-95-8883-1_11

Keywords

Virulent microorganisms · Plant immunity · Climatic change · Pathogenicity · Disease spread

1 Introduction

Agriculture stands as the cornerstone of human civilization, ensuring not only sustenance, livelihoods, and economic stability but also serving as a fundamental pillar of society. Yet, this vital system faces an escalating threat from a myriad of pathogenic microbes like bacteria, fungi, viruses, and nematodes, which pose severe challenges to crop health, yield, and quality. The recent global climate shifts, characterized by changes in temperature, precipitation patterns, and the upsurge in extreme weather events all propelled by human-induced climate change have significantly reshaped the intricate dynamics between crops and their microbial adversaries (Raza et al., 2019; Singh et al., 2023). These climatic perturbations have not only heightened plant vulnerability but also bolstered pathogen vigor, leading to the emergence and proliferation of diseases in regions that were once considered immune to such threats (Timmusk et al., 2017; Rudgers et al., 2020).

The confluence of environmental stressors and increased pathogenicity disrupts the existing defense mechanisms of plants, rendering them more susceptible to infections. Factors such as erratic temperature variations and heightened humidity profoundly impact both the pathogen life cycles and the defense signaling pathways in plants, ultimately compromising host resistance. This disruption not only weakens the plants' ability to fend off invasive species but also creates a breeding ground for the augmentation of novel and more aggressive strains of plant pathogens.

The indirect effects of climate change on the soil microbiome, especially in the rhizosphere where pathogenic and symbiotic organisms coexist, have been the subject of several studies. Beneficial microorganisms may become less dominant under climate-induced stress, which would allow phytopathogens to flourish (Chaudhary et al., 2021a, b, c, d; Khati et al., 2024). Changes in soil temperature and moisture content, for example, can increase the survival of opportunistic pathogens while suppressing plant growth-promoting rhizobacteria (PGPR). According to the interaction between microbial communities and nanomaterials, some nano-compounds have the ability to alter microbial balance, which may lead to the development of novel mitigation techniques (Khati et al., 2017, 2018, 2019a, b; Chaudhary et al., 2021a, b, c, d).

The arrival, establishment, and spread of a pathogen across borders are becoming a global phenomenon, driven by climate-induced changes in pathogen ecology and evolution (Raza et al., 2019; Rudgers et al., 2020). In this context, developing integrated, multisource, and real-time predictive modeling frameworks for infectious diseases combining with meteorological, genomic, and ecological data is quite critical for modernizing surveillance and improving outbreak, along with public health and agricultural policies. Such interdisciplinary efforts blending plant pathology, climatology, and microbiology will inform adaptive strategies for real-time disease monitoring and control (Timmusk et al., 2017).

2 Pathogenic Microbes in Agriculture

The presence of pathogenic microorganisms in the agricultural sector such as bacteria, fungi, viruses, and nematodes presents a great threat to the productivity and quality of crops. These pathogens infect crops, interfere with the physiological functioning, and may result in massive epidemics causing economic losses and food security. The environmental stressors like climate change add to their effect. Being a consistent issue in the agricultural industry, pathogenic microorganisms are starting to impact the present-day climate even more evidently. Their general groups are as follows: the viruses, fungus, bacteria, and small beings such as nematodes. Bacterial pathogens are recognizable as pseudomonas and xanthomonas among many others. The *Pseudomonas* spp. cause bacterial blight and the *Xanthomonas* spp. cause the bacterial leaf spots. Sundin et al. (2016) point out that these diseases are transmitted by contaminated water, equipment, or plant material after they have entered into the plant using natural holes or wounds. They have developed ways of using stressed plant tissues, penetrating them through natural holes or injuries, and spreading through environmental vectors (Raza et al., 2019; Burdon and Zhan, 2020). This interactive relation between environmental values and bacterial virulence should be a topic of a combined study of microbiology and climatology.

Fungi constitute one of the most varied and destructive causes of plant diseases globally. *Fusaria* and *Alternaria* are fungi that cause vascular wilt and leaf spot, respectively. Their capacity to reproduce in diverse humidity and temperature conditions, which are increasingly occurring with climate change, and to rapidly adapt to genetic changes through the use of wind and water droplets indicates the need to consider the issue of fungal infections in crop protection programs (Singh et al., 2023). Their capacity to disperse by using wind and water droplets coupled with their capacity to rapidly adapt to genetic changes makes it more obvious that it is high time to consider the problem of fungal infections in crop protection programs.

Plant viruses pose a significant threat to global agriculture, disrupting plant growth and causing severe losses in both yield and quality. According to Timmusk et al. (2017), viruses like the tomato spotted wilt virus (TSWV) are heavily reliant on insect vectors, particularly thrips, whose populations thrive and multiply in warmer climates. Symptoms such as chlorosis, necrosis, and severe developmental distortions result in substantial economic damage. Furthermore, rising temperatures and milder winters have facilitated the spread of these viral diseases into new geographical regions. Consequently, controlling these pathogens requires a holistic approach that merges virology, ecology, and climate.

Simultaneously, plant-parasitic nematodes, such as those from the *Meloidogyne* genus, cause critical damage by forming root galls, which severely disrupt water and nutrient intake. These soil-borne pathogens are highly resilient, capable of surviving unfavorable conditions, and are severely exacerbated by rising soil temperatures, which accelerate their reproductive cycles.

The enhanced threat posed by both viruses and nematodes under changing climatic conditions underscores the urgent need for targeted integrated pest management (IPM) strategies. Effective mitigation depends on combining ecological, biological, and, when necessary, chemical controls to combat these evolving agricultural threats.

3 Economic Implications

Pathogenic outbreaks have an impact on quality and subvert agricultural economies, but their economic effects go beyond lost harvests. Agro-biodiversity losses brought on by pathogen-induced monoculture vulnerabilities may have a domino impact on trade balances and food supplies. To mitigate these losses, proactive investments in disease management techniques are crucial (Singh et al., 2023) (Fig. 1).

4 The Intersection of Climate Change and Agricultural Pathogens

Pathogen metabolism is improved by higher ambient temperatures, which speeds up life cycles and raises the risk of infection (Raza et al., 2019). Rise in temperature has direct and indirect effects on pathogen dynamics. Higher temperatures boost microbial metabolic rates and replication, but they also diminish host plant defenses, thereby facilitating the expansion of pathogens into regions historically characterized by cooler temperatures.

Similarly, both excessive rainfall and extended drought periods critically affect plant health. Waterlogged soils resulting from heavy precipitation favor the rapid spread of soilborne pathogens, while drought-induced plant stress heightens susceptibility to infections. The resulting microclimatic variability challenges traditional farming practices, prompting adaptive management strategies to mitigate these risks (Timmusk et al., 2017).

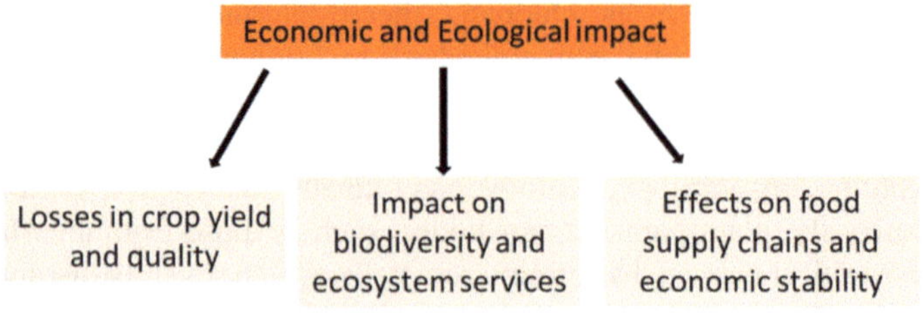

Fig. 1 Ecological and economic implications

Extreme weather events, particularly flooding, enable the long-distance transmission of soil-borne pathogens, effectively breaching geographic barriers that formerly contained them (Rudgers et al., 2020). Concurrently, more frequent heatwaves have been linked to increased pathogen virulence and faster life cycles, complicating regional biosecurity. As climate zones shift, the expanding ranges of plant viruses, combined with intensified global trade, have led to the emergence of *Xylella fastidiosa* and new strains of wheat rust in previously rare areas. These patterns underscore the urgent need for harmonized international biosecurity frameworks and proactive agricultural policies.

Modern farming methods are changing to mitigate the negative impacts of plant diseases brought on by climate change. Changing the dates of planting, adjusting irrigation techniques, and creating crop types that are more resistant to biotic and abiotic stressors are examples of adaptive methods. The use of integrated pest management (IPM) systems, which include targeted chemical interventions, cultural practices, and biological control, is growing. A diverse approach to climate issues is made possible by emerging technologies like precision agriculture and remote sensing, which provide farmers more ability to track and anticipate pathogen outbreaks.

Predictive disease risk maps are currently produced by sophisticated models that include real-time meteorological data, historical climate records, and pathogen life cycle characteristics.

5 Case Studies of Pathogen Emergence and Spread due to Climate Change

One of the most damaging fungal diseases in banana farming is fusarium wilt, also referred to as Panama disease. *Fusarium oxysporum* is the pathogen that causes wilting, chlorosis, and ultimately plant death by invading the vascular tissues. The Cavendish variety, which had offered reliable resistance for several decades, was later adopted after the broad effects of fusarium wilt caused the "Gros Michel" cultivar to fail (Raza et al., 2019). However, additional weaknesses have been shown by the introduction of Tropical Race 4 (TR4), putting Cavendish bananas in jeopardy. The mitigation strategies include the use of microbial biocontrol agent in IPM strategy, breed resistant cultivars, and others.

In another case, *Magnaporthe oryzae*, the fungus that causes rice blast, is still one of the most major destructive diseases that threaten rice production globally. This disease significantly lowers yield and quality by causing lesions on leaves, stems, and panicles. According to recent epidemiological data, climate change-induced increases in temperature and humidity have led rice blast to appear in areas that were thought to be resistant (Rudgers et al., 2020). While extended periods of high humidity and the resulting leafwetness from changed rainfall patterns promote persistent infections, elevated temperatures hasten the fungal life cycle and spore germination (Burdon & Zhan, 2020).

Mitigation strategies for rice blast are multifaceted; they include the development of rice cultivars with multiple layers of resistance via conventional breeding and genomic interventions and the implementation of integrated disease management practices such as optimized planting schedules, precise irrigation management, and rigorous field sanitation.

The third case study examines a notable increase in plant virus outbreaks, focusing on tomato spotted wilt virus (TSWV) and potato virus Y (PVY). These viruses exhibit rapid mutation rates and require efficient insect vectors for transmission. Milder winters and extended growing seasons have contributed to an increase in vector populations (e.g., thrips and aphids), thereby intensifying virus transmission (Díaz et al., 2021). Warmer climates not only enhance vector survival and reproduction but also create favorable microhabitats due to modified precipitation patterns that support elevated vector activity. This case underscores the necessity for robust surveillance systems that integrate climatic forecasts with vector monitoring, alongside policy interventions such as enhanced integrated pest management (IPM) and rapid dissemination of research findings to support timely adaptations.

6 Policy Implications

The findings emphasize how food security is threatened by both the geographic spread of viruses and the increase in speed of disease severity. The results highlight the need for coordinated action, which must include extension services, development of policy, and multiple research projects. Adopting innovative agronomic techniques, creating resilient cultivars, and putting in place reliable surveillance and early warning systems are all examples of proactive steps that are essential.

Furthermore, international collaboration is required to establish harmonized policies that address transboundary disease dynamics. Integrating traditional agricultural knowledge with modern technological advances will be pivotal to enhancing resilience across all levels, from local farms to global food systems.

7 Agricultural Adaptation and Resilience Mechanisms

1. Methods of integrated pest management (IPM) in a changing environment IPM is an ecosystem-based approach that uses chemical, biological, and cultural strategies to provide long-term pest management while reducing threats to the environment, economy, and human health. To disrupt pest life cycles, altered planting schedules, crop diversification, and habitat alteration have been implemented; real-time surveillance and locally targeted studies have also been employed to increase resilience (Leng & Huang, 2017; Foyer & Kranner, 2023; Kamakaula, 2024).
2. Advanced biotechnological techniques like clustered regularly interspaced short palindromic repeats (CRISPR)-mediated gene editing, which has demonstrated promise in precisely targeting susceptibility genes, are increasingly used to

supplement traditional breeding. To keep up with the rapid evolution of diseases, these genetic advancements enable accelerated breeding cycles; however, they also bring up ethical and ecological issues that need to be assessed (Bailly & Roldán, 2023; Rani et al., 2023).

3. Resilient agriculture still relies heavily on traditional approaches. While the use of cover crops enhances soil nitrogen and fosters beneficial microbial communities, crop rotation breaks the life cycles of diseases unique to a given host. Similarly modern irrigation methods like drip irrigation assist in controlling moisture to discourage water-dependent diseases, while practices like organic additions, reduced tillage, and selective composting improve soil microbial diversity (Sun et al., 2022; Tiwari, 2024).
4. Bacterial and fungal antagonists, including strains of *Bacillus*, *Pseudomonas*, and *Trichoderma*, are important components of integrated pathogen management because they lessen the need for chemical interventions and increase plant vigor (Webb et al., 2017).
5. Farmers may more efficiently monitor environmental conditions, predict pest outbreaks, and carry out timely interventions thanks to mobile-based decision support systems and precision agricultural technologies, which improve local adaptive capacity (Kgosikoma et al., 2018; Costa et al., 2019).
6. Policies that support extension services, finance research, and encourage climate-smart practices are shaped in part by agencies like the FAO, WHO, and IPCC. To protect global food security, quarantine and trade laws also work to limit the transboundary spread of infections (Alfizar & Nasution, 2024; Selvaraju et al., 2011).
7. Real-time environmental and pathogen data are essential to provide the temporal and spatial resolution for abrupt changes in crop conditions that could be signs of new disease threats (Zhou et al., 2023; Naik et al., 2024; Goyal et al., 2024). These parameters are crucial indicators of the conditions which favor the proliferation and spread of fungal, bacterial, and viral pathogens (Ashraf et al., 2023). Data collected by these sensors are instrumental in the calibration of early warning algorithms, providing the nuanced evidence required to forecast outbreaks before they reach epidemic levels. Furthermore, an essential component of modern surveillance strategies is the participatory approach. Farmers, extension agents, and citizen scientists are increasingly being integrated into formal monitoring frameworks, supplementing automated data collection systems. This synergistic approach to data integration is essential for creating a thorough and responsive monitoring system that can adjust to quick changes in the environment, as shown by Korošak et al. (2019) and Adilah et al. (2023). Lu et al. (2020a, b), for instance, have shown that predictive models that incorporate both historical and current environmental data can be used to quantify possible risk areas using probabilistic maps in addition to forecasting disease outbreaks.
8. These platforms enable agricultural policymakers and local communities to implement timely, focused interventions by giving them easily comprehensible information. These systems are made to combine information from sensor networks, satellite imagery, and user feedback, turning large, complicated

databases into useful insights. In this light, the work of Lu et al. (2020a, b) and Purwoko et al. (2023) backs up the idea that these integrated systems are essential to contemporary agricultural risk management.

8 Global Cooperation and Policy

Global challenges in agriculture now require a coordinated international response of Food and Agriculture Organization (FAO), World Health Organization (WHO), and the Intergovernmental Panel on Climate Change (IPCC), which are instrumental in formulating guidelines and technical standards designed to support adaptive agricultural practices. As described by Alfizar and Nasution (2024), these institutions play a pivotal role in establishing national research programs, building surveillance infrastructure, and promoting adaptive strategies that align local efforts with global sustainability goals (Fig. 2).

Policies developed at the international level are very critical for addressing transboundary challenges attached with the spread of agricultural pathogens causing serious diseases. Effective policy frameworks support the equitable distribution of resources. This will further ensure that developing nations receive the technical assistance necessary to implement advanced surveillance and early warning systems. In this regard, Selvaraju et al. (2011) emphasize that national policies should be designed to promote climate-smart agriculture, provide targeted funding to vital research and infrastructure projects, and incentivize the adoption of new technological solutions by local farming communities.

At a broader level, international cooperation must also converge with national policy measures aimed at mitigating the negative impacts of climate change on agriculture. Recent studies indicate that countries with well-established policy frameworks exhibit greater resilience when confronted with unexpected pathogen outbreaks (Lewis & Rudnick, 2019). Moreover, the effectiveness of these policies

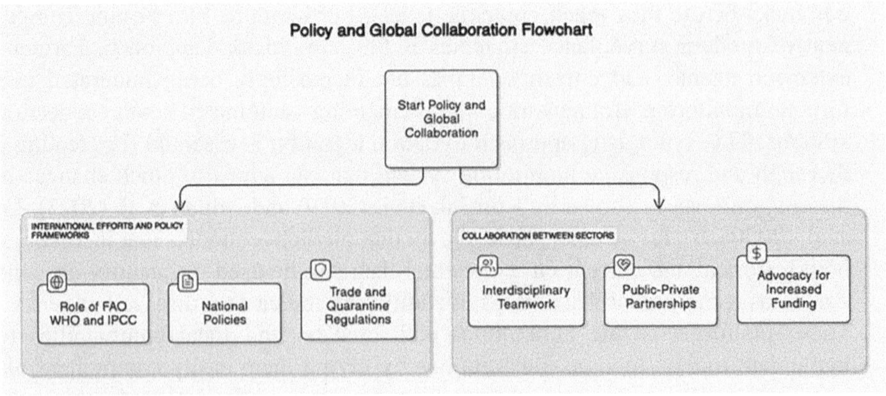

Fig. 2 Global Cooperation Policy flowchart

relies on continuous knowledge exchange and capacity-building efforts between nations. For instance, Mekyassi and Кизилdeniz (2024) argue that nutrition-sensitive climate-smart agriculture (CSA) is fundamental to ensuring food security, thereby highlighting the interrelated nature of agricultural practices, environmental health, and policy frameworks on the global stage.

International agreements such as the Paris Agreement provide a regulatory context within which individual nations must operate to reduce emissions and increase resilience in food production systems. As the urgency to achieve these global targets increases, policymakers are increasingly leveraging climate-smart agriculture (CSA) principles to develop comprehensive plans that simultaneously address adaptation and mitigation goals (Goyal et al., 2024). Alfizar and Nasution (2024) also suggest that such cross-sector collaborations are essential for pooling expertise and resources, thereby driving innovative solutions that address the multi-faceted nature of agricultural challenges.

Collaboration between sectors bridges the gap between high-level scientific research and practical on-the-ground farming practices. Jebory et al. (2024) demonstrate that linking consumer-grade sensor technologies with advanced remote data analytics can revolutionize how farmers manage crop health, while facilitating rapid responses to environmental stressors. These collaborations not only accelerate the adoption of cutting-edge technologies, such as machine learning-based predictive models and precision agriculture devices, but also facilitate ongoing refinements through feedback from end users (Naik et al., 2024). Collaborative efforts between international organizations, nongovernmental organizations (NGOs), and local governments have proven essential in disseminating best practices in climate-smart agriculture. As such, it is vital that policymaking processes include stakeholders from diverse fields, ensuring that agricultural policies are both scientifically robust and contextually relevant (Lewis & Rudnick, 2019; Goyal et al., 2024; Gardezi et al., 2021; Zougmoré et al., 2016).

9 Conclusion

The synergy between monitoring, early warning systems, and policy frameworks underscores the importance of a coordinated, interdisciplinary approach. The challenges posed by climate change require robust surveillance systems capable of detecting early signs of pathogen emergence, coupled with analytics to detect potential outbreaks. At the same time, the development of effective policies and the fostering of international, cross-sector collaborations ensure that these technological solutions are successfully implemented on the ground. By integrating advanced remote sensing, extensive sensor networks, participatory data collection, and adaptive predictive models with supportive policy frameworks and broad-based collaboration, the agricultural community can enhance both preparedness and resilience in the face of an unpredictable climate.

The rapid evolution of climatic conditions and pathogen threats has fundamentally reshaped agriculture. The integration of traditional agronomic practices with advanced remote sensing, sensor networks, and predictive analytics now forms the backbone of comprehensive surveillance and early warning systems. These specific strategies, when combined with robust policy frameworks and interdisciplinary collaborations, are essential for mitigating immediate crop losses and ensuring long-term food security (Lu et al., 2020a, b; Ashraf et al., 2023). Climate change accelerates the emergence and spreads of fungal, bacterial, viral, and nematode pathogens, as illustrated by the fusarium wilt outbreaks in bananas, the geographic expansion of rice blast, and increased plant virus incidences.

10 Future Prospects

Future research priorities include refining predictive models, enhancing real-time data integration, and advancing molecular genetic tools to develop pathogen-resistant crops. Strengthening international partnerships and aligning policy frameworks with localized adaptation needs will be vital in fostering innovations to meet these challenges. It is imperative that stakeholders from researchers and policymakers to agronomists and farming communities invest in sustainable practices and resilient technologies. The deployment of integrated pest management strategies, climate-smart agriculture, and early warning systems must be prioritized to safeguard global food security in the face of an unpredictable climate (Alfizar & Nasution, 2024; Jebory et al., 2024).

References

Adilah, N., Zaki, I., Ismail, I., Awal, M., Munajat, N., & Wahy, A. (2023). Soil monitoring for agriculture activity using low power wide area network. *Journal of Advanced Research in Applied Sciences and Engineering Technology, 33*(1), 219–230.

Alfizar, A., & Nasution, M. (2024). The explosion of pests and diseases due to climate change. *IOP Conference Series: Earth and Environmental Science, 1297*, 012072.

Ashraf, A., Ahmad, L., Ferooz, K., Ramzan, S., Ashraf, I., Khan, J., & Nazir, S. (2023). Remote sensing as a management and monitoring tool for agriculture: Potential applications. *International journal of environment and climate change, 13*(8), 324–343.

Bailly, C., & Roldán, M. (2023). Impact of climate perturbations on seeds and seed quality for global agriculture. *Biochemical Journal, 480*(3), 177–196.

Burdon, J. J., & Zhan, J. (2020). Climate change and disease in plant communities. *PLoS Biology, 18*(3), e3000949.

Chaudhary, P., Khati, P., Chaudhary, A., Gangola, S., Kumar, R., & Sharma, A. (2021a). Bioinoculation using indigenous Bacillus spp. improves growth and yield of Zea mays under the influence of nanozeolite. *3 Biotech, 11*, 11.

Chaudhary, P., Khati, P., Chaudhary, A., Maithani, D., Kumar, G., & Sharma, A. (2021b). Cultivable and metagenomic approach to study the combined impact of nanogypsum and *Pseudomonas taiwanensis* on maize plant health and its rhizospheric microbiome. *PLoS One, 16*(4), e0250574.

Chaudhary, P., Khati, P., Gangola, S., Kumar, A., Kumar, R., & Sharma, A. (2021c). Impact of nanochitosan and Bacillus spp. on health, productivity and defence response in *Zea mays* under field condition. *3 Biotech, 11*(5), 237.

Chaudhary, P., Sharma, A., Chaudhary, A., Khati, P., Gangola, S., & Maithani, D. (2021d). Illumina-based high-throughput analysis of microbial diversity of maize rhizosphere treated with nanocompounds and Bacillus sp. *Applied Soil Ecology, 159*, 103386.

Costa, J., Silva, J., Pinheiro, C., Barón, M., Mylona, P., Centritto, M., & Oliveira, M. (2019). Opportunities and limitations of crop phenotyping in Southern European countries. *Frontiers in Plant Science, 10*, 1125.

Díaz, J., et al. (2021). Climate change and plant viruses: Impacts and management. *Plant Pathology Journal, 37*(4), 289–300.

Foyer, C., & Kranner, I. (2023). Plant adaptation to climate change: A biochemistry perspective. *Biochemical Journal, 480*(2), 237–249.

Gardezi, S., et al. (2021). Prioritizing climate-smart agriculture: An organizational and temporal review. *Wiley Interdisciplinary Reviews: Climate Change, 12*, e755.

Goyal, A., Singh, A., Raghuraman, M., Ghosh, P., & Jadhav, A. (2024). Unveiling trends in forecasting models for crop pest and disease outbreaks: A systematic and scientometric analysis.

Jebory, A., Miry, A., & Aramice, G. (2024). Agriculture based on wireless sensor network: Literature survey. *Wasit Journal of Engineering Sciences, 12*(4), 171–190.

Kamakaula, Y. (2024). Ethnoecology and climate change adaptation in agriculture. *Global, 2*(2), 473–485.

Kgosikoma, K., Lekota, P., & Kgosikoma, O. (2018). Agro-pastoralists' determinants of adaptation to climate change. *International Journal of Climate Change Strategies and Management, 10*(3), 488–500.

Khati, P., Chaudhary, P., Bhatt, P., Nisha, Kumar, R., & Sharma, A. (2018). Effect of nanozeolite and plant growth promoting rhizobacteria on maize. 3. *Biotech, 8*(141), 1–12.

Khati, P., Chaudhary, P., Gangola, S., & Sharma, A. (2019a). Influence of nanozeolite on plant growth promotory bacterial isolates recovered from nanocompound-infested agriculture field. *Environment and Ecology, 37*(2), 521–527.

Khati, P., Chaudhary, P., Gangola, S., & Sharma, A. (2017). Nanochitosan-induced growth of Zea mays with soil health maintenance. *3 Biotech, 7*(81), 1–9.

Khati, P., Mishra, P. K., & Kant, L. (2024). Isolation, screening and characterization of drought tolerant plant growth promoting bacteria from Indian Himalayas. *Pantnagar Journal of Research, 22*(2).

Khati, P., Sharma, A., Chaudhary, P., Singh, A. K., Gangola, S., & Kumar, R. (2019b). High-throughput sequencing approach to assess the impact of nanozeolite treatment on species richness and evenness of soil metagenome. *Biocatalysis and Agricultural Biotechnology, 20*, 101249.

Korošak, Ž., Suhadolnik, N., & Pleteršek, A. (2019). The implementation of a low power environmental monitoring and soil moisture measurement system based on UHF RFID. *Sensors, 19*(24), 5527.

Leng, L., & Huang, F. (2017). Crop yield response to climate change varies with crop spatial distribution pattern. *Scientific Reports, 7*, 15999.

Lewis, P., & Rudnick, D. (2019). The policy enabling environment for climate-smart agriculture: A case study of California. *Frontiers in Sustainable Food Systems, 3*, 31.

Lu, S., Shang, Y., & Zhang, H. (2020a). Evaluation on early drought warning system in the Jinghui channel irrigation area. *International Journal of Environmental Research and Public Health, 17*(1), 374.

Lu, Y., et al. (2020b). Enhancing early warning systems for agricultural diseases by integrating climate data and machine learning. *Agricultural Systems, 179*, 102748.

Mekyassi, A., & Kizildeniz, İ. (2024). The role of nutrition-sensitive climate-smart agriculture in ensuring global food security. In *Bio Web of Conferences*.

Naik, A., Jogi, M., & Shreenivas, B. (2024). Assessing the impact of climate change on global crop yields and farming practices. *Archives of Current Research International, 24*(5), 696–712.

Purwoko, J., Wingardi, T., & Soewito, B. (2023). Smart agriculture water system using crop water stress index and weather prediction. *Commit: Communication and Information Technology Journal, 17*(1), 61–70.

Rani, K., Kumar, K., Navpute, G., Panwar, A., Kaur, A., & Kaushik, P. (2023). Microbiome interactions affect plant physiology and immunity to environmental stresses.

Raza, A., et al. (2019). Impact of climate change on plant pathology and crop production. *Journal of Agricultural Science, 157*(3), 301–315.

Rudgers, J. A., et al. (2020). The interplay of climate and pathogen evolution in agricultural ecosystems. *Global Change Biology, 26*(6), 3452–3464.

Selvaraju, R., Gommes, R., & Bernardi, M. (2011). Climate science in support of sustainable agriculture and food security. *Climate Research, 47*(1), 95–110.

Singh, B., Delgado-Baquerizo, M., Egidi, E., Guirado, E., Leach, J., Liu, H., & Trivedi, P. (2023). Climate change impacts plant pathogens, food security and paths forward. *Nature Reviews Microbiology, 21*(10), 640–656.

Sun, X., Lyu, J., & Ge, C. (2022). Knowledge and farmers' adoption of green production technologies: An empirical study on IPM adoption intention in major indica-rice-producing areas in the Anhui Province of China. *International Journal of Environmental Research and Public Health, 19*(21), 14292.

Sundin, G. W., Castiblanco, L. F., Yuan, X., Zeng, Q., & Yang, C. H. (2016). Bacterial disease management: Challenges, experience, innovation and future prospects. *Molecular Plant Pathology, 17*(9), 1506–1518.

Timmusk, S., et al. (2017). Advances in predictive plant pathology: The effects of climate change on vector-borne plant viruses. *Environmental Research Letters, 12*(4), 045001.

Tiwari, G. (2024). Insect pests in agriculture: Identifying and overcoming challenges through IPM. *Archives of Current Research International, 24*(3), 124–130.

Webb, N., Marshall, N., Stringer, L., Reed, M., Chappell, A., & Herrick, J. (2017). Land degradation and climate change: Building climate resilience in agriculture. *Frontiers in Ecology and the Environment, 15*(8), 450–459.

Zhou, Z., He, S., & He, B. (2023). Research on lossless continuous transmission method of agricultural low-altitude remote sensing image data. *Remote Sensing, 15*(7), 1245.

Zougmoré, L., et al. (2016). Toward climate-smart agriculture in West Africa: A review of climate change impacts, adaptation strategies and policy developments for the livestock, fishery and crop production sectors. *Agriculture & Food Security, 5*, 19.

Microbial Adaptation Strategies in Relation to Climate Change: Genetic and Phenotypic Changes

Meena Rathore, Shivani Kaushik, Pushpendra Sharma, and Rajeev Kaushik

Abstract

Microorganisms play a vital role in various ecosystems and are constantly challenged by the impacts of climate change. To survive and maintain their ecological functions, they have developed diverse adaptation mechanisms. This chapter delves into these mechanisms, examining both phenotypic and genotypic responses. Phenotypic adaptations enable microbes to respond immediately to environmental changes. These include producing heat-shock proteins to combat thermal stress, accumulating osmoprotectants to cope with salinity and drought, and forming biofilms for protection. Additionally, microbes exhibit metabolic flexibility, utilizing alternative carbon sources and optimizing energy metabolism. Behavioural adaptations, such as altered motility and quorum sensing, allow microbes to navigate dynamic habitats and coordinate survival strategies. Genotypic adaptations involve heritable changes in genetic material. Point mutations and indels enhance stress tolerance, while horizontal gene transfer (HGT) facilitates the acquisition of beneficial traits, including novel metabolic pathways and antibiotic resistance. Genome rearrangements streamline functions for resource efficiency, and epigenetic modifications regulate stress-responsive gene expression. Microbial adaptation is evident in diverse environments. For example, Arctic Ocean bacteria exhibit increased HGT rates under warming conditions, acquiring thermal tolerance genes. Similarly, *Sphingobium chlorophenolicum* evolved the ability to degrade the pollutant pentachlorophenol, highlighting the microbial potential for bioremediation. Understanding microbial adaptation is crucial for predicting ecosystem responses to climate change and developing mitigation strategies. By elucidating the mechanisms of adaptation, researchers can better anticipate shifts in microbial community dynamics and

M. Rathore · S. Kaushik · P. Sharma · R. Kaushik (✉)
Division of Microbiology, ICAR-Indian Agricultural Research Institute, New Delhi, India

G. Bhandari et al. (eds.), *Microbial Resilience*,
https://doi.org/10.1007/978-981-95-8883-1_12

their ecosystem impacts. Future studies should integrate multi-omic approaches and environmental monitoring to unlock microbial potential for climate resilience. Microbial adaptation serves a dual role: as a natural response to climate stress and as a tool for sustainable climate solutions. Harnessing microbial strategies enables the development of innovative solutions to combat climate change while maintaining ecological balance. Therefore, incorporating microbial research into comprehensive climate action frameworks is essential to promote a sustainable future.

Keywords

Climate change · Microbial adaptation · Microbial resilience · Microbial mechanisms · Horizontal gene transfer

1 Introduction

Climate change is one of the most pressing environmental challenges of the twenty-first century, with far-reaching effects on ecosystems, biodiversity, and human societies. Its main causes are human activities such as deforestation, industrialization, urbanization, and overuse of natural resources (Tilman & Lehman, 2001; Yue & Gao, 2018; Webster et al., 2018). These actions increase greenhouse gas emissions, such as CO_2 (carbon dioxide), CH_4 (methane), and N_2O (nitrous oxide), trapping heat and driving global warming (Heshmati, 2020). While the climate remained stable for over 12,000 years, recent decades have seen rapid and significant changes. Global temperatures have already risen by about 0.7 °C in the past century, with projections indicating a rise of 1.1–6.4 °C by the century's end (Qin et al., 2014; Sintayehu, 2018; Muluneh, 2021), potentially disrupting weather patterns, sea levels, and ecosystems.

Microorganisms are among the most resilient and adaptable life forms on Earth, capable of thriving in diverse and extreme environments. They drive key biogeochemical cycles, including carbon and nitrogen processing, soil fertility, and GHG emissions (Fig. 1). For example, heterotrophs emit CO_2 through organic matter decomposition, while photosynthetic microbes sequester carbon. Their ecological roles are deeply influenced by climate variables. However, the rapid pace of climate change, driven by increasing greenhouse gas emissions, global warming, and altered precipitation patterns, poses significant challenges to microbial communities (Cavicchioli et al., 2019). In response, microbes employ a range of genetic and phenotypic adaptation strategies to survive and maintain ecosystem functions under shifting environmental conditions (Singh et al., 2010). Microbial adaptation mechanisms can be broadly categorized into genetic changes, such as mutations, horizontal gene transfer (HGT), and plasmid acquisition, and phenotypic changes, including metabolic flexibility, biofilm formation, and stress-responsive gene expression (Ferenjian et al., 2021) (Table 1). These adaptations not only ensure microbial survival but also influence biogeochemical cycles, plant–microbe

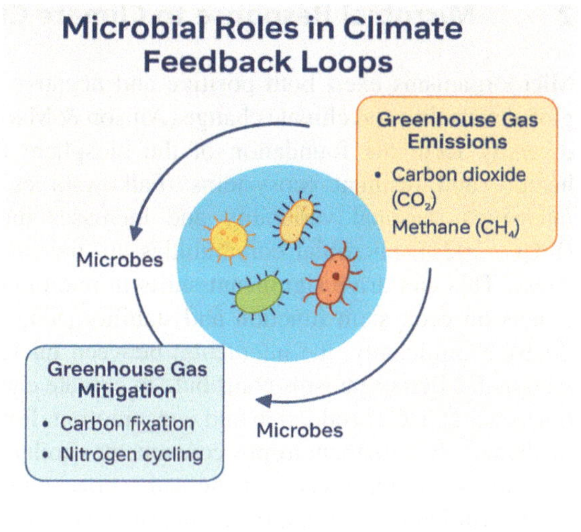

Fig. 1 Microbial roles in climate feedback loops, showing both greenhouse gas emissions and mitigation pathways through carbon fixation, nitrogen cycling, and microbial adaptation

Table 1 Climate change factors and their corresponding microbial responses and affected ecological processes

Climate factor	Microbial response	Affected processes
↑ Temperature	↑ Metabolism, enzyme activity, respiration	Soil respiration, carbon cycling
↑ CO_2 concentration	Ocean acidification, pH shifts	Marine productivity, carbon fixation
↓ pH	Enzyme inhibition, altered gene expression	Nitrification, denitrification
↑ Salinity	Osmotic stress, selection for halotolerant strains	Microbial diversity, survival
↓ Oxygen availability	Anaerobic metabolism, microbial shifts	Methanogenesis, sulfate reduction
↑ Climate variability	Stress responses, gene regulation	Community composition, adaptation dynamics

Note: ↑ represents an increase and ↓ represents a decrease.

interactions, and the overall stability of ecosystems (Jansson & Hofmockel, 2020; Khati et al., 2024). For instance, some bacteria evolve enhanced carbon utilization pathways in response to increased atmospheric CO_2, while others develop heat-shock proteins to cope with rising temperatures (López-García et al., 2022).

Understanding microbial adaptation strategies is crucial for predicting ecosystem responses to climate change and developing mitigation strategies, such as bioengineering climate-resistant microbiomes (Trivedi et al., 2022). This chapter explores the genetic and phenotypic mechanisms underpinning microbial resilience, highlighting recent research on evolutionary dynamics, microbial ecology, and potential biotechnological applications in a changing climate.

2 Microbial Response to Climate Change

Microorganisms exert both positive and negative feedback effects in response to global warming and climate change (Allison & Martiny, 2008). Their abundance and diversity form the foundation of the biosphere and are essential for sustaining healthy and resilient ecosystems (Falkowski et al., 2008). As climate change intensifies seasonal variability and increases the frequency of extreme events (IPCC, 2021), microbial communities are increasingly exposed to environmental stress. This can drive significant shifts in microbial biodiversity, with downstream effects on ecosystem function and stability (Singh et al., 2010; Cavicchioli et al., 2019). Significantly, the interaction between microbes and climate change is bidirectional. Microorganisms contribute to climate change through their roles in greenhouse gas (GHG) production and consumption. For example, methanogens produce methane, while methanotrophs consume it; similarly, nitrifiers and denitrifiers influence nitrous oxide levels. At the same time, microbes are also central to climate change mitigation through processes such as carbon sequestration, nutrient cycling, and bioremediation. Moreover, microbial biodiversity is itself highly sensitive to changing environmental conditions, including temperature rise, altered precipitation patterns, and soil chemistry shifts, which may alter microbial-mediated functions critical to ecosystem resilience (Schimel & Schaeffer, 2012). Altogether, microbial responses to climate change are multifaceted, complex, and challenging to quantify, yet they play a pivotal role in shaping global climate feedback loops as shown in Fig. 1 (Graham et al., 2016).

2.1 Changes in Population Dynamics and Community Structure

Climate change significantly alters microbial population dynamics, diversity, and community structure by modifying key environmental variables such as temperature, moisture, and pH. These effects are often complex and variable, influenced by both deterministic factors (e.g., climate gradients) and stochastic events (e.g., extreme weather). Microbial metabolism, growth rate, and species richness generally increase with rising temperatures. For instance, bacterial and fungal diversity tends to decrease with increasing latitude, paralleling the decline in ambient temperature (Lozupone & Knight, 2007). Temperature-driven changes can lead to geographic range shifts, altered phenology, and modified species abundance (Bellard et al., 2012). While some microbial taxa thrive in warmer conditions, others decline, leading to disrupted community composition and shifts in ecosystem functions. Precipitation changes further complicate microbial dynamics. Drought reduces soil moisture, favoring drought-resistant taxa and suppressing moisture-dependent microbes. This often results in reduced microbial diversity, diminished functional redundancy, and altered nutrient cycling, ultimately impairing soil fertility and plant productivity. Conversely, excessive rainfall and flooding create anoxic conditions that favor denitrifiers and methanogens, enhancing the production of potent greenhouse gases such as nitrous oxide (N_2O) and methane (CH_4). Such conditions can profoundly affect biogeochemical cycling in both terrestrial and aquatic systems. Altogether, shifts in microbial population dynamics and community structure under changing climatic regimes have profound implications for ecosystem resilience, nutrient availability, and climate feedback loops.

2.2 Shifts in Metabolic Activity and Biogeochemical Cycling

Rising global temperatures significantly influence microbial metabolism and biogeochemical cycling, with cascading effects on ecosystem functioning and climate feedback loops. One of the most immediate consequences is the acceleration of microbial respiration, leading to enhanced decomposition of organic matter in soils and aquatic ecosystems. This results in increased emissions of carbon dioxide (CO_2) and methane (CH_4), intensifying global warming through positive feedback mechanisms. Temperature affects microbial biodiversity and metabolism through several pathways. Warmer conditions typically increase metabolic rates, which shorten population doubling times and elevate the rates of ecological and evolutionary processes such as mutation, speciation, and interspecies interactions, including parasitism, mutualism, and competition (Gillooly et al., 2001; Cavicchioli et al., 2019). In theory, this can lead to more diverse and active microbial communities in warming ecosystems such as tropical forests or alpine soils, enhancing decomposition, carbon sequestration, and nutrient cycling (Lozupone & Knight, 2007). However, rising temperatures can also act as environmental filters, selecting for heat-adapted species while reducing stochastic diversity and dispersal, potentially limiting the temporal stability of microbial communities. The overall impact depends heavily on functional group traits and phylogenetic lineages (Graham et al., 2016).

In Arctic and sub-Arctic regions, where warming occurs at nearly twice the global rate, permafrost thawing exposes long-frozen organic carbon to microbial degradation. Methanogenic archaea, which thrive under anaerobic conditions, rapidly metabolize this material, releasing substantial quantities of CH_4, i.e., a potent greenhouse gas. This process could significantly accelerate climate change beyond current model projections. In marine ecosystems, rising sea surface temperatures disrupt microbial plankton communities, affecting primary production and carbon export to deeper ocean layers. Additionally, the absorption of atmospheric CO_2 leads to ocean acidification, which negatively affects calcium carbonate-forming microorganisms such as coccolithophores and foraminifera. These changes reduce the efficiency of carbon fixation, alter nutrient dynamics, and threaten symbiotic associations, such as those between corals and their microbial partners, contributing to coral bleaching and ecosystem degradation.

2.3 Range Shifts and Alterations in Microbial Biogeographic Distributions

Microbial species exhibit remarkable adaptability to changing environmental conditions, leading to shifts in their geographic distributions. Conversely, sensitive microbial populations may decline or migrate to more favorable habitats, altering ecosystem functions. Climate change-induced variations in precipitation patterns, soil moisture, and temperature drive microbial dispersal and survival, influencing plant–microbe interactions, soil health, and decomposition processes. In marine environments, shifts in ocean currents, salinity, and temperature gradients profoundly affect microbial biogeography, impacting primary production, nutrient cycling, and food web dynamics (Picazo et al., 2020).

Additionally, climate change and anthropogenic pressures can restructure microbial community composition, leading to functional redundancy or the loss of specialized metabolic capabilities that are critical for ecosystem resilience. Certain microbial groups, such as thermophilic cyanobacteria and extremophilic archaea, are projected to expand their ranges poleward, while sensitive taxa such as symbiotic mycorrhizal fungi may face significant range contractions, threatening plant biodiversity. Moreover, microbial shifts are not purely passive responses; they actively reshape biogeochemical cycles, with potential feedbacks to climate systems, such as increased methane emissions from warming wetlands and altered carbon sequestration dynamics. Human activities, including land-use changes, pollution, and habitat fragmentation, further interact with climate drivers, accelerating changes in microbial distributions. Predictive models suggest that continued environmental change could lead to a homogenization of microbial communities across biomes, reducing global microbial diversity and ecosystem multifunctionality (Cavicchioli et al., 2019). In terrestrial ecosystems, soil warming favors the proliferation of thermophilic microbes, often at the expense of native taxa, leading to altered decomposition rates and disruptions in carbon and nitrogen cycling (Graham et al., 2014; Crowther et al., 2016). Likewise, changes in precipitation and soil moisture influence microbial dispersal and community assembly, which in turn impacts plant–microbe interactions, soil fertility, and organic matter turnover. For example, drought conditions often reduce microbial diversity, selecting drought-tolerant species with limited functional redundancy, while excessive rainfall can lead to anaerobic conditions that enhance denitrification and methane production (Fierer & Jackson, 2006; Evans & Wallenstein, 2012).

2.4 Specific Examples of Microbial Responses to Different Climate Change Factors

Microbial communities are highly sensitive and adaptable to various climate change drivers, including temperature shifts, ocean acidification, extreme weather events, and elevated atmospheric CO_2 levels. These environmental changes trigger complex microbial responses that directly impact global biogeochemical cycles, ecosystem stability, and climate feedback mechanisms. Understanding specific microbial adaptations and their ecological consequences is crucial for predicting future ecosystem resilience under ongoing climate change (Table 2).

(a) *Temperature*

Temperature changes exert both direct and indirect effects on microbial communities by regulating metabolism, growth, survival, and shaping community composition through environmental filtering (Allison & Treseder, 2008; Bradford, 2013). Warmer conditions increase microbial metabolic rates and enzyme activities up to a thermal optimum, beyond which cellular processes destabilize. For instance, rising temperatures accelerate microbial respiration and population doubling times, contributing to increased CO_2 and CH_4 emissions, thereby intensifying climate change feedback loops, especially in sensitive regions such as the Arctic where permafrost thaw releases large quantities of organic carbon (Schuur et al., 2015; Natali et al., 2019). Cold-adapted ecosystems are particularly vulnerable; warming favors methanogenic archaea, substantially boosting greenhouse gas emissions.

Microorganisms have evolved multiple strategies to cope with extreme temperatures. Psychrophilic species surviving in cryospheric environments use biochemical adaptations and antioxidant systems, while thermophiles thrive in near-boiling hot springs (Margesin & Miteva, 2011; Bakermans, 2015). Bacterial endospores provide heat resistance, enabling persistence in extreme environments. In newly exposed glacial and high-altitude soils, pioneering microbial communities influence early soil development and biogeochemical cycles, acting as indicators of environmental change (Nemergut et al., 2007).

Furthermore, microbial adaptation to warming includes metabolic shifts, changes in respiration rates, and reductions in average cell size. For example, bacterial respiration increases with warming, while cold-adapted marine viruses have evolved cold-shock proteins to tolerate low temperatures (Liu et al., 2022; Zhou et al., 2016).

(b) *Ocean acidification*

The rising concentration of atmospheric carbon dioxide (CO_2) is leading to ocean acidification, characterized by a decrease in surface ocean pH and alterations in seawater carbonate chemistry, with potentially profound effects on marine biogeochemical processes (Keeling et al., 2010; Brewer & Peltzer, 2009). Microbial processes such as carbon dioxide production and nitrification influence pH dynamics, while warming-induced stratification reduces dissolved oxygen (O_2) levels, further constraining aerobic microbial activity in low-O_2 regions (Keeling et al., 2010). Ocean acidification disrupts microbial interactions by impairing biofilm formation, symbiotic associations, and nutrient uptake efficiency and affects microbial carbonate metabolism, thereby influencing phytoplankton productivity and carbon sequestration (Joint et al., 2011). Variability exists among phytoplankton responses: diazotrophs, diatoms, and dinoflagellates often exhibit enhanced growth under elevated CO_2, while coccolithophores and cyanobacteria such as *Synechococcus* and

Table 2 Microbial responses to climate stressors

Climate change factor	Microbial response	Ecological impact	Specific examples
Temperature	Increased microbial respiration Shifts in microbial community composition Adaptations to extreme temperatures	Enhanced greenhouse gas emissions intensify global warming (feedback loop) Loss of cold-adapted species Altered soil carbon storage and nutrient dynamics	Higher CO_2 and CH_4 emissions from thawing permafrost (Schuur et al., 2015; Natali et al., 2019) Thermophiles thriving in hot springs (Margesin & Miteva, 2011) Psychrophiles adapting to glacial melt zones (Margesin & Miteva, 2011) Cold shock proteins in marine viruses (Zhou et al., 2016)
Ocean acidification	Disruption of calcifying microorganisms Changes in phytoplankton community composition Impaired microbial interactions and biofilm formation	Reduced primary production Weakened marine food webs Decreased oceanic carbon sequestration efficiency	Negative effects on coccolithophores, foraminifera (Hutchins & Fu, 2017; Jiao et al., 2010) Mixed phytoplankton responses (Riebesell et al., 2000; Fu et al., 2007) Impaired biofilm formation (Joint et al., 2011)
Extreme weather events	Stress-tolerant taxa selection Dormancy and biofilm formation under drought Successional shifts after flooding	Loss of microbial diversity Reduced soil fertility and ecosystem stability Increased methane emissions during floods Permanent alteration of microbial-driven nutrient cycles	Microbial dormancy during drought (Fuchslueger et al., 2014) Flood-induced community shifts (Johnstone & Lucy, 2020) Synergistic effects from compound events (Canarini et al., 2021)
Increased CO_2 levels	Altered carbon and nitrogen cycling Favoring of acidophilic and anaerobic microbes Ocean acidification effects on microbial diversity	Altered soil respiration rates Higher greenhouse gas emissions (N_2O, CH_4) Decreased deep-sea microbial diversity Risk of ecosystem destabilization near CO_2 leakage zones	Enhanced microbial carbon fixation (Wang et al., 2016) Increased N_2O emissions (van Groenigen et al., 2011) Microbial community shifts in CO_2 seep zones (Yanagawa et al., 2013) Metal mobilization affecting deep-sea microbes (Basallote et al., 2014)

Prochlorococcus display mixed or negligible responses (Riebesell et al., 2000; Fu et al., 2007; Tortell et al., 1997). Some studies even report increased calcification in certain coccolithophore strains under high CO_2 conditions (Langer et al., 2006; Iglesias-Rodriguez et al., 2008), though strain variation and experimental differences may explain conflicting results (Ridgwell et al., 2009; Rost et al., 2008). In addition to acidification, elevated ocean temperatures will intensify stratification, limit nutrient flux, and alter light penetration, all of which affect microbial community structure and function (Steinacher et al., 2010). While surface CO_2 and pH levels naturally fluctuate, particularly in coastal and freshwater systems, the adaptability of microbial communities to future conditions remains a crucial research focus. Despite concerns for calcifying organisms and specific photosynthetic taxa, many marine microbes may exhibit resilience, and assumptions of catastrophic shifts in microbial-driven biogeochemical processes should be validated with further evidence.

(c) *Extreme weather events and microbial resilience*

Climate change is projected to increase the intensity, frequency, and duration of extreme weather events, such as droughts and floods, which significantly impact ecosystems and microbial communities (Johnstone & Lucy, 2020). These disturbances alter the physicochemical properties of soil and disrupt microbial structures and functions, raising concerns about the resilience and recovery of ecosystems (Philippot et al., 2021). Repeated disturbances can have cumulative effects that reduce microbial stability and impair ecological function. Microbial communities often adapt to stress via physiological changes such as modifications in membrane structure or transcriptional regulation (Eberlein et al., 2018). Additionally, ecological filtering favors disturbance-tolerant taxa, while evolutionary processes such as hypermutability may enhance community resilience (Eberlein et al., 2018).

Drought, anticipated to increase with global warming, limits microbial substrate diffusion, reduces microbial abundance and may induce dormancy (Fuchslueger et al., 2014). Microorganisms can mitigate drought impacts through protective strategies such as biofilm formation and osmolyte accumulation, which support osmotic balance and survival during desiccation (Schimel et al., 2007). Upon rewetting, microbes rapidly release osmolytes to reestablish homeostasis. Some studies have shown that exposure to repeated drought events enhances microbial resistance, preserving community composition and functional activity (de Vries et al., 2012; Meisner et al., 2018). However, other research suggests that not all microbial taxa exhibit legacy responses; for instance, methanotrophs in lake sediments and paddy soils showed no adaptive benefits from repeated desiccation (Ho et al., 2016). Microbial resilience under drought appears context-dependent, varying with soil type, land use, and initial community composition (Evans et al., 2014; Canarini et al., 2017).

Flooding also imposes selective pressure on microbial communities. Johnstone and Lucy (2020) observed that summer floods had more significant impacts on soil microbiomes than winter floods, with longer inundation (8 weeks) causing persistent changes in community composition, particularly in arable and grassland soils. These floods increased methane emissions while decreasing nitrous oxide flux, with the magnitude and direction of change influenced by land management practices. Disturbances such as flooding often trigger secondary succession, altering microbial diversity and structure based on the community's successional stage and the nature of the disturbance (Röhl et al., 2022). Röhl and colleagues found that while terrestrial microbial communities in flooded mesocosms were resilient, aquatic communities underwent lasting shifts due to nutrient influx from suspended soil particles, which disproportionately altered aquatic microbial dynamics. Furthermore, compound extreme events such as sequential droughts followed by flooding can have unique, synergistic effects that differ from isolated disturbances. These cascading events may permanently alter microbial composition and function, affecting biogeochemical cycling and long-term ecosystem stability (Canarini et al., 2021). As climate variability increases, so does the likelihood of irreversible transformations in soil microbial communities.

(d) *Increased CO_2 levels*

As atmospheric CO_2 concentrations rise far above pre-industrial levels, microbial communities are undergoing profound transformations (Petit et al., 1999). Microbes, though microscopic, are essential to processes such as carbon cycling, nitrogen transformation, and ecosystem stability. Figure 2 displays how these changes are reshaping life beneath our feet and the waves and also highlights how high CO_2 flux can mobilize toxic metals in sediments, impacting deep-sea microbial health (Basallote et al., 2014).

On land, elevated CO_2 alters plant–microbe interactions, affects soil respiration, and changes nutrient cycling. Certain microbes enhance carbon fixation and boost soil carbon storage (Wang et al., 2016), while others increase nitrous oxide emissions, a potent greenhouse gas (van Groenigen et al., 2011). Moreover, high CO_2 concentrations can select for acidophilic and anaerobic microbes, reshaping entire microbial ecosystems (Beulig et al., 2015).

In marine systems, dissolved CO_2 leads to acidification, threatening microbial diversity and activity, especially around high-CO_2 seep zones (Yanagawa et al., 2013). Acidification enhances methanogen activity, increasing greenhouse gas emissions, while impairing critical processes such as organic matter degradation (Rastelli et al., 2015).

High CO_2 fluxes can also mobilize toxic metals from sediments, impacting deep-sea microbial health (Basallote et al., 2014). Furthermore, emerging carbon capture and storage (CCS) technologies, while promising for climate mitigation, could alter soil and marine microbial communities if leaks occur (de Coninck & Benson, 2014).

Figure 2 illustrates how increased atmospheric CO_2 levels impact microbial communities across terrestrial and marine ecosystems. Elevated CO_2 contributes to ocean acidification, leading to reduced microbial diversity, impaired nutrient cycling, and increased mobilization of toxic metals. In soil systems, higher CO_2 levels influence plant–microbe interactions, enhance methane and nitrous oxide

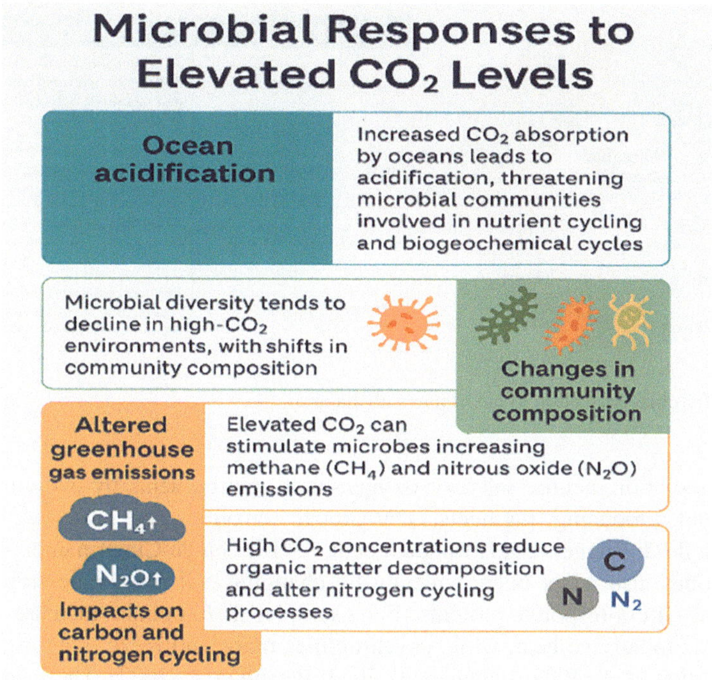

Fig. 2 Microbial responses to elevated CO_2

emissions and shift microbial community composition toward acidophilic and anaerobic taxa. Carbon capture and storage (CCS) strategies, while mitigating atmospheric CO_2, pose ecological risks if leakage occurs, potentially disrupting microbial functions. These microbial responses have cascading effects on ecosystem stability and global biogeochemical cycles.

3 Microbial Adaptation in Response to Climate Change

Microbial evolution and adaptation are essential for understanding how ecosystems respond to climate change (Fig. 3). Microorganisms such as bacteria, archaea, fungi, and viruses play major roles in nutrient and carbon cycling. As global temperatures rise and rainfall patterns change, microbial communities shift, influencing whether ecosystems release or absorb carbon (Cavicchioli et al., 2019).

Soil microbes help control the global carbon cycle. Their response to climate change affects ecosystem stability and feedback to the atmosphere (Thakur & Geisen, 2019; Wang et al., 2021). Microbes adapt to environmental stress through genetic and physical changes, including mutations, horizontal gene transfer, and altered gene regulation (Ahamad, 2022). Extremophiles—microbes that thrive in extreme conditions—show how powerful microbial adaptation can be. They survive

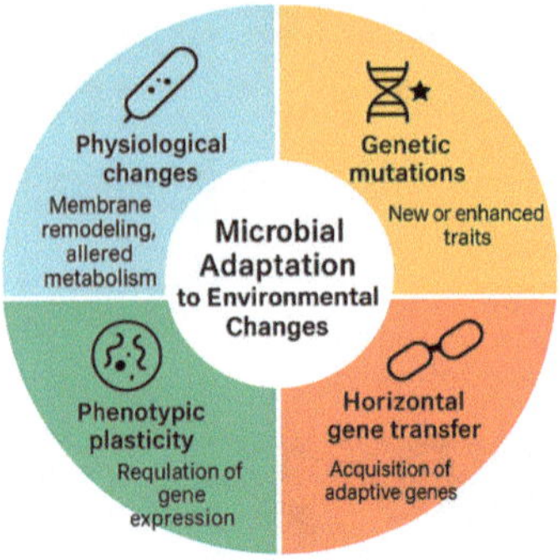

Mechanism	Description
Physiological changes	Alterations in membrane composition, protein exprresssion, and metabolic activity in response to abiotic factors
Genetic mutations	Mutations that arise during replication, generating new talts or enhancing existing traits
Horizontal gene transfer	Acquisition of adaptive genes from other microorganisms through transformation, or conjugition
Phenotypic plasticity	Regulation of gene expression in response to environmental stimuli, enabling rapid acclimation

Fig. 3 Microbial adaptation to environmental changes

in high radiation, acidic, salty, or oxygen-poor environments by evolving special genes and producing pigments, DNA-repair enzymes, or protective biofilms (Canganella & Wiegel, 2011; Raddadi et al., 2015; Gunde-Cimerman et al., 2018).

Microbes also adapt phenotypically by changing cell membrane structure and producing stress-response proteins. For example, thermophiles increase saturated fatty acids to survive heat, while psychrophiles make antifreeze proteins to handle cold (Bartlett et al., 2007; Lewin et al., 2013; Parrilli et al., 2022). These adaptations help microbes maintain ecosystem functions and influence the carbon and nitrogen cycles (Hutchins et al., 2019; Jansson & Hofmockel, 2020). As climate change progresses, understanding microbial adaptation will be crucial for predicting ecosystem responses and leveraging microbial capabilities in biotechnological innovations aimed at environmental sustainability and climate mitigation (Burckhardt, 2023; Malhi et al., 2020; Cavicchioli et al., 2019).

3.1 Genetic Adaptation Mechanism

Microbial adaptation to climate change is largely governed by genotypic changes driven by diverse molecular mechanisms. Microorganisms exhibit extraordinary genetic plasticity, allowing them to respond rapidly to environmental pressures such as rising temperatures, salinity, oxidative stress, heavy metal toxicity, and increased atmospheric CO_2. This adaptability is facilitated by mutation, horizontal gene transfer, gene duplication, and selective pressures that enable the evolution of novel traits such as metabolic flexibility, stress resistance, and the ability to colonize extreme environments (Ahamad, 2022). Such traits are essential for survival in dynamic ecosystems and contribute to key nutrient cycling, which directly influences global climate feedback (Liu et al., 2018).

The general mechanism by which microbes detect stress and activate response pathways is often mediated by two-component systems and sigma factors (Fig. 4). For example, in response to heat stress, microbes synthesize heat shock proteins

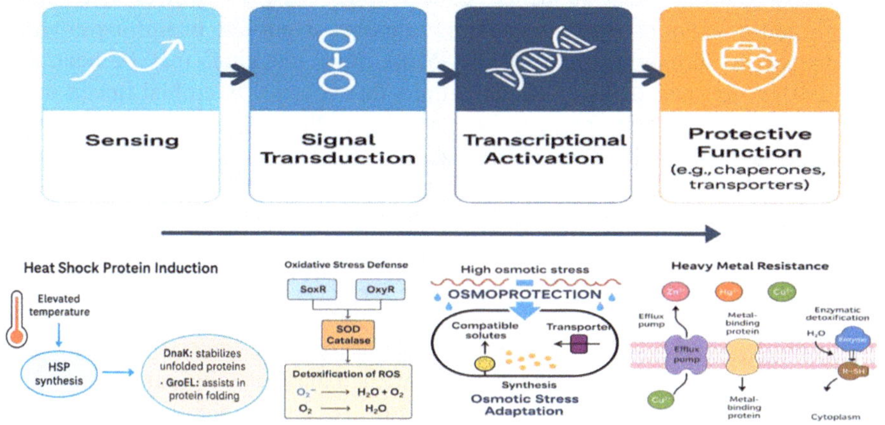

Fig. 4 General stress response pathway in microbes

Table 3 Compatible solutes in osmotic stress

Solute	Microbes	Function
Trehalose	*Escherichia coli*, *Bacillus* spp.	Stabilizes proteins/membranes
Glycine/betaine	*Halomonas* spp., *Vibrio*	Maintains osmotic balance
Proline	*Rhizobium*, *Pseudomonas*	Prevents dehydration

Table 4 Microbial heavy metal resistance mechanisms

Metal	Resistance gene	Mechanism
Zinc	*czcA*	Efflux pump
Copper	*copA*	ATP-driven efflux
Mercury	*merA*	Reductive detoxification
Arsenic	*arsC*	Reduction and efflux

(HSPs) such as DnaK, GroEL, and HtpG, which act as molecular chaperones to refold denatured proteins (Fig. 4). Similarly, during oxidative stress, regulatory proteins such as OxyR and SoxR detect reactive oxygen species (ROS) and trigger the expression of detoxifying enzymes such as catalase and superoxide dismutase (Fig. 4). Microbes accumulate compatible solutes such as trehalose, proline, glycine, and betaine, stabilizing cellular structures to counteract osmotic stress (Fig. 4 and Table 3). For heavy metal resistance, microbes employ efflux pumps, metal sequestration proteins, and enzymatic detoxification systems such as merA for mercury resistance (Fig. 4 and Table 4).

(i) *Mutation*: It serves as a fundamental adaptation strategy for microbes in response to climate change by generating genetic diversity essential for evolutionary processes. These genetic alterations—ranging from point mutations (insertions, deletions, duplications, and substitutions) to larger chromosomal rearrangements—can influence critical traits such as growth rate, stress response, and reproduction, enabling microbes to survive in shifting environmental conditions (Carline, 2011). While many mutations may be neutral or deleterious, some confer beneficial traits that enhance microbial fitness under climate-induced stressors, such as temperature fluctuations or oxidative stress (Ahamad, 2022). Microorganisms, particularly bacteria, possess short generation times and large population sizes, which facilitate the rapid accumulation of advantageous mutations and allow swift adaptation (Horton & Taylor, 2023). These mutations may affect protein function or modify gene regulatory elements, altering gene expression patterns to better suit new environmental demands. For instance, small changes in regulatory regions can lead to significant phenotypic shifts that improve thermal tolerance, nutrient utilization, or toxin resistance. The adaptive potential of mutation is further amplified by horizontal gene transfer, enhancing resilience across microbial populations. Notably, mutation-driven adaptations have been observed in microbes such as *E. coli* and *Daphnia*, where specific genetic changes confer thermal resilience—a key trait under global warming scenarios (Table 5) (Ramsay et al., 1988; Scheffer et al., 2022). Over generations, such beneficial mutations accumulate, supporting microbial survival, driving ecosystem-level feedback, and posing challenges in areas such as antimicrobial resistance and pathogen evolution (Rodríguez-Beltrán et al., 2021).

(ii) *Horizontal gene transfer (HGT)*: It plays a pivotal role in microbial evolution by enabling organisms to acquire new functions that enhance adaptability to changing environments rapidly. The three primary mechanisms include transformation, transduction, and conjugation. *Transformation* involves the uptake of free DNA fragments from the environment, which is then incorporated into the recipient genome through recombination. While *Transduction* is mediated by bacteriophages, which inadvertently package host DNA during viral replication and introduce it into new host cells upon infection. This method allows for gene transfer even between distant species under certain conditions. *Conjugation*, arguably the most efficient and direct method, involves the transfer of plasmids or chromosomal DNA through cell-to-cell contact, typically via a conjugative pilus. This mechanism enables the rapid spread of functional traits, such as antibiotic resistance or metabolic pathways, particularly in dense microbial communities. In extreme environments, additional vectors such as membrane vesicles and nanotubes have also been implicated in facilitating HGT. Smillie et al. (2011) demonstrated that shared ecological niches, specifically oxygen tolerance and habitat type, are more predictive of gene transfer patterns than evolutionary lineage, highlighting the environment's strong selective pressure in shaping genomic exchange.

Organisms with larger genomes tend to incorporate and retain a greater number of horizontally transferred genes, potentially enhancing metabolic versatility and ecological plasticity (Cordero & Hogeweg, 2009; Guieysse & Wuertz, 2012). In contrast, microbes with streamlined genomes, such as *Candidatus Pelagibacter ubique*, often purge non-essential genes to maintain efficiency, thereby limiting the extent of gene acquisition (Giovannoni et al.,

Table 5 Mutation-driven adaptation mechanisms in microbes

Type of mutation	Effect on microbial traits	Microbial example	Adaptation outcome	References
Point mutations (substitution)	Alters amino acid sequence, modifies protein function	*Escherichia coli*	Improved thermal tolerance	Ramsay et al. (1988)
Insertion mutations	Frameshift or functional change in regulatory or coding regions	*Daphnia magna* (microbial symbionts)	Enhanced temperature resilience	Scheffer et al. (2022)
Deletion mutations	Removal of non-essential or deleterious genes	*Pseudomonas* spp.	Streamlined genome for resource efficiency	Ahamad (2022)
Gene duplication	Increased gene dosage, diversification of function	*Staphylococcus aureus*	Enhanced oxidative stress resistance	Horton and Taylor (2023)
Chromosomal rearrangements	Creation of novel gene linkages, regulatory network shifts	*Salmonella enterica*	Adaptation to nutrient limitation	Carline (2011)
Regulatory region mutations	Altered gene expression without coding changes	*Bacillus subtilis*	Rapid response to oxidative or osmotic stress	Ahamad (2022)
Mutations enhanced by Horizontal Gene Transfer	Spreading adaptive mutations across populations	*Acinetobacter baumannii*	Increased multidrug and environmental resistance	Horton and Taylor (2023)

2014). For instance, *Methanosarcina mazei* has exhibited gene loss compared to its relatives, likely due to genome size constraints that restrict retention of accessory genes (Garushyants et al., 2015).

Ecological adaptation through HGT is exemplified in *Prochlorococcus marinus*, where distinct clades have acquired genes related to phosphate acquisition to survive in oligotrophic oceanic environments, underscoring HGT's role in niche specialization (Martiny et al., 2009). Moreover, the acquisition of stress-resistance genes from extremophiles allows microbes to circumvent the slower pace of vertical evolution. A striking example includes the transfer of genes from symbiotic *Actinobacteria* such as *Saccharothrix* sp. to the red algae *Pyropia haitanensis*, enhancing the host's resistance to drought, thermal, and osmotic stress (Wang et al., 2024). This inter-domain gene flow promotes the emergence of novel traits, improving microbial resilience and ecological performance (Nelson-Sathi et al., 2015). For instance, the horizontal transfer of genes from methanogenic archaea to methanotrophic bacteria has enhanced methane metabolism, a critical adaptation for carbon cycling in anaerobic ecosystems under climate pressure. These examples underscore HGT as a

Table 6 HGT-driven microbial traits enhancing climate change resilience

Climate-resilient trait	Gene(s)	Donor/recipient organism(s)	Function/adaptation	References
Heat tolerance	*groES*, *groEL*	*Thermophilic Archaea* → *E. coli*	Chaperonins aiding in protein folding under heat stress	Boto (2010)
Acid tolerance	*cadA*, *cadB*	*Lactobacillus* spp. → *E. coli*	Acid resistance through lysine decarboxylase system	van Hoek et al. (2011)
Salt tolerance	*betA*, *betB*, *proU*	*Halomonas* spp. → *Bacillus subtilis*	Osmoprotectant synthesis for halotolerance	Oren (2008)
Oxidative stress resistance	*katG*, *sodA*, *ahpC*	*Bacillus* spp. → *Pseudomonas* spp.	Detoxification of ROS under drought and high UV	Gogarten and Townsend (2005)
Desiccation resistance	*otsA*, *otsB*	*Deinococcus radiodurans* → *E. coli*	Trehalose biosynthesis for water stress	Csonka and Hanson (2000)
Heavy metal resistance (e.g., arsenic)	*arsC*, *arsB*, *arsR*	*Pseudomonas* spp. → *Cupriavidus metallidurans*	Arsenate detoxification	Silver and Phung (2005)
Methane metabolism	*mmoX*, *mmoY*, *mmoZ*	*Methylocystis* spp. → *Various methanotrophs*	Methane oxidation for carbon utilization	Semrau et al. (2010)
Nitrogen fixation under stress	*nifH*, *nifD*, *nifK*	*Rhizobium* spp. → *Free-living soil bacteria*	Enhanced nitrogen fixation under temperature/acid stress	Raymond et al. (2004)
Cold adaptation	*cspA*, *desK*, *desR*	*Psychrophilic bacteria* → *E. coli*	Cold-shock proteins and membrane fluidity modulation	Phadtare (2004)
UV resistance	*phr*, *uvrA*, *recA*	*Deinococcus* spp. → *Soil bacteria*	DNA repair and protection mechanisms	Battista (1997)

central mechanism of microbial innovation and adaptability (Table 6). By enabling rapid functional diversification, HGT supports microbial communities in maintaining ecosystem functions and resilience in the face of global climate change.

(iii) *Gene duplication*: It is a fundamental evolutionary process that enhances microbial adaptability, particularly in the face of environmental stressors associated with climate change. This mechanism involves the replication of genomic segments containing one or more genes, resulting in paralogous copies. The evolutionary significance of gene duplication lies in its capacity to introduce genetic redundancy. One gene copy can maintain its ancestral function while the other is liberated to diverge, potentially acquiring new or specialized roles that contribute to the organism's survival in fluctuating environments.

Table 7 Examples of gene duplication enhancing climate stress tolerance in microbes

Organism	Duplicated gene(s)	Environmental stress	Functional outcome	References
Escherichia coli	*recA*, *katG*, *sodA*	Heat/oxidative stress	Improved DNA repair and ROS detoxification	Kondrashov (2012)
Pseudomonas aeruginosa	EPS synthesis genes, *algD*, *pslA*	Drought	Enhanced biofilm formation and water retention	Kondrashov (2012)
Deinococcus radiodurans	*recA*, *uvrA*, *phr*	Radiation/UV stress	DNA damage repair and tolerance to genotoxic stress	Slade and Radman (2011)
Halobacterium salinarum	*betA*, *proVWX*	Salinity	Osmoprotection via increased osmolyte synthesis	Oren (2008)
Buchnera aphidicola	Reduction in DNA repair genes	Heat (via host buffering)	Energy conservation; host-dependent stress mitigation	Moran (2002)
Streptomyces spp.	Antibiotic biosynthesis genes	Competitive/ nutrient stress	Increased metabolic versatility and antimicrobial defense	Sayer et al. (2017)

In microbial systems, gene duplication is closely tied to stress response mechanisms. Duplicated genes often encode proteins such as heat shock proteins, membrane transporters, antioxidant enzymes, and components of biofilm formation pathways. These duplications can be upregulated in response to abiotic stresses such as heat, drought, salinity, or oxidative stress, enhancing microbial resilience (Table 7). For example, *Escherichia coli* demonstrates the duplication of DNA repair and antioxidant genes during exposure to high temperatures, supporting enhanced stress recovery (Kondrashov, 2012). Likewise, *Pseudomonas aeruginosa* increases gene copy numbers related to exopolysaccharide (EPS) synthesis and biofilm matrix formation under drought stress, aiding in the retention of water and nutrients in dry environments. Conversely, genome streamlining in symbiotic bacteria such as *Buchnera aphidicola* involves the loss of duplicated genes, especially those involved in DNA repair. This reflects an energy saving strategy aligned with its reliance on host organisms for stress mitigation (Moran, 2002).

3.2 Phenotypic Adaptation Mechanisms

Microorganisms are among the most adaptable life forms on Earth, capable of rapid phenotypic changes in response to environmental fluctuations. Phenotypic adaptations refer to non-genetic modifications in an organism's physiology, morphology, and behavior that enhance survival under environmental stress (Table 8). Climate change, driven by rising global temperatures, altered precipitation patterns,

Table 8 Phenotypic adaptations across stress types

Adaptation type	Mechanism	Example organism/system	Environmental stress	Citations
Physiological	Isozyme production	*Thermus aquaticus*	High temperature	Feler and Gerday (2003)
	Membrane lipid alteration	*Pseudomonas* spp.	Cold/heat	Russell (2002)
	Trehalose/ osmolyte synthesis	*Halomonas elongata*, desert plants	Salinity, drought	Yancey (2005)
Morphological	Filamentation	*E. coli*	DNA damage, antibiotics	Justice et al. (2008)
	Biofilm formation	*P. aeruginosa*, *S. aureus*	Nutrient limitation, disinfectants	Hall-Stoodley et al. (2004)
	Succulence and cuticle thickening	*Salicornia*, Xerophytes	Drought, salinity, UV	
Behavioral	Chemotaxis and thermotaxis	*E. coli*	Nutrient/ temperature gradients	Porter et al. (2011)
	Quorum sensing modulation	*Vibrio fischeri*, *Pseudomonas* spp.	Oxidative stress, host defense	Ng and Bassler (2009)
	Aggregation and sporulation	*Dictyostelium discoideum*	Starvation	Kessin (2001)

and increased atmospheric CO_2 levels, exerts significant selective pressures on microbial communities (Cavicchioli et al., 2019). Phenotypic plasticity—the ability of a single genotype to produce different phenotypes under varying environmental conditions—plays a crucial role in microbial survival and evolution (Fierer et al., 2020). This adaptive strategy allows microbes to adjust their metabolic pathways, morphology, and behavior without genetic mutations, ensuring resilience in dynamic ecosystems.

Various examples from diverse environments, including soil, aquatic systems, and extreme habitats, illustrate how phenotypic plasticity enhances microbial survival under climate-induced stressors. Together, these phenotypic adaptations underscore the crucial role of microbial plasticity in maintaining ecological stability amid climate change. Whether through adjusting metabolism, forming protective communities, entering dormant states, or altering cellular architecture, microbes continuously reshape their physiology to meet environmental demands. These responses not only ensure individual survival but also influence broader ecosystem processes such as nutrient cycling, carbon sequestration, and greenhouse gas emissions. Understanding microbial phenotypic plasticity thus provides valuable insight into the resilience of ecosystems and the feedback mechanisms driving global climate dynamics.

Phenotypic plasticity enables microorganisms to respond swiftly to environmental changes, providing a short-term survival advantage before long-term genetic adaptations can evolve (Schlichting & Smith, 2002). Unlike mutations, which occur over generations, phenotypic adjustments happen rapidly and reversibly, making them crucial for survival during sudden climatic shifts such as temperature spikes, water scarcity, or nutrient imbalances (Ghalambor et al., 2007). For instance, in prolonged droughts, soil microbial communities shift toward species with high phenotypic plasticity. Arid-soil bacteria such as *Streptomyces* modulate secondary metabolite production to increase drought resistance (Barka et al., 2016). Similarly, mycorrhizal fungi expand their hyphal networks to reach deeper soil layers, facilitating water acquisition and enhancing host plant resilience in dry climates (Hawkes et al., 2020).

3.2.1 Physiological Changes

One of the most immediate forms of adaptation is physiological change, which involves a range of biochemical and cellular responses.

(a) *Altered enzyme activity and metabolic pathways*
Altered enzyme activity and the reconfiguration of metabolic pathways are common strategies. Physiological responses are often the first line of adaptation, allowing organisms to adjust internal processes to external fluctuations. One such mechanism is the alteration of enzyme activity and metabolic pathways. Organisms may express temperature- or pH-optimized isozymes to maintain metabolic function under thermal or acidic stress (Somero, 2010). Rising global temperatures influence microbial metabolic rates, with many species altering their enzymatic activities to maintain homeostasis. Moreover, microbes adjust their respiratory metabolism depending on oxygen availability—a condition heavily influenced by climate. In flooded or waterlogged environments where oxygen becomes scarce, methanogenic archaea shift from aerobic to anaerobic respiration. In hypoxic or anoxic conditions, microorganisms such as *Saccharomyces cerevisiae* switch from oxidative phosphorylation to fermentative metabolism to maintain ATP production (DeRisi et al., 1997). These microbes metabolize carbon substrates into methane, a potent greenhouse gas. This shift has been notably observed in Arctic permafrost regions, where thawing due to climate change releases trapped organic matter. As temperatures rise, methanogenic archaea become more active, contributing significantly to atmospheric methane levels (Tveit et al., 2019; Conrad, 2009).

(b) *Changes in membrane composition and fluidity*
Membrane composition and fluidity also undergo stress-induced changes. The composition and fluidity of cellular membranes can change due to environmental stress. In response to temperature changes, membranes adjust their lipid composition: cold stress typically increases unsaturated fatty acids to keep membranes fluid, whereas heat stress increases saturated fatty acids to maintain structural stability (Russell, 2002). For example, Psychrophilic (cold-adapted) and mesophilic (moderate-temperature) microbes exhibit metabolic flexibility

by adjusting lipid membrane composition to preserve fluidity under thermal stress (Chintalapati et al., 2004). For instance, psychrophilic bacteria such as *Pseudomonas syringae*, which inhabit cold environments, increase the proportion of unsaturated fatty acids in their cell membranes. This alteration prevents the membrane from becoming too rigid at low temperatures, ensuring proper nutrient transport and enzymatic activity (Mansilla et al., 2004). In contrast, thermophilic bacteria such as *Thermus aquaticus*, found in hot springs, respond to high temperatures by increasing saturated fatty acids in their membranes, thus maintaining membrane integrity and functionality under thermal stress (Brock, 1985). Some extremophiles, such as archaea, have ether-linked lipids with cyclopentane rings that maintain membrane integrity under extreme salinity and temperature (Koga, 2012).

(c) *Production of stress protectants*
Organisms produce a range of stress protectants to maintain cellular homeostasis. Heat-shock proteins (HSPs) serve as molecular chaperones that refold misfolded proteins and prevent aggregation during heat and oxidative stress (Lindquist & Craig, 1988). Osmoprotectants such as proline, trehalose, glycine betaine, and ectoine stabilize macromolecules and counteract dehydration or salt-induced osmotic pressure (Yancey, 2005). These compounds are critical in desiccation-tolerant bacteria and salt-tolerant plants (halophytes). In Thermophiles, they utilize regulatory molecules such as osmolytes (e.g., mannosylglycerate, polyolphosphodiesters) to stabilize cellular proteins by shifting thermodynamic equilibrium toward the native state, with negatively charged osmolytes balanced by positive counterions such as K^+ (Faria et al., 2004). Heat shock proteins (e.g., GroES, GroEL, DnaK) assist in protein folding, and Clp proteases prevent the accumulation of misfolded proteins, ensuring cellular stability in extreme conditions (Large et al., 2009).

3.2.2 Morphological Changes

Microbes can also undergo morphological changes to adapt to environmental pressures. These changes often involve alterations in cell size, shape, or motility, enhancing the microbe's ability to survive or exploit limited resources.

(a) *Cell size and shape alterations*
Structural modifications provide longer term phenotypic plasticity to cope with environmental changes. Alterations in cell size and shape often occur under stress. Smaller cell sizes, with higher surface-area-to-volume ratios, are beneficial in nutrient-poor environments by enhancing uptake efficiency (Young, 2006). Bacteria under genotoxic stress may undergo filamentation—growing longer cells without dividing—which provides a temporary survival advantage by delaying cell division and increasing the time available for DNA repair (Justice et al., 2008). In marine ecosystems, *Prochlorococcus*, one of the most abundant photosynthetic organisms in the ocean, responds to nutrient-poor conditions by reducing its cell size. Smaller cells have a higher surface-area-to-volume ratio, which facilitates more efficient uptake of scarce nutrients such

as nitrogen and phosphorus (Partensky et al., 1999). These morphological shifts highlight the nuanced ways in which microbes adjust their physiology to maintain homeostasis in changing environments.

In response to intense or prolonged stress, many microorganisms adopt dormancy strategies like sporulation. This behavior is especially common among bacteria and fungi exposed to harsh conditions, including heat, drought, or nutrient depletion. For example, Bacillus subtilis produces highly resistant endospores when environmental conditions turn unfavorable. These spores can survive extreme desiccation, radiation, and nutrient scarcity for extended periods, reactivating only when conditions improve (Setlow, 2006). Similarly, the fungus *Aspergillus niger* produces robust asexual spores when exposed to high temperatures or oxidative stress. These spores enable dispersal over long distances and germination under more favorable conditions, ensuring the species' persistence in variable climates (Krijgsheld et al., 2013).

(b) *Biofilm formation and its role in stress tolerance*
Structured microbial communities encased in extracellular polymeric substances (EPS) are a key phenotypic adaptation to environmental stressors (Flemming & Wingender, 2010). Climate change-induced droughts and nutrient limitations promote biofilm formation, enhancing microbial resistance to desiccation and predation (Rillig et al., 2019). Biofilms serve as protective barriers against a range of stressors, including desiccation, UV radiation, antimicrobial agents, and temperature fluctuations. For example, *Pseudomonas aeruginosa*, a common soil bacterium, forms dense biofilms under high-salinity and elevated temperature conditions. The EPS matrix retains moisture and nutrients, thereby improving microbial survival in arid or saline soils (Toyofuku et al., 2012). In aquatic environments, diatoms and other microalgae respond to ocean acidification—a consequence of elevated atmospheric CO_2 by increasing EPS production. This helps stabilize their communities and maintain physiological processes despite declining pH levels (Passow, 2002). A particularly striking example is *Emiliania huxleyi*, a marine coccolithophore that contributes significantly to global carbon cycling. Under low pH conditions, certain strains reduce calcification to maintain metabolic efficiency, demonstrating a form of phenotypic adaptation that enables continued survival in acidifying oceans (Meyer & Riebesell, 2015).

3.3 Behavioral Changes

Behavioral adaptations are also key in managing environmental fluctuations. These are dynamic and often involve movement or communal interactions in response to stimuli.

(a) *Altered motility and chemotaxis*
Microbial motility and chemotaxis are often modulated to move toward more favorable conditions or escape harmful stimuli. For instance, *Escherichia coli*

can adjust the rotation of its flagella in response to temperature gradients, allowing it to migrate toward optimal thermal niches (Porter et al., 2011). Illustrative examples of behaviorally adaptive mechanisms include heterocyst formation in nitrogen-starved cyanobacteria. These specialized cells isolate nitrogenase from oxygen to enable nitrogen fixation (Flores & Herrero, 2010). In aquatic systems, diatoms and algae exhibit diel vertical migration, rising toward the surface during daylight to maximize photosynthesis and sinking at night to access nutrients.

(b) *Changes in quorum sensing and communication*
Quorum sensing (QS) is another critical behavioral adaptation, particularly in bacteria. QS is a cell-to-cell communication system that allows populations to coordinate gene expression based on cell density. This can occur by the exchange of small organic molecules or proteins, as well as the transmission of electrical signals (Prindle et al., 2015). Environmental stressors such as oxidative stress or nutrient limitation can modulate QS signals, leading to enhanced collective behaviors such as biofilm formation, production of protective enzymes, or expression of virulence factors (Ng & Bassler, 2009). These behavioral shifts, although subtle, significantly enhance an organism's fitness by optimizing group responses to environmental cues. QS regulates gene expression for group behaviors such as bioluminescence, virulence, and biofilm formation in response to environmental cues (Ng & Bassler, 2009).

4 Case Studies: Microbial Adaptations to Climate Change

Climate change imposes significant selective pressures on microbial populations, driving rapid evolutionary changes, physiological adaptations, and shifts in community structure. Here, we present three detailed case studies illustrating microbial responses to key climate change factors: rising temperatures, ocean acidification, and drought stress.

Example 1: Evolution of Thermal Tolerance in a Bacterial Species due to Rising Temperatures
Recent studies demonstrate that bacterial species can rapidly evolve to tolerate higher temperatures. For instance, in a study by Bennett et al. (1992), *E. coli* populations were cultured over thousands of generations under elevated temperatures (42 °C). Over time, the bacteria exhibited improved growth rates at high temperatures, reflecting the evolution of thermal tolerance. Genomic analysis revealed mutations in genes associated with heat shock proteins (e.g., dnaK, groEL) and alterations in membrane composition that enhanced protein stability and membrane integrity (Tenaillon et al., 2012).

In another study, *Pseudomonas fluorescens* showed that within a few hundred generations, populations exposed to elevated temperatures developed mutations enhancing membrane stability, heat-shock protein expression, and enzyme efficiency under thermal stress (Ketola et al., 2013). Experimental evolution approaches revealed that bacterial strains subjected to progressive temperature increases (from 28 to 37 °C) exhibited enhanced fitness, altered fatty acid compositions in their membranes, and changes in growth rates. These adaptations not only improved survival under heat stress but also had ecological implications, such as shifts in competitive interactions with other microbes and altered nutrient cycling rates (Bennett & Lenski, 2007).

Example 2: Adaptation of a Marine Microbial Community to Ocean Acidification
Marine microbial communities are increasingly exposed to lower pH conditions due to rising atmospheric CO_2. A case study at natural CO_2 seep sites near volcanic vents off Ischia Island, Italy, revealed that microbial assemblages adapted to long-term acidification by altering their taxonomic and functional composition (Taylor et al., 2014). The community showed a dominance of acidophilic and metal-resistant bacterial taxa, such as *Epsilonproteobacteria* and *Gammaproteobacteria*, which replaced more sensitive species. Functional genes related to nitrogen cycling, sulfur metabolism, and organic matter decomposition were also enriched under low pH conditions (Kerfahi et al., 2014).

A field-based mesocosm study conducted in the Arctic Ocean revealed significant compositional and functional shifts in microbial communities exposed to low pH conditions (~7.7) (Bunse et al., 2016). In particular, certain groups of proteobacteria and flavobacteria increased in abundance, while others such as cyanobacteria declined. Functional metagenomics showed upregulation of genes involved in proton extrusion, carbon acquisition (e.g., bicarbonate transporters), and extracellular polymer production, suggesting a phenotypic adaptation to maintain pH homeostasis and energy balance (Liu et al., 2010). These findings indicate that microbial communities can reassemble and modulate their metabolic pathways under acidified conditions, potentially buffering ecosystem-level functions such as carbon cycling and nutrient turnover under future climate scenarios.

Example 3: Phenotypic Plasticity in a Soil Microbe in Response to Drought
Soil microbial communities exhibit remarkable phenotypic plasticity to survive and function under drought stress. One detailed case involves *Streptomyces* spp., filamentous soil bacteria known for their resilience during desiccation. When exposed to prolonged drought, *Streptomyces* altered their physiological traits by producing protective osmolytes (such as trehalose), thickening their cell walls, increasing spore formation rate, and reducing hyphal extension to minimize water loss (Chater et al., 2010; Berard et al., 2015). A study by Narsing et al. (2022) showed that drought-stressed *Streptomyces coelicolor* upregulated genes for trehalose biosynthesis, osmoprotectant production, and oxidative stress defense enzymes such as catalase and superoxide dismutase. These physiological adaptations enhanced survival in low-moisture soils and allowed the bacterium to resume metabolic activity quickly upon rehydration. Additionally, soil microcosm experiments revealed that drought-adapted strains had an enhanced ability to form stable biofilms and resist desiccation-induced DNA damage. This case illustrates how microbial phenotypic plasticity

contributes to ecosystem resilience under water-limiting conditions, a scenario increasingly common in arid and semi-arid regions due to climate change.

In a field experiments demonstrated that after drought events, *Streptomyces* quickly resumed metabolic activity, indicating a high degree of functional plasticity. This rapid recovery is essential for maintaining soil health, as *Streptomyces* contribute to organic matter decomposition, antibiotic production, and nutrient cycling. However, extreme or repeated droughts could still lead to community shifts favoring more stress-tolerant but less functionally diverse taxa, impacting long-term soil ecosystem functions (de Vries et al., 2012).

5 Conclusion

Microbial adaptation plays a pivotal role in shaping ecosystem responses to climate change. As primary drivers of essential processes such as carbon sequestration, nutrient cycling, and soil formation, microbes are not only sensitive indicators of environmental change but also active agents influencing Earth's future climate trajectory. Microbial adaptation is vital for understanding how ecosystems respond to climate change. As key players in processes like carbon sequestration, nutrient cycling, and soil formation, microbes act both as sensitive environmental indicators and as active contributors to Earth's future climate. The ability of microbial communities to evolve, adapt functions, and show phenotypic plasticity under stressors such as rising temperatures, ocean acidification, and drought highlights their resilience and also reveals potential vulnerabilities.

Despite growing insights from case studies, many aspects of microbial adaptation remain poorly understood, particularly regarding long-term ecological consequences and feedbacks to global biogeochemical cycles. Interdisciplinary research is urgently needed to deepen our understanding of microbial responses across diverse environments, to predict ecosystem resilience or tipping points, and to inform climate change mitigation and adaptation strategies. As the invisible architects of life on Earth, microbes must be central to our scientific and policy frameworks addressing the climate crisis.

6 Future Prospects

Microbial adaptation to climate change holds profound and far-reaching implications for ecological stability, biogeochemical cycling, human and animal health, and the advancement of biotechnological solutions. As foundational components of all ecosystems, microbes drive essential processes such as organic matter decomposition, nutrient cycling, primary production, and the regulation of greenhouse gases. Their metabolic versatility and rapid evolutionary potential make them highly responsive to environmental shifts, which can in turn produce cascading effects throughout ecological networks and Earth system processes.

One of the most critical implications lies in the impact of microbial adaptation on biogeochemical cycling, particularly with respect to nitrogen, carbon, and sulfur dynamics. Microbial-mediated processes such as nitrification, denitrification, methanogenesis, and sulfate reduction are highly sensitive to environmental variables such as temperature, moisture, and pH. As global temperatures rise and precipitation patterns become more erratic, the composition and functional traits of microbial communities are likely to shift. For example, warming can favor thermophilic and fast-growing microbial taxa, potentially accelerating the decomposition of soil organic matter. This, in turn, may increase soil respiration and carbon dioxide emissions, particularly in carbon-rich ecosystems such as tundra and boreal forests, thereby contributing to positive feedback loops in climate change (Crowther et al., 2016).

Furthermore, shifts in the abundance and activity of nitrifying and denitrifying bacteria under warming and hydrological stress can alter nitrogen availability in both terrestrial and aquatic systems. Such changes not only affect plant productivity and food web dynamics but also risk increasing emissions of nitrous oxide (N_2O), i.e., a greenhouse gas with approximately 300 times the global warming potential of CO_2 over a 100-year period (Butterbach-Bahl et al., 2013). Similarly, the response of methanogenic archaea to thawing permafrost and anoxic conditions may intensify methane emissions, another potent greenhouse gas, further exacerbating global warming.

In marine ecosystems, microbial primary producers, including phytoplankton, cyanobacteria, and other photoautotrophs, are central to oceanic food webs and the biological carbon pump, which sequesters atmospheric CO_2 by transporting organic matter to the deep sea. However, ocean warming, stratification, nutrient limitation, and acidification are altering phytoplankton community structure, reducing net primary productivity in some regions and potentially compromising carbon export efficiency (Bopp et al., 2013). These changes could diminish the ocean's role as a carbon sink and disrupt trophic dynamics, with far-reaching consequences for global climate regulation and marine biodiversity.

Climate-driven changes in microbial ecology also intersect significantly with public health. Warming temperatures, increased humidity, and more frequent extreme weather events are expanding the geographic ranges and seasons of pathogenic microbes and their vectors. For instance, the proliferation of *Vibrio* spp. in temperate coastal waters, driven by rising sea surface temperatures, is increasing the incidence of gastrointestinal and wound infections in regions previously unaffected (Vezzulli et al., 2016). Similarly, shifts in precipitation can influence waterborne disease outbreaks and vector-borne illnesses, while melting permafrost may release ancient or dormant pathogens, raising new biosecurity concerns.

On the innovation front, microbial adaptation presents promising opportunities in biotechnology and climate change mitigation. Microorganisms that thrive under extreme environmental conditions, such as high salinity, acidity, temperature, or radiation, can be harnessed for bioremediation of contaminated environments, including degraded soils, wastewater, and oil-polluted marine areas (Ghosh et al., 2013). Furthermore, engineered microbial consortia or naturally adapted strains can be deployed to improve carbon sequestration, for instance, through soil amendments with beneficial microbes that stabilize organic carbon or promote plant growth.

The production of biofuels from algal biomass, powered by photosynthetic microbes, represents another avenue for reducing reliance on fossil fuels, especially if integrated into circular bioeconomy models. Additionally, studying microbial stress responses, such as thermostable enzymes, osmoprotectant synthesis, quorum sensing, and biofilm formation, offers blueprints for designing novel biomaterials, pharmaceuticals, and industrial processes that remain functional under harsh conditions (Cavicchioli et al., 2019). These innovations can help build climate-resilient technologies and support sustainable development.

Despite these exciting prospects, significant knowledge gaps remain in our understanding of how microbial communities interact with climate dynamics across spatial and temporal scales. Many Earth system models still lack adequate representation of microbial processes, leading to underestimation or oversimplification of critical feedback mechanisms that influence atmospheric greenhouse gas concentrations and ecosystem responses (Wieder et al., 2015). The complexity of microbial ecosystems characterized by high diversity, functional redundancy, and context-dependent behavior poses a challenge for predictive modeling.

To address this, there is a pressing need for integrative, interdisciplinary research frameworks that bridge microbial ecology, climate science, molecular biology, and systems modeling. Key priorities include long-term ecological monitoring, multi-omic approaches (e.g., metagenomics, transcriptomics, metabolomics), mesocosm and field experiments simulating future climate scenarios, and the development of trait-based models that link microbial functions to environmental variables. Emphasis should also be placed on data standardization, open-access databases, and cross-sector collaboration to enhance reproducibility and global applicability.

Ultimately, incorporating microbial processes and adaptive responses into climate models, biodiversity conservation frameworks, and climate policy instruments is essential to produce more accurate forecasts and implement effective mitigation and adaptation strategies. Recognizing microbes as both regulators and responders in the Earth's climate system is not only scientifically imperative but also crucial for informing holistic, evidence-based approaches to global sustainability.

References

Ahamad, A. (2022). *Microbial adaptation mechanisms in response to climate change*. Springer.

Allison, S. D., & Martiny, J. B. H. (2008). Resistance, resilience, and redundancy in microbial communities. *Proceedings of the National Academy of Sciences, 105*(Suppl 1), 11512–11519.

Allison, S. D., & Treseder, K. K. (2008). Warming and drying suppress microbial activity and carbon cycling in boreal forest soils. *Global Change Biology, 14*(12), 2898–2909.

Bakermans, C. (2015). Microbial evolution under extreme conditions. *Life, 5*(1), 163–190.

Barka, E. A., Vatsa, P., Sanchez, L., Gaveau-Vaillant, N., Jacquard, C., Klenk, H. P., Clément, C., Ouhdouch, Y., & van Wezel, G. P. (2016). Taxonomy, physiology, and natural products of Actinobacteria. *Microbiology and Molecular Biology Reviews, 80*(1), 1–43.

Bartlett, D. H., Kato, C., & Horikoshi, K. (2007). High-pressure adaptation of piezophilic bacteria and archaea. *Trends in Microbiology, 15*(9), 475–483.
Basallote, M. D., Rodríguez-Romero, A., Blasco, J., & Riba, I. (2014). Mobilization of metals in marine sediments under simulated CO_2 leak scenarios. *Science of the Total Environment, 493*, 416–423.
Battista, J. R. (1997). Against all odds: The survival strategies of *Deinococcus radiodurans*. *Annual Review of Microbiology, 51*(1), 203–224.
Bellard, C., Bertelsmeier, C., Leadley, P., Thuiller, W., & Courchamp, F. (2012). Impacts of climate change on the future of biodiversity. *Ecology Letters, 15*(4), 365–377.
Bennett, A. F., Lenski, R. E., & Mittler, J. E. (1992). Evolutionary adaptation to temperature. I. Fitness responses of Escherichia coli to changes in its thermal environment. *Evolution, 46*(1), 16–30.
Bennett, A. F., & Lenski, R. E. (2007). An experimental test of evolutionary trade-offs during temperature adaptation. *Proceedings of the National Academy of Sciences, 104*(1), 8649–8654.
Berard, A., Sassi, M. B., Kaisermann, A., & Renault, P. (2015). Soil microbial community responses to heat wave components: drought and high temperature. *Climate Research, 66*(3), 243–264.
Beulig, F., Røy, H., Glombitza, C., & Jørgensen, B. B. (2015). Controls on microbial carbon turnover in high-temperature sediment. *Nature Communications, 6*, 7407.
Bopp, L., Resplandy, L., Orr, J. C., Doney, S. C., Dunne, J. P., Gehlen, M., Halloran, P., Heinze, C., Ilyina, T., Seferian, R. and Tjiputra, J., Vichi, M. (2013). Multiple stressors of ocean ecosystems in the 21st century: projections with CMIP5 models. *Biogeosciences, 10*(10), 6225–6245.)
Boto, L. (2010). Horizontal gene transfer in evolution: facts and challenges. *Proceedings of the Royal Society B: Biological Sciences, 277*(1683), 819–827.
Bradford, M. A. (2013). Thermal adaptation of decomposer communities in warming soils. *Frontiers in Microbiology, 4*, 333.
Brewer, P. G., & Peltzer, E. T. (2009). Limits to marine life. *Science, 324*(5925), 347–348.
Brock, T. D. (1985). *Life at high temperatures*. Springer.
Bunse, C., Lundin, D., Karlsson, C. M., Vila-Costa, M., Palovaara, J., Akram, N., et al. (2016). Response of marine bacterioplankton pH homeostasis gene expression to elevated CO_2. *Nature Climate Change, 6*(5), 457–461.
Burckhardt, R. (2023). *Microbial biotechnology for environmental sustainability*. Elsevier.
Butterbach-Bahl, K., Baggs, E. M., Dannenmann, M., Kiese, R., & Zechmeister-Boltenstern, S. (2013). Nitrous oxide emissions from soils: How well do we understand the processes and their controls? *Philosophical Transactions of the Royal Society of London. Series B, Biological Sciences, 368*(1621), 20130122.
Canarini, A., Kaiser, C., Merchant, A., Richter, A., & Wanek, W. (2017). Root exudation of primary metabolites: Mechanisms and their roles in plant responses to environmental stimuli. *Frontiers in Plant Science, 8*, 157.
Canarini, A., Schmidt, H., Fuchslueger, L., Martin, V., Herbold, C. W., Zezula, D., Gündler, P., Hasibeder, R., Jecmenica, M., Bahn, M., & Richter, A. (2021). Ecological memory of recurrent drought modifies soil processes via changes in soil microbial community. *Nature communications, 12*(1), 5308.
Canganella, F., & Wiegel, J. (2011). Extremophiles: From abyssal to terrestrial ecosystems and possibly beyond. *Naturwissenschaften, 98*(4), 253–279.
Carline, R. (2011). Mutation and microbial adaptation: A review. *Microbial Biotechnology, 4*(1), 1–13.
Cavicchioli, R., Ripple, W. J., Timmis, K. N., Azam, F., Bakken, L. R., Baylis, M., et al. (2019). Scientists' warning to humanity: Microorganisms and climate change. *Nature Reviews. Microbiology, 17*(9), 569–586.
Chater, K. F., Biró, S., Lee, K. J., Palmer, T., & Schrempf, H. (2010). The complex extracellular biology of Streptomyces. *FEMS Microbiology Reviews, 34*(2), 171–198.

Chintalapati, S., Kiran, M. D., & Shivaji, S. (2004). Role of membrane lipid fatty acids in cold adaptation. *Cellular and Molecular Biology, 50*(5), 631–642.

Conrad, R. (2009). The global methane cycle: Recent advances in understanding the microbial processes involved. *Environmental Microbiology Reports, 1*(5), 285–292.

Cordero, O. X., & Hogeweg, P. (2009). The impact of long-distance horizontal gene transfer on prokaryotic genome size. *Proceedings of the National Academy of Sciences, 106*(51), 21748–21753.

Crowther, T. W., Todd-Brown, K. E. O., Rowe, C. W., Wieder, W. R., Carey, J. C., Machmuller, M. B., et al. (2016). Quantifying global soil carbon losses in response to warming. *Nature, 540*(7631), 104–108.

Csonka, L. N., & Hanson, A. D. (2000). Prokaryotic osmoregulation: Genetics and physiology. *Annual Review of Microbiology, 54*(1), 349–381.

de Coninck, H., & Benson, S. M. (2014). Carbon dioxide capture and storage: Issues and prospects. *Annual Review of Environment and Resources, 39*(1), 243–270.

DeRisi, J. L., Iyer, V. R., & Brown, P. O. (1997). Exploring the metabolic and genetic control of gene expression on a genomic scale. *Science, 278*(5338), 680–686.

de Vries, F. T., Griffiths, R. I., Knight, C. G., Nicolitch, O., & Williams, A. (2012). Harnessing rhizosphere microbiomes for drought-resilient crop production. *Science, 368*(6488), 270–274.

Eberlein, C., et al. (2018). Stress resilience in microbial ecosystems. *Frontiers in Microbiology, 9*, 2032.

Evans, S. E., & Wallenstein, M. D. (2012). Soil microbial community response to drying and rewetting stress: Does historical precipitation regime matter? *Biogeochemistry, 109*(1), 101–116.

Evans, S. E., Wallenstein, M. D., & Burke, I. C. (2014). Is bacterial moisture niche a good predictor of shifts in community composition under long-term drought? *Ecology, 95*(1), 110–122.

Falkowski, P. G., Fenchel, T., & Delong, E. F. (2008). The microbial engines that drive Earth's biogeochemical cycles. *Science, 320*(5879), 1034–1039.

Feler, G., & Gerday, C. (2003). Psychrophilic enzymes: Hot topics in cold adaptation. *Nature Reviews. Microbiology, 1*(3), 200–208.

Ferenjian, A., Lloyd, K. G., & Mayhew, L. E. (2021). Microbial metabolic adaptations to environmental stress. *Annual Review of Earth and Planetary Sciences, 49*(1), 363–389.

Fierer, N., & Jackson, R. B. (2006). The diversity and biogeography of soil bacterial communities. *Proceedings of the National Academy of Sciences, 103*(3), 626–631.

Fierer, N., Ladau, J., Clemente, J. C., Leff, J. W., Owens, S. M., Pollard, K. S., et al. (2020). Reconstructing the microbial diversity and function of pre-agricultural tallgrass prairie soils. *Science, 367*(6484), 676–680.

Flemming, H. C., & Wingender, J. (2010). The biofilm matrix. *Nature reviews microbiology, 8*(9), 623–633.

Flores, E., & Herrero, A. (2010). Compartmentalized function through cell differentiation in filamentous cyanobacteria. *Nature Reviews Microbiology, 8*(1), 39–50.

Fu, F. X., Place, A. R., Garcia, N. S., & Hutchins, D. A. (2007). CO_2 and phosphate availability control the toxicity of the harmful bloom cyanobacterium *Microcystis aeruginosa*. *Environmental Microbiology, 9*(2), 512–520.

Fuchslueger, L., Bahn, M., Fritz, K., Hasibeder, R., & Richter, A. (2014). Experimental drought reduces the transfer of recently fixed plant carbon to soil microbes and alters the bacterial community composition in a mountain meadow. *The New Phytologist, 201*(3), 916–927.

Garushyants, S. K., Kazanov, M. D., & Gelfand, M. S. (2015). Horizontal gene transfer and genome evolution in Methanosarcina. *Genome Biology and Evolution, 7*(2), 293–309.

Ghalambor, C. K., McKay, J. K., Carroll, S. P., & Reznick, D. N. (2007). Adaptive versus non-adaptive phenotypic plasticity and the potential for contemporary adaptation in new environments. *Functional Ecology, 21*(3), 394–407.

Ghosh, S. K., Pal, S., & Ray, S. (2013). Study of microbes having potentiality for biodegradation of plastics. *Environmental Science and Pollution Research, 20*(7), 4339–4355.

Gillooly, J. F., Brown, J. H., West, G. B., Savage, V. M., & Charnov, E. L. (2001). Effects of size and temperature on metabolic rate. *Science, 293*(5538), 2248–2251.

Giovannoni, S. J., Cameron Thrash, J., & Temperton, B. (2014). Implications of streamlining theory for microbial ecology. *The ISME Journal, 8*(8), 1553–1565.

Gogarten, J. P., & Townsend, J. P. (2005). Horizontal gene transfer, genome innovation and evolution. *Nature Reviews. Microbiology, 3*(9), 679–687.

Graham, E. B., Wieder, W. R., Leff, J. W., Weintraub, S. R., Townsend, A. R., Cleveland, C. C., & Nemergut, D. R. (2014). Do we need to understand microbial communities to predict ecosystem function? A comparison of statistical models of nitrogen cycling processes. *Soil Biology and Biochemistry, 68*, 279–282.

Graham, E. B., et al. (2016). Microbes as engines of ecosystem function: When does community structure enhance predictions of ecosystem processes? *Frontiers in Microbiology, 7*, 214.

Guieysse, B., & Wuertz, S. (2012). Metabolically versatile large-genome prokaryotes. *Current Opinion in Biotechnology, 23*(3), 467–473.

Gunde-Cimerman, N., Plemenitaš, A., & Oren, A. (2018). Strategies of adaptation of microorganisms of the three domains of life to high salt concentrations. *FEMS Microbiology Reviews, 42*(3), 353–375.

Hall-Stoodley, L., Costerton, J. W., & Stoodley, P. (2004). Bacterial biofilms: From the natural environment to infectious diseases. *Nature Reviews. Microbiology, 2*(2), 95–108.

Hawkes, C. V., Keitt, T. H., & Peay, K. G. (2020). Effects of climate change on soil microbial communities and feedbacks to plants. *Annual Review of Ecology, Evolution, and Systematics, 51*(1), 83–106.

Heshmati, B. (2020). Global warming and climate change: Causes, consequences, and mitigation measures. *International Journal of Environmental Research and Public Health, 17*(6), 1969.

Ho, A., Angel, R., Veraart, A. J., Daebeler, A., Jia, Z., & Schleper, C. (2016). Biotic interactions in the root microbiome of agricultural crops drive soil nitrogen cycling. *Frontiers in Microbiology, 7*, 1704.

Horton, J. S., & Taylor, C. M. (2023). Evolutionary adaptation to climate extremes. *Evolutionary Applications, 16*(2), 358–374.

Hutchins, D. A., & Fu, F. (2017). Microorganisms and ocean global change. *Nature Microbiology, 2*(6), 17058.

Hutchins, D. A., Jansson, J. K., & Trivedi, P. (2019). Microbiome influence on plant and soil health under climate change. *The ISME Journal, 13*(10), 2342–2359.

Iglesias-Rodriguez, M. D., Halloran, P. R., Rickaby, R. E. M., Hall, I. R., Colmenero-Hidalgo, E., Gittins, J. R., et al. (2008). Phytoplankton calcification in a high-CO_2 world. *Science, 320*(5874), 336–340.

IPCC. (2021). *Climate change 2021: The physical science basis. Contribution of Working Group I to the Sixth Assessment Report of the Intergovernmental Panel on Climate Change*. Cambridge University Press.

Jansson, J. K., & Hofmockel, K. S. (2020). The soil microbiome—From metagenomics to metaphenomics. *Current Opinion in Microbiology, 43*, 162–168.

Jiao, N., Herndl, G. J., Hansell, D. A., Benner, R., Kattner, G., Wilhelm, S. W., et al. (2010). Microbial production of recalcitrant dissolved organic matter: Long-term carbon storage in the global ocean. *Nature Reviews. Microbiology, 8*(8), 593–599.

Johnstone, J. F., & Lucy, B. (2020). Disturbance, climate change, and ecological resilience in terrestrial ecosystems. *Frontiers in Ecology and the Environment, 18*(10), 547–556.

Joint, I., Doney, S. C., & Karl, D. M. (2011). Will ocean acidification affect marine microbes?. *The ISME journal, 5*(1), 1–7.

Justice, S. S., Hunstad, D. A., Cegelski, L., & Hultgren, S. J. (2008). Morphological plasticity as a bacterial survival strategy. *Nature Reviews. Microbiology, 6*(2), 162–168.

Keeling, R. F., Körtzinger, A., & Gruber, N. (2010). Ocean deoxygenation in a warming world. *Annual Review of Marine Science, 2*(1), 199–229.

Kerfahi, D., Tripathi, B. M., Takahashi, K., Lee, J., Edwards, D. P., & Adams, J. M. (2014). The impact of ocean acidification on the functional diversity of soil microbial communities. *Frontiers in Microbiology, 5*, 632.

Kessin, R. H. (2001). *Dictyostelium: evolution, cell biology, and the development of multicellularity* (Vol. 38). Cambridge University Press.

Ketola, T., Mikonranta, L., Laakso, J., & Mappes, J. (2013). Evolution of bacterial temperature tolerance in fluctuating environments. *Proceedings of the Royal Society B: Biological Sciences, 280*(1760), 20131454.

Khati, P., Mishra, P. K., & Kant, L. (2024). Isolation, screening and characterization of drought tolerant plant growth promoting bacteria from Indian Himalayas. *Pantnagar Journal of Research, 22*(2), 261.

Koga, Y. (2012). Thermal adaptation of the archaeal and bacterial lipid membranes. *Archaea, 2012*, 789652.

Kondrashov, F. A. (2012). Gene duplication as a mechanism of genomic adaptation to a changing environment. *Proceedings of the Royal Society B: Biological Sciences, 279*(1749), 5048–5057.

Krijgsheld, P., Bleichrodt, R., van Veluw, G. J., Wang, F., Müller, W. H., Dijksterhuis, J., & Wösten, H. A. B. (2013). Development in Aspergillus. *Studies in Mycology, 74*(1), 1–29.

Langer, G., Geisen, M., Baumann, K. H., Kläs, J., Riebesell, U., Thoms, S., & Young, J. R. (2006). Species-specific responses of calcifying algae to changing seawater carbonate chemistry. *Geochemistry, Geophysics, Geosystems, 7*(9). https://doi.org/10.1029/2005GC001227

Large, A., Goldberg, S., & Mokry, R. (2009). Heat shock proteins and their role in plant stress response. *Molecular Plant, 2*(3), 381–393.

Lewin, G. R., Carlos, C., Chevrette, M. G., Horn, H. A., McDonald, B. R., Stankey, R. J., et al. (2013). Evolution and ecology of Actinobacteria and their bioenergy applications. *Annual Review of Microbiology, 67*, 471–493.

Lindquist, S., & Craig, E. A. (1988). The heat-shock proteins. *Annual review of genetics, 22*(1), 631-677.

Liu, Y., He, M., Uchimiya, M., Wang, S., Sposito, G., & Zhu, Y. G. (2010). Environmental fate of microbial DNA in natural environments: Implications for horizontal gene transfer. *Frontiers in Ecology and the Environment, 8*(8), 437–444.

Liu, J., Fu, B., Yang, H., & Zhao, M. (2018). Role of microbes in regulating global biogeochemical cycles. *Frontiers in Microbiology, 9*, 3257.

Liu, Q., Makhalanyane, T. P., & Cowan, D. A. (2022). Evolutionary strategies of cold-adapted microorganisms. *Frontiers in Microbiology, 13*, 860359.

López-García, P., Zivanovic, Y., Deschamps, P., & Moreira, D. (2022). Eukaryogenesis and gene transfers across the tree of life. *Cold Spring Harbor Perspectives in Biology, 14*(6), a040675.

Lozupone, C. A., & Knight, R. (2007). Global patterns in bacterial diversity. *Proceedings of the National Academy of Sciences, 104*(27), 11436–11440.

Malhi, Y., Franklin, J., Seddon, N., Solan, M., Turner, M. G., Field, C. B., & Knowlton, N. (2020). Climate change and ecosystems: Threats, opportunities and solutions. *Philosophical Transactions of the Royal Society of London. Series B, Biological Sciences, 375*(1794), 20190104.

Mansilla, M. C., Cybulski, L. E., Albanesi, D., & de Mendoza, D. (2004). Control of membrane lipid fluidity by molecular thermosensors. *Journal of bacteriology, 186*(20), 6681–6688.

Margesin, R., & Miteva, V. (2011). Diversity and ecology of psychrophilic microorganisms. *Research in Microbiology, 162*(3), 346–361.

Martiny, A. C., Coleman, M. L., & Chisholm, S. W. (2009). Phosphate acquisition genes in Prochlorococcus ecotypes: Evidence for genome-wide adaptation. *Proceedings of the National Academy of Sciences, 103*(4), 12552–12557.

Meisner, A., Bååth, E., & Rousk, J. (2018). Microbial legacy effects in response to drying and rewetting cycles. *The ISME Journal, 12*(4), 1004–1015.

Meyer, J., & Riebesell, U. (2015). Reviews and syntheses: Responses of coccolithophores to ocean acidification—A meta-analysis. *Biogeosciences, 12*(6), 1671–1682.

Moran, N. A. (2002). Microbial minimalism: Genome reduction in bacterial pathogens. *Cell, 108*(5), 583–586.

Muluneh, A. (2021). Climate change and variability impacts on biodiversity and food security. *Environmental Sustainability, 4*(2), 202–212.

Narsing Rao, M. P., Lohmaneeratana, K., Bunyoo, C., & Thamchaipenet, A. (2022). Actinobacteria–plant interactions in alleviating abiotic stress. *Plants, 11*(21), 2976.

Natali, S. M., Schuur, E. A. G., & Rubin, R. L. (2019). Permafrost thaw and soil moisture driving CO_2 and CH_4 release from upland tundra. *Journal of Geophysical Research. Biogeosciences, 124*(4), 1050–1064.

Nelson-Sathi, S., Sousa, F. L., Roettger, M., Lozada-Chávez, N., Thiergart, T., Janssen, A., Bryant, D., Landan, G., Schönheit, P., Siebers, B., Martin, W. F., & McInerney, J.O. (2015). Origins of major archaeal clades correspond to gene acquisitions from bacteria. *Nature, 517*(7532), 77–80.

Nemergut, D. R., Anderson, S. P., Cleveland, C. C., Martin, A. P., Miller, A. E., Seimon, A., & Schmidt, S. K. (2007). Microbial community succession in an unvegetated, recently deglaciated soil. *Microbial Ecology, 53*(1), 110–122.

Ng, W. L., & Bassler, B. L. (2009). Bacterial quorum-sensing network architectures. *Annual Review of Genetics, 43*, 197–222.

Oren, A. (2008). Microbial life at high salt concentrations: Phylogenetic and metabolic diversity. *Saline Systems, 4*(1), 2.

Parrilli, E., De Vizio, D., Schiraldi, C., & Tutino, M. L. (2022). Adaptations of extremophilic bacteria to harsh environments and biotechnological potential. *Microorganisms, 10*(1), 82.

Passow, U. (2002). Transparent exopolymer particles (TEP) in aquatic environments. *Progress in Oceanography, 55*(3–4), 287–333.

Partensky, F., Hess, W. R., & Vaulot, D. (1999). Prochlorococcus, a marine photosynthetic prokaryote of global significance. *Microbiology and molecular biology reviews, 63*(1), 106–127.

Phadtare, S. (2004). Recent developments in bacterial cold-shock response. *Current issues in molecular biology, 6*(2), 125–136.

Petit, J. R., Jouzel, J., Raynaud, D., Barkov, N. I., Barnola, J. M., Basile, I., et al. (1999). Climate and atmospheric history of the past 420,000 years from the Vostok ice core, Antarctica. *Nature, 399*(6735), 429–436.

Philippot, L., Raaijmakers, J. M., Lemanceau, P., & van der Putten, W. H. (2021). Going back to the roots: The microbial ecology of the rhizosphere. *Nature Reviews. Microbiology, 19*(7), 414–426.

Picazo, A., Domaizon, I., Boucher, D., & Debroas, D. (2020). Continental and oceanic plankton communities in a changing climate. *Frontiers in Microbiology, 11*, 535.

Porter, S. L., Wadhams, G. H., & Armitage, J. P. (2011). Signal processing in complex chemotaxis pathways. *Nature Reviews. Microbiology, 9*(3), 153–165.

Qin, D., Stocker, T. F., Plattner, G. K., Tignor, M., Allen, S. K., Boschung, J., et al. (2014). *Climate change 2014: Synthesis report*. IPCC.

Raddadi, N., Giacomucci, L., Totaro, S., Fava, F., & Daffonchio, D. (2015). Potential of extremophilic microorganisms for biotechnological processes. *New Biotechnology, 32*(6), 749–758.

Ramsay, J. A., Kowalchuk, G. A., & Leung, K. T. (1988). Mutation and adaptation in microorganisms. *Journal of Bacteriology, 170*(8), 3453–3459.

Rastelli, E., Corinaldesi, C., Dell'Anno, A., & Middelburg, J. J. (2015). Evidence of efficient carbon processing in the hadal trench benthic food web. *Progress in Oceanography, 137*, 38–48.

Raymond, J., Siefert, J. L., Staples, C. R., & Blankenship, R. E. (2004). The natural history of nitrogen fixation. *Molecular Biology and Evolution, 21*(3), 541–554.

Ridgwell, A., Zondervan, I., Hargreaves, J. C., Bijma, J., & Lenton, T. M. (2009). Assessing the potential long-term increase of oceanic carbon uptake by CO_2-induced enhancement of marine $CaCO_3$ production. *Biogeosciences, 6*(7), 1429–1440.

Riebesell, U., Zondervan, I., Rost, B., Tortell, P. D., Zeebe, R. E., & Morel, F. M. (2000). Reduced calcification of marine plankton in response to increased atmospheric CO_2. *Nature, 407*(6802), 364–367.

Rillig, M. C., Aguilar-Trigueros, C. A., Bergmann, J., Verbruggen, E., Veresoglou, S. D., & Lehmann, A. (2019). Plant root and mycorrhizal fungal traits for understanding soil aggregation. *The New Phytologist, 223*(2), 554–564.

Rodríguez-Beltrán, J., DelaFuente, J., Leon-Sampedro, R., MacLean, R. C., & San Millan, A. (2021). Beyond horizontal gene transfer: the role of plasmids in bacterial evolution. *Nature Reviews Microbiology, 19*(6), 347–359.

Röhl, O., Graupner, N., Trumbore, S., Dörfer, C., & Amelung, W. (2022). Carbon cycle feedbacks in terrestrial ecosystems in response to extreme precipitation. *Global Change Biology, 28*(9), 2953–2966.

Rost, B., Zondervan, I., & Wolf-Gladrow, D. A. (2008). Sensitivity of phytoplankton to future changes in ocean carbonate chemistry: Current knowledge, contradictions and research directions. *Marine Ecology Progress Series, 373*, 227–237.

Russell, N. J. (2002). Bacterial membranes: The effects of chill stress. *Current Opinion in Microbiology, 5*(3), 209–214.

Sayer, E. J., Oliver, A. E., Fridley, J. D., Askew, A. P., Mills, R. T., & Grime, J. P. (2017). Links between soil microbial communities and plant traits in a species-rich grassland under long-term climate change. *Ecology and Evolution, 7*(3), 855–862.

Scheffer, M., Bascompte, J., Brock, W. A., Brovkin, V., Carpenter, S. R., Dakos, V., et al. (2022). Early-warning signals for critical transitions. *Nature, 499*(7457), 295–300.

Schimel, J., & Schaeffer, S. M. (2012). Microbial control over carbon cycling in soil. *Frontiers in Microbiology, 3*, 348.

Schimel, J., Balser, T. C., & Wallenstein, M. (2007). Microbial stress-response physiology and its implications for ecosystem function. *Ecology, 88*(6), 1386–1394.

Schlichting, C. D., & Smith, H. (2002). Phenotypic plasticity: Linking molecular mechanisms with evolutionary outcomes. *Evolutionary Ecology, 16*(3), 189–211.

Schuur, E. A. G., McGuire, A. D., Schadel, C., Grosse, G., Harden, J. W., Hayes, D. J., et al. (2015). Climate change and the permafrost carbon feedback. *Nature, 520*(7546), 171–179.

Semrau, J. D., DiSpirito, A. A., & Yoon, S. (2010). Methanotrophs and copper. *FEMS microbiology reviews, 34*(4), 496–531.

Setlow, P. (2006). Spores of *Bacillus subtilis*: Their resistance to and killing by radiation, heat and chemicals. *Journal of Applied Microbiology, 101*(3), 514–525.

Silver, S., & Phung, L. T. (2005). Genes and enzymes involved in bacterial oxidation and reduction of inorganic arsenic. *Applied and Environmental Microbiology, 71*(2), 599–608.

Singh, B. K., Bardgett, R. D., Smith, P., & Reay, D. S. (2010). Microorganisms and climate change: Terrestrial feedbacks and mitigation options. *Nature Reviews. Microbiology, 8*(11), 779–790.

Sintayehu, D. W. (2018). Impact of climate change on biodiversity and associated key ecosystem services in Africa: A systematic review. *Ecosystem Health and Sustainability, 4*(9), 225–239.

Slade, D., & Radman, M. (2011). Oxidative stress resistance in Deinococcus radiodurans. *Microbiology and Molecular Biology Reviews, 75*(1), 133–191.

Smillie, C. S., Smith, M. B., Friedman, J., Cordero, O. X., David, L. A., & Alm, E. J. (2011). Ecology drives a global network of gene exchange connecting the human microbiome. *Nature, 480*(7376), 241–244.

Somero, G. N. (2010). The physiology of climate change: how potentials for acclimatization and genetic adaptation will determine 'winners' and 'losers'. *Journal of Experimental Biology, 213*(6), 912–920.

Steinacher, M., Joos, F., Frolicher, T. L., Plattner, G. K., & Doney, S. C. (2010). Imminent ocean acidification projected with the independent effects of greenhouse gases and aerosols. *Nature Climate Change, 1*(5), 302–305.

Taylor, J. D., Cunliffe, M., & Green, D. H. (2014). Low pH reduces the diversity and adaptability of the marine nitrogen cycle. *Environmental Microbiology, 16*(4), 1182–1194.

Tenaillon, O., Rodríguez-Verdugo, A., Gaut, R. L., McDonald, P., Bennett, A. F., Long, A. D., & Gaut, B. S. (2012). The molecular diversity of adaptive convergence. *Science, 335*(6067), 457–461.

Thakur, M. P., & Geisen, S. (2019). Trophic regulations of the soil microbiome. *Trends in Microbiology, 27*(9), 771–780.

Tilman, D., & Lehman, C. (2001). Human-caused environmental change: Impacts on plant diversity and evolution. *Proceedings of the National Academy of Sciences, 98*(10), 5433–5440.

Tortell, P. D., Morel, F. M. M., & Legendre, L. (1997). The influence of phytoplankton community structure on biogeochemical cycling. *Limnology and Oceanography, 42*(8), 1208–1214.

Toyofuku, M., Roschitzki, B., Riedel, K., & Eberl, L. (2012). Identification of proteins associated with the Pseudomonas aeruginosa biofilm extracellular matrix. *Journal of proteome research, 11*(10), 4906-4915.

Trivedi, P., Delgado-Baquerizo, M., Jeffries, T. C., Trivedi, C., Anderson, I. C., & Singh, B. K. (2022). The functional potential of microbial communities in the context of climate change. *Global Change Biology, 28*(1), 263–275.

Tveit, A., Schwacke, R., Svenning, M. M., & Urich, T. (2019). Organic carbon transformations in high-Arctic peat soils: Key functions and microorganisms. *The ISME Journal, 13*(7), 1831–1847.

van Groenigen, K. J., Osenberg, C. W., & Hungate, B. A. (2011). Increased soil emissions of potent greenhouse gases under increased atmospheric CO_2. *Nature, 475*(7355), 214–216.

Van Hoek, A. H., Mevius, D., Guerra, B., Mullany, P., Roberts, A. P., & Aarts, H. J. (2011). Acquired antibiotic resistance genes: an overview. Frontiers in microbiology, 2, 203.

Vezzulli, L., Grande, C., Reid, P. C., Hélaouët, P., Edwards, M., Höfle, M. G., & Pruzzo, C. (2016). Climate influence on Vibrio and associated human diseases during the past half-century in the coastal North Atlantic. *Proceedings of the National Academy of Sciences, 113*(34), E5062–E5071.

Wang, S., Tang, J., Chen, H. Y., & Yin, Y. (2016). Elevated atmospheric CO_2 affects soil microbial communities. *Soil Biology and Biochemistry, 98*, 63–72.

Wang, Y., et al. (2021). Soil microbial responses to climate change. *Soil Biology and Biochemistry, 152*, 108058.

Wang, W., Ge, Q., Wen, J., Zhang, H., Guo, Y., Li, Z., Xu, Y., Ji, D., Chen, C., Guo, L Xu, M., & Xie, C. (2024). Horizontal gene transfer and symbiotic microorganisms regulate the adaptive evolution of intertidal algae, Porphyra sense lato. *Communications Biology, 7*(1), 976.

Webster, P. J., Holland, G. J., Curry, J. A., & Chang, H. R. (2018). Changes in tropical cyclone number, duration, and intensity in a warming environment. *Science, 309*(5742), 1844–1846.

Wieder, W. R., Bonan, G. B., & Allison, S. D. (2015). Global soil carbon projections are improved by modelling microbial processes. *Nature Climate Change, 5*(1), 56–60.

Yanagawa, K., Morono, Y., Kuramoto, S., Terada, T., Kubo, Y., Tanikawa, W., et al. (2013). Geological CO_2 sequestration impacts on deep-sea microbial ecosystems. *Scientific Reports, 3*, 1034.

Yancey, P. H. (2005). Organic osmolytes as compatible, metabolic and counteracting cytoprotectants in high osmolarity and other stresses. *Journal of experimental biology, 208*(15), 2819–2830.

Young, K. D. (2006). The selective value of bacterial shape. *Microbiology and molecular biology reviews, 70*(3), 660–703.

Yue, X., & Gao, Q. (2018). Responses of soil microbial communities to climate change. *Global Change Biology, 24*(2), 559–569.

Zhou, J., et al. (2016). Microbial adaptation to environmental changes. *Annual Review of Ecology, Evolution, and Systematics, 47*(1), 217–241.

Bioremediation and Climate Change: Microbial Strategies for Environmental Clean-Up

Swati Lohani, Abhiruchi Uniyal, Amir Khan, Priyanka Khati, and Shaohua Chen

Abstract

The escalating environmental impact of climate change, including rising global temperatures and intensified extreme weather events, has led to increased pollution from oil spills, heavy metals, plastics, and greenhouse gases, threatening ecosystems and human health. An integrated approach to bioremediation, encompassing both in situ and ex situ strategies, is necessary to address these problems. Bioremediation harnesses the natural capabilities of organisms, including plants, bacteria, fungi, and algae, to break down pollutants and adapt to extreme environments. Advances in microbial genomics, synthetic biology, and nano-biotechnology have paved the way for developing optimized microbial strains and engineered microbes tailored to specific pollutants and environments. By integrating bioremediation with broader environmental management strategies and sustainable development goals, and utilizing tools such as artificial intelligence and bioaugmentation, this field holds immense potential as a long-term, sustainable solution to combat climate change and environmental

S. Lohani (✉)
ICAR-VPKAS, Almora, Uttarakhand, India

A. Uniyal
Wildlife Institute of India, Dehradun, Uttarakhand, India

A. Khan
CSIR-Central Institute of Medicinal & Aromatic Plants (CIMAP), Research Centre, Pantnagar, Uttarakhand, India

P. Khati
Crop Production Division, ICAR-VPKAS, Almora, Uttarakhand, India

S. Chen
College of Plant Protection and the Integrative Microbiology Research Centre, South China Agricultural University (SCAU), Guangzhou, China

G. Bhandari et al. (eds.), *Microbial Resilience*,
https://doi.org/10.1007/978-981-95-8883-1_13

degradation. Continued innovation, community engagement, and policy support are essential to maximizing its global impact. Additionally, this chapter examines the role of microbial diversity and activity in fluctuating environmental conditions, focusing on applications such as oil spill remediation in warming oceans, heavy metal removal in altered rainfall patterns, and plastic degradation in elevated temperatures. Addressing the challenges of scale-up, environmental variability, and regulatory frameworks, this chapter provides insights into the future of bioremediation as a vital tool in global environmental management strategies.

Keywords

Bioremediation · In situ and ex situ bioremediation · Biodegradation · Bioaccumulation · Biosorption · Bioaugmentation · Biostimulation · Microbial genomics · Nano-biotechnology · Artificial intelligence · Sustainable development goals (SDGs)

1 Introduction

Climate change is one of the major challenges of the twenty-first century, as it is intimately entwined with the health and functioning of the biosphere. It has placed immense stress on various global systems, including agriculture, leading to shifts in feeding patterns. Additionally, climate change has caused biodiversity loss by disrupting the integrity and survival of many species due to changes in optimal temperature ranges. Another alarming consequence is its role in accelerating antimicrobial resistance—a growing threat to human health due to the increasing incidence of resistant pathogenic infections (Abbass et al., 2022).

Anthropogenic activities such as industrialization, toxic waste production, greenhouse gas emissions, and pollution have exacerbated climate change, triggering events such as climate variability, ocean acidification, global warming, and rising atmospheric carbon dioxide levels. These activities have also contributed to ecological degradation, defaunation, and ecosystem disintegration (Malhi et al., 2020). Addressing climate change is of paramount importance, as this global threat demands a unified commitment to tackle its severe consequences and ensure the sustainability of our planet.

Nature itself is capable of mitigating and adapting to climate change. From an ecological perspective it is essential to explore and quantify the mechanisms, potential, and limitations of these nature-based solutions to effectively handle climate challenges. One such solution is bioremediation, a process that incorporates microbes and their metabolic activity to degrade contaminants. This technique employs naturally occurring biological organisms such as microbes, fungi, green plants, or their enzymes to break down hazardous substances which are present in natural environment, such as soil, sediments, air, and water, into less toxic or non-toxic forms by the process of reduction, elimination, alteration, and transformation of contaminants (Hlihor et al., 2017).

The removal of pollutants is commonly tackled by traditional remediation methods, which include chemical and physical approaches. However, these techniques are mostly expensive and non-specific, resulting in less effective treatment and potential harm to non-target areas. Furthermore, their frequent uses lead to secondary pollution, ecological harm and resource depletion. On the contrary, bioremediation is an eco-friendly, cost-effective, and sustainable practice. In addition to these benefits, it offers targeted treatment, facilitating more precise and effective remediation. Ongoing research and advancements in microbial technology ensure that bioremediation methods continue to evolve and become increasingly effective. By harnessing the innate abilities of microorganisms, bioremediation offers a holistic and sustainable approach to combating environmental pollution and plays a crucial role in tackling climate change challenges (Kumar et al., 2020; Bhatt et al., 2015a, b, 2016, 2019).

Plant growth-promoting rhizobacteria (PGPR), when combined with nanoformulations, such as nanozeolite and nanochitosan, can enhance soil quality, microbial diversity, and plant productivity in changing climates (Khati et al., 2017, 2018, 2019a, 2024). Such integrated approaches support both bioremediation and climate resilience by improving microbial stability and promoting nutrient cycling. Metagenomic analyses have further confirmed the influence of these treatments on rhizospheric microbial richness and ecosystem health (Chaudhary et al., 2021; Khati et al., 2019b).

2 Introduction to Bioremediation

Bioremediation involves the use of bacteria, fungi, and plants to degrade, remove, alter, immobilize, or detoxify various chemical and physical wastes in the environment. Microorganisms act as biocatalysts through their enzymatic pathways to facilitate biochemical reactions that degrade pollutants. The efficiency of bioremediation depends on several factors, including the chemical nature and concentration of pollutants, environmental conditions (such as temperature, pH, and oxygen levels), and the bioavailability of pollutants to microorganisms. Successful bioremediation requires adapting microorganisms to specific pollutants, continuous monitoring, and employing various degradation mechanisms to restore balance to contaminated ecosystems effectively (El Fantroussi & Agathos, 2005).

2.1 Historical Development and Major Milestones in Bioremediation

The historical development of bioremediation research spans over a century, beginning with early observations of microbial decomposition in the pre-twentieth century and formalizing in the 1970s with the coining of the term "bioremediation." The 1989 Exxon Valdez oil spill demonstrated the practical application of bioremediation on a large scale (Bragg et al., 1994). In the 1990s, significant advancements in genetic and molecular tools enabled a deeper understanding of microbial degradation pathways. In the twenty-first century, the integration of omics, synthetic biology, and nanotechnology has further advanced the field, allowing for the development of engineered microbes and innovative techniques such as nanobioremediation. Today, bioremediation is recognized as a sustainable and eco-friendly solution for addressing environmental pollution and mitigating the impacts of climate change.

2.2 Classification of Bioremediation

Bioremediation is categorized on its mode of application (Fig. 1).

2.2.1 In Situ Bioremediation

In situ remediation involves treating contaminants directly at the site using bioremediators. This process targets pollutant removal or attenuation by implementing microbial metabolic potential under natural environmental conditions (Farhadian et al., 2008).

Types of In Situ Bioremediation

1. *Intrinsic Bioremediation*: Intrinsic Bioremediation, also known as natural attenuation, involves a passive approach to remediating polluted areas, relying solely on native microorganisms for treatment. Considering the fact that native microorganisms are already adapted to the local environmental conditions, they are highly effective in degrading pollutants. This effectiveness is further supported by the microbe/hydrocarbon relationship that has developed over the course of evolution (Kumar et al., 2018).

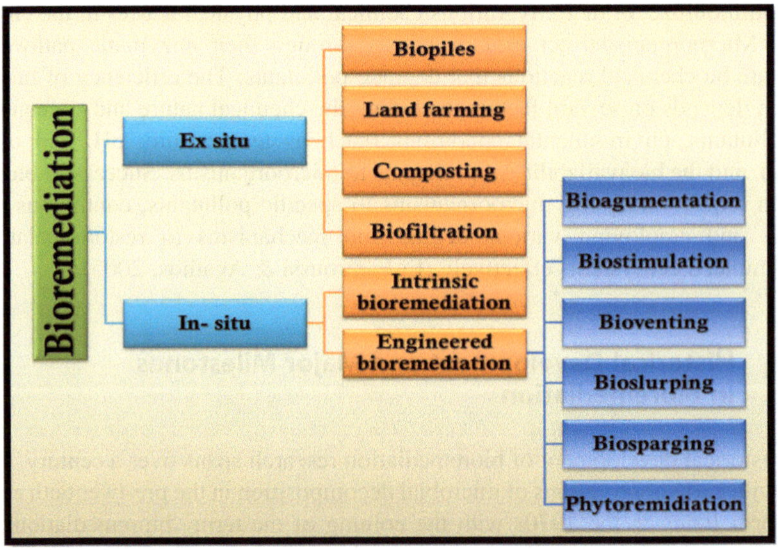

Fig. 1 Flow diagram to show types of bioremediation

2. *Engineered Bioremediation*: Engineered or accelerated bioremediation uses engineered systems to boost microbial activity for cleaning environmental pollutants. Substrates or nutrients are added to the environment to enhance microbial growth and speed up biodegradation by optimizing physical and chemical conditions. This method is often preferred over intrinsic bioremediation due to time and liability considerations (Sales da Silva et al., 2020). Engineered bioremediation includes the following techniques:
   (i) *Bioaugmentation*: Involves enhancing the autochthonous microflora of a polluted site by adding previously selected indigenous or genetically modified bioremediator species to support the degradation process (Woźniak-Karczewska et al., 2019).
   (ii) *Bio-stimulation*: This process includes injecting specific nutrients such as phosphorus, nitrogen, oxygen, or carbon at the site to stimulate the activity of indigenous microorganisms for bioremediation (Zeneli et al., 2019).
   (iii) *Bioventing*: Controlled stimulation of airflow, providing oxygen to enhance the activity of bioremediators and consequently improve the bioremediation process (Brown & Ulrich, 2014).
   (iv) *Bioslurping*: Combines vacuum-enhanced pumping, soil vapor extraction, and bioventing. It is effective for remediating semi-volatile and volatile organic compounds from contaminated soils and removing light non-aqueous phase liquids (LNAPLs) from the water table (Kumar et al., 2018).
   (v) *Bio-sparging*: Involves injecting air and gas-phase nutrients into the saturated zone under pressure to promote aerobic biodegradation. This method effectively remediates aerobically degradable organic contaminants, including petroleum products such as oils, diesel, gasoline, and jet fuels (Azubuike et al., 2016).
   (vi) *Phytoremediation*: Utilizes plants and their products to clean or stabilize contaminants and metals in soil or water. Certain plants absorb heavy metals through their roots and concentrate them in their stems, shoots, and leaves, using genes that control metal uptake and distribution. Phytoremediation techniques include phytodegradation, phytostimulation/rhizodegradation, phytovolatilization, phytoextraction, rhizofiltration, and phytostabilization (Flathman & Lanza, 1998).

2.2.2 Ex Situ Bioremediation

Ex situ bioremediation or off-site bioremediation involves removing and transporting contaminants from a polluted site to another location for treatment. Techniques vary based on the type and level of contaminants, pollution degree, site topography, and treatment cost (Atlas & Philp, 2005). This method effectively addresses a wide range of hydrocarbon pollutants, including aliphatic chains, lubricating oils, carbon compounds, ketones, phenols, cresols, polycyclic aromatic hydrocarbons, and semi-volatile organic compounds (SVOCs).

Types of Ex Situ Bioremediation

1. *Biopiles*: also referred to as biocells, bioheaps, biomounds, and compost piles. It involves the enhancement techniques through above-ground piling of excavated soil contaminated with aerobically remediable hydrocarbons (Delille et al., 2008).
2. *Land farming*: a simple bioremediation process where polluted soil is excavated and placed above ground for aerobic biodegradation by autochthonous microorganisms, stimulated by tilling, nutrient amendments, aeration, and irrigation (Al-Awadhi et al., 1996).
3. *Composting*: an age-old, self-heating process for managing organic waste, resulting in a nutrient-rich product beneficial for agriculture and gardening (Lin et al., 2022).
4. *Biofiltration*: a biological process that uses microbial populations to treat biodegradable waste and purify air contaminated with volatile organic compounds (Flemming, 2000).

2.3 Mechanisms Involved in Bioremediation

Bioremediation relies on several key mechanisms to eradicate or neutralize pollutants from the environment. These mechanisms include:

(i) *Biodegradation*: Biodegradation refers to any biologically driven process that alters or breaks down a substrate. This process is facilitated by degraders, which are microorganisms such as fungi, bacteria, and algae. These organisms thrive on decomposing organic matter, a natural component of ecosystems. Biodegradation plays a vital role in reducing pollution by breaking down organic waste, making it an essential environmental mechanism for maintaining ecological balance (Razza & Innocenti, 2012). Biodegradation operates through different mechanisms based on environmental conditions and the type of microorganisms involved. Aerobic biodegradation occurs in the presence of oxygen, where microorganisms such as Pseudomonas bacteria convert organic pollutants into carbon dioxide (CO_2) and water (H_2O), exemplified by their role in breaking down oil spills. Anaerobic biodegradation takes place in the absence of oxygen, utilizing electron acceptors such as nitrate or sulfate to decompose pollutants, as seen with anaerobic bacteria that degrade trichloroethylene (TCE). Co-metabolism involves the degradation of pollutants by microorganisms while they metabolize another substrate; for example, methanotrophic bacteria break down TCE while consuming methane. These biodegradation processes are crucial for mitigating environmental pollution (Doukani et al., 2022).

(ii) *Bioaccumulation*: Bioaccumulation is a process in which living organisms, such as plants and animals, absorb and concentrate contaminants (e.g., heavy metals or organic compounds) from their environment into their tissues. This occurs because pollutants are absorbed and stored in cells or tissues, often due to their inability to be metabolized or excreted. The mechanism is utilized in applications such as phytoremediation, where plants accumulate pollutants, and microbial-based metal recovery, where microorganisms extract valuable metals from contaminated environments (Daley, 2013).

(iii) *Biosorption*: Biosorption is the passive adsorption of pollutants, such as heavy metals or dyes, onto the surface of biological materials, including microbial cell walls, algae, or plant biomass. This process occurs as pollutants bind to functional groups, such as carboxyl, hydroxyl, or amino groups, on the surface of these materials through physical or chemical interactions. The mechanism relies on the affinity between the pollutants and the functional groups present on the biological surfaces, enabling the capture and immobilization of contaminants.

3 Microorganisms in Bioremediation

Microorganisms play a vital role in bioremediation, utilizing their metabolic capabilities to combat environmental pollution by breaking down, transforming, or immobilizing harmful substances such as petroleum hydrocarbons, heavy metals, pesticides, solvents, and industrial chemicals. These tiny yet powerful microscopic organisms—including bacteria, fungi, and algae—produce specialized enzymes that enable them to degrade pollutants, often converting toxic compounds into less harmful substances such as water and carbon dioxide through a process known as mineralization (Mahmoud, 2021). Their exceptional adaptability allows them to thrive in extreme environments and utilize a wide range of pollutants as carbon sources, making them highly effective in bioremediation efforts (Kour et al., 2022). For example, acidophiles flourish in acidic conditions, psychrophiles adapt to cold climates, and halophiles thrive in saline regions, demonstrating how microbial diversity enhances their bioremediation potential (Perera & Hemamali, 2022). Due to their resilience and widespread presence, these microbes play a vital role in detoxifying contaminated environments, restoring ecosystems, and contributing to a cleaner, safer planet (Fig. 2).

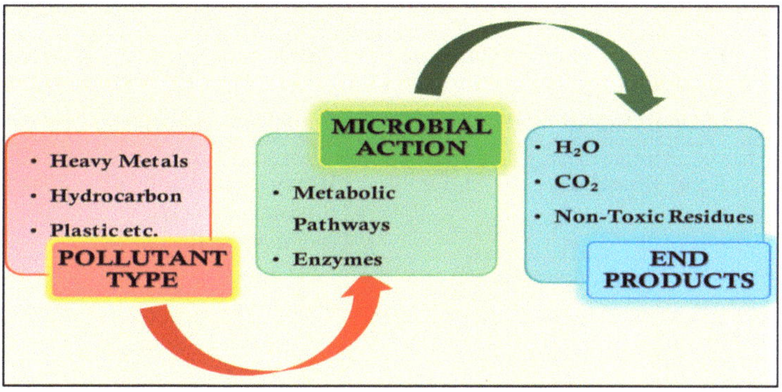

Fig. 2 Microbial bioremediation overview

3.1 Types of Microorganisms Used in Bioremediation

There are several types of microorganisms involved in the process of bioremediation, which are listed in Table 1.

Table 1 Types of microorganisms involved in bioremediation

Microbial group	Microorganisms	Target pollutants	References
Bacteria	*Bacillus sp., Rhodopirellula sp., Rhodovibrio sp., Formosa sp.*	Hydrocarbons	Machado et al. (2019)
	P. cepacia, B. coagulans, B. cereus, Serratia ficaria	Diesel oil	Miri et al. (2022)
	Pseudoalteromonas sp., Agarivorans sp.	Hydrocarbons	Dell'Anno et al. (2021)
	Pseudomonas aeruginosa, Aeromonas sp.	Heavy metals (Chromium, uranium, nickel, copper)	Gaur et al. (2022)
	Escherichia coli	Hexavalent chromium	Mohamed et al. (2020)
	Microbacterium sp., Micrococcus sp., Bacillus sp., Shigella sp.	Uranium, arsenic	Bhakat et al. (2019)
	P. aeruginosa	Crude oil	Mukjang et al. (2022)
	Dehalococcoides sp.	Chloroethenes	Dutta et al. (2022)
	Bacillus sp., Staphylococcus sp.	Endosulfans	Liu et al. (2018)
	P. aeruginosa, B. subtilis, Corynebacterium propinquum, Alcaligenes odorans	Phenol	Gaur and Narasimhulu (2018)
Fungi	*Phanerochaete chrysosporium*	Benzene, toluene, organochlorines	Singh et al. (2020)
	Saccharomyces cerevisiae	Arsenic	Verma et al. (2022)
	Aspergillus sp., Penicillium sp.	Hydrocarbons, chlorophenols	Li et al. (2020)
	Trichoderma sp., Penicillium sp., Aspergillus sp.	Cobalt, copper	Dusengemungu et al. (2020)
Algae	*Chlamydomonas reinhardtii*	Chromium, cadmium	Nowicka et al. (2020)
	Fucus vesiculosus	Heavy metals (Chromium, nickel, cadmium, lead)	Moreira et al. (2019)
	Chlorella sp., Spirulina sp.	Lead, nickel, dichromate	Geetha et al. (2021)

3.2 Microbial Mechanisms in Pollutant Degradation

Microorganisms eliminate pollutants through two primary mechanisms: mobilization and immobilization (Verma & Kuila, 2019). Mobilization involves enzymatic oxidation, bioleaching, biostimulation, bioaugmentation, and enzymatic reduction, while immobilization encompasses biosorption, bioaccumulation, and precipitation (Tak et al., 2012; Ayangbenro et al., 2019).

3.2.1 Mobilization Mechanisms

- *Enzymatic Oxidation*: This process mitigates pollutant toxicity by oxidizing compounds, leading to electron loss in heavy metals, thereby reducing their harmful effects. Microbial oxidoreductase enzymes play a crucial role in degrading persistent pollutants such as dyes and phenols (Unuofin et al., 2019). Laccase, an essential oxidoreductase, catalyzes the oxidation of aromatic amines (Gangola et al., 2018), while phenols and polyphenols facilitate oxygen reduction to water (Kushwaha et al., 2018; Sahay, 2021). Notably, *Pycnoporus* and *Leptosphaerulina* species are known for laccase production, aiding in heavy metal degradation (Copete-Pertuz et al., 2018; Tian et al., 2020).
- *Enzymatic Reduction*: This mechanism transforms pollutants into less oxidized and insoluble forms, primarily through the action of obligate and facultative anaerobes. It is particularly effective in bioremediating pollutants such as polychlorinated dibenzo-p-dioxins, dibenzofurans, and hexavalent chromium. The enzyme chrome reductase facilitates the reduction of hexavalent chromium to its less toxic trivalent state (Zacharia, 2019; Saxena et al., 2020).
- *Bioaugmentation*: This technique accelerates bioremediation by introducing external microbial strains into contaminated environments. In certain cases, genetically engineered microorganisms are utilized to enhance efficiency (Goswami et al., 2018).
- *Biostimulation*: This method boosts the activity of native microbes by supplying essential nutrients to expedite bioremediation. However, excessive nutrient input may lead to environmental issues such as eutrophication (Sayed et al., 2021).
- *Bioleaching*: Acidophilic microbes, such as *Acidithiobacillus thiooxidans* and *Penicillium* species, facilitate the solubilization of heavy metals, enhancing their extraction from contaminated sites (Sun et al., 2019).

3.2.2 Immobilization Mechanisms

- *Biosorption*: This mechanism involves the removal of heavy metals through ion displacement, chelation, and electrostatic interactions on microbial cell surfaces (Mahmoud, 2021). Microbes such as *Rhodococcus erythropolis* and *Bacillus anthracis* facilitate this process, while valuable metals such as gold, zinc, and copper can be recovered via desorption (Mathew & Krishnamurthy, 2018; Medfu Tarekegn et al., 2020; Gangola et al., 2015). Additionally, microbial extracellular substances enhance complexation, further aiding in metal removal (Ayangbenro & Babalola, 2017; Xie et al., 2020). Biochar is also instrumental in extracting metals such as copper and chromium (Xiao et al., 2019). In iron-deficient

conditions, microbial siderophores chelate metals, increasing solubility and promoting degradation. Cyanobacteria such as *Anabaena* and *Synechococcus* play significant roles in bioremediation (Khan et al., 2018).
- *Bioaccumulation*: This process occurs when microbes absorb pollutants at a rate higher than their expulsion, leading to intracellular accumulation. Heavy metals enter microbial cells via carrier-mediated transport, protein channels, or ion pumps (Mir-Tutusaus et al., 2018). Notable bioaccumulators include *Rhizopus arrhizus* (mercury), *Pseudomonas putida* (cadmium), and *Aspergillus niger* (thorium) (Sharma et al., 2022).
- *Precipitation*: This mechanism converts heavy metals into insoluble precipitates, reducing their toxicity. It plays a vital role in biogeochemical cycling, leading to the formation of metal deposits such as iron, manganese, and silver through enzymatic activity (Sharma et al., 2022). Sulfate-reducing bacteria contribute by converting organo-phosphate to ortho-phosphate under alkaline conditions (Pratush et al., 2018). Additionally, *Bacillus subtilis* and *Oceanobacillus indicireducens* aid in heavy metal precipitation within the environment (Maity et al., 2019).

3.3 Biotechnological Advances in Microbial Bioremediation

- *Enzymatic Role in Pollution Degradation*: Microbial enzymes facilitate pollutant breakdown through oxidation, reduction, and elimination (Bhandari et al., 2021). However, enzyme-based bioremediation faces challenges such as stability and sourcing, making it costly (Narayanan et al., 2023).
- *Molecular Techniques in Microbial Bioremediation*: Advanced molecular tools, including sequencing, protein profiling, and bioinformatics, enable the identification of microbial species and their metabolic pathways (Sharma et al., 2022). Omics approaches such as metagenomics and metabolomics provide deeper insights, while Next-Generation Sequencing (NGS) has revolutionized gene analysis (Eisenhofer et al., 2019).
- *Microbial Byproducts in Pollution Cleanup*: Microorganisms produce biosurfactants, organic acids, and polymeric substances that enhance pollutant removal. Organic acids increase metal solubility, biosurfactants improve pollutant bioavailability, and polymeric substances stabilize metals (Ojuederie & Babalola, 2017).
- *Emerging Technologies in Microbial Bioremediation*: Microbial glycoconjugates aid pollutant removal by reducing surface tension and enhancing bioavailability (Bhatt et al., 2021). Species such as *Scedosporium sp.* and *Acinetobacter sp.* are effective in petroleum hydrocarbon degradation (Atakpa et al., 2022). Bioelectrochemical systems, which integrate biological and electrochemical

processes, have shown promise in hydrocarbon remediation by leveraging microbial interactions (Ambaye et al., 2023). Certain bacteria, including *Pseudomonas sp.*, *Ralstonia sp.*, *Rhodococcus sp.*, and *Thauera sp.*, efficiently degrade phenanthrene in polluted soils (Sharma et al., 2020).

4 Climate Change, Microbial Responses, and Bioremediation

Climate change exacerbates environmental challenges by intensifying the spread and impact of pollutants such as oil spills, heavy metals, plastics, and greenhouse gases. Rising temperatures and extreme weather events increase oil spill risks, accelerate heavy metal release from melting ice and permafrost, and slow plastic degradation. Additionally, greenhouse gases fuel global warming, creating a feedback loop that worsens pollution. Understanding this interplay is crucial for effective mitigation strategies to protect ecosystems and public health (Tiedje et al., 2022).

Environmental degradation, driven by human activities and natural processes, leads to air, water, and soil contamination, threatening biodiversity and ecosystem stability. Industrial waste, oil spills, and plastic pollution reduce resource quality, emphasizing the need for bioremediation—a sustainable method using microorganisms, plants, or enzymes to neutralize pollutants. As climate change accelerates degradation, implementing bioremediation becomes essential for ecosystem restoration (Arora, 2018). Climate change alters microbial diversity and activity by shifting environmental factors such as temperature, moisture, and pH. Rising temperatures favor heat-tolerant microbes while reducing sensitive species, disrupting nutrient cycling and decomposition (Mutambara, 2024). In aquatic ecosystems, warming and acidification affect microbes vital for carbon and nitrogen processing, impacting ecosystem stability. Extreme weather events such as droughts, floods, and heatwaves further disrupt microbial communities by altering habitats. Droughts limit soil moisture, reducing microbial activity and biodiversity, while floods wash away communities and introduce pathogens. Heatwaves accelerate metabolism but may stress heat-sensitive species, impairing nutrient cycling and carbon storage (Jha & Dev, 2024).

Microbial adaptability is crucial for bioremediation success. Changes in temperature, pH, moisture, and pollutant levels influence microbial growth and degradation efficiency. Some microbes activate stress-response mechanisms, while extremophiles thrive in harsh conditions, offering potential for remediation in challenging environments. Understanding microbial responses helps optimize bioremediation strategies for effective pollutant removal. Recognizing the interconnectedness of climate change, pollution, and microbial ecosystems is key to developing sustainable environmental protection and remediation strategies. Integrating climate action, pollution control, and bioremediation will help safeguard ecosystems and public health for future generations (Mishra et al., 2022).

5 Case Studies: Successful Microbial Bioremediation in the Context of Climate Change

5.1 Oil Spill Remediation

The ability of indigenous microorganisms to degrade petroleum hydrocarbons in marine environments is well documented. These microbes naturally break down oil pollutants, helping to cleanse affected ecosystems. Bioremediation techniques such as bioaugmentation (introducing specific microbial strains) and biostimulation (adding nutrients to enhance microbial activity) have been successfully employed in major oil spill incidents, including the Exxon Valdez spill in Alaska (1989) and the Deepwater Horizon spill in the Gulf of Mexico (2010) (Zapata-Peñasco et al., 2023). These cases demonstrate the effectiveness of microbial bioremediation in mitigating the environmental impact of oil spills.

5.2 Heavy Metal Remediation

Unlike organic pollutants, heavy metals cannot be biodegraded; however, microbial and plant-based bioremediation techniques can transform them into less toxic or more manageable forms. Strategies such as bioaugmented rhizoaccumulation, rhizofiltration, and phytoaccumulation have been used to remediate contaminated soils and waters. Studies have reported the removal of 19–65% of arsenic (As), copper (Cu), lead (Pb), and zinc (Zn); 99.3% of iron (Fe), Zn, cadmium (Cd), Cu, boron (B), and chromium (Cr); 50–100% of Pb; 25–60% of nickel (Ni); and 20–70% of silver nanoparticles (Mesa et al., 2015). Specific plants such as willows, sunflowers, and Indian mustard have been effective in phytoextraction (Yang et al., 2022), while bioaugmentation with rhizobacteria has enhanced metal uptake (Tiecher et al., 2016). These results highlight the potential of microbial and plant-assisted strategies in reducing heavy metal contamination.

5.3 Plastic Degradation

Plastic pollution is a growing environmental concern, with microplastics posing risks to ecosystems and human health. Some microbial species have shown promise in degrading plastics. The bacterium *Ideonella sakaiensis 201-F6* efficiently degrades polyethylene terephthalate (PET) (Yoshida et al., 2016). Additionally, fungi such as *Phanerochaete chrysosporium*, *Aspergillus niger*, and *Penicillium funiculosum* have demonstrated the ability to degrade low-density polyethylene (Weiland et al., 1995). The discovery of these microbial plastic degraders offers a potential solution for mitigating plastic waste in the environment.

5.4 Greenhouse Gas Mitigation

Microorganisms play crucial roles in the carbon cycle and greenhouse gas (GHG) mitigation by influencing carbon dioxide (CO_2), methane (CH_4), and nitrous oxide (N_2O) levels. Methanotrophic bacteria oxidize CH_4, converting it into less harmful compounds, particularly in landfill environments (Chanton & Liptay, 2000). Estimates suggest that these bacteria oxidize between 10% and 100% of landfill-generated methane, making landfills potential CH_4 sinks rather than sources (Bogner, 2006). Additionally, soil microbial communities contribute to CO_2 sequestration by regulating organic matter decomposition and soil respiration. Enhancing the activity of these microbes could be a viable strategy for mitigating climate change impacts.

Microbial bioremediation offers a sustainable and effective approach to addressing environmental challenges related to pollution and climate change. From oil spill cleanup and heavy metal remediation to plastic degradation and GHG mitigation, microbial communities provide natural solutions for restoring ecosystems and reducing anthropogenic impacts on the environment. Continued research and technological advancements in microbial bioremediation will further enhance its application in combating global environmental issues.

6 Adaptation Strategies for Microbial Bioremediation in a Changing Climate

As climate change alters environmental conditions, effective microbial bioremediation requires adaptive strategies to maintain efficiency due to challenges such as low microbial adaptability, slow processing times, reduced competitiveness, and limited bioavailability of target pollutants (Singh et al., 2017).

6.1 Selection and Optimization of Microbial Strains for Specific Environments

Systems biology (SB) and metabolic engineering (ME), widely applied in life sciences, are gaining interest in bioremediation. Systems biology analyzes complex biological systems and their interconnections at molecular, cellular, community, and ecosystem levels, providing insights into biological processes such as gene expression, enzyme activity, biosynthetic pathways, and secondary metabolite production under pollutant-induced stress (Megharaj et al., 2011).

"Omics" techniques (genomics, proteomics, transcriptomics, and metabolomics) are key in systems biology research. These tools offer a deeper understanding of microbial metabolic pathways, characterizing cellular processes, microbial community composition, and metabolic activity, especially under toxic stress. The goal is to eliminate or detoxify harmful chemicals by understanding environmental factors affecting microbial processes and cell-to-cell interactions (Kumar Singh & Shukla, 2015). Choosing the right systems biology tools depends on cost, time, workforce, and objectives. DNA-based approaches such as genomics, metagenomics, and phenomics are ideal for studying microbial community composition. For identifying functional genes and understanding metabolic pathways and proteins, transcriptomics, proteomics, and mass spectrometry are effective (Chakraborty et al., 2012).

For characterizing microbially secreted small molecules, nuclear magnetic resonance (NMR), desorption electrospray ionization (DESI), and matrix-assisted laser desorption/ionization (MALDI) are recommended (El Amrani et al., 2015). Functional genomics microarrays, proteomics, and metabolomics are useful for analyzing enzyme activity in complex microbial communities.

6.2 Metabolic Engineering of Microbes for Enhanced Pollutant Degradation

Advancements in genetics, biochemical pathways, and synthetic biology provide valuable insights for developing effective bioremediation strategies. While many microbes can naturally degrade pollutants, their metabolic pathways are often inefficient for large-scale applications. Bioremediation is limited by factors such as complex pollutant mixtures, prolonged degradation times, and microbial nonbiocompatibility. Metabolic engineering (ME) addresses these challenges by modifying existing pathways or introducing new catabolic pathways in microbial hosts. This approach enhances biomolecule yield, broadens the range of target pollutants, improves enzymatic activity, increases toxic compound bioavailability, and enables novel biodegradation functions absent in native microbes (Kumar & Prasad, 2011).

Beyond bioremediation, microbes are widely used in industries such as pharmaceuticals, agriculture, and food. Metabolic engineering supports industrial applications by increasing product formation, accelerating processes, conserving energy, reducing by-products, and developing stress-resistant strains. Key targets include expanding substrate and product ranges, improving productivity and process efficiency, and enhancing cellular properties (Anesiadis et al., 2008).

6.2.1 Approaches for Metabolic Engineering

6.2.1.1 Heterologous Expression of Entire Gene Clusters

Advancements in synthetic biology and genome sequencing have uncovered gene clusters essential for pollutant degradation and natural product biosynthesis. Despite challenges such as low production levels and limited genetic tools in certain strains, transferring entire gene clusters or metabolic pathways to optimized hosts has proven effective (Vilchez-Vargas et al., 2013). For instance, the polyhydroxybutyrate (PHB) biosynthetic pathway genes from *Ralstonia eutropha* have been successfully transferred into the industrially advantageous organism *Saccharomyces cerevisiae* (Abd-El-haleem et al., 2007).

6.2.1.2 Engineering Regulatory Networks

Metabolic pathway genes in microbes are often grouped into clusters regulated by specific genes that activate or repress their expression. These clusters are expressed only when necessary to avoid metabolic costs. For bioremediation and industrial use, microbes can be engineered to continuously produce activators or suppress repressors, enabling efficient and sustained metabolite production (Madison & Huisman, 1999). For example, continuous expression of SARP regulators in *Streptomyces coelicolor* boosts the production of secondary metabolites such as actinorhodin. Similarly, in *Streptomyces griseus*, inactivation of pathway-specific repressors enhances chromomycin production by enabling uninterrupted synthesis (Nakamura & Whited, 2003).

6.2.1.3 Gene Insertion and Deletion

Expression of heterologous pathways in microbial systems can sometimes cause metabolic imbalances, necessitating the addition or removal of specific genes to achieve desired traits. For instance, deleting multiple genes in the multidrug resistance operon of *Burkholderia cenocepacia* K56-2 using endonuclease I-SceI enhanced its suitability for environmental bioremediation by making it easier to manipulate and less likely to cause infections (Flannagan et al., 2008). Similarly, introducing genes encoding 2-keto acid decarboxylase and alcohol dehydrogenase from *Saccharomyces cerevisiae* into *Escherichia coli* allowed the conversion of 2-keto acids (amino acid biosynthesis intermediates) into higher alcohols, which can be used as biofuels (Kumar & Prasad, 2011).

6.2.1.4 Redirecting Metabolic Pathways

Multiple pathway modifications can direct microbes to produce specific products. For instance, while *E. coli* typically produces succinic acid in small amounts, redirecting metabolic fluxes can increase its production while reducing by-products. Systems biology tools such as transcriptomics and metabolomics, alongside computational models, aid strain optimization and pathway engineering (Wesolowski et al., 2010). Advances in genomics and large biological databases enable *in silico* designs and experimental validation of novel pathways for bioremediation or industrial applications (Verma et al., 2022).

6.2.1.5 Quorum Sensing

Quorum sensing is a bacterial gene regulation mechanism enabling cell-to-cell communication and population-density-based gene transcription control through extracellular chemical signals such as acylhomoserine lactones (AHLs) in Gram-negative bacteria and small peptides in Gram-positive bacteria. It regulates processes such as plasmid transfer, biofilm formation, bioluminescence, and bioactive compound production, with applications in improving recombinant protein yields and bioremediation. Studies have identified quorum-sensing bacteria capable of degrading aromatic compounds, with over 11% of species in the NCBI genome database possessing aromatic ring-hydroxylating dioxygenase (RHD) and AHL/AI-synthetase genes (Huang et al., 2013). For instance, *Sphingomonadales* and

Rhizobiales were isolated as phenanthrene degraders producing AHLs, and quorum sensing genes (*lasI* and *rhlI*) in *Pseudomonas aeruginosa* N6P6 were found to promote biofilm formation and polycyclic aromatic hydrocarbon (PAH) degradation, illustrating the synergistic potential of quorum sensing for ecological and industrial applications (Kumari et al., 2016).

6.2.1.6 Genome Editing

Non-homologous end joining (NHEJ) is a genome editing repair mechanism, with tools such as TALEN and ZFNs using specific DNA-binding domains to introduce edits (Gupta & Shukla, 2017). The CRISPR-Cas9 system, guided by RNA (gRNA), precisely cuts DNA strands and is highly compatible with archaeal and bacterial systems, making it versatile for bioremediation, synthetic biology, and beyond. Unlike single-gene editing, CRISPR-Cas9 enables simultaneous multi-locus editing, allowing the construction of efficient gene circuits (Doudna & Charpentier, 2014). This capability aids in modifying microbial metabolic pathways to degrade pollutants or produce targeted compounds (Selle & Barrangou, 2015). Integrating synthetic biology and metabolic engineering enhances microbial bioremediation by reconstructing metabolic pathways for efficient pollutant degradation. Coordinated efforts, advanced bioinformatics tools, and emerging technologies such as gene shuffling, whole-genome editing, cybernetics, and nanobioremediation can significantly improve environmental clean-up processes.

7 Challenges and Limitations in Bioremediation in a Changing Climate

7.1 Impact of Climate Change on Microbial Bioremediation

Climate change significantly affects microbial bioremediation by altering environmental factors such as temperature, moisture, pH, and nutrient cycles. These changes can either enhance or inhibit microbial activity, directly impacting pollutant degradation. Temperature shifts play a crucial role—excessive heat denatures enzymes, reducing microbial viability, while cold slows enzymatic reactions. Deviations in pH disrupt microbial functions, with acidification and alkalinization reducing metabolic efficiency. Although specialized microbes may adapt, many lose effectiveness outside optimal conditions (Alkorta et al., 2017). Nutrient availability is also impacted by changes in biogeochemical cycles. Reduced nutrient levels slow microbial metabolism and intensify competition, while excess nutrients (eutrophication) lead to oxygen depletion, further hindering bioremediation. Climate change also disrupts microbial community dynamics, increasing competition and reducing pollutant degradation efficiency.

7.2 Challenges of Microbial Degradation in Bioremediation

Microbial degradation is a slow and complex process influenced by pollutant characteristics, environmental conditions, and biological limitations. Many pollutants, such as PAHs, heavy metals, and synthetic chemicals, resist microbial breakdown due to their toxic or complex structures. Bioavailability also limits degradation, as some pollutants bind tightly to soil or exist in hydrophobic forms, making them inaccessible to microbes. Biological challenges include the lack of specific enzymes, competition for resources, and the presence of microbial predators, all of which slow the degradation process. This can lead to pollutants persisting in the environment for years or even decades.

7.3 Implications of Slow Microbial Degradation on Long-Term Sustainability

The slow pace of microbial degradation poses serious environmental, economic, and social challenges. Delayed degradation increases the risk of pollutants spreading through runoff and wind erosion, contaminating land and water and threatening biodiversity. Extended remediation timelines require ongoing monitoring and maintenance, increasing financial burdens. Incomplete degradation can also contribute to climate change, as anaerobic decomposition produces methane, a potent greenhouse gas. Addressing these issues requires advanced bioremediation techniques and sustainable management policies (Boopathy, 2000).

7.4 Ethical and Biosafety Concerns in Using GMOs for Bioremediation

The use of GMOs in bioremediation raises ethical and biosafety concerns. Ecologically, GMOs may disrupt ecosystems by outcompeting native species or transferring genes, potentially causing new problems such as antibiotic resistance. Ethical concerns include the lack of informed consent and the need for fair distribution of risks and benefits. Biosafety risks include the uncontrolled spread of GMOs, loss of genetic stability through mutations, and potential misuse or unintended release. To mitigate these risks, comprehensive assessments, biocontainment strategies (such as kill switches), and strict regulatory oversight are essential for the safe and responsible use of GMOs in bioremediation (Jafari et al., 2012).

8 Integrating Bioremediation with Other Environmental Management Strategies

In a holistic way, bioremediation addresses soil, water, and air contamination while preserving biodiversity and ecosystem functions. It reduces toxic pollutants, improves soil health, and restores water quality, contributing to long-term ecological balance. This approach integrates community engagement, regulatory compliance, and continuous monitoring to ensure environmental and public health protection. When combined with other restoration techniques such as phytoremediation and habitat rehabilitation, bioremediation supports a comprehensive strategy for restoring ecosystems, promotes sustainable land use and climate resilience (Philp et al., 2005).

Integrating bioremediation with physical and chemical treatments enhances pollutant removal by combining their strengths. Physical methods, such as soil excavation and air sparging, reduce contaminant levels and improve accessibility for microbial degradation. Chemical treatments, such as oxidation and surfactant washing, break down complex pollutants into simpler forms for easier microbial processing. Applied sequentially or simultaneously, these methods reduce toxicity, improve conditions for microbial activity, and address residual contamination. This integrated approach increases efficiency, treats a wider range of pollutants, reduces costs, and enhances environmental safety, making it ideal for complex or heavily contaminated environments (Aparicio et al., 2022).

Community and policy engagement is also crucial to the success of large-scale bioremediation projects. Effective community involvement fosters public trust, ensures transparency, and addresses local concerns. Public consultations, community advisory boards, and regular updates allow residents to participate in decision-making and understand the project's goals and potential impacts. Policy engagement ensures compliance with environmental regulations and promotes sustainable practices. Collaborating with government agencies helps secure funding, streamline permits, and enforce safety standards. Furthermore, aligning bioremediation efforts with broader environmental policies supports long-term ecological restoration. This combined approach enhances project efficiency, social acceptance, and environmental accountability (Bilyard et al., 1996).

Bioremediation supports sustainable development goals (SDGs) by using natural processes to clean up environmental pollutants, restoring ecosystems, and conserving resources. It offers a cost-effective and eco-friendly alternative to traditional remediation methods, reducing pollution in soil, water, and air. This process enhances public health, supports biodiversity, and promotes the reuse of contaminated land and water. Additionally, bioremediation helps mitigate climate change by reducing greenhouse gas emissions and sequestering carbon. By fostering cleaner environments and sustainable industrial practices, bioremediation contributes to a healthier planet and a more sustainable future (Kumar et al., 2018).

9 Conclusion

Bioremediation is a crucial and sustainable method for addressing environmental pollution, particularly in the face of climate change. As ecosystems experience mounting pressures from human activities and extreme weather, microorganisms play an essential role in breaking down pollutants. Their adaptability to harsh conditions allows them to degrade contaminants such as oil spills, heavy metals, plastics, and greenhouse gases, making bioremediation a key strategy in environmental restoration. Although microbial approaches have demonstrated effectiveness, climate change brings additional challenges, such as shifts in microbial communities, changes in metabolic processes, and the persistence of pollutants. Advances in metabolic engineering, synthetic biology, and bioaugmentation are enhancing microbial efficiency, leading to more precise and effective remediation methods. Furthermore, integrating bioremediation with physical, chemical, and policy-based environmental strategies increases its impact and scalability. Despite it, bioremediation faces challenges, including environmental limitations, competition among microbes, and biosafety concerns related to genetically modified organisms. However, innovative research in microbial genomics, artificial intelligence, and nanobiotechnology offers new solutions. Strengthening interdisciplinary collaboration and promoting supportive policies will be crucial in overcoming these challenges and unlocking the full potential of bioremediation.

10 Future Prospects

Bioremediation is evolving rapidly through innovations in genomics, artificial intelligence, and nanotechnology. These advancements promise more efficient, targeted, and sustainable approaches to mitigating environmental contamination and addressing climate change.

Emerging genomics and metagenomics technologies such as next-generation sequencing (NGS) allow researchers to identify and characterize microbial communities involved in pollutant degradation. These techniques reveal novel genes and metabolic pathways responsible for breaking down complex and hazardous substances. Future research will focus on engineering microbes using tools such as CRISPR Cas9 to enhance their capacity to degrade persistent pollutants, including microplastics, pharmaceutical residues, and per- and polyfluoroalkyl substances (PFAS). Additionally, metagenomics enables real-time monitoring of microbial dynamics in contaminated sites, improving process efficiency and ecological safety (Gunjal et al., 2023).

Artificial intelligence and machine learning can transform bioremediation by enabling data-driven decision-making and process optimization. These technologies can analyze large metagenomic datasets to identify key microbial players and predict their pollutant degradation capabilities. AI-driven models can simulate bioremediation scenarios, optimizing conditions for microbial activity and enhancing the effectiveness of remediation efforts. Machine learning algorithms also facilitate the discovery of new biodegradation pathways and improve the design of microbial consortia tailored to specific contaminants. In the future, integrating AI with real-time environmental monitoring will allow for adaptive and automated bioremediation processes, increasing both precision and sustainability (Raviteja & Reddy, 2022).

Nano-biotechnology holds great promise in enhancing microbial bioremediation capabilities. Nanoparticles can improve microbial activity by delivering essential nutrients, stimulating microbial growth, or enhancing pollutant bioavailability. For instance, iron nanoparticles can facilitate the reduction of heavy metals, while carbon-based nanomaterials may accelerate the breakdown of organic pollutants. Future research will focus on developing biodegradable and eco-friendly nanomaterials to minimize secondary contamination. Additionally, nano-carriers can be used to deliver engineered microbes precisely to polluted areas, ensuring targeted and efficient bioremediation. This technology will play a key role in addressing complex contamination scenarios in both terrestrial and aquatic environments (Guha et al., 2024).

References

Abbass, K., Qasim, M. Z., Song, H., Murshed, M., Mahmood, H., & Younis, I. (2022). A review of the global climate change impacts, adaptation, and sustainable mitigation measures. *Environmental Science and Pollution Research, 29*(28), 42539–42559.

Abd-El-haleem, D., Amara, A., Zaki, S., Abulhamd, A., & Abulreesh, G. (2007). Biosynthesis of biodegradable polyhydroxyalkanotes biopolymers in genetically modified yeasts. *International journal of Environmental Science and Technology, 4*, 513–520.

Al-Awadhi, N., Al-Daher, R., ElNawawy, A., & Salba, M. T. (1996). Bioremediation of oil-contaminated soil in Kuwait. I. landfarming to remediate oil-contaminated soil. *Soil and Sediment Contamination, 5*(3), 243–260.

Alkorta, I., Epelde, L., & Garbisu, C. (2017). Environmental parameters altered by climate change affect the activity of soil microorganisms involved in bioremediation. *FEMS Microbiology Letters, 364*(19), fnx200.

Ambaye, T. G., Vaccari, M., Franzetti, A., Prasad, S., Formicola, F., Rosatelli, A., et al. (2023). Microbial electrochemical bioremediation of petroleum hydrocarbons (PHCs) pollution: Recent advances and outlook. *Chemical Engineering Journal, 452*, 139372.

Anesiadis, N., Cluett, W. R., & Mahadevan, R. (2008). Dynamic metabolic engineering for increasing bioprocess productivity. *Metabolic Engineering, 10*(5), 255–266.

Aparicio, J. D., Raimondo, E. E., Saez, J. M., Costa-Gutierrez, S. B., Álvarez, A., Benimeli, C. S., & Polti, M. A. (2022). The current approach to soil remediation: A review of physicochemical and biological technologies, and the potential of their strategic combination. *Journal of Environmental Chemical Engineering, 10*(2), 107141.

Arora, N. K. (2018). Bioremediation: A green approach for restoration of polluted ecosystems. *Environmental Sustainability, 1*(4), 305–307.

Atakpa, E. O., Zhou, H., Jiang, L., Ma, Y., Liang, Y., Li, Y., et al. (2022). Improved degradation of petroleum hydrocarbons by co-culture of fungi and biosurfactant-producing bacteria. *Chemosphere, 290*, 133337.

Atlas, R. M., & Philp, J. (2005). *Bioremediation. Applied microbial solutions for real-world environmental cleanup* (p. xi+−366). American Society of Microbiology.

Ayangbenro, A. S., & Babalola, O. O. (2017). A new strategy for heavy metal polluted environments: A review of microbial biosorbents. *International Journal of Environmental Research and Public Health, 14*(1), 94.
Ayangbenro, A. S., Babalola, O. O., & Aremu, O. S. (2019). Bioflocculant production and heavy metal sorption by metal resistant bacterial isolates from gold mining soil. *Chemosphere, 231*, 113–120.
Azubuike, C. C., Chikere, C. B., & Okpokwasili, G. C. (2016). Bioremediation techniques–classification based on site of application: Principles, advantages, limitations and prospects. *World Journal of Microbiology and Biotechnology, 32*, 1–18.
Bhakat, K., Chakraborty, A., & Islam, E. (2019). Characterization of arsenic oxidation and uranium bioremediation potential of arsenic resistant bacteria isolated from uranium ore. *Environmental Science and Pollution Research, 26*, 12907–12919.
Bhandari, S., Poudel, D. K., Marahatha, R., Dawadi, S., Khadayat, K., Phuyal, S., et al. (2021). Microbial enzymes used in bioremediation. *Journal of Chemistry, 2021*(1), 8849512.
Bhatt, P., Negi, G., Gangola, S., Khati, P., Srivastava, A., & Sharma, A. (2015a). Optimization of sulfosulfuron biodegradation through response surface methodology using indigenous bacterial strain isolated from contaminated agriculture field. *International Journal of Current Microbiology and Applied Sciences, 4*(8), 105–112.
Bhatt, P., Sharma, A., Khati, P., Kumar, G., Gangola, S., & Srivastava, A. (2015b). Novel pathway of cypermethrin biodegradation in a *Bacillus* sp. strain SG2 isolated from cypermethrin-contaminated agriculture field. *3 Biotech, 6*, 45.
Bhatt, P., Negi, G., Gangola, S., Khati, P., Srivastava, A., & Sharma, A. (2016). Differential expression and characterization of cypermethrin degrading potential proteins in *Bacillus thuringiensis* strain SG4. *3 Biotech, 6*, 225.
Bhatt, P., Gangola, S., Chaudhary, P., Khati, P., Kumar, G., Sharma, A., & Srivastava, A. (2019). Pesticide induced up-regulation of esterase and aldehyde dehydrogenase in indigenous *Bacillus* spp. *Bioremediation Journal, 23*(1), 42–52.
Bhatt, P., Verma, A., Gangola, S., Bhandari, G., & Chen, S. (2021). Microbial glycoconjugates in organic pollutant bioremediation: Recent advances and applications. *Microbial Cell Factories, 20*(1), 72.
Bilyard, G. R., McCabe, G. H., White, K. A., Gajewski, S. W., Hendrickson, P. L., Jaksch, J. A., et al. (1996). *Legal and social concerns to the development of bioremediation technologies* (No. PNNL–11301). Pacific Northwest National Lab.
Bogner, J. (2006). Garbage and global change. *Waste Management, 26*(5), 451–452.
Boopathy, R. (2000). Factors limiting bioremediation technologies. *Bioresource Technology, 74*(1), 63–67.
Bragg, J. R., Prince, R. C., Harner, E. J., & Atlas, R. M. (1994). Effectiveness of bioremediation for the Exxon Valdez oil spill. *Nature, 368*(6470), 413–418.
Brown, L. D., & Ulrich, A. C. (2014). Bioremediation of oil spills on land. In *Handbook of oil spill science and technology* (pp. 395–406). https://doi.org/10.1002/9781118989982.ch15
Chakraborty, R., Wu, C. H., & Hazen, T. C. (2012). Systems biology approach to bioremediation. *Current Opinion in Biotechnology, 23*(3), 483–490.
Chanton, J., & Liptay, K. (2000). Seasonal variation in methane oxidation in a landfill cover soil as determined by an in situ stable isotope technique. *Global Biogeochemical Cycles, 14*(1), 51–60.
Chaudhary, P., Khati, P., Chaudhary, A., Maithani, D., Kumar, G., & Sharma, A. (2021). Cultivable and metagenomic approach to study the combined impact of nanogypsum and *Pseudomonas taiwanensis* on maize plant health and its rhizospheric microbiome. *PLoS One, 16*(4), e0250574.
Copete-Pertuz, L. S., Plácido, J., Serna-Galvis, E. A., Torres-Palma, R. A., & Mora, A. (2018). Elimination of Isoxazolyl-Penicillins antibiotics in waters by the ligninolytic native Colombian strain *Leptosphaerulina* sp. considerations on biodegradation process and antimicrobial activity removal. *Science of the Total Environment, 630*, 1195–1204.

Daley, J. (2013). Bioamplification as a bioaccumulation mechanism for persistent organic pollutants (POPs) in wildlife. *Reviews of Environmental Contamination and Toxicology, 227*, 107–155.

Delille, D., Duval, A., & Pelletier, E. (2008). Highly efficient pilot biopiles for on-site fertilization treatment of diesel oil-contaminated sub-Antarctic soil. *Cold Regions Science and Technology, 54*(1), 7–18.

Dell'Anno, F., Rastelli, E., Tangherlini, M., Corinaldesi, C., Sansone, C., Brunet, C., & Dell'Anno, A. (2021). Highly contaminated marine sediments can host rare bacterial taxa potentially useful for bioremediation. *Frontiers in Microbiology, 12*, 584850.

Doudna, J. A., & Charpentier, E. (2014). The new frontier of genome engineering with CRISPR-Cas9. *Science, 346*(6213), 1258096.

Doukani, K., Boukirat, D., Boumezrag, A., Bouhenni, H., & Bounouira, Y. (2022). Fundamentals of biodegradation process. In *Handbook of biodegradable materials* (pp. 1–27). Springer.

Dusengemungu, L., Kasali, G., Gwanama, C., & Ouma, K. O. (2020). Recent advances in biosorption of copper and cobalt by filamentous fungi. *Frontiers in Microbiology, 11*, 582016.

Dutta, N., Usman, M., Ashraf, M. A., Luo, G., & Zhang, S. (2022). Efficacy of emerging technologies in addressing reductive dechlorination for environmental bioremediation: A review. *Journal of Hazardous Materials Letters, 3*, 100065.

Eisenhofer, R., Minich, J. J., Marotz, C., Cooper, A., Knight, R., & Weyrich, L. S. (2019). Contamination in low microbial biomass microbiome studies: Issues and recommendations. *Trends in Microbiology, 27*(2), 105–117.

El Amrani, A., Dumas, A. S., Wick, L. Y., Yergeau, E., & Berthomé, R. (2015). "Omics" insights into PAH degradation toward improved green remediation biotechnologies. *Environmental Science & Technology, 49*(19), 11281–11291.

El Fantroussi, S., & Agathos, S. N. (2005). Is bioaugmentation a feasible strategy for pollutant removal and site remediation? *Current Opinion in Microbiology, 8*(3), 268–275.

Farhadian, M., Vachelard, C., Duchez, D., & Larroche, C. (2008). In situ bioremediation of monoaromatic pollutants in groundwater: A review. *Bioresource Technology, 99*(13), 5296–5308.

Flannagan, R. S., Linn, T., & Valvano, M. A. (2008). A system for the construction of targeted unmarked gene deletions in the genus Burkholderia. *Environmental Microbiology, 10*(6), 1652–1660.

Flathman, P. E., & Lanza, G. R. (1998). Phytoremediation: Current views on an emerging green technology. *Journal of soil Contamination, 7*(4), 415–432.

Flemming, H. C. (2000). Biofilms in biofiltration. In *Biotechnology: Environmental processes III* (Vol. 11, pp. 445–455). Wiley-VCH Verlag GmbH.

Gangola, S., Bhatt, P., Khati, P., & Sharma, A. (2015). Mycoremediation of imidacloprid in the presence of different soil amendments using *Trichoderma longibrachiatum* and *Aspergillus oryzae* isolated from pesticide-contaminated agricultural field of Uttarakhand. *Journal of Bioremediation & Biodegradation, 6*, 310.

Gangola, S., Sharma, A., Bhatt, P., Khati, P., & Chaudhary, P. (2018). Presence of esterase and laccase in *Bacillus subtilis* facilitates biodegradation and detoxification of cypermethrin. *Scientific Reports, 8*(1), 12755.

Gaur, N., & Narasimhulu, K. (2018). Recent advances in the bio-remediation of persistent organic pollutants and its effect on environment. *Journal of Cleaner Production, 198*, 1602–1631.

Gaur, V. K., Tripathi, V., & Manickam, N. (2022). Bacterial-and fungal-mediated biodegradation of petroleum hydrocarbons in soil. In *Development in wastewater treatment research and processes* (pp. 407–427). Elsevier.

Geetha, N., Bhavya, G., Abhijith, P., Shekhar, R., Dayananda, K., & Jogaiah, S. (2021). Insights into nanomycoremediation: Secretomics and mycogenic biopolymer nanocomposites for heavy metal detoxification. *Journal of Hazardous Materials, 409*, 124541.

Goswami, M., Chakraborty, P., Mukherjee, K., Mitra, G., Bhattacharyya, P., Dey, S., & Tribedi, P. (2018). Bioaugmentation and biostimulation: A potential strategy for environmental remediation. *Journal of Microbiology & Experimentation, 6*(5), 223–231.

Guha, T., Mitra, S., Rez, S., & Kundu, R. (2024). Advances in nano-biotechnology for bioremediation applications. In *Toxicity of aquatic system and remediation* (pp. 264–279). CRC Press.

Gunjal, A., Gupta, S., Nweze, J. E., & Nweze, J. A. (2023). Metagenomics in bioremediation: Recent advances, challenges, and perspectives. In *Metagenomics to bioremediation* (pp. 81–102). Academic.

Gupta, S. K., & Shukla, P. (2017). Gene editing for cell engineering: Trends and applications. *Critical Reviews in Biotechnology, 37*(5), 672–684.

Hlihor, R. M., Gavrilescu, M., Tavares, T., Favier, L., & Olivieri, G. (2017). Bioremediation: An overview on current practices, advances, and new perspectives in environmental pollution treatment. *Biomed Research International, 2017*, 6327610.

Huang, Y., Zeng, Y., Yu, Z., Zhang, J., Feng, H., & Lin, X. (2013). In silico and experimental methods revealed highly diverse bacteria with quorum sensing and aromatics biodegradation systems—A potential broad application on bioremediation. *Bioresource Technology, 148*, 311–316.

Jafari, M., Danesh, Y. R., Goltapeh, E. M., & Varma, A. (2012). Bioremediation and genetically modified organisms. In *Fungi as bioremediators* (pp. 433–451). Springer.

Jha, M. K., & Dev, M. (2024). Impacts of climate change. In *Smart internet of things for environment and healthcare* (pp. 139–159). Springer.

Khan, A., Singh, P., & Srivastava, A. (2018). Synthesis, nature and utility of universal iron chelator–Siderophore: A review. *Microbiological Research, 212*, 103–111.

Khati, P., Chaudhary, P., Gangola, S., & Sharma, A. (2017). Nanochitosan-induced growth of *Zea mays* with soil health maintenance. *3 Biotech, 7*(81), 1–9.

Khati, P., Chaudhary, P., Bhatt, P., Nisha, Kumar, R., & Sharma, A. (2018). Effect of nanozeolite and plant growth promoting rhizobacteria on maize. *3 Biotech, 8*(141), 1–12.

Khati, P., Chaudhary, P., Gangola, S., & Sharma, A. (2019a). Influence of nanozeolite on plant growth promotory bacterial isolates recovered from nanocompound-infested agriculture field. *Environment and Ecology, 37*(2), 521–527.

Khati, P., Sharma, A., Chaudhary, P., Singh, A. K., Gangola, S., & Kumar, R. (2019b). High-throughput sequencing approach to assess the impact of nanozeolite treatment on species richness and evenness of soil metagenome. *Biocatalysis and Agricultural Biotechnology, 20*, 101249.

Khati, P., Mishra, P. K., & Kant, L. (2024). Isolation, screening and characterization of drought tolerant plant growth promoting bacteria from Indian Himalayas. *Pantnagar Journal of Research, 22*(2), 261.

Kour, D., Khan, S. S., Kour, H., Kaur, T., Devi, R., Rai, P. K., et al. (2022). Microbe-mediated bioremediation: Current research and future challenges. *Journal Of Applied Biology & Biotechnology, 10*(2), 6–24.

Kumar, R. R., & Prasad, S. (2011). Metabolic engineering of bacteria. *Indian Journal of Microbiology, 51*(3), 403–409.

Kumar Singh, P., & Shukla, P. (2015). Systems biology as an approach for deciphering microbial interactions. *Briefings in Functional Genomics, 14*(2), 166–168.

Kumar, A., Devi, S., & Singh, D. (2018). Significance and approaches of microbial bioremediation in sustainable development. In *Microbial bioprospecting for sustainable development* (pp. 93–114). Springer Nature Singapore.

Kumar, V., Saxena, G., & Shah, M. P. (Eds.). (2020). *Bioremediation for environmental sustainability: Approaches to tackle pollution for cleaner and greener society*. Elsevier.

Kumari, S., Mangwani, N., & Das, S. (2016). Synergistic effect of quorum sensing genes in biofilm development and PAHs degradation by a marine bacterium. *Bioengineered, 7*(3), 205–211.

Kushwaha, A., Maurya, S., Pathak, R. K., Agarwal, S., Chaurasia, P. K., & Singh, M. P. (2018). Laccase from white rot fungi having significant role in food, pharma, and other industries. In

Research advancements in pharmaceutical, nutritional, and industrial enzymology (pp. 253–277). IGI Global.

Li, Q., Liu, J., & Gadd, G. M. (2020). Fungal bioremediation of soil co-contaminated with petroleum hydrocarbons and toxic metals. *Applied Microbiology and Biotechnology, 104*(21), 8999–9008.

Lin, C., Cheruiyot, N. K., Bui, X. T., & Ngo, H. H. (2022). Composting and its application in bioremediation of organic contaminants. *Bioengineered, 13*(1), 1073–1089.

Liu, Z., Shao, B., Zeng, G., Chen, M., Li, Z., Liu, Y., et al. (2018). Effects of rhamnolipids on the removal of 2, 4, 2, 4-tetrabrominated biphenyl ether (BDE-47) by *Phanerochaete chrysosporium* analyzed with a combined approach of experiments and molecular docking. *Chemosphere, 210*, 922–930.

Machado, L. F., de Assis Leite, D. C., da Costa Rachid, C. T. C., Paes, J. E., Martins, E. F., Peixoto, R. S., & Rosado, A. S. (2019). Tracking mangrove oil bioremediation approaches and bacterial diversity at different depths in an in situ mesocosms system. *Frontiers in Microbiology, 10*, 2107.

Madison, L. L., & Huisman, G. W. (1999). Metabolic engineering of poly (3-hydroxyalkanoates): From DNA to plastic. *Microbiology and Molecular Biology Reviews, 63*(1), 21–53.

Mahmoud, G. A. E. (2021). Microbial scavenging of heavy metals using bioremediation strategies. In *Rhizobiont in bioremediation of hazardous waste* (pp. 265–289). Springer Singapore.

Maity, J. P., Chen, G. S., Huang, Y. H., Sun, A. C., & Chen, C. Y. (2019). Ecofriendly heavy metal stabilization: Microbial induced mineral precipitation (MIMP) and biomineralization for heavy metals within the contaminated soil by indigenous bacteria. *Geomicrobiology Journal, 36*(7), 612–623.

Malhi, Y., Franklin, J., Seddon, N., Solan, M., Turner, M. G., Field, C. B., & Knowlton, N. (2020). Climate change and ecosystems: Threats, opportunities and solutions. *Philosophical Transactions of the Royal Society B, 375*(1794), 20190104.

Mathew, B. B., & Krishnamurthy, N. B. (2018). Screening and identification of bacteria isolated from industrial area groundwater to study lead sorption: Kinetics and statistical optimization of biosorption parameters. *Groundwater for Sustainable Development, 7*, 313–327.

Medfu Tarekegn, M., Zewdu Salilih, F., & Ishetu, A. I. (2020). Microbes used as a tool for bioremediation of heavy metal from the environment. *Cogent Food & Agriculture, 6*(1), 1783174.

Megharaj, M., Ramakrishnan, B., Venkateswarlu, K., Sethunathan, N., & Naidu, R. (2011). Bioremediation approaches for organic pollutants: A critical perspective. *Environment International, 37*(8), 1362–1375.

Mesa, J., Rodríguez-Llorente, I. D., Pajuelo, E., Piedras, J. M. B., Caviedes, M. A., Redondo-Gómez, S., & Mateos-Naranjo, E. (2015). Moving closer towards restoration of contaminated estuaries: Bioaugmentation with autochthonous rhizobacteria improves metal rhizoaccumulation in native *Spartina maritima*. *Journal of Hazardous Materials, 300*, 263–271.

Miri, S., Rasooli, A., Brar, S. K., Rouissi, T., & Martel, R. (2022). Biodegradation of p-xylene—A comparison of three psychrophilic Pseudomonas strains through the lens of gene expression. *Environmental Science and Pollution Research, 2022*, 1–15.

Mir-Tutusaus, J. A., Baccar, R., Caminal, G., & Sarrà, M. (2018). Can white-rot fungi be a real wastewater treatment alternative for organic micropollutants removal? A review. *Water Research, 138*, 137–151.

Mishra, S., Hora, S., Mishra, R., & Kanaujia, P. K. (2022). Bioremediation-based microorganisms to break down pollutants decelerate due to climate change. In *Plant stress mitigators: Action and application* (pp. 125–143). Springer Nature.

Mohamed, M. S., El-Arabi, N. I., El-Hussein, A., El-Maaty, S. A., & Abdelhadi, A. A. (2020). Reduction of chromium-VI by chromium-resistant *Escherichia coli* FACU: A prospective bacterium for bioremediation. *Folia Microbiologica, 65*, 687–696.

Moreira, V. R., Lebron, Y. A. R., Lange, L. C., & Santos, L. V. S. (2019). Simultaneous biosorption of Cd (II), Ni (II) and Pb (II) onto a brown macroalgae Fucus vesiculosus: Mono-and multi-

component isotherms, kinetics and thermodynamics. *Journal of Environmental Management, 251*, 109587.
Mukjang, N., Chitov, T., Mhuantong, W., Champreda, V., Pathom-Aree, W., Sattayawat, P., & Bovonsombut, S. (2022). Bacterial communities associated with crude oil bioremediation through composting approaches with indigenous bacterial isolate. *Life, 12*(11), 1712.
Mutambara, P. (2024). *Soil microbial response to increasing temperatures: Influence of elevated temperatures on enzymatic activities, bacterial and fungal abundance and diversity*. Doctoral dissertation, Laurentian University Library & Archives.
Nakamura, C. E., & Whited, G. M. (2003). Metabolic engineering for the microbial production of 1, 3-propanediol. *Current Opinion in Biotechnology, 14*(5), 454–459.
Narayanan, M., Ali, S. S., & El-Sheekh, M. (2023). A comprehensive review on the potential of microbial enzymes in multipollutant bioremediation: Mechanisms, challenges, and future prospects. *Journal of Environmental Management, 334*, 117532.
Nowicka, B., Fesenko, T., Walczak, J., & Kruk, J. (2020). The inhibitor-evoked shortage of tocopherol and plastoquinol is compensated by other antioxidant mechanisms in *Chlamydomonas reinhardtii* exposed to toxic concentrations of cadmium and chromium ions. *Ecotoxicology and Environmental Safety, 191*, 110241.
Ojuederie, O. B., & Babalola, O. O. (2017). Microbial and plant-assisted bioremediation of heavy metal polluted environments: A review. *International Journal of Environmental Research and Public Health, 14*(12), 1504.
Perera, I. C., & Hemamali, E. H. (2022). Genetically modified organisms for bioremediation: Current research and advancements. In *Bioremediation of environmental pollutants: Emerging trends and strategies* (pp. 163–186). Springer.
Philp, J. C., Bamforth, S. M., Singleton, I., & Atlas, R. M. (2005). Environmental pollution and restoration: A role for bioremediation. In *Bioremediation: Applied microbial solutions for real-world environmental cleanup* (pp. 1–48). American Society for Microbiology.
Pratush, A., Kumar, A., & Hu, Z. (2018). Adverse effect of heavy metals (As, Pb, Hg, and Cr) on health and their bioremediation strategies: A review. *International Microbiology, 21*, 97–106.
Raviteja, K. V. N. S., & Reddy, K. R. (2022). Application of artificial intelligence, machine learning, and deep learning in contaminated site remediation. In *International conference on trends and recent advances in civil engineering* (pp. 411–425). Springer Nature.
Razza, F., & Innocenti, F. D. (2012). Bioplastics from renewable resources: The benefits of biodegradability. *Asia-Pacific Journal of Chemical Engineering, 7*, S301–S309.
Sahay, R. (2021). Synthetic applications of laccase and its catalytic potentials. *International Journal of Advanced Engineering Research and Science, 8*(6), 112–120.
Sales da Silva, I. G., Gomes de Almeida, F. C., Padilha da Rocha e Silva NM, Casazza, A. A., Converti, A., & Asfora Sarubbo, L. (2020). Soil bioremediation: Overview of technologies and trends. *Energies, 13*(18), 4664.
Saxena, G., Kishor, R., & Bharagava, R. N. (2020). Application of microbial enzymes in degradation and detoxification of organic and inorganic pollutants. In *Bioremediation of industrial waste for environmental safety: Volume I: Industrial waste and its management* (pp. 41–51). Springer.
Sayed, K., Baloo, L., & Sharma, N. K. (2021). Bioremediation of total petroleum hydrocarbons (TPH) by bioaugmentation and biostimulation in water with floating oil spill containment booms as bioreactor basin. *International Journal of Environmental Research and Public Health, 18*(5), 2226.
Selle, K., & Barrangou, R. (2015). Harnessing CRISPR–Cas systems for bacterial genome editing. *Trends in Microbiology, 23*(4), 225–232.
Sharma, M., Nandy, A., Taylor, N., Venkatesan, S. V., Kollath, V. O., Karan, K., & Gieg, L. M. (2020). Bioelectrochemical remediation of phenanthrene in a microbial fuel cell using an anaerobic consortium enriched from a hydrocarbon-contaminated site. *Journal of Hazardous Materials, 389*, 121845.

Sharma, P., Dutta, D., Udayan, A., Nadda, A. K., Lam, S. S., & Kumar, S. (2022). Role of microbes in bioaccumulation of heavy metals in municipal solid waste: Impacts on plant and human being. *Environmental Pollution, 305*, 119248.

Singh, P., Jain, R., Srivastava, N., Borthakur, A., Pal, D. B., Singh, R., et al. (2017). Current and emerging trends in bioremediation of petrochemical waste: A review. *Critical Reviews in Environmental Science and Technology, 47*(3), 155–201.

Singh, R. K., Tripathi, R., Ranjan, A., & Srivastava, A. K. (2020). Fungi as potential candidates for bioremediation. In *Abatement of environmental pollutants* (pp. 177–191). Elsevier.

Sun, L., Guo, D., Liu, K., Meng, H., Zheng, Y., Yuan, F., & Zhu, G. (2019). Levels, sources, and spatial distribution of heavy metals in soils from a typical coal industrial city of Tangshan, China. *Catena, 175*, 101–109.

Tak, H. I., Ahmad, F., & Babalola, O. O. (2012). Advances in the application of plant growth-promoting rhizobacteria in phytoremediation of heavy metals. *Reviews of Environmental Contamination and Toxicology, 223*, 33–52.

Tian, Q., Dou, X., Huang, L., Wang, L., Meng, D., Zhai, L., et al. (2020). Characterization of a robust cold-adapted and thermostable laccase from Pycnoporus sp. SYBC-L10 with a strong ability for the degradation of tetracycline and oxytetracycline by laccase-mediated oxidation. *Journal of Hazardous Materials, 382*, 121084.

Tiecher, T. L., Ceretta, C. A., Ferreira, P. A., Lourenzi, C. R., Tiecher, T., Girotto, E., et al. (2016). The potential of *Zea mays* L. in remediating copper and zinc contaminated soils for grapevine production. *Geoderma, 262*, 52–61.

Tiedje, J. M., Bruns, M. A., Casadevall, A., Criddle, C. S., Eloe-Fadrosh, E., Karl, D. M., & Zhou, J. (2022). Microbes and climate change: A research prospectus for the future. *MBio, 13*(3), e00800–e00822.

Unuofin, J. O., Okoh, A. I., & Nwodo, U. U. (2019). Aptitude of oxidative enzymes for treatment of wastewater pollutants: A laccase perspective. *Molecules, 24*(11), 2064.

Verma, S., & Kuila, A. (2019). Bioremediation of heavy metals by microbial process. *Environmental Technology and Innovation, 14*, 100369.

Verma, S., Kour, S., & Pathak, R. K. (2022). In silico approaches in bioremediation research and advancements. In *Bioremediation of environmental pollutants: Emerging trends and strategies* (pp. 221–238). Springer.

Vilchez-Vargas, R., Geffers, R., Suárez-Diez, M., Conte, I., Waliczek, A., Kaser, V. S., & Pieper, D. H. (2013). Analysis of the microbial gene landscape and transcriptome for aromatic pollutants and alkane degradation using a novel internally calibrated microarray system. *Environmental Microbiology, 15*(4), 1016–1039.

Weiland, M., Daro, A., & David, C. (1995). Biodegradation of thermally oxidized polyethylene. *Polymer Degradation and Stability, 48*(2), 275–289.

Wesolowski, J., Hassan, R. Y. A., Reinhardt, K., Hodde, S., & Bilitewski, U. (2010). Antifungal compounds redirect metabolic pathways in yeasts: Metabolites as indicators of modes of action. *Journal of Applied Microbiology, 108*(2), 462–471.

Woźniak-Karczewska, M., Lisiecki, P., Białas, W., Owsianiak, M., Piotrowska-Cyplik, A., Wolko, Ł., & Chrzanowski, Ł. (2019). Effect of bioaugmentation on long-term biodegradation of diesel/biodiesel blends in soil microcosms. *Sci Total Environ, 671*, 948–958.

Xiao, F., Cheng, J., Cao, W., Yang, C., Chen, J., & Luo, Z. (2019). Removal of heavy metals from aqueous solution using chitosan-combined magnetic biochars. *Journal of Colloid and Interface Science, 540*, 579–584.

Xie, Q., Liu, N., Lin, D., Qu, R., Zhou, Q., & Ge, F. (2020). The complexation with proteins in extracellular polymeric substances alleviates the toxicity of Cd (II) to Chlorella vulgaris. *Environmental Pollution, 263*, 114102.

Yang, L., Wang, J., Yang, Y., Li, S., Wang, T., Oleksak, P., et al. (2022). Phytoremediation of heavy metal pollution: Hotspots and future prospects. *Ecotoxicology and Environmental Safety, 234*, 113403.

Yoshida, S., Hiraga, K., Takehana, T., Taniguchi, I., Yamaji, H., Maeda, Y., et al. (2016). A bacterium that degrades and assimilates poly (ethylene terephthalate). *Science, 351*(6278), 1196–1199.

Zacharia, J. T. (2019). Degradation pathways of persistent organic pollutants (POPs) in the environment. In *Persistent organic pollutants* (pp. 17–30). IntechOpen.

Zapata-Peñasco, I., Avelino-Jiménez, I. A., Mendoza-Pérez, J., Guevara, M. V., de Guevara, M. G. L., Valadez-Martínez, M., et al. (2023). Environmental stressor assessment of hydrocarbonoclastic bacteria biofilms from a marine oil spill. *Biotechnology Reports, 42*, e00834.

Zeneli, A., Kastanaki, E., Simantiraki, F., & Gidarakos, E. (2019). Monitoring the biodegradation of TPH and PAHs in refinery solid waste by biostimulation and bioaugmentation. *Journal of Environmental Chemical Engineering, 7*(3), 103054.

Microbial Resource Management for Climate Change Mitigation

Devendra Singh, M. Saritha, and Mahipal Choudhary

Abstract

Microbial Resource Management (MRM) has emerged as a promising eco-friendly approach to mitigate climate change by leveraging the natural capabilities of microbes in regulating key biogeochemical processes. This chapter explores the scientific basis and applications of microbial consortia for greenhouse gas (GHG) reduction, carbon sequestration, and sustainable agriculture. It emphasizes the role of nitrogen-fixing bacteria, methanotrophs, denitrifiers, and plant growth-promoting rhizobacteria (PGPR) in reducing carbon emissions and enhancing soil fertility. Key microbial strategies such as biofertilizers, bioremediation, and stress-resilient crop production are discussed with a focus on climate-resilient agriculture. Recent advancements in metagenomics, microbial engineering, and large-scale bioformulations for field application are also highlighted. While acknowledging the challenges in mass multiplication and field stability, the chapter underscores the need for integrated policy frameworks and region-specific microbial solutions to support global climate mitigation efforts.

Keywords

Microbial resource management · Climate change mitigation · Microbial consortia · Plant growth-promoting rhizobacteria · Mirobial biofertilizers

D. Singh (✉) · M. Saritha · M. Choudhary
ICAR-Central Arid Zone Research Institute, Jodhpur, Rajasthan, India

G. Bhandari et al. (eds.), *Microbial Resilience*,
https://doi.org/10.1007/978-981-95-8883-1_14

1 Introduction

Climate change, driven primarily by human activities such as the burning of fossil fuels, deforestation, and industrial agriculture, poses a significant threat to ecological balance, economic stability, and global food and health systems. The unchecked emission of greenhouse gases (GHGs)—especially carbon dioxide (CO_2), methane (CH_4), and nitrous oxide (N_2O)—has led to a rise in global temperatures, causing erratic weather patterns, prolonged droughts, frequent flooding, melting glaciers, and rising sea levels (IPCC, 2023). These climatic shifts not only endanger biodiversity and ecosystem services but also compromise crop productivity, water availability, and public health (Scott et al., 2024). As highlighted by the Intergovernmental Panel on Climate Change (IPCC), limiting global warming to 1.5 °C above pre-industrial levels is imperative to prevent irreversible environmental damage (IPCC, 2023). While global efforts have largely focused on technological solutions and policy reforms, there is growing recognition of the critical role that biological systems—especially microorganisms—can play in mitigating climate change. Microorganisms play a pivotal dual role in both adapting to and mitigating the impacts of climate change (Fig. 1).

Microorganisms are fundamental to Earth's biogeochemical cycles and serve as key regulators of carbon and nitrogen dynamics. Through processes such as organic

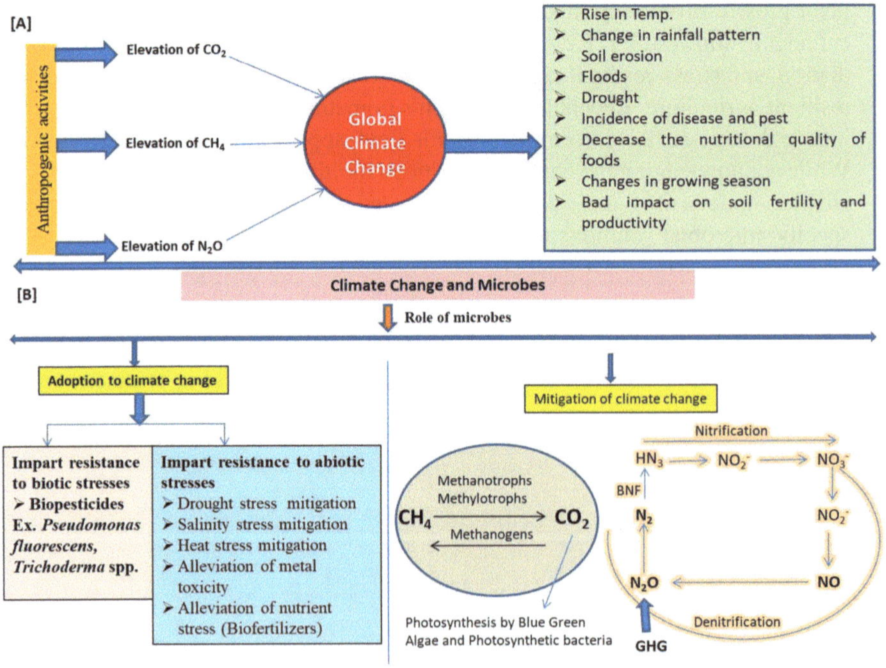

Fig. 1 Interconnection between anthropogenic emissions, climate change impacts, and microbial interventions

matter decomposition, carbon sequestration, methane oxidation, nitrogen fixation and denitrification, microbes influence the flux and fate of GHGs in terrestrial and aquatic ecosystems (Falkowski et al., 2008; Wu et al., 2024). Harnessing this potential through Microbial Resource Management (MRM) offers a sustainable, nature-based solution to climate change mitigation. MRM involves the strategic selection, optimization and application of beneficial microbial strains or consortia to enhance ecosystem resilience and reduce environmental impacts (Bender et al., 2016; Mousa et al., 2024; Panday et al., 2024; Tomer et al., 2024). For instance, soil microbes contribute to long-term carbon storage by stabilizing organic matter, while methanotrophic bacteria oxidize methane into less harmful forms (Bhattacharyya et al., 2022). Similarly, denitrifying microbes can convert nitrous oxide—a potent greenhouse gas into inert nitrogen gas, thereby mitigating emissions from agricultural systems (Saud et al., 2022).

Moreover, beneficial microbes can help build climate-resilient agriculture by improving plant stress tolerance against drought, salinity and heat (Giri et al., 2025; Khati et al., 2024). Plant growth-promoting rhizobacteria (PGPR), mycorrhizal fungi, and endophytes enhance nutrient uptake, root architecture and hormonal balance, reducing the need for chemical fertilizers and improving soil health (Ahmed et al., 2025; Bhat et al., 2023; Singh et al., 2023; Khati et al., 2019a; Kumari et al., 2020). Nano-enabled microbial strategies, such as those involving nanozeolite and nanochitosan in combination with PGPR, have shown promise in improving crop performance and microbial community structure under field conditions (Chaudhary et al., 2021a, b, c, d; Khati et al., 2017, 2019a, b). Despite this immense potential, the application of microbial solutions at scale remains limited due to challenges such as variability in field performance, environmental heterogeneity and limited awareness or policy support. MRM seeks to overcome these barriers by integrating microbial ecology, genomics, bioinformatics and applied microbiology into holistic strategies for environmental and agricultural sustainability. MRM seeks to overcome these barriers by integrating microbial ecology, genomics, bioinformatics, and applied microbiology into holistic strategies for environmental and agricultural sustainability (Parihar et al., 2025; Chaudhary et al., 2021a, b, c, d).

This chapter provides an in-depth exploration of microbial roles in mitigating climate change, with emphasis on their ecological mechanisms, practical applications and integration into sustainable development frameworks. It also highlights current research trends, successful case studies, and the potential of microbial technologies to support global climate action goals. By fostering a deeper understanding of microbial functions and promoting their practical deployment, MRM can play a transformative role in achieving a more climate-resilient and sustainable future.

2 Microbial Processes and Climate Change Mitigation

Soil microorganisms play a pivotal role in climate change mitigation through their influence on GHG dynamics, particularly involving CO_2, CH_4, and N_2O. These microbial processes can be broadly classified into those that (i) directly influence GHG transformation (e.g., methanotrophs, denitrifiers), (ii) contribute to carbon sequestration, and (iii) indirectly mitigate emissions through enhanced soil health and crop productivity.

2.1 GHG Mitigation

Methanotrophs, a specialized group of bacteria, oxidize methane and other C1 carbon compounds such as methanol, formate, and methylamines into carbon dioxide, thereby reducing atmospheric CH_4 levels (Kolb, 2009). Although methanotrophs account for only ~5% of the global atmospheric methane sink, they are highly efficient within soil ecosystems—oxidizing up to 90% of methane produced in soils before it escapes into the atmosphere (Oremland & Culbertson, 1992). Several methanotroph and methylotroph strains have shown potential as plant growth promoters (Jeyajothi et al., 2014; Jhala et al., 2014; Zúniga et al., 2011). These microbes offer a dual benefit: GHG mitigation and resource recycling.

2.2 Soil Carbon Storage

Soils act as major carbon sinks, and their carbon sequestration capacity mediated by microorganisms has far-reaching implications for climate regulation. Soil organic matter (SOM) stores vast quantities of carbon, and microbial processes determine both the sequestration of carbon in SOM and its release into the atmosphere as CO_2 through decomposition. Microorganisms are essential drivers of the carbon cycle, contributing to both the stabilization and release of carbon. In addition to CO_2, soil microbes influence the fluxes of other potent greenhouse gases such as methane (CH_4) and N_2O, which are significantly more potent than CO_2 in terms of global warming potential (Schils et al., 2008).

Soil microorganisms facilitate carbon stabilization by converting dead organic matter into simpler carbon compounds that can be protected from further degradation and sequestered in stable forms (Six et al., 2006; Grandy & Wickings, 2010). Fungi, particularly arbuscular mycorrhizal fungi (AMF), play a significant role in soil carbon sequestration. AMF form symbiotic associations with plant roots, enhancing nutrient uptake for the plant in exchange for plant-derived sugars. This mutualistic interaction boosts fungal biomass in the soil, thereby increasing soil carbon content (Rillig et al., 2001). Moreover, AMF produce glomalin, a sticky glycoprotein that binds soil particles together, contributing to soil structure stability and carbon protection (Rillig, 2004). Wilson et al. (2009) found a positive correlation between AMF colonization, soil aggregation, and carbon sequestration.

These microbial degradation processes are essential for microbial biomass production and the formation of stable soil carbon structures (Six et al., 2006). Managing soils to support microbial communities by providing diverse plant residues, minimizing soil disturbance (e.g., reduced tillage), and avoiding excessive chemical inputs can enhance soil carbon sequestration.

2.3 Indirect Effect of Microbes

Microbial inoculants such as PGPR and biocontrol agents can increase nutrient availability, suppress pathogens, and improve drought resistance, contributing to yield stability under climate stress. These organisms increase land-use efficiency and reduce emissions associated with fertilizer use, thereby contributing to sustainable, low-emission agriculture (Mahmuti et al., 2009).

Furthermore, microbes hold potential in bioenergy production. Methanogenic archaea can generate methane from organic waste, while yeasts such as *Saccharomyces cerevisiae* convert molasses and cellulose into bioethanol. Although cellulose-based bioethanol production is still under development, it represents a sustainable energy pathway. Likewise, green algae can produce biodiesel using sunlight, though technical challenges remain (Patil et al., 2011; Pfromm et al., 2011). Soil microbes also contribute to climate change buffering by stabilizing ecosystem processes. During the decomposition of SOM, microbes release CO_2, CH_4, and N_2O. Increased microbial diversity contributes to ecosystem resilience, enabling agricultural systems to better withstand environmental stress.

Microbes, particularly biocontrol agents, can stabilize plant and herbivore populations, making them more resistant to climate variability (Hoover & Newman, 2004). Thus, soil microorganisms not only mitigate GHG emissions but also enhance ecological stability and adaptive capacity in the face of climate change.

3 Microbial Biofertilizers and Biostimulants for Sustainable Agriculture

Microbial biofertilizers and bio-stimulants are eco-friendly solutions for sustainable agriculture, enhancing crop growth, yield, and soil health while reducing reliance on chemical inputs (Santoyo et al., 2021; Vessey, 2003; Oleńska et al., 2020).

3.1 PGPR and Sustainable Agriculture

PGPR function as both biofertilizers (enhancing nutrient availability) and biostimulants (stimulating plant growth and stress resilience). Tables 1 and 2 show the role of PGPR in sustainable agriculture and their role in crop production. Their dual role makes them integral to sustainable agricultural practices. Soil with low fertility and extreme climatic conditions requires higher amounts of chemical

Table 1 Various PGPR strains function as biofertilizers, producing siderophores and plant growth-promoting hormones in different crops

Type of biofertilizer	PGPR strains	Host crop	Mode of beneficial action	References
Nitrogen-fixing microorganisms (NFMs)	*Rhizobium meliloti, Sinorhizobium melilotii, Rhizobium legumninosarum biovar, Rhizobium phaseoli, Rhizobium tropici, Rhizobium leguminosarum biovar trifolii, Mesorhizobium loti, R. leguminosarum biovar viceae, Bradyrhizobium japonicum, Bradyrhizobium elkanii, Rhizobium fredii, Azorhizobium caulinodans,* and *Burkholderia*	Legumes, i.e., soybean, chickpea, alfalfa (*Medicago* sp.), pigeon pea, beans, clover, lotus, peas, sesbania)	Symbiotic nitrogen fixation	Zahran (1999), Elliott et al. (2007), and Guo et al. (2023)
	Azospirillum brasilense, Pseudomonas stutzeri, Bacillus, Paenibacillus Burkholderia, Azoarcus Herbaspirillum, and *Gluconobacter*	Maize, wheat, rice	Associative nitrogen fixation	Bashan and de Bashan (2010), Nadarajah and Abdul Rahman (2021)
	Azotobacter chroococcum, Azotobacter vinelandii, and *Paenibacillus sabinae*	Cereals, vegetables	Free-living nitrogen fixation	Subba Rao (1982), Pankievicz et al. (2019) and Soumare et al. (2020)
	Frankia spp.	Actinorhizal plants (e.g., alder)	Symbiotic nitrogen fixation	Benson and Silvester (1993)
	Gluconacetobacter diazotrophicus	Sugarcane, rice	Endophytic nitrogen fixation	Cavalcante and Döbereiner (1988)
	Burkholderia vietnamiensis	Rice, sugarcane		Govindarajan et al. (2008)
	Herbaspirillum seropedicae	Maize, rice, sugarcane		Baldani et al. (2002)
	Beijerinckia indica	Forest and cultivated soils	Free-living nitrogen fixation	Verma et al. (2011)

	Anabaena, Nostoc flagelliforme, Trichormus, Calothrix, Chlorogloeopsis. Aulosira fertilissima, Cylindrospermum	Rice, sugarcane	Symbiotic nitrogen fixation in root nodules	Singh et al. (2016)
Phosphate solubilizing and mobilizing microorganisms (PSMMs)	Bacterial strains: *Alcaligenes* sp., *Aerobactor aerogenes, Achromobacter* sp., *Actinomadura oligospora, Agrobacterium* sp., *Azospirillum brasilense, Bacillus* sp., *B. megaterium, B. circulans, B. cereus, B. fusiformis, B. pumils, B. mycoides, B. polymyxa, B. coagulans B. chitinolyticus, B. subtilis, Bradyrhizobium* sp., *Brevibacterium* sp., *Citrobacter* sp., *Pseudomonas* sp., *P. putida, P. striata, P. fluorescens, P. calcis, Flavobacterium* sp., *Nitrosomonas* sp., *Klebsiella variicola, Erwinia* sp., *Micrococcus* sp., *Escherichia intermedia, Serratia phosphoticum, Nitrobacter* sp., *Thiobacillus ferroxidans, Thiobacillus thioxidans, Rhizobium meliloti, Xanthomonas* sp., *Gluconacetobacter* sp. *Burkholderia* sp.	Rice, wheat, maize, legumes, vegetables, pulses, horticultural crops, etc.	Solubilize insoluble phosphate via organic acid production, i.e., oxalic acid, malic acid, formic acid, acetic acid, tartaric acid, gluconic acid, citric acid, lactic acid, propionic acid, etc.	Saharan and Nehra (2011), Sharma et al. (2013), Babalola and Glick (2012), Nehra and Choudhary (2015), Alori et al. (2017), Biswas et al. (2018), Nacoon et al. (2020), Rawat et al. (2021), Siedliska et al. (2021) and Elhaissoufi et al. (2022)

(continued)

Table 1 (continued)

Type of biofertilizer	PGPR strains	Host crop	Mode of beneficial action	References
	Fungal strains: *Aspergillus awamori, Aspergillus niger, Aspergillus tereus, Aspergillus flavus, Aspergillus nidulans, Aspergillus foetidus, Aspergillus wentii. Fusarium oxysporum, Alternaria teneius, Achrothcium* sp., Arthrobotrys, *Penicillium digitatum, Penicillium lilacinium, Penicillium balaji, Penicillium funicolosum, Cephalosporium* sp., *Cladosprium* sp., *Curvularia lunata,* Cunninghamella, *Candida* sp., *Chaetomium globosum,* Glomus, *Humicola inslens, Humicola lanuginosa, Helminthosporium* sp., *Paecilomyces fusisporous, Pythium* sp., *Phoma* sp., *Populospora mytilina, Myrothecium roridum, Morteirella* sp., *Micromonospora* sp., *Oideodendron* sp., *Rhizoctonia solani, Rhizopus* sp., *Mucor* sp., *Trichoderma viridae, Torula thermophila, Schwanniomyces occidentalis,*			Ichriani et al. (2018), Adhikari and Pandey (2019), Mercl et al. (2020), Kiprotich et al. (2023) and Pan and Cai (2023)

	Sclerotium rolfsii. Torula, Trichoderma, and Yarrowia			
	Yeast strains: *Yarrowia lipolytica, Schizosaccharomyces pombe and Pichiafermentans, Rhodotorula* sp., *Candida rugosa, Saccharomyces cerevisiae,* and *Saccharomyces rouxii*			Xiao et al. (2013)
	Actinobacterial strains: *Actinobispora yunnanensis, Actinomodura citrea, Microtetrospora astidiosa, Micromonospora echinospora, Sacchromonospora viridis, Saccharopolyspora hirsute, Streptomyces albus, Streptoverticillium album, Streptomyces cyaneus, Thermonospora mesophila*			Hamim et al. (2019) and Chouyia et al. (2022)
	Algal strains: *Cyanobacterium* sp., *Chlorella* sp., *Lyngbya* sp., *Anabena* sp., *Calothrix braunii, Nostoc* sp., *Scytonema* sp.			Yang et al. (2017)
Potassium solubilizing microorganisms (KSMs)	*Agrobacterium tumefaciens, Bacillus mucilaginosus, Bacillus licheniformis, Bacillus edaphicus, Bacillus*		Solubilize potassium from waste mica and increase potassium availability in soil; Acidification, complexation	Basak and Biswas (2010), Parmar and Sindhu (2013), Meena et al. (2014), Etesami

(continued)

Table 1 (continued)

Type of biofertilizer	PGPR strains	Host crop	Mode of beneficial action	References
	circulans, Acidothiobacillus ferrooxidans, Pseudomonas azotoformans, Paenibacillus spp., *Rhizobium pusense* and *Flavobacterium anhuiense* and fungal strains (*Aspergillus* spp. and *Aspergillus terreus*)		Lipo-chitooligosaccharides production Mica	et al. (2017) and Olaniyan et al. (2022)
Sulphur solubilizing microorganisms (SSMs)	*Thiobacillus ferrooxidans, Thiobacillus denitrificans, Thiobacillus thiooxidans, Thiobacillus thioparus, Xanthobacter, Alcaligens, Bacillus, Pseudomonas,* and *Arthrobacter* and fungi (*Aspergillus, Penicillium*), *Chlorbium, Chromatiu, Rhodopseudomonas, Chromatium purpuratum, Thiocapsa pfennigii, Klebsiella oxytoca, Enterobacter ludwigii*		Oxidizing organic sulphur compounds into plant available sulphate (SO_4^{2-}) form.	Prasad and Shivay (2016), Naorem et al. (2017), Chaudhary et al. (2022, 2023) and Shinde et al. (2022)
Siderophore production	*Azotobacter* sp., *A. chroococcum, Acinetobacter calcoaceticus, Alcaligenes* sp., *Bacillus amyloliquefaciens, B. subtilis, B. siamensis, B. brevis, B. anthracis, Escherichia coli, Vibrio cholera, Klebsiella*	Rice, wheat, soybean, maize, mungbean, groundnut, beans, radish, cucumber, potato, onion etc	Producing of different type of siderophores, i.e., Azotobactin, bacillibactin, ferrichrome, desferrioxamine B (deferoxamine), desferrioxamine E, fusarinine C, ornibactin, ferribactin, gonobactin,	Muthuselvan and Balagurunathan (2013), Sayyed et al. (2013), Amaya-Gómez et al. (2020), Dimopoulou et al. (2021), Shen et al. (2022), Miller et al. (2023), Benaissa (2024), Deb

	pneumoniae, Kosakonia radicincitans, Lysobacter enzymogenes, Pseudomonas aeruginosa, Pseudomonas fluorescens, Pseudomonas aureofaciens, Pseudomonas putida, R. meliloti, Ustilagosphaerogena, Streptomyces pilosus, Streptomyces coelicolor, Fusarium roseum, Burkholderiacepacia, Neisseria gonorrhoeae, Paracoccus denitrificans, Agrobacterium tumefaciens, Staphylococcus hyicus, Yersinia pestis		nocobactin, enterobactin, vibriobactin, parabactin, agrobactin, rhizobactin, staphyloferrin A, pseudobactin, pyoverdine, spermidine (Lysochelin), yersiniabactin	and Tatung (2024) and Chaudhary and Sindhu (2025)
Phytohormones producing microorganisms (PPMs)	*Acinetobacter* sp., *Azospirillum* sp., *Bacillus* sp., *B. thuringiensis, Bacillus subtilis* LK14, *Rhizobium leguminosarum, Mesorhizobium cicero, R. phaseoli, Azospirillum brasilense, Psedomonas* sp., *P. putida, P. aeruginosa, Rhizobium* sp., *Ensifer meliloti strain RD64, Serratia marcescens*	Rice, wheat, potato, tomato, mustard, Maize, pepper, lettuce, carrot, *Lactuca sativa, Arabidopsis thaliana*	Production of indole acetic acid (IAA)	Kumar et al. (2017), Defez et al. (2019) and Zhang et al. (2022)
	Pseudomonas sp., *P. putida, Azospirillum lipoferum,*	Maize, soybean, apple, pear, tomato, pepper	Production of gibberellic acid (GA)	Kang et al. (2019) and Nett et al. (2022)

(continued)

Table 1 (continued)

Type of biofertilizer	PGPR strains	Host crop	Mode of beneficial action	References
	Sphingomonas sp., *Bacillus* sp., *Bradyrhizobium diazoefficiens, Bacillus tequilensis* SSB07, *Leifsonia xyli* SE13,			
	Bacillus subtilis, Bacillus sp., *Azotobacter* spp., *Rhizobium* spp., *Pantoea agglomerans, Rhodospirillum rubrum, Pseudomonas fluorescens, Bacillus subtilis,* and *Paenibacillus polymyxa, A. brasilense* Sp245	Tomato, cucumber	Production of cytokinin (CK)	Glick (2012), Carvajal et al. (2017) and Mekureyaw et al. (2022)
	Azospirillum lipoferum, A. brasilense, Phyllobacterium brassicacearum strain STM196, *B. subtilis* IB-22, *Pseudomonas mandelii* IB-Ki14, *Bacillus licheniformis* SA03	*Maize, wheat, Chinese cabbage, Arabidopsis*	Production of abscisic acid (ABA)	Armada et al. (2015), Iqbal et al. (2017), Zhou et al. (2017), Pan et al. (2019) and He et al. (2024)

Table 2 Plant Growth-Promoting Rhizobacteria (PGPR) and their mechanisms of bio-stimulation and biocontrol through induced systemic resistance and antibiotics for managing plant pathogens

	PGPR strains	Mechanism	Pathogen	References
Induced systemic resistance (ISR)	*Bacillus* sp. *B. amyloliquefaciens,* *B. amyloliquefaciens,* *B. subtilis*	Strengthening of cell wall	*F. oxysporum* f. sp. *Radices-lycopersici* Zymoseptoria tritici, *Sclerotinia sclerotiorum*	Le Mire et al. (2018), Farzand et al. (2019) and Chandwani et al. (2023)
	Pseudomonas sp. *P. fluorescens WCS417r,* *P. aeruginosa,* *P. chlororaphis* *P. fluorescens* *Pseudomonas* sp.	Cell wall thickening, deposition of callose and accumulation of phenolic compounds, Accumulation of PR proteins namely β-1,3 glucanases and endochitinases	*Pythium ultimum,* *F. oxysporum* f. sp. *Pisi,* *Leptosphaeria maculans,* *Tobacco mosaic virus,* *Botrytis cinerea* *Magnaporthe oryzae*	Pieterse et al. (2014), Kong et al. (2018), Omoboye et al. (2019), Chae et al. (2020) and Monnier et al. (2020)
	Trichoderma spp.	Jasmonic acid and Ethylene singnaling, Ferricrocin	*Pythium ultimum,* *F. oxysporum,* *Cochliobolus heterostrophus*	Mukherjee et al. (2018)
	Piriformospora indica, AM fungi, Penicillium simplicissimum, Phoma sp.	By secretion of salicylic acid and malic acid	*F. oxysporum, Pythium ultimum*	
	Streptomyces pulveraceus	Ferrioxamines	*Cylindrocarpon olidum*	Armin et al. (2021)
Biocontrol	*Agrobacterium radiobacter*	Antibiosis and competition	*Agrobacterium tumefaciens*	Kerr and Bullard (2020)
	Azotobacter salinestris	Antifungal substances, HCN, siderophores	*Fusarium* sp.	Chennappa et al. (2018)
	Actinobacteria, Pseudomonas, Bacillus, Staphylococcus, Enterobacter, Pantoae, Alcallgenes	Antagonism and growth promotion	*Aspergillus niger, Penicillium chrysogenum, Microsporum gypseum, Fusarium* sp., *Pythium ultimum, Alternaria alternate, Botrytis cinerea, Phytophtoracryptogea*	
	Achromobacter sp.	Secretion of volatile metabolites	*Magnaporthe oryzae*	Sahu et al. (2021)

(continued)

Table 2 (continued)

	PGPR strains	Mechanism	Pathogen	References
	Bacillus amyloliquefaciens, Bacillus cereus, Brevibacteriumlaterosporus, Pseudomonas fluorescens, Serratiamarcescens, Bacillus subtilis,	Antibiosis (lipopeptides) induced resistance, siderophores, induction of systemic resistance (ISR)	*Pythium, clubroot; yellow rust Pythium ultimum, F. oxysporum* f. sp. *Cubense*	Reiss and Jørgensen (2017), Miljaković et al. (2020) and Verma et al. (2023)
	Bacillus velezensis	Antimicrobial substances, IAA, protease and cellulase	*Botrytis cinerea*	Xue et al. (2023)
	Pseudomonas chlororaphis	Extracellular antibiotics, production of volatiles, siderophores, effective root colonization	*Macrophomina phaseolina* (charcoal rot of sorghum)	Das et al. (2013)
	Purpureocillium lilacinum	Production of siderophore and phosphate solubilization	*Fusarium oxysporum, Rhizoc tonia solani and Botrytis cinerea*	Herrera et al. (2022)
	Trichoderma asperellum	Induced defense-related enzymes such as peroxidase and polyphenol oxidase, and cell wall-degrading enzymes β—1,3-glucanase and chitinase	*Corynespora cassiicola and Curvularia aeria*	Baiyee et al. (2019)

fertilizers compared to soils rich in organic carbon and vibrant microbial ecosystems (Bender et al., 2016; Backer et al., 2018). Soil microbial activity is a key factor in managing the application of organic nutrient sources (Zeiner et al., 2024). Certain microbes, functioning as biofertilizers, improve plant nutrient uptake through diverse mechanisms, including increasing root surface area, fixing nitrogen, solubilizing phosphorus, potassium mobilization, producing siderophores, and synthesizing Hydrocyanic acid (HCN) (Kumar et al., 2022a, b; Solomon et al., 2024).

3.1.1 Nitrogen Fixation

Nitrogen fixation is a vital biological process where specific microorganisms transform atmospheric nitrogen gas (N_2) into ammonia (NH_3) in the presence of nitrogenase enzyme, and a compound plant can absorb and utilize in the form of either nitrate or ammonia. This process enriches soil fertility and increases nutrient availability for plants (Barney, 2020; de Andrade et al., 2023; Sharma et al., 2023). Nitrogen-fixing microorganisms are divided into two categories: symbiotic and non-symbiotic. Members of the Rhizobiaceae family, which establish symbiotic relationships with leguminous plants, are classified as symbiotic nitrogen fixers, i.e., *Rhizobium* (Kumar et al., 2020a, b). Rhizobium-based biofertilizers are applied to soil or seeds to foster symbiotic relationships with leguminous crops such as soybeans, peas, and alfalfa (Samantaray et al., 2024). In contrast, non-symbiotic nitrogen fixers, including *Azotobacter* (Barman et al., 2019), *Azospirillum* (Bourscheidt et al., 2023), and *Cyanobacteria* species such as *Anabaena* (Zulkefli & Hwang, 2020; Zhang et al., 2025), *Nostoc* (Joshi et al., 2020), *Spirulina* (Alghamdi et al., 2023), and *Aulosira* (Mohan & Kumar, 2019), occur as free-living or endophytic microbes.

3.1.2 Phosphorus Solubilization

Phosphorus-solubilizing biofertilizers (PSBs) are applied to improve the availability and uptake of phosphorus, thereby promoting plant growth and development. PSBs, also known as phospho-bacterin, facilitate the conversion of insoluble phosphate compounds, such as di-calcium phosphate ($CaHPO_4$), tri-calcium phosphate ($Ca_3(PO_4)_2$), hydroxyapatite ($Ca_{10}(PO_4)_6(OH)_2$), and rock phosphates ($Ca_5(PO_4)_3(F, OH, Cl)$), into forms that plants can readily absorb (Fig. 2) (Ghimirey et al., 2024). Various bacteria, including *Pseudomonas, Bacillus, Rhizobium, Penicillium, Burkholderia, Achromobacter, Agrobacterium, Micrococcus, Aerobacter, Flavobacterium, Aspergillus,* and *Erwinia*, are effective in solubilizing and mobilizing phosphorus, contributing to improved soil fertility and plant health (Alori et al., 2017; Narwal et al., 2023; de Oliveira-Paiva et al., 2024; Chen et al., 2024; Wani et al., 2025; Xie et al., 2025).

3.1.3 Potassium Mobilization

Potassium-solubilizing bacteria (KSB) play a vital role in enhancing soil fertility by breaking down potassium-containing minerals, transforming insoluble potassium into soluble forms that plants can readily absorb (Parmar & Sindhu, 2013; Meena

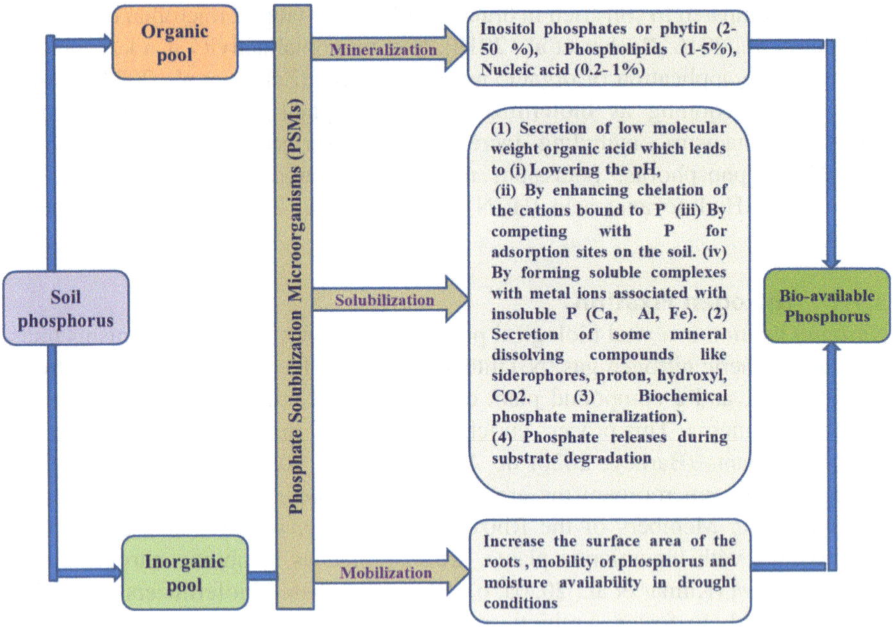

Fig. 2 Mechanism showing phosphorus transformation in soil through microorganisms

et al., 2014; Baghel et al., 2020). Various bacterial species, including *Acidothiobacillus ferrooxidans*, *Paenibacillus* spp., *Bacillus mucilaginosus, Bacillus edaphicus*, *Bacillus circulans*, *Bacillus polymyxa*, *Priestia megaterium,* and *Burkholderia* sp. demonstrate the ability to solubilize potassium from minerals, such as biotite, feldspar, illite, muscovite, orthoclase, and mica (Basak & Biswas, 2010; Meena et al., 2015, 2016; Etesami et al., 2017; Olaniyan et al., 2022). The KSB facilitate the release of potassium through mechanisms, such as the production of organic and inorganic acids, acidolysis, polysaccharide secretion, complexolysis, chelation, and ion-exchange reactions (Jini et al., 2023; Sharma et al., 2024).

3.1.4 Sulphur Solubilizing and Transformation

Sulphur is a vital macronutrient for the crop's growth and development (Supikova et al., 2022). The cycling of sulphur in soil primarily results from microbial activity, though some chemical transformations, such as the oxidation of iron sulphide, also occur (Wani et al., 2025). The key transformations in the sulphur cycle include mineralization, immobilization, and oxidation. Sulphur oxidation is primarily mediated by specialized bacteria, particularly species of Thiobacillus (e.g., Thiobacillus ferrooxidans, Thiobacillus denitrificans, Thiobacillus thiooxidans, and Thiobacillus thioparus), which are obligate chemolithotrophs. Additionally, heterotrophic bacteria such as *Paracoccus*, *Xanthobacter*, *Alcaligenes*, *Bacillus*, *Pseudomonas*, and *Arthrobacter*, as well as fungi (*Aspergillus*, *Penicillium*) and certain actinomycetes, contribute to sulphur oxidation (Chaudhary et al., 2018;

Shinde et al., 2022). In aquatic environments, photolithotrophic green and purple bacteria (*Chlorobium*, *Chromatium*, *Rhodopseudomonas*) also oxidize sulphur compounds (Chaudhary et al., 2023). Conversely, dissimilatory sulphate reduction is carried out by obligate anaerobic bacteria such as Desulfobulbus, Desulfobacter, Desulfococcus, Desulfosarcina, Desulfovibrio, Desulfobacterium, Desulfonema, Thermodesulfobacterium, Desulfomonas, and Desulfotomaculum (Fuentes-Lara et al., 2019).

3.1.5 Siderophores Production

Siderophores are produced by a range of microorganism such as bacteria, fungi, and yeasts, and certain monocotyledonous plants in response to low iron availability have a high affinity for Fe^{3+} (Dwivedi & Johri, 2003; Hider & Kong, 2010; Deb & Tatung, 2024; Soares, 2022). These compounds display significant structural variation and are categorized into four types: carboxylates, hydroxamates, catecholates, and mixed types (Arora & Verma, 2017; Timofeeva et al., 2022). Siderophores producing micro-organisms as diverse as *Bacillus amyloliquefaciens, Bacillus subtilis, Rhizobium, Rhizopus microspores, Staphylococcus* sp., *Pseudomonas, Agrobacterium, Azotobacter* sp., *A. chroococcum, Acinetobacter calcoaceticus, Escherichia coli, Klebsiella pneumoniae, Kosakonia radicincitans, Paracoccus denitrificans,* and many fungi produce a wide range of iron-chelating compounds (Ghosh et al., 2020; Khasheii et al., 2021; Kalyan et al., 2022).

3.1.6 Plant Hormones Production

Phytohormones (cytokinins, gibberellins, auxins, abscisic acid, and ethylene) are organic compounds synthesized by plants as well as various microorganisms, which play a crucial role in regulating plant growth and development (Khan et al., 2020a, b; Orozco-Mosqueda et al., 2023; Choudhary et al., 2018; Bag et al., 2022; Khoso et al., 2024; Nagel et al., 2018; Keswani et al., 2020). They also act as molecular messengers, helping plants respond to environmental stresses that can restrict growth or become fatal if not properly managed (Usman et al., 2022; Vejan et al., 2016; Mok, 2019; Eichmann et al., 2021; Stefen et al., 2022; Hershey et al., 2014; Mia, 2022; Singh et al., 2022; Kaur & Pandove, 2023; Vaghela & Gohel, 2023).

Some hormones such as abscisic acid (ABA) are responsible for the mitigation of drought stress by triggering stomatal closure (Karnwal et al., 2023; Mariotti et al., 2021). Reduced level of ethylene is another indicator of stress alleviation (Dehghanian et al., 2022; Gamalero et al., 2023).

3.1.7 Induced Systemic Resistance

Certain beneficial rhizosphere bacteria increase a plant's resistance to a variety of diseases and insect herbivores by causing induced systemic resistance (ISR), a crucial defence mechanism (Choudhary & Varma, 2016; Mitra et al., 2025). By priming the plant's immune system, these bacteria enable a quicker and more powerful defence response in the event of an actual attack rather than immediately triggering energy-intensive defence responses (Pieterse et al., 2014; Ashra & Nair,

2022). Furthermore, microorganisms that induce ISR may modify the plant's defence in ways that promote their own colonization and survival.

These microbial compounds help in stimulating the plant's immune response. They trigger major accumulation of PR proteins (pathogenesis-related proteins) or acid-like jasmonic, ethylene (ET), salicylic acid and malic acid, synthesis of phyto-alexin, production of hydrolytic enzymes, such as chitinases and β-1,3 glucanases, calcium calmodulin, and other secondary metabolites against the challenge pathogen (Kour et al., 2024). All these responses, including those from salicylic acid, jasmonic acid (JA), and ethylene (ET), converge at a central regulator called NPR1 (Non-expresser of Pathogenesis-Related Genes 1). The activation of NPR1 leads to increased production of defense-related proteins, especially pathogenesis-related (PR) proteins, which strengthen the plant's ability to resist infections. Finally, this results in two major systemic defence mechanisms: induced systemic resistance (ISR), primarily regulated by JA and ET, and systemic acquired resistance (SAR), mainly regulated by SA. Both ISR and SAR improve the plant's defensive capacity and prepare it for future attacks by harmful organisms.

3.2 Mycorrhizal Fungi for Nutrient Cycling and Soil Health

Mycorrhizal fungi form mutualistic symbiotic associations with the roots of most terrestrial plants (>80%), playing a pivotal role in nutrient cycling and soil health (Ma et al., 2022). These fungi extend the root system through their hyphal networks, enhancing nutrient and water uptake while improving soil structure and microbial diversity. In sustainable agriculture, mycorrhizal fungi are key components of microbial biofertilizers and bio stimulants, reducing reliance on chemical inputs and supporting ecosystem resilience (Parihar et al., 2020b; Akter et al., 2024). Mycorrhizal fungi are broadly classified into two major types that are particularly important in agriculture. (a) Arbuscular mycorrhizal fungi (AMF): These fungi belong to the phylum Glomeromycota, with common genera including Rhizophagus irregularis and Funneliformis mosseae. The AMF form specialized structures called arbuscules and vesicles within the root cortical cells, which facilitate efficient nutrient exchange between the fungus and the plant. (b) Ectomycorrhizal fungi (EMF): These fungi belong to the phyla *Basidiomycota* and *Ascomycota*, with common representatives including *Pisolithus tinctorius* and *Laccaria bicolor*. EMF forms a sheath or mantle around plant roots and develops a Hartig net, which is a network of hyphae that penetrates between root cortical cells to facilitate nutrient exchange.

Mycorrhizal fungi enhance nutrient cycling and soil health through several mechanisms:

3.2.1 Phosphorus Acquisition

AMF extend hyphae up to 10–20 cm beyond roots, accessing insoluble phosphorus (e.g., rock phosphate) in soil micropores. They produce phosphatases and organic acids to solubilize phosphorus, increasing plant uptake by 20–60% compared to non-mycorrhizal plants (Smith & Read, 2010; Battini et al., 2017; Parihar et al., 2020a, 2025).

3.2.2 Nitrogen Cycling and Uptake

The AMF acquire organic and inorganic nitrogen from soil, transferring it to host plants. The EMF degrade organic matter using enzymes such as laccases, releasing nitrogen for plant use in forest soils. The AMF can enhance nitrogen fixation in legumes by improving phosphorus supply to *Rhizobium* nodules, increasing nodulation by 15–30% (Scheublin & Van Der Heijden, 2006).

3.2.3 Carbon Cycling

AMF contribute predominantly to soil organic matter by creating a sink demand for plant carbon and distributing to below-ground hyphal biomass (Parihar, Rakshit, et al., 2020b). Plants allocate 10–20% of photosynthates to mycorrhizal fungi, which use this carbon to build hyphal networks and produce glomalin—a glycoprotein that stabilizes soil aggregates and increases water holding capacity (Andrino et al., 2021; Hawkins et al., 2023).

3.2.4 Nutrient Exchange Efficiency

Mycorrhizal fungi mediate bidirectional nutrient exchange, providing nutrients to plants in return for carbon. This symbiosis optimizes resource use, reducing nutrient leaching by 20–40% compared to non-mycorrhizal systems.

3.2.5 Soil Health

Mycorrhizal fungi enhance soil health through physical, chemical, and biological improvements: soil structure and aggregation, water retention and drought tolerance, soil microbial diversity, and organic matter decomposition.

3.3 Role of Biofertilizers in Reducing Synthetic Fertilizer Use and Emissions

Biofertilizers are eco-friendly inputs that contain living microorganisms, which promote plant growth by enhancing the availability of essential nutrients through natural processes. These include nitrogen fixation, phosphorus solubilization, potassium mobilization, and the production of plant growth-promoting substances. The widespread use of synthetic fertilizers in modern agriculture has led to environmental concerns, such as soil degradation, water pollution, and GHG emissions (Xu et al., 2023; Tomer et al., 2024). In this context, biofertilizers have emerged as an effective and sustainable alternative that can reduce reliance on synthetic fertilizers while maintaining or even improving crop productivity (Rai & Shukla, 2020).

3.3.1 Reduction in Synthetic Fertilizer Use

Biofertilizers contribute significantly to nutrient supply by enhancing nutrient availability in the soil. By utilizing these microorganisms, farmers can reduce the application rates of synthetic nitrogen, phosphorus, and potassium fertilizers (Fasusi et al., 2021; Tomer et al., 2024). This not only lowers input costs but also reduces the negative impacts associated with over-fertilization.

3.3.2 Mitigation of Greenhouse Gas Emissions

The production and application of synthetic fertilizers, especially nitrogen-based ones, are significant sources of greenhouse gases such as carbon dioxide (CO_2) and nitrous oxide (N_2O). Biofertilizers help mitigate these emissions in several ways. Biofertilizers promote efficient nitrogen use by plants, reducing the excess nitrogen in soils that could otherwise be lost as N_2O through denitrification and volatilization (Kumar et al., 2022a, b).

3.3.3 Improvement in Soil Health and Sustainability

Beyond reducing emissions and fertilizer use, biofertilizers have positive effects on overall soil health (Samantaray et al., 2024). Regular use of biofertilizers maintains a healthy microbial ecosystem in the rhizosphere, which is critical for nutrient cycling and disease suppression. Microbial secretions, such as exopolysaccharides, improve soil aggregation, porosity, and water retention capacity.

3.3.4 Environmental and Economic Benefits

The overuse of synthetic fertilizers often leads to nutrient runoff into water bodies, causing eutrophication. Biofertilizers minimize this risk by ensuring nutrient release is in synchrony with plant demand (Khoshru et al., 2023; Panday et al., 2024). Biofertilizers are generally cheaper than synthetic fertilizers, making them economically viable for small and marginal farmers. Moreover, their production relies on renewable biological resources. Adoption of biofertilizers not only supports climate-smart agriculture but also ensures long-term productivity and ecological balance.

4 Bioremediation and Soil Health Restoration

Bioremediation is a biological process that uses living organisms, mainly microorganisms and plants, to degrade, detoxify, or remove pollutants from contaminated soils, water, and environments (Sonawane et al., 2022). This eco-friendly and cost-effective method plays a crucial role in restoring soil health by breaking down harmful substances, such as heavy metals, petroleum hydrocarbons, pesticides, and industrial waste residues. Microbial bioremediation involves bacteria and fungi that possess the enzymatic ability to transform toxic compounds into less harmful or non-toxic forms. For instance, species such as *Achromobacter, Alcaligenes, Xanthobacter, Arthrobacter, Pseudomonas, Bacillus, Mycobacterium, Corynebacterium, Flavobacterium,* and *Nitrosomonas,* are widely used to degrade organic pollutants and immobilize heavy metals in soil (Sharma et al., 2021). Phytoremediation, on the other hand, uses specific plants to absorb, stabilize, or degrade contaminants, contributing to soil recovery. Restoring soil health through bioremediation not only reduces toxicity but also improves soil structure, nutrient availability, microbial diversity, and overall fertility. It helps revive the biological activity of degraded soils, making them suitable again for agricultural and ecological use (Rebello et al., 2021).

4.1 Microbial Degradation of Pollutants and Xenobiotics

Microbial degradation is a natural process in which microorganisms such as bacteria, fungi, and actinomycetes break down harmful pollutants and xenobiotics into simpler, non-toxic compounds (Phale et al., 2019). Xenobiotics are synthetic chemicals that are foreign to natural ecosystems, including pesticides, herbicides, plastics, pharmaceuticals, and industrial chemicals. These compounds often persist in the environment and can negatively affect soil, water, and living organisms. Microorganisms use a variety of enzymatic pathways to degrade these substances (Imam et al., 2022). For example, microorganisms such as *Alcaligenes*, *Cellulosimicrobium*, *Microbacterium*, *Micrococcus*, *Methanospirillum*, *Aeromonas*, *Sphingobium*, *Flavobacterium*, *Rhodococcus*, *Aspergillus*, *Penicillium*, *Trichoderma*, *Streptomyces*, *Rhodotorula*, *Candida*, and *Aureobasidium* have been isolated, characterized and have exhibited an excellent ability to biodegrade a variety of xenobiotic pollutants (Miglani et al., 2022; Mishra et al., 2021a, b). The *fungi*, such as *Phanerochaete chrysosporium*, also play a significant role by producing lignin-degrading enzymes that can break down persistent organic pollutants. This microbial activity not only reduces the toxicity of contaminated environments but also contributes to nutrient recycling and soil detoxification. Factors such as temperature, pH, oxygen availability, and the presence of co-metabolites influence the efficiency of microbial degradation.

4.2 Management of Saline and Degraded Soils Using Microbial Inoculants

Saline and degraded soils are major constraints to agricultural productivity due to high salt content, poor soil structure, nutrient imbalances, and reduced microbial activity (Priya et al., 2025). The use of microbial inoculants offers an eco-friendly and sustainable approach to improve the health and fertility of such soils. Microbial inoculants consist of beneficial microorganisms such as plant growth-promoting rhizobacteria (PGPR), arbuscular mycorrhizal fungi (AMF), nitrogen-fixing bacteria, and phosphate-solubilizing bacteria. In saline environments, these inoculants often include an abundance of halophilic microorganisms, predominantly belonging to the genera Bacillus, Pseudomonas, Micrococcus, and Alcaligenes (Kumawat et al., 2022). These microbes enhance plant tolerance to salt stress by producing growth-promoting substances, such as phytohormones and osmolytes, and by improving root growth and nutrient uptake. Some microbes also produce exopolysaccharides, which help in soil aggregation and reduce the impact of salinity by binding toxic ions. Mycorrhizal fungi form symbiotic associations with plant roots, enhancing water and nutrient absorption, especially phosphorus, and increasing the plant's resistance to saline conditions (Kumar et al., 2020a, b). Additionally, microbial inoculants improve soil biological activity, enhance organic matter decomposition, and support the restoration of degraded land.

5 Microbial Consortia in Climate-Resilient Crop Production

Microbial consortia, multi-community members of plant beneficial microorganisms, are becoming a valuable instrument towards fostering climate-resilient plant cultivation. In contrast to single-strain inoculants, the consortia act synergistically. They enable plants to resist environmental stresses in a number of direct and indirect ways. They facilitate better nutrient uptake by fixing nitrogen, solubilizing phosphorus, and enhancing the availability of micronutrients (Bargaz et al., 2018). Besides, microbial consortia regulate plant hormone balances, enhance root growth, and trigger plant tolerance through the activation of stress-responsive genes and antioxidant pathways (Nunes et al., 2024). Under water-limited and saline conditions, they enhance water retention and keep the ionic balance in plant tissues in check (Muhammad et al., 2024). Through these processes, microbial consortia enable increased crop yields under stress conditions. In addition, they also make inputs to sustainable soil fertility. The use of microbial consortia in agriculture is an environmentally friendly way of constructing robust and productive farming systems.

5.1 Mechanisms of Stress Tolerance by Microbial Consortia

Microbial consortia enhance plant resilience to environmental stresses through a combination of physiological, biochemical, and molecular interactions. These mechanisms are particularly effective in addressing major abiotic stressors such as drought, salinity, and high temperature (Table 3).

5.1.1 Drought Stress Tolerance

Microorganisms, including microbial consortia, are able to promote drought stress tolerance in plants predominantly by triggering antioxidant systems, enhancing water and nutrient uptake, and root development (Shaffique et al., 2022). Plant-growth-promoting rhizobacteria such as *Bacillus* sp., *Pseudomonas* sp. *Azospirillum*, and *Actinobacteria*, endophytes such as *Gluconacetobacter diazotrophicus*, *Streptomyces laurentii*, and phyllospheric microorganisms such as *Bacillus altitudinis* have been reported to be involved in drought stress tolerance or alleviation (Bano et al., 2013; Filgueiras et al., 2020; Kour et al., 2020; Devarajan et al., 2021; Khati et al., 2024). Microbial ACC deaminase production reduces ethylene levels, alleviating the adverse effects of drought stress on plant growth. In addition, microbes assist in osmotic adjustment through the production of

Table 3 Role of microorganisms in abiotic stress mitigation

Stress type	Mechanism	Microbial role	References
Drought	Enhanced root architecture	Stimulate root growth and branching	Hartman and Tringe (2019)
Drought	Production of exopolysaccharides	Improve soil water retention	Morcillo and Manzanera (2021)
Drought	Hormonal regulation (ABA modulation)	Balance plant water usage through hormone control	Chattaraj et al. (2025)
Drought	Induced systemic tolerance	Trigger drought-responsive genes	Kumar et al. (2024)
Salinity	Ionic homeostasis (Na^+/K^+ regulation)	Prevent toxic salt accumulation in plant tissues	Kumar et al. (2020a, b)
Salinity	Osmoprotectant production (proline, betaine)	Maintain osmotic balance and protect cell structures	Mishra et al. (2021a, b)
Salinity	Antioxidant defense activation (SOD, CAT)	Reduce oxidative damage caused by salt stress	Kumar et al. (2020a, b)
Salinity	Improved nutrient uptake in saline soils	Enhance availability and absorption of nutrients	Hidri et al. (2016)
High temperature	Induction of heat shock proteins (HSPs)	Stabilize cellular proteins during heat stress	Ahmad et al. (2023)
High temperature	Ethylene modulation via ACC deaminase	Lower ethylene levels to delay senescence	Acuña et al. (2019)
High temperature	Membrane stability and fluidity maintenance	Preserve membrane function under high temperatures	Mitra et al. (2021)
High temperature	Preservation of photosynthetic efficiency	Maintain chlorophyll and photosynthetic enzyme activity	Dastogeer et al. (2022)

osmolytes such as proline, glycine betaine, and trehalose, which support cellular activities (Shaffique et al., 2022). Exopolysaccharides made by microbes develop biofilms that surround plant roots, shielding them from dehydration and improving soil water retention (Morcillo & Manzanera, 2021). In addition, microbial volatile organic compounds (VOCs) and antioxidant mechanisms induce plant stress responses, aiding in their response to oxidative stress (Ali et al., 2024). These mechanisms combined allow crops to withstand adverse environmental conditions and increase overall resilience against climate change. Al Mamun et al. (2024) have demonstrated that a consortium of *Bacillus amyloliquefaciens*, Arbuscular Mycorrhizal Fungi and *Pseudomonas brassicacearum* was very efficient at improving drought tolerance and phosphate nutrition in potato plants under drought stress conditions. They aided in enhanced root growth, higher production of antioxidants, improved osmotic adjustment, and improved stress signaling by phytohormones such as ABA and IAA (Al Mamun et al., 2024). Likewise, inoculation of *Bacillus megaterium* strains JTB1, MT22, and their consortia significantly enhanced root growth and sugarcane development under moderate and severe drought stress by increasing the expression of auxin transporters and photosynthetic gene expression (Buqori et al., 2025).

5.1.2 Salinity Stress Tolerance

Microbial consortia are critical in regulating salinity stress by improving plant tolerance via various synergistic processes. These involve ionic balance maintenance by controlling sodium (Na^+) and potassium (K^+) uptake, osmoprotectant accumulation such as proline and glycine betaine, and the induction of antioxidant enzymes to minimize oxidative damage due to salt-induced reactive oxygen species (Kumar et al., 2020a, b). Microbes also synthesize phytohormones such as auxins and cytokinins that are beneficial for root growth, resulting in increased nutrient uptake from saline soils. They also enhance soil structure and microbial activity by producing exopolysaccharides. Inoculation of tomato plants with microbial consortia made up of *Bacillus*, *Delftia*, *Enterobacter*, and *Achromobacter* species greatly mitigated the harmful impacts of salinity stress by improving root growth, biomass accumulation, mineral uptake, and chlorophyll contents (Kapadia et al., 2021). A PGPR multi-strain consortium (*Ensifer adhaerens* BK-30, *Pseudomonas fluorescens* SN5, and *Bacillus megaterium* SN15) greatly promoted wheat growth and physiological characteristics under salinity stress by minimizing electrolyte leakage and enhancing chlorophyll content, RWC, and K^+/Na^+ ratio (Khan et al., 2022). Likewise, research by Sathya et al. (2024) reported that a microbial consortium of *Chryseobacterium indologenes* ICKM4 and *Stenotrophomonas maltophilia* ICKM15 greatly promoted growth, nodulation, chlorophyll content, antioxidant gene expression, and tolerance to stress in chickpea under salinity. These investigations prove that co-inoculation of compatible strains with varying growth-promoting properties is a superior approach compared to single-strain applications.

5.1.3 High Temperature Stress Tolerance

The multi-strain populations, usually consisting of plant growth-promoting rhizobacteria (PGPR), arbuscular mycorrhizal fungi (AMF), and endophytes, improve heat tolerance by increasing antioxidant enzyme activity, minimizing oxidative stress, stabilizing membrane integrity, and enhancing water retention and nutrient uptake (Carillo, 2025). They also regulate stress-responsive hormones such as abscisic acid (ABA), salicylic acid (SA), and ethylene, ensuring improved photosynthetic efficiency and biomass accumulation under high temperatures (Dastogeer et al., 2022). The lower contents of malondialdehyde (MDA) and hydrogen peroxide (H_2O_2) in inoculated plants indicated decreased oxidative damage, which was correlated with increased antioxidant enzyme activity (Dastogeer et al., 2022). The bacterial consortium containing three *Bacillus* sp. and one *Pseudomonas* sp. was seen to increase the heat tolerance in maize seedlings (Ahmad et al., 2023). The consortium stimulated maize seedling development through enhanced antioxidant enzyme activity and upregulation of major heat shock proteins under heat stress (Ahmad et al., 2023).

These stress tolerance mechanisms demonstrate the potential of microbial consortia as an eco-friendly, sustainable solution to facilitate crop growth in the face of climate variability. Their multifunctional roles not only help plants survive but also maintain yield and quality under challenging environmental conditions.

5.2 Case Studies on Microbial Applications in Arid Regions

Microbial technologies have shown promising outcomes in enhancing crop resilience and soil fertility in arid and semi-arid regions, where harsh conditions such as drought, high salinity, and poor soil structure prevail.

5.2.1 Thar Desert (India): Dual Inoculation Improves Pomegranate Yield

Aseri et al. (2008) carried out a five-year field experiment in the Thar Desert to evaluate the effect of inoculating pomegranate plants with *Azotobacter chroococcum* and the Arbuscular Mycorrhizal Fungus *Glomus mosseae*. Inoculated cuttings sprouted 8–10 days in advance and had enhanced branch number, leaf area (up to 57.6%), shoot biomass (up to 36%), and chlorophyll content. Inoculated plants also had enhanced biochemical parameters, such as greater reducing sugars, amino nitrogen, and phenols, and lower oxidative stress. Soil microbial processes (dehydrogenase, phosphatase, nitrogenase, and FDA hydrolysis) and AM fungal colonization were significantly greater in treated plants. Uptake of nutrients (N, P, K, Ca, Mg, and micronutrients) was observed to increase considerably, the maximum effect being observed with dual inoculation. Multidisciplinary field trials across several years validated uniform improvements in plant height, canopy size, pruned biomass, and fruit yield, confirming the efficacy of microbial consortia as a viable method for enhancing pomegranate performance in arid and nutrient-poor conditions.

5.2.2 Atacama Desert (Chile): Enhancing Wheat Growth Under Salt Stress

Acuña et al. (2019) examined ACC deaminase-producing rhizobacteria from *Parastrephia quadrangularis*, a native species of the Chilean Andean Altiplano, for their plant growth-promoting potential under salt stress. Among 30 bacterial isolates, 10 had ACC deaminase activity and were characterized as predominantly *Serratia* and *Klebsiella* species. They survived in high salinity (2.5–7.5% NaCl) and showed principal PGPR characteristics such as the production of auxin and siderophores. Inoculation with two *Klebsiella* strains selected from native, salt-tolerant plants resulted in a 45–62% increase in plant biomass and an 18–35% enhancement in root SOD activity even under salt stress. These results indicate the utility of rhizobacteria isolated from indigenous, salinity-tolerant vegetation as bioinoculants for enhancing salt tolerance in plants.

5.2.3 Negev Desert (Israel): PGPR Enhances Maize Growth

A bacterial isolate, *Dietzia cinnamea* 55, was isolated by Khan et al. (2020a, b) from the rhizosphere of *Zygophyllum dumosum*. The isolate showed salt, drought (PEG), and pH stress tolerance. Genome analysis of the bacterium indicated genes for trehalose biosynthesis, heat shock proteins, osmoprotectant transporters, and additional stress-regulated adaptations. Greenhouse and field trials showed that *D. cinnamea* 55 significantly promoted corn growth by increasing shoot and root growth, dry biomass, and yield.

5.2.4 Chihuahuan Desert (Mexico): PGPR Enhance Crop Growth Under Salinity

Palacio-Rodríguez et al. (2017) isolated halophilic PGPR strains, *Bacillus* sp. and *Pseudomonas lini* from *Distichlis spicata* roots, a salt-tolerant grass, in the Chihuahuan Desert. These strains showed plant growth-promoting characteristics such as IAA and siderophore production, and ACC deaminase activity. Inoculation of *Arabidopsis*, cucumber, and watermelon plants with these strains under saline conditions led to enhanced biomass and lateral root growth, highlighting their potential in promoting crop growth under salinity stress. These case studies demonstrate that location-specific microbial consortia can play a vital role in sustaining agriculture in arid regions through eco-friendly and low-input approaches.

6 Technological Innovations in Microbial Resource Management

Microbial resources find utility in agricultural, environmental, and industrial applications. Improved technological capabilities have greatly enhanced the roles of researchers and practitioners in discovering, manipulating, and applying microbial communities. This section comprehends the significant progress attained in three principal areas: metagenomics and next-generation sequencing, gene editing technologies, and bioformulations for large-scale production.

6.1 Metagenomics and Next-Generation Sequencing for Microbial Identification

Meta-omics strategies, including metagenomics, metatranscriptomics, metaproteomics, and metabolomics, provide holistic information on the structure, function, and dynamics of plant-associated microbiomes, especially under stress (Meena et al., 2017; Khati et al., 2019b). Metagenomics provides profiling of microbial community composition and functional potential through sequencing of environmental DNA, which indicates the occurrence of stress-responsive genes, such as those related to nutrient solubilization, phytohormone production, and abiotic stress resistance (Garg et al., 2024). This method enables detailed insights into ecological roles of microorganisms without the need for culturing, which is crucial as most microbes are unculturable (Suenaga, 2012). Metatranscriptomics builds on this by revealing which microbial genes are being actively expressed under a particular stress, thereby detecting functional changes and microbial response in real-time (Aguiar-Pulido et al., 2016). Metaproteomics fills this gap by detecting and quantifying the proteins being expressed by the microbial community, connecting gene expression with functional protein activity, such as enzymes in osmoprotection and antioxidant defense (Kleiner, 2019). Furthermore, metabolomics uncovers the end-products of these biological processes, such as signaling molecules and metabolites that facilitate plant-microbe interactions and plant stress responses

(Aguiar-Pulido et al., 2016; Meena et al., 2017). Together, these meta-omics techniques combined disaggregate the intricate, multilayered interactions between plants and their microbiomes, allowing key microbial taxonomic groups and functions responsible for plant resilience under abiotic and biotic stress to be identified. These findings open up new avenues for developing targeted microbial consortia and bioinoculants to promote sustainable agriculture in climate-stress environments.

Next Generation Sequencing (NGS) platforms such as Illumina and 454 Roche have improved the resolution and speed of metagenomic analysis, allowing for the rapid sequencing of millions of DNA fragments simultaneously (Satam et al., 2023). Nkongolo and Narendrula-Kotha (2020) suggested combining NGS with PLFA (phosphor-lipid fatty acid profile) for a comprehensive view of microbial structure and function, while Nannipieri et al. (2020) recommended integrating metagenomics, gene expression, and culturomics for accurate functional analysis. In addition, the combination of metagenomics with bioinformatics tools allows predictive modeling of microbial interactions and their ecological functions (Hiraoka et al., 2016). This has provided new avenues in designing synthetic microbial consortia specific to particular agricultural or environmental issues. Metagenomic information also guides strain selection for microbial inoculants, improving their efficacy and ecological compatibility.

6.2 Gene Editing Tools for Enhancing Microbial Efficiency

Genome editing technologies utilizing CRISPR/Cas are broadly employed in engineering bacterial genomes and metabolism (Lone et al., 2018). These tools are now crucial in biotechnological applications for optimizing microbial traits. To efficiently knock out or insert target genes, researchers can select appropriate Cas enzymes and DNA repair mechanisms tailored to their experimental goals (Liu et al., 2020). Gene editing can be used to enhance desirable functions by targeted activation or suppression of genes in microorganisms. Such modifications can enhance microbial colonization proficiency, metabolic yield, and stress tolerance and consequently their efficacy as bioinoculants. Gene editing can also be employed to suppress undesirable characteristics such as antibiotic resistance (Moitra et al., 2024). Bioengineering, in synthetic biology, could be used to design advanced biological systems for producing beneficial products (Karataş & Ayaz, 2025). This has made significant progress in recent years, with applications in human health, industry, and agriculture, driven by advances in DNA sequencing, synthesis, and metabolic engineering.

6.3 Bioformulations and Large-Scale Application Strategies

Microbial resources need to be scaled up and translated from the lab to the field through strong bioformulations and large-scale delivery systems. Bioformulations constitute preparations of microbial inoculants in a stable, viable, and effective state. Such formulations can consist of carrier materials (peat, compost, alginate beads), protectants, and adjuvants to provide protection to the microbes during storage, transportation, and application (Khan et al., 2023). The recent developments in controlled-release systems, nanotechnology, and microencapsulation have contributed greatly towards the shelf life and field performance of microbial products. For example, encapsulating useful microbes within biopolymer matrices can insulate them against environmental stresses, including UV radiation, drying, and temperature changes (Pinaeva & Noskov, 2024). Targeted delivery to plant roots or targeted niches in soil can also be provided by nanocarriers. Large-scale application techniques are also paramount. These range from seed coating, foliar sprays, and soil drenches to drip irrigation systems that are suitable for mechanized farming. Technologies for precision delivery, including drones and automated dispensers, are being investigated for improving coverage while reducing labor inputs. Microbial products also have the potential of being integrated in circular bioeconomy models. The utilization of agro-waste as substrate for microbial growth and setting up of production units of biofertilizers co-operatively near farms could save the costs of logistics and carbon emissions.

Microbial resource management has evolved into a precision-guided, data-driven technology. The advanced tools not only enable us to identify and engineer beneficial microbes but also facilitate their safe upscaling in agricultural systems. As global challenges such as climate change, soil degradation, and food insecurity intensify, these innovations could prove to be sustainable, resilient, and bio-based solutions.

7 Conclusion

As the global community seeks sustainable and scalable solutions to combat the multifaceted challenges of climate change, microbial resource management emerges as a nature-based, scientifically grounded, and cost-effective approach. Microorganisms, with their intrinsic ability to regulate biogeochemical processes and promote plant resilience, represent an untapped reservoir of climate mitigation potential. By harnessing their capabilities through targeted interventions such as biofertilizers, microbial consortia, and stress-tolerant inoculants, it is possible to reduce greenhouse gas emissions, enhance soil carbon storage, and build climate-resilient agricultural systems. However, realizing this potential requires interdisciplinary collaboration, technological innovation, and supportive policy frameworks. This chapter aims to bridge the gap between microbial ecology and practical climate action, laying the foundation for integrated, microbe-based strategies that support both environmental sustainability and agricultural productivity in a warming world.

8 Future Prospects

The future of Microbial Resource Management (MRM) lies in harnessing cutting-edge technologies such as synthetic biology, metagenomics, and AI-driven microbial modeling to design highly efficient and resilient microbial consortia tailored for specific agro-climatic zones. Advances in scalable bioformulation techniques and precision delivery systems will enhance field stability and effectiveness of microbial applications. Integration of MRM into national climate action plans, supported by robust policy frameworks and public-private partnerships, will be crucial for widespread adoption. Furthermore, interdisciplinary research linking microbial ecology, soil science, and climate modeling will open new pathways for optimizing carbon sequestration and reducing greenhouse gas emissions. As global climate challenges intensify, MRM holds strong potential to become a cornerstone of sustainable and climate-resilient agriculture.

References

Acuña, J. J., Campos, M., de la Luz Mora, M., Jaisi, D. P., & Jorquera, M. A. (2019). ACCD-producing rhizobacteria from an Andean Altiplano native plant (*Parastrephia quadrangularis*) and their potential to alleviate salt stress in wheat seedlings. *Applied Soil Ecology, 136*, 184–190.

Adhikari, P., & Pandey, A. (2019). Phosphate solubilization potential of endophytic fungi isolated from *Taxus wallichiana* Zucc. roots. *Rhizosphere, 9*, 2–9.

Aguiar-Pulido, V., Huang, W., Suarez-Ulloa, V., Cickovski, T., Mathee, K., & Narasimhan, G. (2016). Metagenomics, metatranscriptomics, and metabolomics approaches for microbiome analysis. *Evolutionary Bioinformatics Online, 12*(Suppl 1), 5–16. https://doi.org/10.4137/EBO.S36436

Ahmad, M., Imtiaz, M., Nawaz, M. S., Ahmad, M. I., Ahmad, Z., Rehman, S. U., Ullah, S., Sohail, M. I., Noman, A., & Shah, M. Z. (2023). Thermotolerant PGPR consortium B3P modulates physio-biochemical and molecular machinery for enhanced heat tolerance in maize during early vegetative growth. *Annales de Microbiologie, 73*, 34.

Ahmed, N., Li, J., Li, Y., Deng, L., Deng, L., Chachar, M., Chachar, Z., Chachar, S., Hayat, F., Raza, A., & Umrani, J. H. (2025). Symbiotic synergy: How arbuscular mycorrhizal fungi enhance nutrient uptake, stress tolerance, and soil health through molecular mechanisms and hormonal regulation. *IMA Fungus, 16*, e144989.

Akter, S., Kamruzzaman, M., Sarder, M. P., Amin, M. S., Joardar, J. C., Islam, M. S., et al. (2024). Mycorrhizal fungi increase plant nutrient uptake, aggregate stability and microbial biomass in the clay soil. *Symbiosis, 93*(2), 163–176.

Al Mamun, M., Neumann, G., Moradtalab, N., Ahmed, A., Dupuis, B., Darbon, G., et al. (2024). Microbial consortia versus single-strain inoculants as drought stress protectants in potato affected by the form of N supply. *Horticulturae, 10*(1), 102.

Alghamdi, S. A., Alharby, H. F., Abdelfattah, M. A., Mohamed, I. A., Hakeem, K. R., Rady, M. M., & Shaaban, A. (2023). *Spirulina platensis*-inoculated humified compost boosts rhizosphere soil hydro-physico-chemical properties and *Atriplex nummularia* forage yield and quality in an arid saline calcareous soil. *Journal of Soil Science and Plant Nutrition, 23*(2), 2215–2236.

Ali, S., Akhtar, M. S., Siraj, M., & Zaman, W. (2024). Molecular communication of microbial plant biostimulants in the rhizosphere under abiotic stress conditions. *International Journal of Molecular Sciences, 25*(22), 12424.

Alori, E. T., Glick, B. R., & Babalola, O. O. (2017). Microbial phosphorus solubilization and its potential for use in sustainable agriculture. *Frontiers in Microbiology, 8*, 971.

Amaya-Gómez, C. V., Porcel, M., Mesa-Garriga, L., & Gómez-Álvarez, M. I. (2020). A framework for the selection of plant growth-promoting rhizobacteria based on bacterial competence mechanisms. *Applied and Environmental Microbiology, 86*(14), e00760–e00720.

Andrino, A., Guggenberger, G., Sauheitl, L., Burkart, S., & Boy, J. (2021). Carbon investment into mobilization of mineral and organic phosphorus by arbuscular mycorrhiza. *Biology and Fertility of Soils, 57*, 47–64.

Armada, E., Barea, J. M., Castillo, P., Roldán, A., & Azcón, R. (2015). Characterization and management of autochthonous bacterial strains from semiarid soils of Spain and their interactions with fermented agrowastes to improve drought tolerance in native shrub species. *Applied Soil Ecology, 96*, 306–318.

Armin, R., Zühlke, S., Grunewaldt-Stöcker, G., Mahnkopp-Dirks, F., & Kusari, S. (2021). Production of siderophores by an apple root-associated *Streptomyces ciscaucasicus* strain GS2 using chemical and biological OSMAC approaches. *Molecules, 26*, 3517.

Arora, N. K., & Verma, M. (2017). Modified microplate method for rapid and efficient estimation of siderophore produced by bacteria. *3 Biotech, 7*(6), 1–9. https://doi.org/10.1007/s13205-017-1008-y

Aseri, G. K., Jain, N., Panwar, J., Rao, A. V., & Meghwal, P. R. (2008). Biofertilizers improve plant growth, fruit yield, nutrition, metabolism and rhizosphere enzyme activities of pomegranate (*Punica granatum* L.) in Indian Thar Desert. *Scientia Horticulturae, 117*, 130–135.

Ashra, H., & Nair, S. (2022). Trait plasticity during plant-insect interactions: From molecular mechanisms to impact on community dynamics. *Plant Science, 317*, 111188.

Babalola, O. O., & Glick, B. R. (2012). The use of microbial inoculants in African agriculture: Current practice and future prospects. *Journal of Food, Agriculture and Environment, 10*, 540–549.

Backer, R., Rokem, J. S., Ilangumaran, G., Lamont, J., Praslickova, D., Ricci, E., Subramanian, S., & Smith, D. L. (2018). Plant growth-promoting rhizobacteria: Context, mechanisms of action, and roadmap to commercialization of biostimulants for sustainable agriculture. *Frontiers in Plant Science, 9*, 1473.

Bag, S., Mondal, A., & Banik, A. (2022). Exploring tea (*Camellia sinensis*) microbiome: Insights into the functional characteristics and their impact on tea growth promotion. *Microbiological Research, 254*, 126890.

Baghel, V., Thakur, J. K., Yadav, S. S., Manna, M. C., Mandal, A., Shirale, A. O., et al. (2020). Phosphorus and potassium solubilization from rock minerals by endophytic *Burkholderia* sp. strain FDN2-1 in soil and shift in diversity of bacterial endophytes of corn root tissue with crop growth stage. *Geomicrobiology Journal, 37*(6), 550–563.

Baiyee, B., Ito, S. I., & Sunpapao, A. (2019). *Trichoderma asperellum* T1 mediated antifungal activity and induced defense response against leaf spot fungi in lettuce (*Lactuca sativa* L.). *Physiological and Molecular Plant Pathology, 106*, 96–101.

Baldani, J. I., Videira, S. S., Reis, V. M., & Baldani, V. L. D. (2002). The art of isolating nitrogen-fixing bacteria from non-leguminous plants using N-free semi-solid media: A practical guide for microbiologists. *Plant and Soil, 356*(1–2), 237–246.

Bano, Q., Ilyas, N., Bano, A., Zafar, N., Akram, A., & Hassan, F. (2013). Effect of *Azospirillum* inoculation on maize (*Zea mays* L.) under drought stress. *Pakistan Journal of Botany, 45*, 13–20.

Bargaz, A., Lyamlouli, K., Chtouki, M., Zeroual, Y., & Dhiba, D. (2018). Soil microbial resources for improving fertilizers efficiency in an integrated plant nutrient management system. *Frontiers in Microbiology, 9*, 1606.

Barman, S., Das, S., & Bhattacharya, S. S. (2019). The prospects of bio-fertilizer technology for productive and sustainable agricultural growth. In *New and future developments in microbial biotechnology and bioengineering* (pp. 233–253). Elsevier.

Barney, B. M. (2020). Aerobic nitrogen-fixing bacteria for hydrogen and ammonium production: Current state and perspectives. *Applied Microbiology and Biotechnology, 104*(4), 1383–1399.

Basak, B. B., & Biswas, D. R. (2010). Co-inoculation of potassium solubilizing and nitrogen fixing bacteria on solubilization of waste mica and their effect on growth promotion and nutrient acquisition by a forage crop. *Biology and Fertility of Soils, 46*, 641–648.

Bashan, Y., & de Bashan, L. E. (2010). How the plant growth-promoting bacterium *Azospirillum* promotes plant growth—A critical assessment. *Advances in Agronomy, 108*, 77–136.

Battini, F., Grønlund, M., Agnolucci, M., Giovannetti, M., & Jakobsen, I. (2017). Facilitation of phosphorus uptake in maize plants by mycorrhizosphere bacteria article. *Scientific Reports, 7*, 1–11.

Benaissa, A. (2024). Rhizosphere: Role of bacteria to manage plant diseases and sustainable agriculture—A review. *Journal of Basic Microbiology, 64*(3), 2300361.
Bender, S. F., Wagg, C., & van der Heijden, M. G. (2016). An underground revolution: Biodiversity and soil ecological engineering for agricultural sustainability. *Trends in Ecology & Evolution, 31*(6), 440–452.
Benson, D. R., & Silvester, W. B. (1993). Biology of *Frankia* strains, actinomycete symbionts of actinorhizal plants. *Microbiological Reviews, 57*(2), 293–319.
Bhat, M. A., Mishra, A. K., Jan, S., Bhat, M. A., Kamal, M. A., Rahman, S., Shah, A. A., & Jan, A. T. (2023). Plant growth promoting rhizobacteria in plant health: A perspective study of the underground interaction. *Plants, 12*(3), 629.
Bhattacharyya, S. S., Ros, G. H., Furtak, K., Iqbal, H. M., & Parra-Saldívar, R. (2022). Soil carbon sequestration—An interplay between soil microbial community and soil organic matter dynamics. *Science of the Total Environment, 815*, 152928.
Biswas, J. K., Banerjee, A., Rai, M., Naidu, R., Biswas, B., Vithanage, M., Dash, M. C., Sarkar, S. K., & Meers, E. (2018). Potential application of selected metal resistant phosphate solubilizing bacteria isolated from the gut of earthworm (*Metaphire posthuma*) in plant growth promotion. *Geoderma, 330*, 117–124.
Bourscheidt, M. L., Gomes, F. J., Pedreira, C. G., Boote, K. J., Hoogenboom, G., Pereira, D. H., & Pedreira, B. C. (2023). Highlighting the benefits of biological nitrogen fixation on agronomic, physiological, and nutritive value traits of brachiariagrass. *European Journal of Agronomy, 143*, 126730.
Buqori, D. M. A. I., Sugiharto, B., Suherman, T. A. S., & Hariyono, K. (2025). Mitigating drought stress by application of drought-tolerant *Bacillus* spp. enhanced root architecture, growth, antioxidant and photosynthetic genes expression in sugarcane. *Scientific Reports, 15*, 5259.
Carillo, P. (2025). Can biostimulants enhance plant resilience to heat and water stress in the Mediterranean hotspot? *Plant Stress, 16*, 100802.
Carvajal, F., Palma, F., Jiménez-Muñoz, R., Jamilena, M., Pulido, A., & Garrido, D. (2017). Unravelling the role of abscisic acid in chilling tolerance of zucchini during postharvest cold storage. *Postharvest Biology and Technology, 133*, 26–35.
Cavalcante, V. A., & Döbereiner, J. (1988). A new acid-tolerant nitrogen-fixing bacterium associated with sugarcane. *Plant and Soil, 108*, 23–31.
Chae, D. H., Kim, D. R., Cheong, M. S., Lee, Y. B., & Kwak, Y. S. (2020). Investigating the induced systemic resistance mechanism of 2, 4-diacetylphloroglucinol (DAPG) using DAPG hydrolase-transgenic *Arabidopsis*. *Plant Pathology Journal, 36*, 255.
Chandwani, S., Dewala, S., Chavan, S. M., Paul, D., Pachaiappan, R., Gopi, M., & Amaresan, N. (2023). Complete genome sequencing of *Bacillus subtilis* (CWTS 5), a siderophore-producing bacterium triggers antagonistic potential against *Ralstonia solanacearum*. *Journal of Applied Microbiology, 134*, lxad066.
Chattaraj, S., Samantaray, A., Ganguly, A., & Thatoi, H. (2025). Employing plant growth-promoting rhizobacteria for abiotic stress mitigation in plants: With a focus on drought stress. *Discovery Applied Science, 7*, 68.
Chaudhary, S., & Sindhu, S. S. (2025). Iron sensing, signalling and acquisition by microbes and plants under environmental stress: Use of iron-solubilizing bacteria in crop biofortification for sustainable agriculture. *Plant Science, 356*, 112496.
Chaudhary, S., Tanvi, V., Verma, N., & Goyal, S. (2018). Response of sulphur oxidizing bacterial inoculation on growth and yield parameters of mustard (*Brassica juncea* L.). *International Journal of Chemical Studies, 6*(6), 2452–2457.
Chaudhary, P., Khati, P., Chaudhary, A., Gangola, S., Kumar, R., & Sharma, A. (2021a). Bioinoculation using indigenous *Bacillus* spp. improves growth and yield of *Zea mays* under the influence of nanozeolite. *3 Biotech, 11*, 11.
Chaudhary, P., Khati, P., Chaudhary, A., Maithani, D., Kumar, G., & Sharma, A. (2021b). Cultivable and metagenomic approach to study the combined impact of nanogypsum and *Pseudomonas taiwanensis* on maize plant health and its rhizospheric microbiome. *PLoS One, 16*(4), e0250574.

Chaudhary, P., Khati, P., Gangola, S., Kumar, A., Kumar, R., & Sharma, A. (2021c). Impact of nanochitosan and *Bacillus* spp. on health, productivity and defence response in *Zea mays* under field condition. *3 Biotech, 11*(5), 237.

Chaudhary, P., Sharma, A., Chaudhary, A., Khati, P., Gangola, S., & Maithani, D. (2021d). Illumina-based high-throughput analysis of microbial diversity of maize rhizosphere treated with nanocompounds and *Bacillus* sp. *Applied Soil Ecology, 159*, 103386.

Chaudhary, S., Dhanker, R., Singh, K., Brar, B., & Goyal, S. (2022). Characterization of sulfur-oxidizing bacteria isolated from mustard (*Brassica juncea* L.) rhizosphere having the capability of improving sulfur and nitrogen uptake. *Journal of Applied Microbiology, 133*(5), 2814–2825.

Chaudhary, S., Sindhu, S. S., Dhanker, R., & Kumari, A. (2023). Microbes-mediated sulphur cycling in soil: Impact on soil fertility, crop production and environmental sustainability. *Microbiological Research, 271*, 127340.

Chen, X., Zhao, Y., Huang, S., Peñuelas, J., Sardans, J., Wang, L., & Zheng, B. (2024). Genome-based identification of phosphate-solubilizing capacities of soil bacterial isolates. *AMB Express, 14*, 85.

Chennappa, G., Sreenivasa, M. Y., & Nagaraja, H. (2018). *Azotobacter salinestris*: A novel pesticide-degrading and prominent biocontrol PGPR bacteria. In *Microorganisms for green revolution: Volume 2: Microbes for sustainable agro-ecosystem* (pp. 23–43). Springer.

Choudhary, D. K., & Varma, A. (Eds.). (2016). *Microbial-mediated induced systemic resistance in plants* (pp. 147–162). Springer.

Choudhary, M., Yadav, R. P., Mondal, T., Ghasal, P. C., Meena, V. S., & Bisht, J. K. (2018). Towards the plant beneficiary rhizobacteria (PBR) and agricultural sustainability. In V. S. Meena, P. K. Mishra, J. K. Bisht, & A. Pattanayak (Eds.), *Role of rhizospheric microbes in soil*. Springer.

Chouyia, F., Ventorino, V., & Pepe, O. (2022). Diversity, mechanisms and beneficial features of phosphate-solubilizing *Streptomyces* in sustainable agriculture: A review. *Frontiers in Plant Science, 13*, 1035358.

Das, A. J., Kumar, M., & Kumar, R. (2013). Plant growth promoting rhizobacteria (pgpr): An alternative of chemical fertilizer for sustainable, environment friendly agriculture. *Research Journal of Agriculture and Forestry Sciences, 4*, 21–23.

Dastogeer, K. M. G., Zahan, I. M., Rhaman, M. S., Sarker, M. S. A., & Chakraborty, A. (2022). Microbe-mediated thermotolerance in plants and pertinent mechanisms—A meta-analysis and review. *Frontiers in Microbiology, 13*, 833566.

de Andrade, L. A., Santos, C. H. B., Frezarin, E. T., Sales, L. R., & Rigobelo, E. C. (2023). Plant growth-promoting rhizobacteria for sustainable agricultural production. *Microorganisms, 11*(4), 1088.

de Oliveira-Paiva, C. A., Bini, D., de Sousa, S. M., Ribeiro, V. P., Dos Santos, F. C., de Paula Lana, U. G., de Souza, F. F., Gomes, E. A., & Marriel, I. E. (2024). Inoculation with *Bacillus megaterium* CNPMS B119 and *Bacillus subtilis* CNPMS B2084 improve P-acquisition and maize yield in Brazil. *Frontiers in Microbiology, 15*, 1426166.

Deb, C. R., & Tatung, M. (2024). Siderophore producing bacteria as biocontrol agent against phytopathogens for a better environment: A review. *South African Journal of Botany, 165*, 153–162.

Defez, R., Andreozzi, A., Romano, S., Pocsfalvi, G., Fiume, I., & Esposito, R. (2019). Bacterial IAA-delivery into *Medicago* root nodules triggers a balanced stimulation of C and N metabolism leading to a biomass increase. *Microorganisms*. https://doi.org/10.3390/microorganisms7100403

Dehghanian, Z., Habibi, K., Sardrodi, M. M., Balilashaki, K., Lajayer, B. A., & Astatkie, T. (2022). Revivification of rhizobacteria-promoting plant growth for sustainable agricultural development. In *New and future developments in microbial biotechnology and bioengineering* (pp. 353–368). Elsevier.

Devarajan, A. K., Muthukrishanan, G., Truu, J., Truu, M., Ostonen, I., Panneerselvam, P., et al. (2021). The foliar application of rice phyllosphere bacteria induces drought-stress tolerance in *Oryza sativa* (L.). *Plants, 10*, 387. https://doi.org/10.3390/plants10020387

Dimopoulou, A., Theologidis, I., Benaki, D., Koukounia, M., Zervakou, A., Tzima, A., Diallinas, G., Hatzinikolaou, D. G., & Skandalis, N. (2021). Direct antibiotic activity of bacillibactin broadens the biocontrol range of *Bacillus amyloliquefaciens* MBI600. *mSphere, 6*(4), e0037621.

Dwivedi, D., & Johri, B. N. (2003). Antifungals from fluorescent *Pseudomonads*: Biosynthesis and regulation. *Current Science, 85*(12), 1693–1703.

Eichmann, R., Richards, L., & Schafer, P. (2021). Hormones as go-betweens in plant microbiome assembly. *The Plant Journal, 105*, 518–541.

Elhaissoufi, W., Ghoulam, C., Barakat, A., Zeroual, Y., & Bargaz, A. (2022). Phosphate bacterial solubilization: A key rhizosphere driving force enabling higher P use efficiency and crop productivity. *Journal of Advanced Research, 38*, 13–28.

Elliott, G. N., Chen, W. M., Chou, J. H., Wang, H. C., Sheu, S. Y., Perin, L., Reis, V. M., Moulin, L., Simon, M. F., Bontemps, C., et al. (2007). *Burkholderia phymatum* is a highly effective nitrogen-fixing symbiont of *Mimosa* spp. and fixes nitrogen ex planta. *The New Phytologist, 173*, 168–180.

Etesami, H., Emami, S., & Alikhani, H. A. (2017). Potassium solubilizing bacteria (KSB): Mechanisms, promotion of plant growth, and future prospects a review. *Journal of Soil Science and Plant Nutrition, 17*(4), 897–911.

Falkowski, P. G., Fenchel, T., & Delong, E. F. (2008). The microbial engines that drive Earth's biogeochemical cycles. *Science, 320*(5879), 1034–1039.

Farzand, A., Moosa, A., Zubair, M., Khan, A. R., Massawe, V. C., Tahir, H. A. S., Sheikh, T. M. M., Ayaz, M., & Gao, X. (2019). Suppression of *Sclerotinia sclerotiorum* by the induction of systemic resistance and regulation of antioxidant pathways in tomato using fengycin produced by *Bacillus amyloliquefaciens* FZB42. *Biomolecules, 9*, 613.

Fasusi, O. A., Cruz, C., & Babalola, O. O. (2021). Agricultural sustainability: Microbial biofertilizers in rhizosphere management. *Agriculture, 11*, 163.

Filgueiras, L., Silva, R., Almeida, I., Vidal, M., Baldani, J. I., & Meneses, C. H. S. G. (2020). *Gluconacetobacter diazotrophicus* mitigates drought stress in *Oryza sativa* L. *Plant and Soil, 451*, 57–73.

Fuentes-Lara, L., Medrano-Macías, J., Pérez-Labrada, F., Rivas-Martínez, E., García-Enciso, E., González-Morales, S., Juarez-Maldonado, A., Rincon-Sanchez, F., & Benavides-Mendoza, A. (2019). From elemental sulfur to hydrogen sulfide in agricultural soils and plants. *Molecules, 24*, 2282.

Gamalero, E., Lingua, G., & Glick, B. R. (2023). Ethylene, ACC, and the plant growth-promoting enzyme ACC deaminase. *Biology, 12*(8), 1043.

Garg, D., Patel, N., Rawat, A., & Rosado, A. S. (2024). Cutting edge tools in the field of soil microbiology. *Current Research in Microbial Sciences, 6*, 100226.

Ghimirey, V., Chaurasia, J., Acharya, N., Dhungana, R., & Chaurasiya, S. (2024). Biofertilizers: A sustainable strategy for enhancing physical, chemical, and biological properties of soil. *Innovations in Agriculture, 7*, 1–11.

Ghosh, S. K., Bera, T., & Chakrabarty, A. M. (2020). Microbial siderophore—A boon to agricultural sciences. *Biological Control*. https://doi.org/10.1016/j.biocontrol.2020.104214

Giri, S., Panwar, H., Singh, K., & Kumar, M. (2025). Microbial contributions to crop adaptation: Innovation for climate-resilient agriculture. In *Cutting edge technologies for developing future crop plants* (pp. 349–364). Springer Nature Singapore.

Glick, B. R. (2012). Plant growth-promoting bacteria: Mechanisms and applications. *Scientifica, 2012*, 963401.

Govindarajan, M., Kwon, S. W., & Weon, H. Y. (2008). Isolation, molecular characterization and growth-promoting activities of endophytic *Burkholderia* spp. *World Journal of Microbiology and Biotechnology, 24*, 2457–2464.

Grandy, A. S., & Wickings, K. (2010). Biological and biochemical pathways of litter decomposition and soil carbon stabilization. *Geochimica et Cosmochimica Acta, 74*(12), A351–A351.

Guo, K., Yang, J., Yu, N., Luo, L., & Wang, E. (2023). Biological nitrogen fixation in cereal crops: Progress, strategies, and perspectives. *Plant Communications, 4*, 100499.

Hamim, A., Boukeskasse, A., Ouhdouch, Y., Farrouki, A., Barrijal, S., Miche, L., Mrabet, R., Duponnois, R., & Hafidi, M. (2019). Phosphate solubilizing and PGR activities of ericaceous shrubs microorganisms isolated from Mediterranean forest soil. *Biocatalysis and Agricultural Biotechnology, 19*, 101128.

Hartman, K., & Tringe, S. G. (2019). Interactions between plants and soil shaping the root microbiome under abiotic stress. *The Biochemical Journal, 476*(19), 2705–2724. https://doi.org/10.1042/BCJ20180615

Hawkins, H. J., Cargill, R. I., Van Nuland, M. E., Hagen, S. C., Field, K. J., Sheldrake, M., et al. (2023). Mycorrhizal mycelium as a global carbon pool. *Current Biology, 33*(11), R560–R573.

He, S., Li, L., Lv, M., Wang, R., Wang, L., Yu, S., et al. (2024). PGPR: Key to enhancing crop productivity and achieving sustainable agriculture. *Current Microbiology, 81*(11), 377.

Herrera, H., Fuentes, A., Ortiz, J., Soto, J., da Silva Valadares, R. B., Salas Eljatib, C., & Arriagada, C. (2022). Root-associated endophytes isolated from juvenile *Ulex europaeus* (Fabaceae) plants colonizing rural areas in South-Central Chile. *Plant and Soil, 474*, 181–193.

Hershey, D. M., Lu, X., Zi, J., & Peters, R. J. (2014). Functional conservation of the capacity for *ent*-kaurene biosynthesis and an associated operon in certain rhizobia. *Journal of Bacteriology, 196*, 100–106.

Hider, R. C., & Kong, X. (2010). Chemistry and biology of siderophores. *Natural Product Reports, 27*(5), 637–657.

Hidri, I. R., Barea, J. M., Mahmoud, M. B., & Azcon, A. R. (2016). Impact of microbial inoculation on biomass accumulation by *Sulla carnosa* provenances, and in regulating nutrition, physiological and antioxidant activities of this species under non-saline and saline conditions. *Journal of Plant Physiology, 201*, 28–41.

Hiraoka, S., Yang, C.-C., & Iwasaki, W. (2016). Metagenomics and bioinformatics in microbial ecology: Current status and beyond. *Microbes and Environments, 31*(3), 204–212. https://doi.org/10.1264/jsme2.ME16024

Hoover, J. K., & Newman, J. A. (2004). Tritrophic interactions in the context of climate change: A model of grasses, cereal aphids and their parasitoids. *Global Change Biology, 10*, 1197–1208.

Ichriani, G. I., Nuraini, Y., & Handayanto. (2018). Formulation of biocharcompost and phosphate solubilizing fungi from oil palm empty fruit bunch to improve growth of maize in an ultisol of Central Kalimantan. *Journal of Ecological Engineering, 19*, 45–55.

Imam, A., Suman, S. K., Kanaujia, P. K., & Ray, A. (2022). Biological machinery for polycyclic aromatic hydrocarbons degradation: A review. *Bioresource Technology, 343*, 126121.

IPCC. (2023). Summary for policymakers. In *Climate change 2023: Synthesis report. Contribution of Working Groups I, II and III to the Sixth Assessment Report of the Intergovernmental Panel on Climate Change*. IPCC.

Iqbal, N., Khan, N. A., Ferrante, A., Trivellini, A., Francini, A., & Khan, M. I. R. (2017). Ethylene role in plant growth, development and senescence: Interaction with other phytohormones. *Frontiers in Plant Science, 8*, 475.

Jeyajothi, R., Subbalakshmi, L., & Nalliah, D. (2014). Effect of PPFM application on microbial population in transplanted rice. *Trends in Biosciences, 7*, 3573–3574.

Jhala, Y. K., Rajababu, V. V., Shelat Harsha, N., Patel, H. K., Patel, H. K., & Patel, K. T. (2014). Isolation and characterization of methane utilizing bacteria from wetland paddy ecosystem. *World Journal of Microbiology and Biotechnology, 30*(6), 1845–1860.

Jini, D., Ganga, V. S., Greeshma, M. B., Sivashankar, R., & Thirunavukkarasu, A. (2023). Sustainable agricultural practices using potassium-solubilizing microorganisms (KSMs) in coastal regions: A critical review on the challenges and opportunities. *Environment, Development and Sustainability, 26*, 13641–13664.

Joshi, H., Shourie, A., & Singh, A. (2020). Cyanobacteria as a source of biofertilizers for sustainable agriculture. In *Advances in cyanobacterial biology* (pp. 385–396). Academic.

Kalyan, V. S. R. K., Meena, S., Karthikeyan, S., & Jawahar, D. (2022). Isolation, screening, characterization, and optimization of bacteria isolated from calcareous soils for siderophore production. *Archives of Microbiology, 204*, 721.

Kang, S.-M., Khan, A. L., Waqas, M., Asaf, S., Lee, K.-E., Park, Y.-G., et al. (2019). Integrated phytohormone production by the plant growth promoting rhizobacterium *Bacillus tequilensis* SSB07 induced thermotolerance in soybean. *Journal of Plant Interactions, 14*(1), 416–423.

Kapadia, C., Sayyed, R. Z., El Enshasy, H. A., Vaidya, H., Sharma, D., et al. (2021). Halotolerant microbial consortia for sustainable mitigation of salinity stress, growth promotion, and mineral uptake in tomato plants and soil nutrient enrichment. *Sustainability, 13*(15), 8369.

Karataş, P., & Ayaz, F. (2025). Synthetic biology and application areas. *Discovery Biotechnology, 2*, 3.

Karnwal, A., Shrivastava, S., Al-Tawaha, A. R. M. S., Kumar, G., Kumar, A., & Kumar, A. (2023). PGPR-mediated breakthroughs in plant stress tolerance for sustainable farming. *Journal of Plant Growth Regulation, 43*(9), 1–17.

Kaur, J., & Pandove, G. (2023). Understanding the beneficial interaction of plant growth promoting rhizobacteria and endophytic bacteria for sustainable agriculture: A bio-revolution approach. *Journal of Plant Nutrition, 46*(12), 1–29.

Kerr, A., & Bullard, G. (2020). Biocontrol of crown gall by *Rhizobium radiobacter*: Challenges in biopesticide commercialization. *Agronomy, 10*, 1126.

Keswani, C., Singh, S. P., Cueto, L., García-Estrada, C., Mezaache-Aichour, S., Glare, T. R., Borriss, R., Singh, S. P., Blázquez, M. A., & Sansinenea, E. (2020). Auxins of microbial origin and their use in agriculture. *Applied Microbiology and Biotechnology, 104*, 8549–8565.

Khan, N., Martínez-Hidalgo, P., Humm, E. A., Maymon, M., Kaplan, D., & Hirsch, A. M. (2020a). Inoculation with a microbe isolated from the Negev Desert enhances corn growth. *Frontiers in Microbiology, 11*, 1149. https://doi.org/10.3389/fmicb.2020.01149

Khan, N., Bano, A., & Ali, S. (2020b). Crosstalk amongst phytohormones from planta and PGPR under biotic and abiotic stress. *Plant Growth Regulation, 90*, 189–203.

Khan, M. Y., Nadeem, S. M., Sohaib, M., Waqas, M. R., Alotaibi, F., Ali, L., Zahir, Z. A., & Al-Barakah, F. N. I. (2022). Potential of plant growth promoting bacterial consortium for improving the growth and yield of wheat under saline conditions. *Frontiers in Microbiology, 13*, 958522. https://doi.org/10.3389/fmicb.2022.958522

Khan, A., Singh, A. V., Gautam, S. S., Agarwal, A., Punetha, A., & Upadhayay, V. K. (2023). Microbial bioformulation: A microbial assisted biostimulating fertilization technique for sustainable agriculture. *Frontiers in Plant Science, 14*, 1270039. https://doi.org/10.3389/fpls.2023.1270039

Khasheii, B., Mahmoodi, P., & Mohammadzadeh, A. (2021). Siderophores: Importance in bacterial pathogenesis and applications in medicine and industry. *Microbiological Research, 250*, 126790.

Khati, P., Chaudhary, P., Gangola, S., & Sharma, A. (2017). Nanochitosan-induced growth of *Zea mays* with soil health maintenance. *3 Biotech, 7*(81), 1–9.

Khati, P., Chaudhary, P., Gangola, S., & Sharma, A. (2019a). Influence of nanozeolite on plant growth promotory bacterial isolates recovered from nanocompound-infested agriculture field. *Environment and Ecology, 37*(2), 521–527.

Khati, P., Sharma, A., Chaudhary, P., Singh, A. K., Gangola, S., & Kumar, R. (2019b). High-throughput sequencing approach to assess the impact of nanozeolite treatment on species richness and evenness of soil metagenome. *Biocatalysis and Agricultural Biotechnology, 20*, 101249.

Khati, P., Mishra, P. K., & Kant, L. (2024). Isolation, screening and characterization of drought tolerant plant growth promoting bacteria from Indian Himalayas. *Pantnagar Journal of Research, 22*(2), 261.

Khoshru, B., Mitra, D., Nosratabad, A. F., Reyhanitabar, A., Mandal, L., Farda, B., et al. (2023). Enhancing manganese availability for plants through microbial potential: A sustainable approach for improving soil health and food security. *Bacteria, 2*(3), 129–141.

Khoso, M. A., Wagan, S., Alam, I., Hussain, A., Ali, Q., Saha, S., et al. (2024). Impact of plant growth-promoting rhizobacteria (PGPR) on plant nutrition and root characteristics: Current perspective. *Plant Stress, 11*, 100341.

Kiprotich, K., Muoma, J., Omayio, D., Ndombi, T., & Wekesa, C. (2023). Molecular characterization and mineralizing potential of phosphorus solubilizing bacteria colonizing common bean (*Phaseolus vulgaris* L.) rhizosphere in Western Kenya. *International Journal of Microbiology, 2023*, 6668097.

Kleiner, M. (2019). Metaproteomics: Much more than measuring gene expression in microbial communities. *mSystems, 4*(3), e00115–e00119.

Kolb, S. (2009). Aerobic methanol-oxidizing bacteria in soil. *FEMS Microbiology Letters, 300*, 1–10.

Kong, H. G., Shin, T. S., Kim, T. H., & Ryu, C. M. (2018). Stereoisomers of the bacterial volatile compound 2, 3-butanediol differently elicit systemic defense responses of pepper against multiple viruses in the field. *Frontiers in Plant Science, 9*, 90.

Kour, D., Rana, K. L., Kaur, T., Sheikh, I., Yadav, A. N., Kumar, V., et al. (2020). Microbe-mediated alleviation of drought stress and acquisition of phosphorus in great millet (*Sorghum bicolour* L.) by drought-adaptive and phosphorus-solubilizing microbes. *Biocatalysis and Agricultural Biotechnology, 23*, 101501.

Kour, D., Negi, R., Khan, S. S., Kumar, S., Kaur, S., Kaur, T., et al. (2024). Microbes mediated induced systemic response in plants: A review. *Plant Stress, 11*, 100334.

Kumar, G., Arya, K., Verma, A., Bhatt, P., Khati, P., Gangola, S., Kumar, R., Sharma, A., & Singh, H. (2017). Bioremediation of petrol engine oil polluted soil using microbial consortium and wheat crop. *Journal of Pure and Applied Microbiology, 11*(3), 1583–1588.

Kumar, A., Singh, S., Gaurav, A. K., Srivastava, S., & Verma, J. P. (2020a). Plant growth-promoting bacteria: Biological tools for the mitigation of salinity stress in plants. *Frontiers in Microbiology, 11*, 12–16.

Kumar, N., Srivastava, P., Vishwakarma, K., Kumar, R., Kuppala, H., Maheshwari, S. K., & Vats, S. (2020b). The rhizobium–plant symbiosis: State of the art. In *Plant-microbe symbiosis* (pp. 1–20). Springer.

Kumar, M., Ahmad, S., & Singh, R. P. (2022a). Plant growth promoting microbes: Diverse roles for sustainable and ecofriendly agriculture. *Energy Nexus, 7*, 100133.

Kumar, S., Sindhu, S. S., & Kumar, R. (2022b). Biofertilizers: An ecofriendly technology for nutrient recycling and environmental sustainability. *Current Research in Microbial Sciences, 3*, 100094.

Kumar, A., Naroju, S. P., Kumari, N., Arsey, S., Kumar, D., Gubre, D. F., et al. (2024). The role of drought response genes and plant growth promoting bacteria on plant growth promotion under sustainable agriculture: A review. *Microbiological Research, 286*, 127827.

Kumari, S., Sharma, A., Chaudhary, P., & Khati, P. (2020). Management of plant vigour and soil health using two agri-usable nanocompounds and plant growth promotory rhizobacteria in fenugreek. *3 Biotech, 10*(461), 1–11.

Kumawat, C., Kumar, A., Parshad, J., Sharma, S. S., Patra, A., Dogra, P., Yadav, G. K., Dadhich, S. K., Verma, R., & Kumawat, G. L. (2022). Microbial diversity and adaptation under salt-affected soils: A review. *Sustainability, 14*(15), 9280. https://doi.org/10.3390/su14159280

Le Mire, G., Siah, A., Brisset, M. N., Gaucher, M., Deleu, M., & Jijakli, M. H. (2018). Surfactin protects wheat against *Zymoseptoria tritici* and activates both salicylic acid-and jasmonic acid-dependent defense responses. *Agriculture, 8*, 11.

Liu, Z., Dong, H., Cui, Y., Cong, L., & Zhang, D. (2020). Application of different types of CRISPR/Cas-based systems in bacteria. *Microbial Cell Factories, 19*, 172.

Lone, B. A., Karna, S. K. L., Ahmad, F., Shahi, N., & Pokharel, Y. R. (2018). CRISPR/Cas9 system: A bacterial tailor for genomic engineering. *Genetics Research International, 2018*, Article 3797214.

Ma, S., Zhu, L., Wang, J., Liu, X., Jia, Z., Li, C., Liu, J., Zeng, J., & Zhang, J. (2022). Arbuscular mycorrhizal fungi promote *Gleditsia sinensis* Lam. root growth under salt stress by regulating nutrient uptake and physiology. *Forests, 13*, 688.

Mahmuti, M., West, J. S., Watts, J., Gladders, P., & Fitt, B. D. L. (2009). Controlling crop disease contributes to both food security. *Journal of Applied Ecology, 7*, 189–202.

Mariotti, L., Scartazza, A., Curadi, M., Picciarelli, P., & Toffanin, A. (2021). *Azospirillum baldaniorum* Sp245 induces physiological responses to alleviate the adverse effects of drought stress in purple basil. *Plants, 10*, 1141.

Meena, V. S., Maurya, B. R., & Verma, J. P. (2014). Does a rhizospheric microorganism enhance K + availability in agricultural soils? *Microbiological Research, 169*, 337–347.

Meena, V. S., Maurya, B. R., Verma, J. P., Aeron, A., Kumar, A., Kim, K., & Bajpai, V. K. (2015). Potassium solubilizing rhizobacteria (KSR): Isolation, identification, and K-release dynamics from waste mica. *Ecological Engineering, 81*, 340–347.

Meena, V. S., Maurya, B. R., Verma, J. P., & Meena, R. S. (2016). *Potassium solubilizing microorganisms for sustainable agriculture*. Springer.

Meena, K. K., Sorty, A. M., Bitla, U. M., Choudhary, K., Gupta, P., Pareek, A., et al. (2017). Abiotic stress responses and microbe-mediated mitigation in plants: The omics strategies. *Frontiers in Plant Science, 8*, 172.

Mekureyaw, M. F., Pandey, C., Hennessy, R. C., Nicolaisen, M. H., Liu, F., Nybroe, O., et al. (2022). The cytokinin-producing plant beneficial bacterium *Pseudomonas fluorescens* G20–18 primes tomato (*Solanum lycopersicum*) for enhanced drought stress responses. *Journal of Plant Physiology, 270*, 153629.

Mercl, F., Garcia-Sanchez, M., Kulhanek, M., Kosnar, Z., Szakova, J., & Tlustos, P. (2020). Improved phosphorus fertilisation efficiency of wood ash by fungal strains *Penicillium* sp. PK112 and *Trichoderma harzianum* OMG08 on acidic soil. *Applied Soil Ecology, 147*, 103360.

Mia, M. B. (2022). Enhanced root morphogenesis in non-legumes as induced by rhizobacteria *Bacillus* spp. In *Bacilli in agrobiotechnology: Plant stress tolerance, bioremediation, and bioprospecting* (pp. 151–168). Springer.

Miglani, R., Parveen, N., Kumar, A., Ansari, M. A., Khanna, S., Rawat, G., Panda, A. K., Bisht, S. S., Upadhyay, J., & Ansari, M. N. (2022). Degradation of xenobiotic pollutants: An environmentally sustainable approach. *Meta, 12*(9), 818.

Miljaković, D., Marinković, J., & Balešević-Tubić, S. (2020). The significance of *Bacillus* spp. in disease suppression and growth promotion of field and vegetable crops. *Microorganisms, 8*, 1037.

Miller, M. L., Li, S., Eichhorn, C. D., Zheng, Y., & Du, L. (2023). Identification and biosynthetic study of the siderophore Lysochelin in the biocontrol agent *Lysobacter enzymogenes*. *Journal of Agricultural and Food Chemistry, 7*(19), 7418–7426.

Mishra, P., Mishra, J., & Arora, N. K. (2021a). Plant growth promoting bacteria for combating salinity stress in plants—Recent developments and prospects: A review. *Microbiological Research, 252*, 126861.

Mishra, S., Lin, Z., Pang, S., Zhang, W., Bhatt, P., & Chen, S. (2021b). Recent advanced technologies for the characterization of xenobiotic-degrading microorganisms and microbial communities. *Frontiers in Bioengineering and Biotechnology, 9*, 632059.

Mitra, D., Rodríguez, A. M. D., Cota, F. I. P., Khoshru, B., Panneerselvam, P., Moradi, S., et al. (2021). Amelioration of thermal stress in crops by plant growth-promoting rhizobacteria. *Physiological and Molecular Plant Pathology, 115*, 101679.

Mitra, D., Rani, A., Janeeshma, E., & Khoshru, B. (2025). Editorial: Microbial-mediated induced resistance: Interactive effects for improving crop health. *Frontiers in Microbiology, 16*, 1614435.

Mohan, A., & Kumar, B. (2019). Cyanobacterial diversity in agriculturally fertile soil of Patna and their population density. *Biotechnology and Applied Biochemistry, 5*(1), 15–32.

Moitra, A., Chakraborty, A., & Dam, B. (2024). CRISPR-Cas9 system: A potent tool to fight antibiotic resistance in bacteria. *Microbe, 5*, 100184.

Mok, M. (2019). Cytokinins and plant development—An overview. In *Cytokinins* (pp. 155–166). Springer.

Monnier, N., Cordier, M., Dahi, A., Santoni, V., Guenin, S., Clement, C., Sarazin, C., Penaud, A., Dorey, S., Cordelier, S., & Rippa, S. (2020). Semipurified rhamnolipid mixes protect *Brassica napus* against *Leptosphaeria maculans* early infections. *Phytopathology, 110*, 834–842.

Morcillo, R. J. L., & Manzanera, M. (2021). The effects of plant-associated bacterial exopolysaccharides on plant abiotic stress tolerance. *Meta, 11*(6), 337. https://doi.org/10.3390/metabo11060337

Mousa, W. K., Ghemrawi, R., Abu-Izneid, T., Al Ramadan, N., & Al Sheebani, F. (2024). The design and development of EcoBiomes: Multi-species synthetic microbial consortia inspired by natural desert microbiome to enhance the resilience of climate-sensitive ecosystems. *Heliyon, 10*(16), e36548.

Muhammad, M., Waheed, A., Wahab, A., Majeed, M., Nazim, M., Liu, Y.-H., Li, L., & Li, W.-J. (2024). Soil salinity and drought tolerance: An evaluation of plant growth, productivity, microbial diversity, and amelioration strategies. *Plant Stress, 11*, 100319.

Mukherjee, P. K., Hurley, J. F., Taylor, J. T., Puckhaber, L., Lehner, S., Druzhinina, I., Schumacher, R., & Kenerley, C. M. (2018). Ferricrocin, the intracellular siderophore of *Trichoderma virens*, is involved in growth, conidiation, gliotoxin biosynthesis and induction of systemic resistance in maize. *Biochemical and Biophysical Research Communications, 505*, 606–611.

Muthuselvan, I., & Balagurunathan, R. (2013). Siderophore production from *Azotobacter* sp. and its application as biocontrol agent. *International Journal of Current Research and Review, 5*(11), 30–42.

Nacoon, S., Jogloy, S., Riddech, N., Mongkolthanaruk, W., Kuyper, T. W., & Boonlue, S. (2020). Interaction between phosphate solubilizing bacteria and arbuscular mycorrhizal fungi on growth promotion and tuber inulin content of *Helianthus tuberosus* L. *Scientific Reports, 10*, 1–10.

Nadarajah, K., & Abdul Rahman, N. S. N. (2021). Plant-microbe interaction: Aboveground to belowground, from the good to the bad. *International Journal of Molecular Sciences, 22*(19), 10388.

Nagel, R., Bieber, J. E., Schmidt-Dannert, M. G., Nett, R. S., & Peters, R. J. (2018). A third class: Functional gibberellin biosynthetic operon in beta-proteobacteria. *Frontiers in Microbiology, 9*, 2916.

Nannipieri, P., Ascher-Jenull, J., Ceccherini, M. T., Pietramellara, G., Renella, G., & Schloter, M. (2020). Beyond microbial diversity for predicting soil functions: A mini review. *Pedosphere, 30*, 5–17.

Naorem, A. K., Udayana, S. K., Kumar, G., & Mukherjee, A. K. (2017). Intervention of multi-tasking PGPR in pulse production under stress conditions. *Innovation in Farming, 2*(1), 41–44.

Narwal, E., Choudhary, J., Jat, N. K., Meena, A. L., Ghasal, P. C., Dutta, D., Mishra, R. P., Saritha, M., Meena, L. K., Bhanu, C., Singh, R., Kumar, G. C., Panwar, A. S., & Choudhary, M. (2023). Plant growth-promoting microbes: Key players in organic agriculture. In R. Mawar, R. Z. Sayyed, S. K. Sharma, & K. S. Sattiraju (Eds.), *Plant growth promoting rhizobacteria (PGPR): Prospects for sustainable agriculture* (Vol. 22, pp. 139–160). Springer Singapore.

Nehra, V., & Choudhary, M. (2015). A review on plant growth promoting rhizobacteria acting as bioinoculants and their biological approach towards the production of sustainable agriculture. *Journal of Applied and Natural Science, 7*, 540–556.

Nett, R. S., Bender, K. S., & Peters, R. J. (2022). Production of the plant hormone gibberellin by rhizobia increases host legume nodule size. *The ISME Journal, 16*(7), 1809–1817.

Nkongolo, K. K., & Narendrula-Kotha, R. (2020). Advances in monitoring soil microbial community dynamic and function. *Journal of Applied Genetics, 61*(2), 249–263. https://doi.org/10.1007/s13353-020-00549-5

Nunes, P. S. O., Lacerda-Junior, G. V., Mascarin, G. M., Guimarães, R. A., Medeiros, F. H. V., Arthurs, S., & Bettiol, W. (2024). Microbial consortia of biological products: Do they have a future? *Biological Control, 188*, 105439.

Olaniyan, F. T., Alori, E. T., Adekiya, A. O., Ayorinde, B. B., Daramola, F. Y., Osemwegie, O. O., & Babalola, O. O. (2022). The use of soil microbial potassium solubilizers in potassium nutrient availability in soil and its dynamics. *Annales de Microbiologie, 72*(1), 45.

Oleńska, E., Grab, K., Kaczmarek, A., Kiersnowska, M., & Kozłowska, M. (2020). Beneficial features of plant growth-promoting rhizobacteria for improving plant growth and health in challenging conditions: A methodical review. *Sci Total Environ, 743*, 140682.

Omoboye, O. O., Oni, F. E., Batool, H., Yimer, H. Z., De Mot, R., & Hofte, M. (2019). *Pseudomonas* cyclic lipopeptides suppress the rice blast fungus *Magnaporthe oryzae* by induced resistance and direct antagonism. *Frontiers in Plant Science, 10*, 901.

Oremland, R. S., & Culbertson, C. W. (1992). Importance of methane-oxidizing bacteria in the methane budget as revealed by the use of a specific inhibitor. *Nature, 356*, 421–423.

Orozco-Mosqueda, M. C., Santoyo, X., & Glick, B. R. (2023). Recent advances in the bacterial phytohormone modulation of plant growth. *Plants, 12*, 606.

Palacio-Rodríguez, R., Coria-Arellano, J. L., López-Bucio, J., et al. (2017). Halophilic rhizobacteria from *Distichlis spicata* promote growth and improve salt tolerance in heterologous plant hosts. *Symbiosis, 73*, 179–189.

Pan, L., & Cai, B. (2023). Phosphate-solubilizing bacteria: Advances in their physiology, molecular mechanisms and microbial community effects. *Microorganisms, 11*(12), 2904.

Pan, W., Lu, Q., Xu, Q.-R., Zhang, R.-R., Li, H.-Y., Yang, Y.-H., et al. (2019). Abscisic acid-generating bacteria can reduce Cd concentration in pakchoi grown in Cd-contaminated soil. *Ecotoxicology and Environmental Safety, 177*, 100–107.

Panday, S. C., Parihar, M., Meena, R. P., Chaudhary, M., Meena, V. S., Mondal, T., Khati, P., Singh, A. K., Bisht, J. K., Kant, L., & Singh, R. D. (2024). Long-term fertilization strategies for improving productivity, profitability and water use efficiency of soybean–wheat cropping system. *The Journal of Agricultural Science, 162*(5), 1–14.

Pankievicz, V. C. S., Irving, T. B., Maia, L. G. S., & Ane, J. M. (2019). Are we there yet? The long walk towards the development of efficient symbiotic associations between nitrogen-fixing bacteria and nonleguminous crops. *BMC Biology, 17*, 99.

Parihar, M., Chitara, M., Khati, P., Kumari, A., Mishra, P. K., Rakshit, A., & Jatav, S. S. (2020a). Arbuscular mycorrhizal fungi: Abundance, interaction with plants and potential biological applications. In *Advances in plant microbiome and sustainable agriculture: Diversity and biotechnological applications* (pp. 105–143). Springer Nature Singapore.

Parihar, M., Rakshit, A., Meena, V. S., Gupta, V. K., Rana, K., Choudhary, M., Tiwari, G., Mishra, P. K., Pattanayak, A., & Bisht, J. K. (2020b). The potential of arbuscular mycorrhizal fungi in C cycling: A review. *Archives of Microbiology, 202*, 1581–1596.

Parihar, M., Mishra, P. K., Paschapur, A., Kumar, A., Khati, P., Joshi, D., Bisht, J., & Kant, L. (2025). Elevation gradients and soil characteristics shape arbuscular mycorrhizal fungal communities in the Indian mid-Himalaya. *Nordic Journal of Botany, e04760*, 1–13.

Parmar, P., & Sindhu, S. S. (2013). Potassium solubilization by rhizosphere bacteria: Influence of nutritional and environmental conditions. *Journal of Microbiological Research, 3*(1), 25–31.

Patil, P. D., Gude, V. G., Mannarswamy, A., Deng, S., Cooke, P., Munson-McGee, S., Rhodes, I., Lammers, P., & Nirmalakhandan, N. (2011). Optimization of direct conversion of wet algae to biodiesel under supercritical methanol conditions. *Bioresource Technology, 102*, 118–122.

Pfromm, P. H., Amanor-Boadu, V., & Nelson, R. (2011). Sustainability of algae derived biodiesel: A mass balance approach. *Bioresource Technology, 102*, 1185–1193.

Phale, P. S., Sharma, A., & Gautam, K. (2019). Microbial degradation of xenobiotics like aromatic pollutants from the terrestrial environments. In *Pharmaceuticals and personal care products: Waste management and treatment technology* (pp. 259–278). Butterworth-Heinemann.

Pieterse, C. M., Zamioudis, C., Berendsen, R. L., Weller, D. M., Van Wees, S. C., & Bakker, P. A. (2014). Induced systemic resistance by beneficial microbes. *Annual Review of Phytopathology, 52*, 347–375.

Pinaeva, L. G., & Noskov, A. S. (2024). Biodegradable biopolymers: Real impact to environment pollution. *Science of the Total Environment, 947*, 174445.

Prasad, R., & Shivay, Y. S. (2016). Sulphur in soil in plant and human nutrition. *Proceedings of the National Academy of Sciences, India, Section B: Biological Sciences, 88*(2), 429–434.

Priya, R., Rath, P., Sharma, D., Suratramka, A., & Bhardwaj, L. K. (2025). Microbial inoculants: Application in the management of metal stress. In *Microbial inoculants* (pp. 95–109). Academic.

Rai, S., & Shukla, N. (2020). Biofertilizer: An alternative of synthetic fertilizers. *Plant Archives, 20*(2), 1374–1379.

Rawat, P., Das, S., Shankhdhar, D., & Shankhdhar, S. C. (2021). Phosphate-solubilizing microorganisms: Mechanism and their role in phosphate solubilization and uptake. *Journal of Soil Science and Plant Nutrition, 21*(1), 49–68.

Rebello, S., Nathan, V. K., Sindhu, R., Binod, P., Awasthi, M. K., & Pandey, A. (2021). Bioengineered microbes for soil health restoration: Present status and future. *Bioengineered, 12*(2), 12839–12853.

Reiss, A., & Jørgensen, L. N. (2017). Biological control of yellow rust of wheat (*Puccinia striiformis*) with Serenade® ASO (*Bacillus subtilis* strain QST713). *Crop Protection, 93*, 1–8.

Rillig, M. C. (2004). Arbuscular mycorrhizae, glomalin, and soil aggregation. *Canadian Journal of Soil Science, 84*(4), 355–363.

Rillig, M. C., Wright, S. F., Nichols, K. A., Schmidt, W. F., & Torn, M. S. (2001). Large contribution of arbuscular mycorrhizal fungi to soil carbon pools in tropical forest soils. *Plant and Soil, 233*(2), 167–177.

Saharan, B. S., & Nehra, V. (2011). Plant growth promoting rhizobacteria: A critical review. *Life Science and Medical Research, 21*, 1–30.

Sahu, K. P., Kumar, A., Patel, A., Kumar, M., Gopalakrishnan, S., Prakash, G., Rathour, R., & Gogoi, R. (2021). Rice blast lesions: An unexplored phyllosphere microhabitat for novel antagonistic bacterial species against *Magnaporthe oryzae*. *Microbial Ecology, 81*, 731–745.

Samantaray, A., Chattaraj, S., Mitra, D., Ganguly, A., Kumar, R., Gaur, A., & Thatoi, H. (2024). Advances in microbial based bio-inoculum for amelioration of soil health and sustainable crop production. *Current Research in Microbial Sciences, 7*, 100251.

Santoyo, G., Moreno-Hagelsieb, G., del Monte-Chávez, P., & Glick, B. R. (2021). Rhizosphere colonization determinants by plant growth-promoting rhizobacteria (PGPR). *Biology, 10*(6), 475.

Satam, H., Joshi, K., Mangrolia, U., Waghoo, S., Zaidi, G., Rawool, S., et al. (2023). Next-generation sequencing technology: Current trends and advancements. *Biology (Basel), 12*(7), 997.

Sathya, A., Rehman, V., Srinivas, V., Kudapa, H., & Gopalakrishnan, S. (2024). Efficacy of two different microbial consortia on salinity tolerance in chickpea: An in-planta evaluation on biochemical, histochemical, and genomic aspects. *3 Biotech, 14*(11), 285.

Saud, S., Wang, D., & Fahad, S. (2022). Improved nitrogen use efficiency and greenhouse gas emissions in agricultural soils as producers of biological nitrification inhibitors. *Frontiers in Plant Science, 13*, 854195.

Sayyed, R. Z., Chincholkar, S. B., Reddy, M. S., Gangurde, N. S., & Patel, P. R. (2013). Siderophore producing PGPR for crop nutrition and phytopathogen suppression. In D. K. Maheshwari (Ed.), *Bacteria in agrobiology: Disease management* (pp. 449–471). Springer.

Scheublin, T. R., & Van Der Heijden, M. G. (2006). Arbuscular mycorrhizal fungi colonize nonfixing root nodules of several legume species. *The New Phytologist, 172*(4), 732–738.

Schils, R., Kuikman, P., Liski, J., Van Oijen, M., Smith, P., Webb, J., Alm, J., Somogyi, Z., Van Den Akker, J., Billett, M., Emmett, B., Evans, C., Lindner, M., Palosuo, T., Bellamy, P., Alm, J., Jandl, R., & Hiederer, R. (Eds.). (2008). *Service contract: Review of existing information on the interrelations between soil and climate change*. Alterra.

Scott, D., Hall, C. M., Rushton, B., & Gössling, S. (2024). A review of the IPCC Sixth Assessment and implications for tourism development and sectoral climate action. *Journal of Sustainable Tourism, 32*(9), 1725–1742.

Shaffique, S., Khan, M. A., Imran, M., Kang, S.-M., Park, Y.-S., Shabir, H. W., & Lee, I.-J. (2022). Research progress in the field of microbial mitigation of drought stress in plants. *Frontiers in Plant Science, 13*, 870626.

Sharma, B. S., Sayyed, R. Z., Trivedi, H. M., & Gobi, A. T. (2013). Phosphate solubilizing microbes: Sustainable approach for managing phosphorus deficiency in agricultural soils. *Springerplus, 2*, 587.

Sharma, P., Pandey, A. K., Kim, S. H., Singh, S. P., Chaturvedi, P., & Varjani, S. (2021). Critical review on microbial community during in-situ bioremediation of heavy metals from industrial wastewater. *Environmental Technology and Innovation, 24*, 101826.

Sharma, P., Sangwan, S., Kaur, H., Patra, A., Anamika, & Mehta, S. (2023). Diversity and evolution of nitrogen fixing bacteria. In N. Singh, A. Chattopadhyay, & E. Lichtfouse (Eds.), *Sustainable agriculture reviews* (Vol. 60, pp. 95–120). Springer.

Sharma, R., Sindhu, S. S., & Glick, B. R. (2024). Potassium solubilizing microorganisms as potential biofertilizer: A sustainable climate-resilient approach to improve soil fertility and crop production in agriculture. *Journal of Plant Growth Regulation, 43*(8), 2503–2535.

Shen, N., Li, S., Li, S., Zhang, S., & Jiang, M. (2022). The siderophore-producing bacterium *Bacillus simensis* Gxun-6, has an antifungal activity against *Fusarium oxysporum* and promotes the growth of banana. *Egyptian Journal Of Biological Pest Control, 32*, 34.

Shinde, A. H., Sharma, A., Doshi, S., Kumar, M. A., & Haldar, S. (2022). Isolation and screening of sulfur-oxidizing bacteria from coast of Bhavnagar, India, and formulation of consortium for bioremediation. *Environmental Science and Pollution Research, 29*(36), 1–14.

Siedliska, A., Piotr, B., Joanna, P., Monika, Z., & Jaromir, K. (2021). Identification of plant leaf phosphorus content at different growth stages based on hyperspectral reflectance. *BMC Plant Biology, 21*, 28.

Singh, J. S., Kumar, A., Rai, A. N., & Singh, D. P. (2016). Cyanobacteria: A precious bio-resource in agriculture, ecosystem, and environmental sustainability. *Frontiers in Microbiology, 7*, 529.

Singh, P., Singh, R. K., Zhou, Y., Wang, J., Jiang, Y., Shen, N., Wang, Y., Yang, L., & Jiang, M. (2022). Unlocking the strength of plant growth promoting *Pseudomonas* in improving crop productivity in normal and challenging environments: A review. *Journal of Plant Interactions, 17*, 220–238.

Singh, D., Thapa, S., Yadav, J., Singh, D., Chakdar, H., Kumar, M., & Saxena, A. K. (2023). Deciphering the mechanisms of microbe mediated drought stress alleviation in wheat. *Acta Physiologiae Plantarum, 45*(6), 81.

Six, J., Frey, S. D., Thiet, R. K., & Batten, K. M. (2006). Bacterial and fungal contributions to carbon sequestration in agroecosystems. *Soil Science Society of America Journal, 70*(2), 555.

Smith, S. E., & Read, D. J. (2010). *Mycorrhizal symbiosis*. Academic.

Soares, E. V. (2022). Perspective on the biotechnological production of bacterial siderophores and their use. *Applied Microbiology and Biotechnology, 106*, 3985–4004.

Solomon, W., Janda, T., & Molnár, Z. (2024). Unveiling the significance of rhizosphere: Implications for plant growth, stress response, and sustainable agriculture. *Plant Physiology and Biochemistry, 206*, 108290.

Sonawane, J. M., Rai, A. K., Sharma, M., Tripathi, M., & Prasad, R. (2022). Microbial biofilms: Recent advances and progress in environmental bioremediation. *Sci Total Environ, 824*, 153843.

Soumare, A., Diedhiou, A. G., Thuita, M., Hafidi, M., Ouhdouch, Y., Gopalakrishnan, S., & Kouisni, L. (2020). Exploiting biological nitrogen fixation: A route towards a sustainable agriculture. *Plants, 9*, 1011.

Stefen, D. L. V., et al. (2022). How phytohormones synthesized by PGPR affect plant growth? In *Secondary metabolites and volatiles of PGPR in plant-growth promotion* (pp. 119–131). Springer.

Subba Rao, N. S. (1982). *Biofertilizers in agriculture*. Oxford/IBH Publishing Co.

Suenaga, H. (2012). Targeted metagenomics: A high-resolution metagenomics approach for specific gene clusters in complex microbial communities. *Environmental Microbiology, 14*(1), 13–22. https://doi.org/10.1111/j.1462-2920.2011.02438.x

Supikova, K., Kosinova, A., Vavrusa, M., Koplikova, L., François, A., Pospisil, J., Zatloukal, M., Wever, R., Hartog, A., & Gruz, J. (2022). Sulfated phenolic acids in plants. *Planta, 255*, 124.

Timofeeva, A. M., Galyamova, M. R., & Sedykh, S. E. (2022). Bacterial siderophores: Classification, biosynthesis, perspectives of use in agriculture. *Plants, 11*(22), 3065.

Tomer, S., Khati, P., Suyal, D. C., Khan, F., Perveen, K., & Barasarathi, J. (2024). Relative multi-beneficial effect of microorganisms on plant health of chickpea (*Cicer arietinum* L. var. PG-186). *Frontiers in Microbiology, 15*, 1452553.

Usman, M., Balsalobre-Lorente, D., Jahanger, A., & Ahmad, P. (2022). Pollution concern during globalization mode in financially resource-rich countries: Do financial development, natural resources, and renewable energy consumption matter? *Renewable Energy, 183*, 90–102.

Vaghela, N., & Gohel, S. (2023). Medicinal plant-associated rhizobacteria enhance the production of pharmaceutically important bioactive compounds under abiotic stress conditions. *Journal of Basic Microbiology, 63*, 308–325.

Vejan, P., Abdullah, R., Khadiran, T., Ismail, S., & Nasrulhaq Boyce, A. (2016). Role of plant growth promoting rhizobacteria in agricultural sustainability—A review. *Molecules, 21*(5), 573.

Verma, J. P., Yadav, J., Tiwari, K. N., & Kumar, A. (2011). Impact of plant growth promoting rhizobacteria on crop production. *International Journal of Agricultural Research, 6*(3), 222–232.

Verma, P., Chandra, P., Rai, A. K., Kumar, A., Prajapat, K., Sundha, P., Basak, N., Mann, A., Saharan, B. S., Bedwal, S., & Yadav, R. K. (2023). Native rhizobacteria suppresses spot blotch disease, improves growth and yield of wheat under salt–affected soils. *Plant Stress, 10*, 100234.

Vessey, J. K. (2003). Plant growth-promoting rhizobacteria as biofertilizers. *Plant and Soil, 255*(2), 571–586.

Wani, A. K., Qadir, F., Elboughdiri, N., Rahayu, F., Pranowo, D., Martasari, C., & Singh, R. (2025). Metagenomics and plant-microbe symbioses: Microbial community dynamics, functional roles in carbon sequestration, nitrogen transformation, sulfur and phosphorus mobilization for sustainable soil health. *Biotechnology Advances, 82*, 108580.

Wilson, G. W. T., Rice, C. W., Rillig, M. C., Springer, A., & Hartnett, D. C. (2009). Soil aggregation and carbon sequestration are tightly correlated with the abundance of arbuscular mycorrhizal fungi: Results from long-term field experiments. *Ecology Letters, 12*(5), 452–461.

Wu, H., Cui, H., Fu, C., Li, R., Qi, F., Liu, Z., Yang, G., Xiao, K., & Qiao, M. (2024). Unveiling the crucial role of soil microorganisms in carbon cycling: A review. *Sci Total Environ, 909*, 168627.

Xiao, C., Chi, R., Pan, X., Liu, F., & He, J. (2013). Rock phosphate solubilization by four yeast strains. *Annales de Microbiologie, 63*, 173–178.

Xie, X., Turatsinze, A. N., Liu, Y., Chen, G., Yue, L., Ye, A., & Wang, R. (2025). Bioinoculant substitution enhances rhizosphere soil quality and maize growth by modulating microbial communities and host gene expression in alkaline soils. *Microbiological Research, 279*, 128194.

Xu, Y., Xu, X., Li, J., Guo, X., Gong, H., Ouyang, Z., & Mathijs, E. (2023). Excessive synthetic fertilizers elevate greenhouse gas emissions of smallholder-scale staple grain production in China. *Journal of Cleaner Production, 430*, 139720.

Xue, Y., Zhang, Y., Huang, K., Wang, X., Xing, M., Xu, Q., & Guo, Y. (2023). A novel biocontrol agent *Bacillus velezensis* K01 for management of gray mold caused by *Botrytis cinerea*. *AMB Express, 13*, 91.

Yang, L., Liu, Y., Cao, X., Zhou, Z., Wang, S., Xiao, J., & Zhou, Y. (2017). Community composition specificity and potential role of phosphorus solubilizing bacteria attached on the different bloom-forming cyanobacteria. *Microbiological Research, 205*, 59–65.

Zahran, H. H. (1999). *Rhizobium*-legume symbiosis and nitrogen fixation under severe conditions and in an arid climate. *Microbiology and Molecular Biology Reviews, 63*(4), 968–989.

Zeiner, C. A., Kisch, M. N., Lynch, E. D., Shrestha, P., & Small, G. E. (2024). Soil microbial activity profiles associated with organic compost fertilizers in an urban garden. *Urban Agriculture & Regional Food Systems, 9*(1), e20059.

Zhang, C., Yu, Z., Zhang, M., Li, X., Wang, M., Li, L., et al. (2022). *Serratia marcescens* PLR enhances lateral root formation through supplying PLR-derived auxin and enhancing auxin biosynthesis in *Arabidopsis*. *Journal of Experimental Botany, 73*(11), 3711–3725.

Zhang, Y., Ding, S., Xue, X., Hu, X., Rehman, O. U., Liu, W., & Huo, S. (2025). Combined application of nitrogen-fixing cyanobacteria enhances rice growth and nutritional quality in saline environments. *Algal Research, 89*, 104061.

Zhou, C., Zhu, L., Xie, Y., Li, F., Xiao, X., Ma, Z., et al. (2017). *Bacillus licheniformis* SA03 confers increased saline-alkaline tolerance in chrysanthemum plants by induction of abscisic acid accumulation. *Frontiers in Plant Science, 8*, 1143.

Zulkefli, N. S., & Hwang, S. J. (2020). Heterocyst development and diazotrophic growth of *Anabaena variabilis* under different nitrogen availability. *Life, 10*(11), 279.

Zúniga, C., Morales, M., Borgne, L. S., & Revah, S. (2011). Production of poly-β-hydroxybutyrate (PHB) by *Methylobacterium organophilum* isolated from a methanotrophic consortium in a two-phase partition bioreactor. *Journal of Hazardous Materials, 190*(1–3), 876–882.

Leveraging Cyanobacteria for Alleviating Climate Change-Induced Abiotic Stress in Crop Plants

Vijaya Rani, Pranita Jaiswal, Himanshu Mahawar, Priyanka Khati, Anju Arora, and Innocent Ojeba Musa

Abstract

Climate change is intensifying abiotic stresses such as drought, salinity, heat, and oxidative stress, which negatively impact agricultural productivity and threaten food security. Cyanobacteria, or blue-green algae, are photosynthetic microorganisms that thrive in extreme environments and exhibit resilience to these stresses. This chapter explores the mechanisms through which cyanobacteria alleviate abiotic stress in plants and their potential agricultural applications. Cyanobacteria enhance drought tolerance by producing extracellular polysaccharides that retain soil water and phytohormones such as auxins and gibberellins that promote plant growth under water-limited conditions. Their nitrogen-fixing ability improves soil fertility, particularly in nutrient-depleted soils under stress. Cyanobacteria also help plants tolerate salinity by reducing ion toxicity, stabilizing cellular structures, and promoting growth. Furthermore, they possess heat-shock proteins and antioxidants that protect plants from oxidative damage caused by heat stress. Continued research is needed to unlock the full potential of cyanobacteria as a sustainable solution for mitigating climate

V. Rani (✉) · P. Jaiswal · A. Arora
CCUBGA, Division of Microbiology, ICAR-Indian Agricultural Research Institute, New Delhi, India

H. Mahawar
Directorate of Weed Research (ICAR), Jabalpur, India

P. Khati
Crop Production Division, ICAR-VPKAS, Almora, Uttarakhand, India

I. O. Musa
Department of Microbiology, Skyline University Nigeria, Kano, Kano State, Nigeria

G. Bhandari et al. (eds.), *Microbial Resilience*,
https://doi.org/10.1007/978-981-95-8883-1_15

change-induced abiotic stress and ensuring resilient agricultural systems for sustained crop production.

Keywords

Cyanobacteria · Climate change · Carbon sequestration · Abiotic stress · N_2-fixation

1 Introduction

Climate change is exacerbating abiotic stresses such as drought, salinity, and heat, significantly impacting agricultural productivity worldwide, with yield losses exceeding 50% for major crops. Addressing these challenges is crucial for ensuring food security for a growing global population, projected to reach 9 billion by 2050 (Tilman et al., 2011). The conventional strategies involving breeding and genetic engineering have faced limitations in producing crops that can withstand the increasing severity and frequency of these environmental stresses (Villalobos-López et al., 2022). The exploration of alternative, sustainable solutions is therefore imperative. The complex interplay of elevated atmospheric CO_2 concentrations, increasing average temperatures, and the heightened incidence of extreme weather phenomena inherent to climate change presents a formidable challenge to global food security, indirectly impacting human health through diminished crop yields and nutritional quality. Innovative and sustainable crop production efforts are required to ensure optimized resilience under climate change conditions. The limited availability of arable land, coupled with soil degradation, water scarcity, and the proliferation of pathogens, further compounds these issues, necessitating a paradigm shift in agricultural practices.

In response, there is a growing focus on nature-based microbial solutions to enhance crop resilience under climate stress. Among these, plant growth-promoting rhizobacteria (PGPR) such as *Bacillus*, *Pseudomonas*, and *Trichoderma* spp. have demonstrated considerable success in improving plant growth, nutrient uptake, stress tolerance, and soil health (Chaudhary et al., 2021a; Khati et al., 2018). PGPR applications have been significantly boosted through the integration of nanomaterials, such as nanozeolite, nanochitosan, and nanogypsum, which enhance microbial colonization and efficiency under field conditions (Chaudhary et al., 2021b, c; Khati et al., 2017, 2019a). Additionally, metagenomic approaches have revealed shifts in rhizospheric microbial diversity in response to these interventions, providing deeper insights into ecosystem-level impacts (Khati et al., 2019b; Chaudhary et al., 2021d). The synergistic effects of these microbes have also been noted in improving plant performance in contaminated or degraded soils through mechanisms such as biodegradation, enzyme production, and nutrient cycling (Bhatt et al., 2016; Gangola et al., 2018; Kumar et al., 2017).

Recent scientific advances highlight the potential of cyanobacteria as a promising bio-inoculant to enhance crop resilience against abiotic stresses. Cyanobacteria's ability to improve plant growth and mitigate stress has recently sparked significant interest. In this chapter, we explore the ecological roles, physiological mechanisms, and biotechnological applications of cyanobacteria in alleviating climate change-induced abiotic stress in crops. With increasing evidence supporting their multifunctionality, cyanobacteria represent a potent, sustainable tool in the ongoing efforts to secure agricultural productivity in the face of a changing climate.

2 Cyanobacteria: Evolutionary Significance and Ecological Importance

Cyanobacteria, also known as blue-green algae, are photosynthetic prokaryotes with a remarkable capacity to thrive in diverse and extreme environments, including saline, arid, and high-temperature conditions. Evolving over 2.5 billion years ago, they were instrumental in the "Great Oxidation Event," fundamentally transforming Earth's atmosphere and paving the way for aerobic life forms. Their metabolic versatility and adaptability have allowed them to colonize nearly every illuminated environment on Earth (Mehdizadeh Allaf & Peerhossaini, 2022). Cyanobacteria are major primary producers in many ecosystems, converting light energy and carbon dioxide into organic matter and releasing oxygen. In marine environments, picocyanobacteria such as *Prochlorococcus* and *Synechococcus* are numerically dominant and contribute significantly to global carbon fixation (Flombaum et al., 2013). In freshwater systems, larger filamentous and colonial forms can dominate the phytoplankton. Beyond their ecological significance, various cyanobacterial strains hold considerable promise for sustainable agriculture, serving as biofertilizers, biostimulants, and agents for soil improvement. Cyanobacteria possess unique capability to fix N_2 and exert beneficial effects on both crops and soil. *Anabaena*, *Aulosira*, *Nostoc*, *Calothrix*, and *Plectonema* are some of the major nitrogen-fixing cyanobacteria used as biofertilizers in the paddy fields and help maintain long-term soil fertility and sustainability. *Anabaena, Aulosira, Nostoc* and *Calothrix* are filamentous cyanobacteria with specialized thickened, non-photosynthetic cells called heterocyst that harbours nitrogenase enzyme essential for N_2- fixation. Members of the *Plectonema* genus donot have heterocysts and require microaerobic conditions for N_2- fixation. Their capacity to fix atmospheric nitrogen, solubilize phosphorus, and produce siderophores enhances nutrient availability in the soil, promoting plant growth and reducing the dependence on synthetic fertilizers, thereby fostering sustainable agricultural practices. In terrestrial ecosystems, particularly in arid and semi-arid regions, cyanobacteria form intricate microbial mats or biological soil crusts. These crusts contribute to soil aggregation, improve water infiltration and retention, and prevent wind and water erosion, playing a vital role in ecosystem stability and soil health (Warren et al., 2019). Numerous genera of cyanobacteria—spanning five orders, Chroococcales, Nostocales, Oscillatoriales, Pleurocapsales, and Synechococcales—are actively involved in the formation of biological soil crusts. Cyanobacteria are recognized as one of the most important and abundant photoautotrophs in many arid land crust formations occurring at or just below the soil surface (Colesie et al., 2016). Some filamentous cyanobacteria species also secrete exopolysaccharide compounds that effectively aggregate soil particles and thus improve the structure and stability of the crust (Rossi & De Philippis, 2015). Cyanobacteria establish beneficial symbiotic relationships with various organisms, including plants (e.g. *Azolla-Anabaena* association in rice paddies), fungi (forming lichens), protists, and even some animals. In these partnerships, cyanobacteria typically provide fixed nitrogen or photosynthetically derived carbohydrates to their hosts (Adams & Duggan, 2008). Their robust nature and ability to fix nitrogen enable cyanobacteria to colonize harsh, barren environments such as newly formed volcanic rock or disturbed lands, initiating primary succession and contributing to soil development.

Their inherent tolerance mechanisms, coupled with their ability to produce a range of beneficial metabolites, make them a compelling candidate for alleviating abiotic stress in crop plants. Cyanobacteria are photosynthetic prokaryotes recognized for their ubiquitous presence and resilience in diverse ecological niches and represent a potentially transformative tool in sustainable agriculture. They harbour a complex repertoire of biochemical pathways enabling the synthesis of phytohormones such as auxins, cytokinins, and gibberellins, all of which play critical roles in modulating plant growth and development. Moreover, cyanobacteria exhibit remarkable adaptability, allowing them to flourish in extreme environments characterized by high salinity, drought, and temperature fluctuations, thus offering a natural solution for improving crop resilience in the face of climate change.

Beyond their nutritional contributions, cyanobacteria produce a variety of bioactive compounds, including exopolysaccharides, osmoprotectants (e.g. glycine betaine, trehalose), and antioxidants (Reddy et al., 2018). These metabolites have been shown to enhance plant growth, improve stress tolerance, and alleviate the detrimental effects of abiotic stress on crops. Cyanobacteria's ability to produce these compounds in response to environmental stress, combined with their nitrogen-fixing capabilities, makes them an attractive tool for promoting crop resilience.

3 Abiotic Stress in Crop Plants

The world's growing population places immense pressure on agricultural systems to produce sufficient food. However, crop production is continuously challenged by a range of environmental stressors, broadly categorized as biotic (living organisms) and abiotic (non-living physical and chemical factors). Among these, abiotic stresses are responsible for significant crop losses globally, often reducing average yields by more than 50% for most major crops. These stresses are becoming more prevalent and severe due to the accelerating pace of climate change, demanding a deeper understanding of their impact on plant physiology and the development of effective mitigation strategies. Abiotic stress, encompassing factors such as drought, salinity, temperature extremes, and nutrient deficiencies, poses a significant threat to global food security by adversely impacting crop yield and plant health. Abiotic stresses are diverse and often occur in combination, leading to complex plant responses.

Drought, or water deficit, is arguably the most widespread and damaging abiotic stress globally. It occurs when the water supply to the plant root system is insufficient to meet transpirational demands, leading to a reduction in water potential, turgor loss, and impaired physiological processes (Chaves et al., 2009). It leads to reduced photosynthesis, oxidative stress, growth inhibition, premature senescence and yield loss (Haghpanah et al., 2024). Salinity stress is caused by high concentrations of soluble salts (primarily NaCl) in the soil and is a major constraint to agricultural productivity, particularly in arid, semi-arid and irrigated lands. It affects plant by creating osmotic stress and ion toxicity. Excessive accumulation of specific ions (e.g. Na^+, Cl^-) in the cytoplasm can inhibit enzyme activity, damage cellular structures, and interfere with nutrient uptake causing ion toxicity. High temperature exceeding the optimal range causes heat stress leading to protein denaturation, membrane damage, increased respiration rate, inhibition of PS-II and reproductive failure (Wahid & Close, 2007). The heat stress in plants induces the synthesis of heat shock proteins, antioxidant enzymes and compatible solutes to maintain membrane stability. Plants are also susceptible to low temperature stress. Low temperatures, including chilling (0–15 °C) and freezing (<0 °C), can cause membrane rigidification, enzyme inactivation, water deficit, and ice crystal formation within cells, leading to severe cellular damage and plant death. Plants require a balanced supply of macro- and micronutrients for optimal growth and development. Nutrient availability is often influenced by soil pH, organic matter content, and microbial activity. Deficiencies in essential nutrients (e.g. nitrogen, phosphorus, potassium, iron, zinc) can impair various physiological processes, leading to stunted growth, chlorosis, necrosis, and reduced yield. Climate change exacerbates abiotic stresses, posing a significant threat to global agriculture. Rising global temperatures, altered precipitation patterns, and increased frequency and intensity of extreme weather events directly impact crop yield and plant health.

Abiotic stress triggers a complex cascade of intracellular signalling events, effectively translating an external threat into an internal cellular response. Through the coordinated action of calcium signaling that involves increase in cytosolic Ca^{2+} concentrations (through calcium-binding proteins, including calmodulins, calmodulin-like proteins, calcineurin B-like proteins, and calcium-dependent protein kinases), ROS signaling (controlled production of reactive oxygen species like H_2O_2), phytohormone regulation (abscisic acid), and osmotic adjustment, plants are able to detect stress quickly and activate protective mechanisms. Upon sensing stress and transducing the signals, plants enact a range of physiological and biochemical adjustments aimed at maintaining cellular homeostasis and mitigating immediate damage. One of the most fundamental responses to drought and salinity is osmotic adjustment. Plants accumulate compatible solutes (osmolytes) such as proline, glycine betaine, various sugars (sucrose, trehalose), and sugar alcohols (mannitol, sorbitol). These molecules help to maintain cell turgor by lowering cellular osmotic potential, thus facilitating water uptake from a drier or saltier environment. Antioxidant defence system comprises enzymatic antioxidants such as superoxide dismutase (SOD) (which converts superoxide radicals to hydrogen peroxide), catalase (CAT) and ascorbate peroxidase (APX) (which detoxify hydrogen peroxide), and glutathione reductase (GR) (which maintains the pool of reduced glutathione). Complementing these are non-enzymatic antioxidants, including ascorbate (vitamin C), glutathione, tocopherols (vitamin E), carotenoids, and flavonoids, which directly scavenge harmful reactive oxygen species (ROS), thus preventing cellular damage. Stomatal regulation is a critical physiological response to drought and heat stress and is tightly regulated by Abscisic acid (ABA), hydraulic signals, and CO_2 levels. While essential for water conservation, prolonged stomatal closure restricts CO_2 uptake, inevitably leading to a reduction in photosynthetic efficiency. Plants also modify their root system architecture to enhance resource acquisition. Under drought, roots tend to grow deeper into the soil to access water, while under nutrient deficiencies, lateral root proliferation might increase in nutrient-rich zones. These morphological changes are crucial for long-term survival in challenging environments. Furthermore, photosynthetic adjustments occur, including non-photochemical quenching (NPQ) to safely dissipate excess light energy as heat, downregulation of key photosynthetic enzymes, and activation of photorespiration or alternative electron transport pathways to consume excess electrons and minimize ROS production in chloroplasts. Cell wall modifications also play a role in maintaining structural integrity under osmotic stress. Beyond morphological and physiological adaptations, plants also make significant adjustments at the protein synthesis and degradation levels. Stress, especially heat, leads to the accumulation of misfolded or damaged proteins. Plants respond by upregulating the synthesis of Heat Shock Proteins (HSPs), which function as molecular chaperones, assisting in protein folding, preventing aggregation, and refolding denatured proteins. Membrane lipid composition adjustment is another crucial molecular response, particularly to temperature extremes. Under cold stress, plants increase the proportion of unsaturated fatty acids in their cell membranes to maintain fluidity and prevent rigidification. Conversely, under heat stress, an increase in saturated fatty acids might occur to stabilize membranes and prevent excessive fluidity.

4 Mechanisms of Cyanobacteria in Alleviating Abiotic Stress

Cyanobacteria respond to the unfavourable environment by producing several secondary metabolites and biomolecules as a part of their defence mechanisms to sustain the adverse environment and enable them to adapt and survive the condition. The role of metabolic pathways in sensing and adapting to prevailing stress conditions becomes very critical. Therefore, understanding the stress tolerance mechanism requires a comprehensive study involving the effect of different stresses on physiological, biochemical, and metabolic changes in cyanobacteria.

4.1 High Light Stress Tolerance Mechanisms

In cyanobacteria, a balance between light harvesting and energy dissipation is crucial for survival. Insufficient light energy can slow down the photosynthetic process and hence cause a shortage of food, while excess energy, if not properly dissipated, can damage the photosynthetic apparatus. Therefore, cyanobacteria require a finely tuned system to manage light capture and dissipation, ensuring optimal photosynthetic performance. Cyanobacteria have evolved several protective mechanisms to counter such harmful consequences of excess light absorption.

Alteration in the Pigment/Phycobilisome Adjustment One of the most common responses to acclimatize to different intensities and duration of light is by alteration in the photosynthetic pigments to cope with changes in light conditions. Several strains of cyanobacteria can adjust the size and composition of their phycobilisomes (harbors colored phycobilliproteins i.e. phycoerythrin, phycocyanin, allophycocyanin), the light-harvesting complexes, to optimize light absorption and minimize photodamage under high light conditions (Rai et al., 2025). Under low light (LL), the levels of phycobilisomes (PBS) increase to enhance light capture. In high light (HL), light-harvesting antenna sizes are reduced to minimize light absorption and protect photosynthetic systems (PS). This reduction in antenna size helps prevent photoinhibition, where excessive light energy can damage photosynthetic machinery (Bag, 2021).

Synthesis of Protective Pigments and Other Metabolites Under high light conditions, cyanobacteria can synthesize carotenoids, which act as antioxidants to neutralize ROS and protect the photosynthetic apparatus.

Similarly, exposure to UV radiation also causes production of ROS and denaturation of DNA, RNA or proteins, which in turn have deleterious effect on the photosynthesis and hence survival of the organisms. Several strains of cyanobacteria like *Nostoc, Anabaena, Scytonema, Lyngbya* and *Synechocystis* are known to counter high light conditions by accumulating carotenoids, antioxidants (ascorbic acid, glutathione, tocopherol), detoxifying enzymes (superoxide dismutase, catalase, and peroxidases) and synthesis of photoprotective compounds such as mycosporine-like amino acids (MAAs) and scytonemin (Häder et al., 2007; Singh et al., 2024). These compounds enable the cyanobacteria to tolerate intense sunlight and oxidative stress.

Photoprotective Mechanisms Cyanobacteria utilize several mechanisms to protect themselves from high light. These include non-photochemical quenching (NPQ), where excess light energy is dissipated as heat to prevent damage to photosystem II. This important function of photoprotection through NPQ of excess light is mainly performed by orange carotenoid proteins (OCP)—a ketocarotenoid, which is of orange colour during the inactive stage (García-Oneto et al., 2024). Under strong blue-green light, it gets converted to the red active form and helps in photoprotection by converting excess light energy to heat (Domínguez-Martín et al., 2022). Under low light (LL) conditions, a 13 kDa fluorescence recovery protein (FRP) converts the active form (red) back to inactive form (orange), restoring normal flow of energy from antenna pigments to PSII (Boulay et al., 2010; Sluchanko et al., 2018).

4.2 Temperature Stress Tolerance Mechanisms

Cyanobacterial species are known to survive in a wide range of temperature (−20 to 70 °C). So, the duration as well as intensity of temperature variations significantly impact cell metabolism and its components, and this impact can vary considerably between different strains or species. Membranes, proteins, and nucleic acids of the cyanobacteria act as thermosensors to detect temperature changes, which in turn affect cellular processes at multiple levels, influencing membrane fluidity, protein structure and function, and gene expression (Gupta et al., 2025).

Membrane Fluidity Regulation To cope with temperature changes, cyanobacteria can alter the fatty acid composition of their membranes. The relative changes in saturated and unsaturated fatty acids of the membrane cause changes in the membrane fluidity (Mironov et al., 2019). For example, they may increase the proportion of unsaturated fatty acids at higher temperatures to maintain membrane fluidity (Gupta et al., 2025). Under low-temperature conditions, cells typically reduce de novo fatty acid synthesis and modify existing fatty acids by introducing double bonds (desaturation), making them more fluid and less rigid for survival.

Heat Shock Proteins (HSPs) High temperatures negatively impact cell membranes by increasing fluidity, altering protein-protein interactions, and causing structural changes that lead to ion leakage. This increased fluidity can disrupt the ordered structure of the membrane, making it easier for ions to pass through and disrupt cellular processes. Additionally, heat can weaken or break down hydrogen bonds and electrostatic interactions between proteins, leading to changes in protein structure and function (Los & Murata, 1998; Yadav et al., 2022). Cyanobacteria produce HSPs in response to heat stress. These proteins assist in the refolding of denatured proteins and the prevention of protein aggregation.

Cold Shock Proteins Cold temperatures induce the expression of a novel family of cold shock chaperones and proteases (cold shock proteins—CSPs), which help stabilize RNA and protein structures. Cold exposure is also known to cause phosphorylation of the ribosomal protein S2, leading to downregulation of overall translation except cold stress-induced mRNA (El-Fahmawi & Owttrim, 2007). Furthermore, cyanobacteria also induce the expression of several gene families in order to adapt to low temperature (Suzuki et al., 2001).

4.3 Salinity Stress Tolerance Mechanisms

Salinity stress lowers the water potential of the external environment as it gets more concentrated with the dissolve salts leading to water loss and an increase in ionic concentration of the cell (Csonka, 1989; Yadav et al., 2022). However, with a variety of physiological mechanisms, cyanobacteria are considered to be more tolerant to salt stress as compared to higher plants (Wei et al., 2006; Whitton & Potts, 2007).

These include ion transport, along with lower Na^+ absorption to regulate the intracellular Na^+ level, followed by the synthesis of compatible solutes.

The cyanobacterial response towards salt stress occurs in three phases. The first phase occurs within seconds and involves rapid Na^+ and Cl^- ions influx and efflux of cytosolic water. K^+ ions replace Na^+ in the second phase, which decreases the intracellular Na+ concentration and its toxic effects. The last phase involves synthesis of compatible solutes, uncharged organic compounds of low molecular weight such as glycine betaine, Glucosyl glycerol/Glucosyl glycerate (GG), proline, sucrose, and trehalose (Karandashova & Elanskaya, 2005).

- *Ion Transporters:* Cyanobacteria utilize ion transporters, such as Na^+/H^+ antiporters, to expel excess sodium ions from the cytoplasm, thus protecting cellular functions from ion toxicity.
- *Compatible Solute Synthesis:* Cyanobacteria maintain osmotic balance by synthesizing compatible solutes such as sucrose, glycerol, and trehalose. These molecules help in maintaining cellular turgor and preventing dehydration. Accumulation of organic osmolytes such as ectoine, glycine betaine, and proline helps stabilize cellular structures under salinity stress (Karandashova & Elanskaya, 2005).

4.4 Drought Stress Tolerance Mechanisms

Some cyanobacteria, particularly those in desert habitats, can tolerate complete desiccation. For example, the survival of dried Nostoc commune samples has been reported for more than a century (Cameron, 1962). They can enter a state of metabolic dormancy, with minimal cellular activity until water availability is restored.

- *Production of Trehalose and Sucrose:* These sugars act as protectants against dehydration, stabilizing proteins and membranes during periods of drought.
- *Exopolysaccharide (EPS) Production:* Cyanobacteria produce EPS, which can form a protective layer around the cells, reducing water loss during drought conditions.

4.5 Oxidative Stress Tolerance Mechanisms

- *Antioxidant Enzymes:* Cyanobacteria produce a variety of antioxidant enzymes, including superoxide dismutase (SOD), catalase (CAT), and peroxidase (POD), to detoxify ROS. These enzymes neutralize the harmful effects of ROS, thereby protecting the cells from oxidative damage.
- *Glutathione and Other Antioxidants:* Cyanobacteria accumulate small-molecule antioxidants, such as glutathione, to scavenge ROS and protect cellular macromolecules.

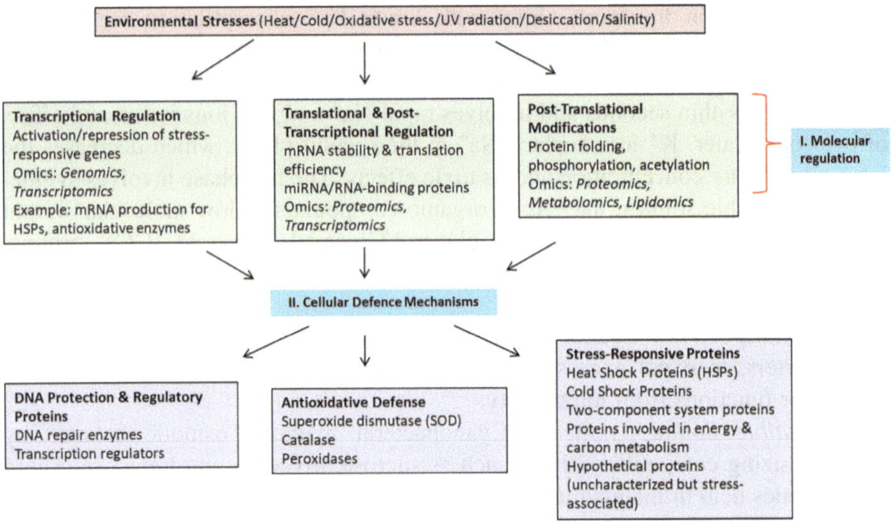

Fig. 1 Molecular regulation and defence mechanism to tolerate environmental stress

- *Metal Chelation:* Cyanobacteria can sequester metal ions such as iron and copper, which participate in the formation of ROS, by synthesizing metal-binding proteins such as ferritin.

5 Molecular Basis of Stress Tolerance

The response of cyanobacteria to different stresses is a result of a complex interaction between physiology, biochemistry, and ecology. Thus, different omics techniques such as genomics, proteomics, transcriptomics, metabolomics, and lipidomics have been used to decipher the basis of cellular responses of cyanobacteria under different stresses. Stress responses are regulated at multiple levels, impacting gene expression, protein activity, and cellular metabolism. These regulations can occur at the transcriptional, translational, and post-translational levels. At the transcriptional level, stress can trigger changes in the production of stress-responsive mRNAs, while post-transcriptional regulation involves controlling the translation of these mRNAs into protein and can also alter their stability. Post-translational modifications either by phosphorylation, methylation, acetylation, protein folding or cleavage can either activate or deactivate proteins which in turn can affect cellular metabolism and facilitate the organism to survive under stress conditions.

The mechanisms used by different cyanobacteria to tolerate abiotic stress vary. Different species and even strains within the same species may employ unique strategies to cope with specific abiotic stresses (Fig. 1). These can broadly be classified as follows:

(i) DNA repair/protection and transcription regulators
(ii) Cellular antioxidative enzymes
(iii) Stress-related proteins
- Heat shock proteins (HSPs)
- Cold shock proteins
- proteins of other cellular metabolisms
- Two-component system proteins and
- Hypothetical proteins

5.1 DNA Repair/Protection and Transcription Regulators

DNA repair and protection are crucial for cyanobacteria to survive during abiotic stress because environmental factors such as radiation and desiccation can cause DNA damage, leading to mutations and potentially disrupting essential cellular processes. Cyanobacteria have evolved various mechanisms to repair and protect their DNA from stresses, ensuring their survival and maintaining genome integrity. Xiong et al. (2015) observed an upregulation of various transcriptional regulators and enzymes related to replication, DNA repair, and modification in Synechococcus, under high light (HL) stress. Specifically, proteins such as ParA, GvrA, and PhrA, as well as NusB exhibited increased expression, suggesting a cellular response to HL-induced DNA damage and stress. Similar response for proteins involved in DNA damage, repair, and modification (polymerases, exonucleases, and methylases) was observed in *Spirulina patensis* (Hongsthong et al., 2009). In cyanobacteria another important transcription regulator is a sigma factor—a protein subunit of the RNA polymerase which plays a very important role during stress conditions (Srivastava et al., 2021a, b). Cyanobacteria utilize "sigma switching" mechanism to cope with the stress condition. This involves replacing primary sigma factors (responsible for transcription of housekeeping genes) with group 2 sigma factors (responsible for transcription of stress-responsive genes) during stress conditions (Rai et al., 2025).

Besides their role in normal cellular processes, DNA-binding proteins (Dps) are also known to play an important role in protecting the cell from oxidative damage and extreme environmental conditions (Rai et al., 2025). For example, among five Dps (NpDps1, NpDps2, NpDps3, NpDps4, and NpDps5) proteins identified in *Nostoc punctiforme* ATCC 29133, one (NpDps2) has been reported to play an important role in H_2O_2—related stress tolerance and another (NpDps5) for iron homeostasis and maintaining PS11/PS1 ratio (Moparthi et al., 2016). Studies have also demonstrated that in many cyanobacteria sRNAs are expressed under stress conditions (Georg et al., 2014; Hu et al., 2017; Hu & Wang, 2018).

5.2 Cellular Antioxidative Enzymes

Cyanobacteria, a ubiquitous group of photosynthetic organisms, can thrive in diverse and even extreme environments due to their adaptation mechanisms. They utilize ROS, which are byproducts of aerobic metabolism, as signalling molecules to regulate stress responses, including antioxidant enzyme production such as superoxide dismutase (SOD), catalase (CAT), peroxidases (POX), glutathione peroxidase (GPX), glutathione reductase (GR), glutathione S-transferases (GST), ascorbate peroxidase (APX), and monodehydroascorbate reductase (DHAR). Overexpression of these enzyme systems has been reported in several cyanobacteria under stress conditions as a defence mechanism (Cameron & Pakrasi, 2010; Fulda et al., 2006; Rai et al., 2021; Waditee-Sirisattha & Kageyama, 2023).

5.3 Stress-Related Proteins

Proteins, including enzymes, are vital not only for cellular function, regulating transcription, and translation but are also crucial for stress response, acclimation, and ultimately, determining tolerance to stressors (Hihara et al., 2001). Furthermore, some of them undergo post-translational modifications (PTMs) in response to abiotic stress signals (Rai et al., 2025). For example, salt stress led to upregulation/accumulation of 55 proteins in *Synechocystis* sp. strain PCC6803 (Fulda et al., 2006). This included salt stress-specific as well as other stress acclimation proteins. At the same time the proteins associated with basic carbohydrate metabolism were deregulated. Similarly, proteins related to the defence system were expressed in *Anabaena* sp. PCC 7120 and *Anabaena doliolum* (Shrivastava et al., 2015), *Synechocystis* sp. PCC 6803 (Gao et al., 2009) following UV-B stress. Babele et al. (2015) reported overexpression of two hypothetical proteins (HPs) following UV-B irradiation, along with several housekeeping proteins involved in carbon, amino acid biosynthesis and energy metabolism, antioxidative enzymes, reductases, etc.

Wang et al. (2013) observed deregulation of 114 proteins under salt stress in *Arthrospira* (*Spirulina*) *plantensis*-YZ. They further reported differential expression of 18 proteins. This included proteins belonging to important metabolic pathways, such as photosynthesis, glucose metabolism, cysteine and methionine metabolism, lysine synthesis, fatty acid metabolism, glutathione metabolism along with stress-related proteins like heat shock protein and ABC transporter proteins, which might have conferred resistance against salt stress. Pandhal and team (2008) used the Isobaric Tags for Relative and Absolute Quantitation (iTRAQ) approach for proteomic analysis and observed that both extremely and moderately halotolerant cyanobacteria exhibit similar salt-stress responses by up-regulating stress-related proteins under identical salinity conditions.

Additionally, two-component systems (TCSs) involving histidine kinases (HKs) and response regulators (RRs) act as molecular switches in cyanobacteria. These play a crucial role in sensing and responding to stresses such as salt, osmotic, and light stress. Exposure of *Spirulina platensis* to low-temperature stress under the light condition has been reported to cause expression of proteins related to two-component response systems, along with those related to DNA repair, molecular chaperones, other stress-induced proteins and proteins associated with secretion systems, nitrogen assimilation (Hongsthong et al., 2008). Two-component response systems proteins, along with those involved in DNA damage and repair systems, molecular chaperones, capsule formation, and unsaturated fatty acid biosynthesis, were up-regulated in *Spirulina platensis* (Hongsthong et al., 2009) under high temperature stress. Kurdrid et al. (2011) reported an important role of signalling proteins (Hik14, Hik26, and Hik28) in cross-talk of cell signalling pathways under low and high temperature stress in *Spirulina platensis.*

6 Enhancing Crop Resilience Through Symbiotic Relationships with Cyanobacteria

Cyanobacteria have shown substantial promise in improving crop growth, productivity, and resilience under a range of abiotic stresses, including drought, salinity, nutrient deficiency, and elevated temperature and CO_2 conditions. These beneficial effects stem from their ability to fix atmospheric nitrogen, produce phytohormones (e.g. auxins, cytokinins, gibberellins), secrete exopolysaccharides (EPS), and synthesize bioactive compounds such as vitamins, amino acids, and siderophores. Mechanistically, this is achieved through antioxidant enzyme activation (e.g. superoxide dismutase, catalase), osmoregulation, and hormonal modulation. Cyanobacteria also improve soil structure and water retention by secreting EPS, forming soil-stabilizing biofilms (Chamizo et al., 2019). Root system architecture and rhizosphere dynamics are enhanced by cyanobacterial inoculation, increasing nutrient uptake efficiency. For instance, *Anabaena* and *Nostoc* inoculation in chrysanthemum and common bean resulted in improved growth metrics under salinity and drought (Bharti et al., 2021; Rady et al., 2018). Phytohormones such as indole acetic acid (IAA), ABA, and cytokinins produced by cyanobacteria further regulate root growth and stomatal conductance, critical under water-deficit conditions.

A growing body of experimental evidence from greenhouse and field studies highlights the successful integration of cyanobacterial symbionts in agricultural systems across cereals, legumes, vegetables, and ornamental crops. One of the most well-documented examples is the *Azolla–Anabaena azollae* symbiosis, extensively used in flooded rice fields across Asia. This symbiosis is characterized by efficient nitrogen fixation and biomass turnover, contributing approximately 30–60 kg N ha^{-1} per cropping season. The regular incorporation of *Azolla* into rice fields enhances soil fertility and reduces the dependency on synthetic nitrogen fertilizers, thereby promoting low-input sustainable rice cultivation. In cereal crops such as wheat and maize, field applications of *Nostoc muscorum* and *Anabaena oryzae* have significantly improved plant growth, yield parameters, and nutrient uptake. In maize, inoculation with multispecies biofilms such as *Anabaena–Nostoc*, *Anabaena–Azotobacter*, and *Anabaena–Trichoderma* combinations resulted in enhanced shoot and root biomass, increased cob size, and improved physiological traits such as chlorophyll content and nitrogen assimilation. These benefits are linked to the simultaneous nitrogen fixation, phosphate solubilization, and plant growth hormone production by the consortia (Prasanna et al., 2016).

Similarly, in leguminous crops such as pea (*Pisum sativum*), application of *Oscillatoria angustissima* and *Nostoc entophytum* improved germination rate, nodule formation, carbohydrate accumulation, and pod yield (Osman et al., 2010). In soybean (*Glycine max*), under elevated CO_2 and temperature conditions simulating future climate scenarios, inoculation with *Anabaena laxa* and a biofilm of *Anabaena torulosa–Bradyrhizobium japonicum* led to significant increases in plant biomass, seed nitrogen content (up to 758 mg N $plant^{-1}$), and nitrogenase activity by 55–72% compared to the uninoculated control (Sanyal et al., 2022). Recent innovations include the use of cyanobacterial inoculants in ornamental and horticultural crops. In *Chrysanthemum indicum*, bioinoculation with *Nostoc* spp. improved flowering traits, total chlorophyll content, relative water content, and shoot dry mass under drought stress, indicating their role in commercial floriculture under suboptimal irrigation regimes (Bharti et al., 2021). Likewise, *Capsicum annuum* treated with cyanobacteria–green algae consortia demonstrated improved plant vigor, nutrient use efficiency, and microbial diversity, thus confirming the broader utility of mixed algal inoculants in vegetable production (Jose et al., 2024).

Additionally, combinations of cyanobacteria with plant growth-promoting rhizobacteria (PGPR), such as *Anabaena–Azotobacter*, *Anabaena–Trichoderma*, and *Anabaena–Providencia* consortia, have been used successfully in maize, significantly enhancing plant height, biomass, nutrient uptake, and soil microbial activity. While the increase in cob yield was statistically on par with controls, improvements in plant physiological traits were notable, confirming their effectiveness as bioenhancers in stress-prone agricultural systems (Prasanna et al., 2016). In peanut cultivation on sandy, nutrient-deficient soils, foliar and soil application of *Nostoc muscorum* in conjunction with 75% recommended nitrogen dose significantly improved pod yield, chlorophyll and carotenoid contents, and enzymatic activities such as urease and dehydrogenase in the rhizosphere (Shaheen et al., 2022). These effects were attributed to the bioavailable nitrogen, EPS-mediated moisture retention, and improved root–soil interactions. In horticulture, foliar application of *Anabaena laxa* and its biofilm with *Trichoderma viride* significantly enhanced fruit fresh weight, nutrient density (Fe and Zn), organic carbon content, and rhizospheric enzyme activities in fruit crops. Such results validate cyanobacteria's multifaceted role in stress adaptation, including antioxidant activation, hormonal regulation, rhizosphere colonization, and resource-use efficiency under fluctuating environmental conditions. Lichen-forming cyanobacteria such as *Nostoc* spp. have also been reported to contribute substantially to ecosystem function in agroforestry and rangeland systems. In these composite organisms, cyanobacteria not only fix nitrogen but also trap dust and mineral particles, secrete weathering agents, and contribute to soil genesis and fertility enhancement (Prasanna et al., 2021).

Field-scale studies conducted across varied agroecological zones have demonstrated that cyanobacterial bioinoculants not only enhance soil fertility and plant productivity but also serve as biological buffers against abiotic stressors. In soybean trials conducted under semi-arid conditions with elevated CO_2 and temperature regimes, biofilms of *Anabaena laxa* and *A. torulosa–Bradyrhizobium japonicum* significantly improved nodulation, nitrogenase activity, plant growth, and seed nitrogen content compared to uninoculated controls. Nitrogenase activity in inoculated plants increased by up to 72%, while nitrogen accumulation per plant rose to 758 mg N $plant^{-1}$ (Sanyal et al., 2022). In maize, inoculation with *Anabaena–Nostoc* biofilms enhanced acetylene reduction activity (a proxy for nitrogenase activity) by threefold, while PLFA analysis revealed increased microbial biomass and higher abundance of beneficial fungal taxa, particularly arbuscular mycorrhizal fungi. The *Anabaena–Trichoderma* biofilm applied to the hybrid variety "Parkash" yielded superior plant height, leaf area, and shoot biomass (Prasanna et al., 2016). Field trials in peanut demonstrated that foliar sprays and soil application of *N. muscorum*, even with reduced nitrogen inputs (75% of the recommended dose), were able to significantly enhance pod yield, root volume, and photosynthetic pigment concentrations (Shaheen et al., 2022). Moreover, trials conducted in *Capsicum annuum* have shown that application of cyanobacteria-based consortia improved root-to-shoot ratios, soil nitrogen availability, and microbial diversity, establishing their role in improving system resilience under abiotic and biotic pressures (Jose et al., 2024). These outcomes were further substantiated by biochemical profiling that demonstrated higher chlorophyll biosynthesis, enhanced antioxidant enzyme activity, and better nutrient uptake efficiency. The use of cyanobacteria as biostimulants, rich in amino acids, vitamins, and EPS, has shown to positively affect plant physiology, especially root development and stress tolerance mechanisms. Their inclusion in integrated nutrient management systems significantly reduces the loss of nitrogen and phosphorus through leaching or volatilization, thereby enhancing fertilizer use efficiency by as much as 20–40% (Dhar et al., 2015; Sinha et al., 2014). Together, these multilocational field studies provide strong evidence for the scalability, adaptability, and ecological benefits of cyanobacteria-based interventions in conventional and climate-resilient agriculture.

7 Application of Cyanobacteria in Crop Stress Management

Cyanobacteria are increasingly employed as bioinoculants to promote crop productivity and mitigate abiotic stress. Their roles in nitrogen fixation, phytohormone production, soil aggregation, and pathogen suppression align with the principles of sustainable agriculture (Rai et al., 2025).

7.1 Cyanobacterial Inoculants: Types and Methods of Application

Cyanobacteria, due to their nitrogen-fixing ability, photosynthetic efficiency, and bioactive compound production, are gaining momentum as sustainable biofertilizers in modern agriculture. Their applications span crop improvement, soil fertility enhancement, and stress mitigation. The classification of cyanobacterial inoculants, along with formulation types and application strategies, determines their efficacy in diverse agroecosystems. Cyanobacterial inoculants are primarily categorized as free-living or symbiotic based on their ecological behaviour:

Free-living cyanobacteria such as *Anabaena oryzae*, *Nostoc muscorum*, and *Calothrix elenkinii* have been used successfully in crops such as rice, peanut, maize, and tomato through soil drenching, seed coating, or foliar application (Shaheen et al., 2022; Prasanna et al., 2016).

Symbiotic cyanobacteria, like *Anabaena azollae* in association with *Azolla*, form mutualistic interactions particularly useful in paddy fields.

Recent phylogenetic and physiological characterization of native heterocystous strains such as *Nostoc* and *Calothrix* from temperate rice fields confirmed their suitability as biofertilizers, exhibiting robust nitrogenase, nitrate reductase, and glutamate synthase activities (Pérez et al., 2012).

7.2 Formulation of Cyanobacterial Inoculants

Various formulation technologies have been developed to ensure microbial viability, ease of application, and extended shelf life.

(a) *Liquid Formulations:* Liquid cyanobacterial fertilizers, including hydroponic applications, show promise due to their ease of absorption and bioactivity enhancement. Cyanobacteria stimulate microbial communities and release phytohormones and enzymes such as IAA, polysaccharides, and antioxidants (Jacob & Kumar, 2020).

(b) *Powder and Granular Formulations:* Powdered inoculants using carriers such as perlite, cocopeat, and vermiculite improve shelf life and microbial colonization. Studies with *Calothrix elenkinii*, *Anabaena laxa*, and biofilms (e.g. An-Tr and An-Bs) showed enhanced chlorophyll content, population stability, and fungicidal activity in the carrier matrix for up to 20 weeks (Prasanna et al., 2020).

(c) *Gel-Based and Encapsulated Inoculants:* Encapsulation in silica matrices can retain biomass and enhance photosynthetic and metabolic activity. Encapsulated *Synechocystis* sp. *PCC 6803* demonstrated altered gene expression and higher metabolic resilience, though further optimization is needed (Dickson & Ely, 2011).

(d) *Biofilm-Based Formulations:* Biofilms composed of *Anabaena torulosa* and *Trichoderma viride* (An-Tr) have proven highly effective. They not only enhance nitrogen fixation and IAA production but also significantly improve seedling vigor, soil enzyme activities, and rhizospheric microbial communities. In tomato, An-Tr inoculants increased microbial biomass carbon by 60–70%, stimulated C–N assimilating enzymes by 45–50%, and improved fruit quality attributes such as carotenoid and lycopene content under elevated CO_2 conditions (Kokila et al., 2022).

7.3 Species Composition of Inoculants

Cyanobacterial formulations are either single-species or multispecies consortia:

- *Single-strain inoculants*, such as *Anabaena laxa*, have shown notable increases in carbonic anhydrase and glutamine synthetase activity, contributing to enhanced nutrient uptake (Kokila et al., 2022).
- *Multispecies consortia*, like *Anabaena–Trichoderma* and *Anabaena–Bradyrhizobium*, promote microbial synergy and stress resilience. An-Tr consortia outperformed single strains in enhancing crop yield, nutrient status, and soil health under both ambient and elevated CO_2 conditions (Sanyal et al., 2022).

7.4 Methods of Application

Cyanobacterial inoculants are deployed through multiple delivery systems, each suited to specific crops and growth stages:

- *Seed Coating and Priming:* Biofilms applied to maize seed improved seedling vigor, enzyme activity, and chlorophyll content. An-Tr biofilms led to three- to fivefold increases in nitrogen fixation and stimulated microbial diversity.
- *Soil Drenching:* Soil application of *Anabaena oryzae* and *Nostoc muscorum* improved nutrient uptake (N, P, K, Ca) and yield in peanut under reduced nitrogen input.
- *Foliar Spray:* Aqueous foliar sprays containing *Anabaena laxa* and An-Tr biofilms increased nitrate reductase and carbonic anhydrase activities by 13–17% and improved tomato yield and soil enzyme activity (Bhardwaj et al., 2024).
- *Root Dip and Seedling Treatment:* Application of *Microcoleus vaginatus* culture filtrates as root dip significantly reduced *Meloidogyne incognita* populations and improved tomato growth parameters (Khan et al., 2005).

8 Conclusion

The increasing frequency and intensity of abiotic stresses pose a significant threat to global food security. Traditional agricultural practices and conventional breeding methods often fall short in providing adequate solutions to these complex challenges. In this context, cyanobacteria offer a sustainable and eco-friendly alternative. Their remarkable adaptability, honed over billions of years of evolution, allows them to thrive in extreme environments and transfer these stress-mitigating benefits to associated plants.

The mechanisms through which cyanobacteria confer stress tolerance are diverse and synergistic. Nitrogen fixation, a hallmark of many cyanobacterial species, directly enriches soil fertility, reducing the reliance on synthetic nitrogen fertilizers, which are energy-intensive to produce and contribute to greenhouse gas emissions.

Beyond nitrogen, cyanobacteria produce a plethora of plant growth-promoting substances, including auxins, cytokinins, and gibberellins, which regulate various physiological processes in plants, such as root development, stomatal function, and overall biomass accumulation.

Furthermore, cyanobacteria synthesize a range of protective compounds. Exopolysaccharides (EPS) secreted by cyanobacteria play a crucial role in soil aggregation, improving soil structure, water retention, and nutrient availability. This is particularly vital in drought-prone regions, where EPS can significantly enhance the soil's capacity to hold moisture, making it more accessible to plant roots. Osmoprotectants such as glycine betaine, trehalose, and proline, accumulated by cyanobacteria under stress, help maintain cellular turgor and protect cellular components from damage due to osmotic imbalance caused by salinity or drought. The robust antioxidant defence systems within cyanobacteria, comprising enzymes such as superoxide dismutase (SOD), catalase (CAT), and peroxidase (POD), along with non-enzymatic antioxidants, enable them to neutralize reactive oxygen species (ROS) generated under stress, thereby preventing oxidative damage to plant cells.

Field-scale studies and greenhouse experiments have consistently demonstrated the efficacy of cyanobacterial inoculants across a wide array of crops, including cereals, legumes, vegetables, and ornamentals. The *Azolla–Anabaena* symbiosis in rice paddies stands as a historical testament to their biofertilizing capabilities. More recent research highlights the success of multispecies consortia, such as *Anabaena–Trichoderma* and *Anabaena–Bradyrhizobium* biofilms, in enhancing crop yield, nutrient uptake, and overall plant health under both ambient and elevated CO_2 conditions. These formulations, applied through various methods, such as seed coating, soil drenching, foliar spray, and root dip, have shown promising results in improving seedling vigour, enzyme activity, chlorophyll content, and ultimately, crop productivity.

The molecular underpinnings of cyanobacterial stress tolerance are complex, involving sophisticated regulatory networks. Transcriptional regulators, DNA repair mechanisms, and the expression of stress-related proteins, including heat shock proteins (HSPs) and cold shock proteins (CSPs), are all finely tuned to enable survival and adaptation. The "sigma switching" mechanism and the role of DNA-binding proteins (Dps) further exemplify their intricate molecular strategies for coping with environmental perturbations.

In essence, cyanobacteria represent a holistic solution for sustainable agriculture. They not only enhance plant resilience to abiotic stresses but also contribute to soil health, nutrient cycling, and reduced environmental footprint associated with conventional farming.

9 Future Prospects

Despite the significant progress in understanding and applying cyanobacteria in agriculture, several avenues for future research and development remain to unlock their full potential. The application of advanced omics technologies (genomics, transcriptomics, proteomics, metabolomics, lipidomics) will be crucial for identifying and characterizing novel cyanobacterial strains with superior stress tolerance mechanisms and plant growth-promoting traits. Genetic engineering and synthetic biology approaches could be employed to enhance specific beneficial pathways, such as increased production of phytohormones, EPS, or stress-responsive proteins, or to improve their colonization efficiency in the rhizosphere.

Developing more stable, viable, and cost-effective and tailored cyanobacterial formulations is paramount for widespread adoption. While the benefits are evident, the precise molecular crosstalk between cyanobacteria and host plants under stress conditions is still not fully elucidated. Future research should delve deeper into the signalling pathways involved, identifying specific plant genes or metabolites that are modulated by cyanobacterial inoculation. Cyanobacterial bio-inoculants should be integrated into broader climate-smart agricultural strategies. This includes their use in conjunction with water-saving irrigation techniques, conservation tillage, and diversified cropping systems.

Scaling up the production of cyanobacterial inoculants from laboratory to industrial levels requires overcoming significant technical and economic hurdles. This includes developing cost-effective mass production methods, ensuring quality control, and navigating regulatory frameworks for their commercialization and use as biofertilizers or biostimulants. Public-private partnerships will be essential to bridge the gap between research and practical application.

References

Adams, D. G., & Duggan, P. S. (2008). Cyanobacteria–bryophyte symbioses. *Journal of Experimental Botany, 59*(5), 1047–1058.

Babele, P. K., Singh, G., Kumar, A., & Tyagi, M. B. (2015). Induction and differential expression of certain novel proteins in *Anabaena* L31 under UV-B radiation stress. *Frontiers in Microbiology, 6*, 133.

Bag, P. (2021). Light harvesting in fluctuating environments: Evolution and function of antenna proteins across photosynthetic lineage. *Plants, 10*(6), 1184.

Bhardwaj, A., Prasanna, R., Bavana, N., Kokila, V., Nivedha, R. M., Gaur Rudra, S., Singh, A. K., & Shivay, Y. S. (2024). Foliar application of cyanobacterial formulations stimulates plant growth and fruit quality in tomato under protected cultivation. *New Zealand Journal of Crop and Horticultural Science, 53*(2), 330–348.

Bharti, A., Prasanna, R., Dantuluri, V. S. R., et al. (2021). Cyanobacterium amended mixes as priming options for stimulating growth and improving nutrient availability in nursery grown *Chrysanthemum* rooted stem cuttings. *Acta Physiologiae Plantarum, 43*, Article 102.

Bhatt, P., Negi, G., Gangola, S., Khati, P., Srivastava, A., & Sharma, A. (2016). Differential expression and characterization of cypermethrin degrading potential proteins in *Bacillus thuringiensis* strain SG4. *3 Biotech, 6*, 225.

Boulay, C., Wilson, A., D'Haene, S., & Kirilovsky, D. (2010). Identification of a protein required for recovery of full antenna capacity in OCP-related photoprotective mechanism in cyanobacteria. *Proceedings of the National Academy of Sciences, 107*(26), 11620–11625.

Cameron, R. E. (1962). Species of *Nostoc vaucher* occurring in the Sonoran Desert in Arizona. *Transactions of the American Microscopical Society, 81*(3), 379–384.

Cameron, J. C., & Pakrasi, H. B. (2010). Essential role of glutathione in acclimation to environmental and redox perturbations in the cyanobacterium *Synechocystis* sp. PCC 6803. *Plant Physiology, 154*(4), 1672–1685.

Chamizo, S., Adessi, A., Mugnai, G., Simiani, A., & De Philippis, R. (2019). Soil type and cyanobacteria species influence the macromolecular and chemical characteristics of the polysaccharidic matrix in induced biocrusts. *Microbial Ecology, 78*(3), 482–493.

Chaudhary, P., Khati, P., Chaudhary, A., Gangola, S., Kumar, R., & Sharma, A. (2021a). Bioinoculation using indigenous *Bacillus* spp. improves growth and yield of *Zea mays* under the influence of nanozeolite. *3 Biotech, 11*, 11.

Chaudhary, P., Khati, P., Chaudhary, A., Maithani, D., Kumar, G., & Sharma, A. (2021b). Cultivable and metagenomic approach to study the combined impact of nanogypsum and *Pseudomonas taiwanensis* on maize plant health and its rhizospheric microbiome. *PLoS One, 16*(4), e0250574.

Chaudhary, P., Khati, P., Gangola, S., Kumar, A., Kumar, R., & Sharma, A. (2021c). Impact of nanochitosan and *Bacillus* spp. on health, productivity and defence response in *Zea mays* under field condition. *3 Biotech, 11*(5), 237.

Chaudhary, P., Sharma, A., Chaudhary, A., Khati, P., Gangola, S., & Maithani, D. (2021d). Illumina-based high-throughput analysis of microbial diversity of maize rhizosphere treated with nanocompounds and *Bacillus* sp. *Applied Soil Ecology, 159*, 103386.

Chaves, M. M., Flexas, J., & Pinheiro, C. (2009). Photosynthesis under drought and salt stress: Regulation mechanisms from whole plant to cell. *Annals of Botany, 103*(4), 551–560.

Colesie, C., Felde, V. J. M. N. L., & Büdel, B. (2016). Composition and macrostructure of biological soil crusts. In *Biological soil crusts: An organizing principle in drylands* (pp. 159–172). https://doi.org/10.1007/978-3-319-30214-0_9

Csonka, L. N. (1989). Physiological and genetic responses of bacteria to osmotic stress. *Microbiological Reviews, 53*(1), 121–147.

Dhar, B. L., Sharma, S., & Verma, S. (2015). Integrated nutrient management for sustainable crop production and soil fertility improvement: A review. *Journal of Applied and Natural Science, 7*(2), 1010–1019.

Dickson, D. J., & Ely, R. L. (2011). Evaluation of encapsulation stress and the effect of additives on viability and photosynthetic activity of *Synechocystis* sp. PCC 6803 encapsulated in silica gel. *Applied Microbiology and Biotechnology, 91*(6), 1633–1646. https://doi.org/10.1007/s00253-011-3517-7

Domínguez-Martín, M. A., Sauer, P. V., Kirst, H., Sutter, M., Bína, D., Greber, B. J., Nogales, E., Polívka, T., & Kerfeld, C. A. (2022). Structures copper, nickel, cobalt and arsenic homeostasis mechanisms. *Life, 4*(4), 865–886. https://doi.org/10.3390/life4040865

El-Fahmawi, B., & Owttrim, G. W. (2007). Cold-stress-altered phosphorylation of EF-Tu in the cyanobacterium *Anabaena* sp. strain PCC 7120. *Canadian Journal of Microbiology, 53*(5), 551–558. https://doi.org/10.1139/W07-030

Flombaum, P., Gallegos, J. L., Gordillo, R. A., Rincón, J., Zabala, L. L., Jiao, N., et al. (2013). Present and future global distributions of the marine cyanobacteria *Prochlorococcus* and *Synechococcus*. *Proceedings of the National Academy of Sciences, 110*(24), 9824–9829.

Fulda, S., Mikkat, S., Huang, F., Huckauf, J., Marin, K., Norling, B., & Hagemann, M. (2006). Proteome analysis of salt stress response in the cyanobacterium *Synechocystis* sp. strain PCC 6803. *Proteomics, 6*(9), 2733–2745.

Gangola, S., Sharma, A., Bhatt, P., Khati, P., & Chaudhary, P. (2018). Presence of esterase and laccase in *Bacillus subtilis* facilitates biodegradation and detoxification of cypermethrin. *Scientific Reports, 8*, 12755.

Gao, Y., Xiong, W., Li, X. B., Gao, C. F., Zhang, Y. L., Li, H., & Wu, Q. Y. (2009). Identification of the proteomic changes in Synechocystis sp. PCC 6803 following prolonged UV-B irradiation. *Journal of experimental botany, 60*(4), 1141–1154. https://doi.org/10.1093/jxb/ern356

García-Oneto, T. M., Moyano-Bellido, C., & Domínguez-Martín, M. A. (2024). Structure and function of the light-protective orange carotenoid protein families. *Current Research in Structural Biology, 7*, 100141. https://doi.org/10.1016/j.crstbi.2024.100141

Georg, J., Dienst, D., Schurgers, N., Wallner, T., Kopp, D., Stazic, D., Kuchmina, E., Klahn, S., Lokstein, H., Hess, W. R., & Wilde, A. (2014). The small regulatory RNA SyR1/PsrR1 controls photosynthetic functions in cyanobacteria. *Plant Cell, 26*(9), 3661–3679.

Gupta, N., Bhattacharjee, S., & Mishra, A. K. (2025). Stress biology and signal perceptions in cyanobacteria. In A. K. Mishra & S. S. Singh (Eds.), *Cyanobacteria metabolisms to molecules* (pp. 383–411). Academic.

Haghpanah, M., Hashemipetroudi, S., Arzani, A., & Araniti, F. (2024). Drought tolerance in plants: physiological and molecular responses. *Plants, 13*(21), 2962. https://doi.org/10.3390/plants13212962

Häder, D. P., Kumar, H. D., Smith, R. C., & Worrest, R. C. (2007). Effects of solar UV radiation on aquatic ecosystems and interactions with climate change. *Photochemical & Photobiological Sciences, 6*(3), 267–285.

Hihara, Y., Kamei, A., Kanehisa, M., Kaplan, A., & Ikeuchi, M. (2001). DNA microarray analysis of cyanobacterial gene expression during acclimation to high light. *Plant Cell, 13*(4), 793–806.

Hongsthong, A., Sirijuntarut, M., Prommeenate, P., Lertladaluck, K., Porkaew, K., Cheevadhanarak, S., et al. (2008). Proteome analysis at the subcellular level of the cyanobacterium *Spirulina platensis* in response to low-temperature stress conditions. *FEMS Microbiology Letters, 288*(1), 92–101.

Hongsthong, A., Sirijuntarut, M., Yutthanasirikul, R., Senachak, J., Kurdrid, P., Cheevadhanarak, S., & Tanticharoen, M. (2009). Subcellular proteomic characterization of the high-temperature stress response of the cyanobacterium *Spirulina platensis*. *Proteome Science, 7*(1), 119.

Hu, J., & Wang, Q. (2018). Regulatory sRNAs in cyanobacteria. *Frontiers in Microbiology, 9*, 2399.

Hu, J., Li, T., Xu, W., Zhan, J., Chen, H., He, C., & Wang, Q. (2017). Small antisense RNA RblR positively regulates RuBisCo in *Synechocystis* sp. PCC 6803. *Frontiers in Microbiology, 8*, 231.

Jacob, S., & Kumar, R. (2020). Sustainable initiative of using cyanobacteria as a liquid fertilizer for hydroponic cultivation: A waste to wealth utilization. *International Journal of Emerging Technologies and Innovative Research, 7*(8), 1430–1461.

Jose, S., Ahmad, M., Renuka, N., Bux, F., & Kumari, S. (2024). Cyanobacteria-green microalgae consortia enhance soil fertility and plant growth by shaping the native soil microbiome of *Capsicum annuum*. *Rhizosphere, 30*, 100892. https://doi.org/10.1016/j.rhisph.2024.100892

Karandashova, I. V., & Elanskaya, I. V. (2005). Genetic control and mechanisms of salt and hyperosmotic stress resistance in cyanobacteria. *Russian Journal of Genetics, 41*(12), 1311–1321.

Khan, Z., Park, S. D., Shin, S. Y., Bae, S. G., Yeon, I. K., & Seo, Y. J. (2005). Management of Meloidogyne incognita on tomato by root-dip treatment in culture filtrate of the blue-green alga, Microcoleus vaginatus. *Bioresource Technology, 96*(12), 1338–1341. https://doi.org/10.1016/j.biortech.2004.11.012

Khati, P., Chaudhary, P., Gangola, S., & Sharma, A. (2017). Nanochitosan-induced growth of *Zea mays* with soil health maintenance. *3 Biotech, 7*(81), 1–9.

Khati, P., Chaudhary, P., Bhatt, P., Nisha, Kumar, R., & Sharma, A. (2018). Effect of nanozeolite and plant growth promoting rhizobacteria on maize. *3 Biotech, 8*(141), 1–12.

Khati, P., Chaudhary, P., Gangola, S., & Sharma, A. (2019a). Influence of nanozeolite on plant growth promotory bacterial isolates recovered from nanocompound-infested agriculture field. *Environment and Ecology, 37*(2), 521–527.

Khati, P., Sharma, A., Chaudhary, P., Singh, A. K., Gangola, S., & Kumar, R. (2019b). High-throughput sequencing approach to assess the impact of nanozeolite treatment on species richness and evenness of soil metagenome. *Biocatalysis and Agricultural Biotechnology, 20*, 101249.

Kokila, V., Prasanna, R., Kumar, A., Nishanth, S., Shukla, J., Gulia, U., Nain, L., Shivay, Y. S., & Singh, A. K. (2022). Cyanobacterial inoculation in elevated CO2 environment stimulates soil C enrichment and plant growth of tomato. *Environmental Technology and Innovation, 26*, 102234.

Kumar, G., Arya, K., Verma, A., Bhatt, P., Khati, P., Gangola, S., Kumar, R., Sharma, A., & Singh, H. (2017). Bioremediation of petrol engine oil polluted soil using microbial consortium and wheat crop. *Journal of Pure and Applied Microbiology, 11*(3), 1583–1588.

Kurdrid, P., Senachak, J., Sirijuntarut, M., Yutthanasirikul, R., Phuengcharoen, P., Jeamton, W., et al. (2011). Comparative analysis of the *Spirulina platensis* subcellular proteome in response to low- and high-temperature stresses: Uncovering cross-talk of signaling components. *Proteome Science, 9*, 39.

Los, D. A., & Murata, N. (1998). Structure and expression of fatty acid desaturases. *BBA – Lipids and Lipid Metabolism, 1394*(1), 3–15.

Mehdizadeh Allaf, M., & Peerhossaini, H. (2022). Cyanobacteria: Model microorganisms and beyond. *Microorganisms, 10*(4), 696.

Mironov, K. S., Sinetova, M. A., Shumskaya, M., & Los, D. A. (2019). Universal molecular triggers of stress responses in cyanobacterium *Synechocystis*. *Life, 9*(3), 67.

Moparthi, V. K., Li, X., Vavitsas, K., Dzhygyr, I., Sandh, G., Magnuson, A., & Stensjö, K. (2016). The two Dps proteins, NpDps2 and NpDps5, are involved in light-induced oxidative stress tolerance in the N2-fixing cyanobacterium *Nostoc punctiforme*. *Biochimica et Biophysica Acta - Bioenergetics, 1857*(11), 1766–1776.

Osman, M. E. H., El-Sheekh, M. M., El-Naggar, A. H., & Gheda, S. F. (2010). Effect of two species of cyanobacteria as biofertilizers on some metabolic activities, growth, and yield of pea plant. *Biology and Fertility of Soils, 46*(8), 861–875.

Pandhal, J., Ow, S. Y., Wright, P. C., & Biggs, C. A. (2008). Comparative proteomics study of salt tolerance between a nonsequenced extremely halotolerant cyanobacterium and its mildly halotolerant relative using in vivo metabolic labeling and in vitro isobaric labeling. *Journal of Proteome Research, 8*(2), 818–828.

Pérez, G., Doldán, S., Borsani, O., & Irisarri, P. (2012). Differential response to moderate UV-B irradiation of two heterocystous cyanobacteria isolated from a temperate rice field. *Advances in Microbiology, 2*(1), 37–47.

Prasanna, R., Kanchan, A., Ranjan, K., Venkatachalam, S., Manjunath, M., Ramakrishnan, B., & Nain, L. (2016). Beneficial cyanobacteria and eubacteria synergistically enhance bioavailability of soil nutrients and yield of okra. *Heliyon, 2*, e00066.

Prasanna, R., Gupta, H., Yadav, V., Gupta, K., Buddhadeo, R., Gogoi, R., Bharti, A., Mahawar, H., & Nain, L. (2020). Prospecting the promise of cyanobacterial formulations developed using soil-less substrates as carriers. *Environmental Technology and Innovation, 18*, 100652.

Prasanna, B. M., Cairns, J. E., Zaidi, P. H., et al. (2021). Beat the stress: Breeding for climate resilience in maize for the tropical rainfed environments. *Theoretical and Applied Genetics, 134*(5), 1729–1752.

Rady, M. M., Taha, S. S., & Kusvuran, S. (2018). Integrative application of cyanobacteria and antioxidants improves common bean performance under saline conditions. *Scientia Horticulturae, 233*, 61–69.

Rai, R., Singh, S., Rai, K. K., Raj, A., Sriwastaw, S., & Rai, L. C. (2021). Regulation of antioxidant defense and glyoxalase systems in cyanobacteria. *Plant Physiology and Biochemistry, 168*, 353–372.

Rai, P., Pathania, R., Bhagat, N., Bongirwar, R., Shukla, P., & Srivastava, S. (2025). Current insights into molecular mechanisms of environmental stress tolerance in cyanobacteria. *World Journal of Microbiology and Biotechnology, 41*(2), 53.

Reddy, M. S., Praveen, R., & Sharma, S. (2018). Cyanobacteria as a tool for enhancing plant growth and combating stress. *Ecological Engineering, 118*, 87–99.

Rossi, F., & De Philippis, R. (2015). Role of cyanobacterial exopolysaccharides in phototrophic biofilms and in complex microbial mats. *Life, 5*(2), 1218–1238.

Sanyal, D., Prasanna, R., Ghosh, P., & Adak, A. (2022). Influence of cyanobacterial inoculation and elevated climate variables on nitrogen fixation and growth of soybean (*Glycine max*). *Journal of Plant Physiology, 270*, 153656.

Shaheen, A., Salem, G., & Ghazal, M. (2022). Impact of cyanobacterial inoculant on growth and productivity of peanut plants in sandy soil. *Biotechnology Journal International, 26*(4), 38–47. https://doi.org/10.9734/bji/2022/v26i4655

Shrivastava, A. K., Chatterjee, A., Yadav, S., Singh, P. K., Singh, S., & Rai, L. (2015). UV-B stress induced metabolic rearrangements explored with comparative proteomics in three *Anabaena* species. *Journal of Proteomics, 127*(Pt A), 122–133.

Singh, P. R., Gupta, A., Singh, A. P., Jaiswal, J., & Sinha, R. P. (2024). Effects of ultraviolet radiation on cellular functions of the cyanobacterium *Synechocystis* sp. PCC 6803 and its recovery under photosynthetically active radiation. *Journal of Photochemistry and Photobiology. B, 252*, 112866.

Sinha, S. K., Singh, R. B., & Kumar, P. (2014). Effect of integrated nutrient management on fertilizer use efficiency, crop yield and soil properties in rice–wheat system. *Indian Journal of Agricultural Sciences, 84*(8), 1015–1019.

Sluchanko, N. N., Slonimskiy, Y. B., Shirshin, E. A., Moldenhauer, M., Friedrich, T., & Maksimov, E. G. (2018). OCP-FRP protein complex topologies suggest a mechanism for controlling high light tolerance in cyanobacteria. *Nature Communications, 9*, 3869.

Srivastava, A., Varshney, R. K., & Shukla, P. (2021a). Sigma factor modulation for cyanobacterial metabolic engineering. *Trends in Microbiology, 29*(3), 266–277.

Srivastava, V., Amanna, R., Rowden, S. J. L., Sengupta, S., Madhu, S., Howe, C. J., & Wangikar, P. P. (2021b). Adaptive laboratory evolution of the fast-growing cyanobacterium *Synechococcus elongatus* PCC 11801 for improved solvent tolerance. *Journal of Bioscience and Bioengineering, 131*(5), 491–500.

Suzuki, L., Kanesaki, Y., Mikami, K., Kanehisa, M., & Murata, N. (2001). Cold-regulated genes under control of the cold sensor Hik33 in *Synechocystis*. *Molecular Microbiology, 40*(2), 235–244.

Tilman, D., Balzer, C., Hill, J., & Befort, B. L. (2011). Global food demand and the sustainable intensification of agriculture. *Proceedings of the National Academy of Sciences of the United States of America, 108*(50), 20260–20264.

Villalobos-López, M. A., Arroyo-Becerra, A., Quintero-Jiménez, A., & Iturriaga, G. (2022). Biotechnological advances to improve abiotic stress tolerance in crops. *International Journal of Molecular Sciences, 23*(19), 12053.

Waditee-Sirisattha, R., & Kageyama, H. (2023). Halotolerance, stress mechanisms, and circadian clock of salt-tolerant cyanobacteria. *Applied Microbiology and Biotechnology, 107*(4), 1–13.

Wahid, A., & Close, T. J. (2007). Expression of dehydrins under heat stress and their relationship with water relations of sugarcane leaves. *Biologia Plantarum, 51*, 104.

Wang, H., Yang, Y., Chen, W., Ding, L., Li, P., Zhao, X., et al. (2013). Identification of differentially expressed proteins of *Arthrospira* (Spirulina) plantensis-YZ under salt-stress conditions by proteomics and qRT-PCR analysis. *Proteome Science, 11*, 6.

Warren, S. D., Stark, L. R., Lewis, L. A., Pombubpa, N., Kurbessoian, T., Stajich, J. E., & Aanderud, Z. T. (2019). Reproduction and dispersal of biological soil crust organisms. *Frontiers in Ecology and Evolution, 7*, 344.

Wei, Y., Xu, X., Tao, H., & Wang, P. (2006). Growth performance and physiological response in the halophyte *Lycium barbarum* grown at salt-affected soil. *The Annals of Applied Biology, 149*(2), 263–269.

Whitton, B. A., & Potts, M. (Eds.). (2007). *The ecology of cyanobacteria: Their diversity in time and space*. Springer.

Xiong, Q., Feng, J., Li, S. T., Zhang, G. Y., Qiao, Z. X., Chen, Z., Wu, Y., Lin, Y., Li, T., Ge, F., & Zhao, J. D. (2015). Integrated transcriptomic and proteomic analysis of the global response of *Synechococcus* to high light stress. *Molecular & Cellular Proteomics, 14*(4), 1038–1053.

Yadav, P., Singh, R. P., Rana, S., Joshi, D., Kumar, D., Bhardwaj, N., Gupta, R. K., & Kumar, A. (2022). Mechanisms of stress tolerance in cyanobacteria under extreme conditions. *Stress, 2*(4), 53.

Nitrogen Cycle Dynamics Under Changing Climatic Conditions: The Impact of Microbes

Atrayee Dey, Somnath Chatterjee, Shubhranil Brahma, and Prasenjit Debbarma

Abstract

Global warming and increasing soil temperature boost the growth and turnover of soil microbial communities. Given the high organic nitrogen content of microbial cell walls, this accelerated microbial activity is expected to drive a faster rate of organic nitrogen cycling in the soil. The quantity of nitrogen sequestered or nitrogen pools within terrestrial soils serves to moderate biogeochemical cycling processes. Environmental factors play a crucial role in regulating both the constitution of soil nitrogen pools and the interconversion of nitrogen forms within these pools. Soil microbes play a significant role in maintaining soil health and fertility through various activities such as moderating the N cycle. Thus, the biological reactions of global N cycling are majorly influenced by the soil microbiome and the ecosystem. Changing climatic conditions may therefore impact the dynamics of the soil N cycle. In this chapter, a sincere effort has been made to establish the correlation between the N cycle and soil microbial communities under changing conditions and their effect on soil health.

A. Dey
Department of Zoology, Banwarilal Bhalotia College, Asansol, West Bengal, India

S. Chatterjee
Department of Zoology, Dr. Bhupendra Nath Dutta Smriti Mahavidyalaya, Burdwan, West Bengal, India

S. Brahma
Department of Zoology, Iswar Chandra Vidyasagar College, Belonia, Tripura, India

P. Debbarma (✉)
Department of Botany, Iswar Chandra Vidyasagar College, Belonia, Tripura, India

G. Bhandari et al. (eds.), *Microbial Resilience*,
https://doi.org/10.1007/978-981-95-8883-1_16

Keywords

Nitrogen cycle · Soil microbiome · Soil health · Climate change

1 Introduction

Climate change, fueled by greenhouse gas emissions from human activities, has led to substantial changes in terrestrial and aquatic ecosystems due to global warming (Priya et al., 2023). Changes in living organisms and the distribution of inorganic and organic compounds are caused by environmental stressors such as rising temperatures and extreme weather events, such as heat waves, storms, floods, and droughts (Mandal & Roy, 2024). For plants, especially those consumed by humans and animals, organic and inorganic compounds are crucial for growth. Nitrogen is a key compound among them (Viancelli & Michelon, 2024). The health of the soil ecosystem heavily relies on nitrogen (N), which is key to producing and transforming soil organic matter and altering the soil's physiochemical and biological characteristics (Sardar et al., 2023). With less than 1% of the world's N present in soils, most is stored in the atmosphere and sedimentary rocks, rendering it inaccessible to plants and impacting forest tree growth (Follett & Hatfield, 2001). Sources of nitrogen in forests include litterfall, biological N fixation (BNF), recycling of tree remnants, and microbial transformations and organisms (Tang et al., 2018). Schmidt et al. (2011) reported that soil organic matter (SOM) serves as the Earth's largest reservoir of organic carbon (C) and nitrogen (N), and its turnover is vital to the cycling of global elements and the climate system. The anticipated rise in global temperatures is expected to hasten the decomposition of SOM by enhancing microbial activity and increasing the expression of enzymes responsible for breaking down high-molecular-weight organic matter. This accelerated process may result in substantial emissions of greenhouse gases, such as CO_2, into the atmosphere, thereby creating a positive feedback loop that intensifies climate change (Bond-Lamberty et al., 2018). Although the processes governing carbon dynamics and their impact on climate change have been extensively studied, soil nitrogen storage and the fate of organic nitrogen have garnered far less attention. This is despite their intrinsic connection to organic carbon and their mutual reliance on temperature (Séneca et al., 2021). The release of nitrogen during the decomposition of soil organic matter directly affects the nitrogen uptake by soil microorganisms, thereby influencing the mineralization rate. Soil microorganisms are pivotal in terrestrial nitrogen cycling (Jansson & Hofmockel, 2020), and research has demonstrated that the availability of nitrogen to these microorganisms can influence the degradation and turnover rates of the various SOM pools (Liu & Greaver, 2010; Knicker, 2011). Within the nitrogen cycle of the soils, the key processes of nitrification, denitrification, and ammonification play crucial roles (Sardar et al., 2023). High-molecular-weight substances such as intracellular and cell wall-bound proteins and nucleic acids with microbial and plant origins make up the organic nitrogen in the soil (Geisseler et al., 2010). Microbes require nitrogen for their growth, and to obtain it, they scavenge the soil, incorporating N into their biomass, thereby rendering it unavailable to plants and other microbes. This process is known as immobilization. The balance between mineralization and immobilization determines the nutrient status of the soil (Li et al., 2021). When mineralization exceeds immobilization, plants and other organisms have a surplus of N. Conversely, if immobilization surpasses mineralization, the soil becomes depleted in N (Robertson & Groffman, 2015).

Microbial residues, including dead microbial cell walls, and their constituents, peptidoglycans, chitin, and cell-wall-bound proteins, contain the majority of organic nitrogen, though, and may account for as much as 80% of the organic nitrogen pool (Wang et al., 2020a, b). In the nitrogen (N) cycle of soils, the processes of nitrification, denitrification, and ammonification are crucial steps. Ammonia (NH_3) is converted into nitrite (NO_2^-) by ammonia-oxidizing archaea and bacteria. Subsequently, nitrite (NO_2^-) is transformed into nitrate (NO_3^-) by nitrite-oxidizing bacteria. Nature and its biodiversity depend on the presence, even in limited quantities, of reactive nitrogen species within the ecosystem. Nitrogen encompasses all nitrogen compounds except N_2, including NH_3, NO_3, NO_2, nitrous oxide (N_2O), amines, and organic nitrogen forms.

Various pieces of research established the fact that soil microbial populations produce hydrolases to decompose plant litter and other organic materials, thereby recycling nutrients and fostering plant growth as part of the nutrient-cycling process in forest environments (Tan et al., 2020; Prasad et al., 2021). As temperatures rise, microbial activity increases, accelerating the decomposition of organic matter. This process releases nitrogen in the form of ammonium (NH_4^+) and nitrate (NO_3^-), which can promote plant growth. However, it can also lead to an increase in nitrous oxide (N_2O) emissions, a powerful greenhouse gas (Mattoo & Suman, 2023). Microbial communities are essential for the nitrogen cycle, and climate change causes major disruptions that can either enhance or reduce their effects through a feedback loop. Microbial communities in the soil can significantly impact soil characteristics, such as nutrient levels and hydrolase enzyme activity, by regulating the decomposition rate of soil organic matter (SOM) (Kooch et al., 2018). Soil microbial populations are greatly impacted by multifaceted soil properties, such as pH, moisture level, and nutrient concentration (Sardar et al., 2023). Therefore, nitrogen fixation, nitrification, denitrification, ammonification, immobilization, and mineralization are all microbial processes that are vital for maintaining soil fertility and supporting plant growth, ultimately contributing to the health and productivity of ecosystems.

Comprehending the interplay between global warming and nitrogen availability is crucial for formulating strategies to mitigate and adapt to climate change. This chapter delves into how global warming affects nitrogen cycling in the environment, the consequent impacts on water quality and ecosystem health, and potential solutions to counteract these adverse effects. Key mitigation strategies include enhancing nitrogen use efficiency in agriculture, embracing sustainable farming practices, and strengthening policies and regulations to manage nitrogen emissions effectively.

2 Global Nitrogen Pools and Climate Change

Nitrogen (N) is a fundamental element in Earth's biogeochemical cycles, essential for biological processes such as protein synthesis and photosynthesis. Nitrogen constitutes approximately 0.3% of the biosphere by weight, ranking as the fourth most abundant element after carbon, oxygen and hydrogen. Conversely, it is a minor constituent of Earth's crust, averaging only 50 ppm in continental crust (Wedepohl, 1995). The global nitrogen pool consists of various reservoirs, including atmospheric nitrogen, soil nitrogen, and oceanic nitrogen. Anthropogenic activities, particularly industrialization and agriculture, have significantly altered nitrogen cycling, leading to environmental consequences such as eutrophication, greenhouse gas emissions, and biodiversity loss.

The following are the major components of the global nitrogen pool:

1. Atmospheric Nitrogen—The atmosphere contains approximately 78% nitrogen in the form of inert N_2 gas. Biological nitrogen fixation and industrial processes convert atmospheric nitrogen into reactive forms.
2. Terrestrial Nitrogen—Soil nitrogen pools include organic nitrogen, ammonium (NH_4^+), nitrate (NO_3^-), and dissolved organic nitrogen (DON). These forms are crucial for plant growth and microbial activity.
3. Aquatic Nitrogen—Oceans and freshwater systems store nitrogen in dissolved inorganic and organic forms. Nitrogen deposition and runoff influence aquatic nitrogen dynamics, affecting marine ecosystems.

The atmosphere is the largest reservoir, primarily composed of dinitrogen (N_2) at 3.7×10^9 Tg, with a minor component of nitrous oxide (N_2O) at 1.4×10^3 Tg (Sorai et al., 2007). Within the Biosphere, marine systems contain 3.0×10^3 Tg (Lerman et al., 2004) and 5.0×10^2 Tg (Sorai et al., 2007), while terrestrial systems hold 5.4×10^3 Tg (Lerman et al., 2004), 2.9×10^4 Tg (Chamiedes & Perdue, 1997), and 7.7×10^3 Tg (Sorai et al., 2007). The Oceanic reservoir includes dissolved N_2 at 1.46×10^6 Tg and N_2O at 0.34 Tg (both calculated from Emerson et al., 2002). Reactive nitrogen forms in the ocean include nitrate (NO_3^-) at 6.0×10^5 Tg (Mackenzie, 1998), particulate organic nitrogen (PON) at 9.0×10^4 Tg (Sorai et al., 2007), and dissolved organic nitrogen (DON) at 8.1×10^5 Tg (Sorai et al., 2007). The geological reservoir represents the longest term storage. Continental crust contains 1.3×10^9 Tg (Wedepohl, 1995), crustal rocks hold 6.4×10^8 Tg (Boyd, 2001), and oceanic crust accounts for 8.9×105 Tg (Li et al., 2007). Sedimentary nitrogen includes coastal sediments at 3.2×10^4 Tg (Lerman et al., 2004) and deep ocean sediments at 2.0×10^9 Tg (Chamiedes & Perdue, 1997). Soil nitrogen, often considered part of the geological or terrestrial long-term reservoir, is reported at 1.4×10^5 Tg (Batjes, 1996) and 2.2×10^4 Tg (Lerman et al., 2004).

Total nitrogen present in the soil ecosystem consists of both organic and inorganic forms that reflect the net balance of all nitrogen entering and leaving the soil. This balance is often influenced by activity in the rhizosphere. Without human intervention, soil nitrogen levels are naturally governed by climatic factors, such as temperature and precipitation, plant N uptake, rhizospheric effects, microbial activities and soil physical and chemical properties (Bytnerowicz et al., 2022; Dubey et al., 2021; Myrold, 2021; Wang et al., 2020a, b). Therefore, anthropogenic activities play a vital role in the alteration of total soil N pools and their cycle in the soil ecosystem.

The projected increase in global temperatures due to climate change is expected to significantly impact nitrogen (N) biogeochemical cycling across various ecosystems, including the soil ecosystem. Nitrogen is a crucial element in ecosystems, cycling between the atmosphere, land, and water through processes such as nitrogen fixation, nitrification, and denitrification. Rising temperatures and changing precipitation patterns influence these processes, leading to shifts in nitrogen availability and distribution. For example, increased temperatures can accelerate microbial activity, enhancing nitrogen mineralization and altering soil nitrogen dynamics (Viancelli & Michelon, 2024). Additionally, climate-induced changes in vegetation patterns can modify nitrogen uptake and storage in ecosystems (Greaver et al., 2016).

3 Biological Reactions of the N Cycle

3.1 Biological N_2 Fixation (BNF)

A vital part of the Earth's ecology, the biological nitrogen cycle involves several activities that change nitrogen (N) into other chemical forms. These transformations maintain life by allowing inert atmospheric nitrogen (N_2) to be transformed into forms that are biologically accessible to living things. The cycle includes several important activities supported by various microbial communities, including nitrogen fixation, ammonification, nitrification, and denitrification.

Nitrogen (N) stands out among the primary soil nutrients because it originates from the atmosphere. The water cycle and biological processes largely govern its transformations and movements within ecosystems. The atmosphere holds a vast, well-mixed reservoir of di-nitrogen (N_2) gas, which is biologically inaccessible. Fowler et al. (2015) established that the atmosphere harbors a vast, well-mixed reservoir of di-nitrogen (N_2) gas, which remains biologically inaccessible. The atmosphere holds approximately 4×10^9 teragrams (Tg) of nitrogen (N), of which a relatively small fraction (473 Tg N) is annually converted into biologically available or reactive nitrogen pools, as calculated by Fowler et al. (2015) and Ladha et al. (2022). The biological fixation of atmospheric N_2 is a critical process in the biosphere, second in importance only to photosynthesis, to maintain life on Earth. Ladha et al. (2022) stated the facts, which were previously established by some investigators, that the key genera of prokaryotic organisms that possess the genetic information required to synthesize nitrogenase and convert gaseous N_2 into NH_3 include *Azotobacter*, *Rhizobium*, *Frankia*, and *Cyanobacteria*. This ammonia can then be biochemically modified to generate various organic forms of nitrogen (Giller, 2001) (Fig. 1). Ammonium (NH_4^+) and nitrate (NO_3^-) are the primary forms

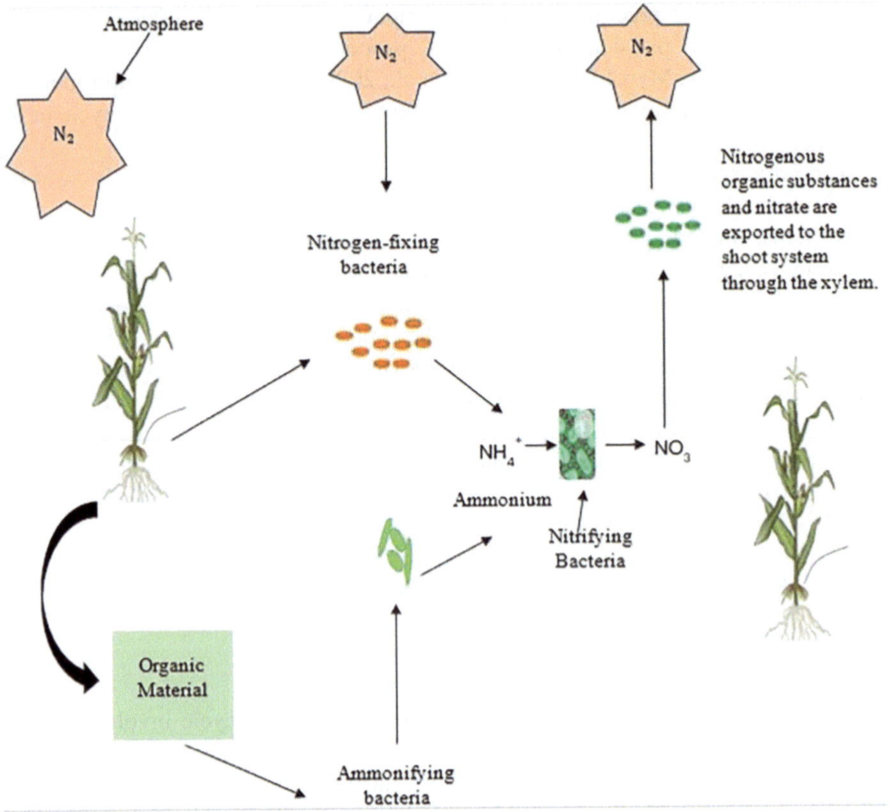

Fig. 1 Schematic diagram depicting biological nitrogen fixation in the soil by microbes

of nitrogen absorbed by plants. Research has demonstrated that a combination of NH_4^+ and NO_3^- is more advantageous for promoting plant growth and enhancing nitrogen uptake (Chen et al., 2023). Biological nitrogen fixation (BNF) is essential for improving soil fertility and addressing climate change by leveraging nitrogen-fixing microbes. These microorganisms transform atmospheric nitrogen (N_2) into forms that plants can use, thereby lessening the dependence on synthetic fertilizers, which are major contributors to greenhouse gas emissions. The subsequent sections will delve into the processes and impacts of BNF in the context of evolving climate conditions (Mattoo & Suman, 2023). Microorganisms collaborate seamlessly, participating in every reaction of the nitrogen cycle to continually transform nitrogen into bio-available forms. As research in this field progresses, it continues to uncover numerous reactions within the nitrogen cycle that remain unknown.

3.2 NH_4 Assimilation

Ammonium (NH_4^+) and nitrate (NO_3^-) are the primary forms of nitrogen absorbed by plants. Research has demonstrated that a combination of NH_4^+ and NO_3^- is more advantageous for promoting plant growth and enhancing nitrogen uptake (Chen et al., 2023). Microbes are critical for ammonium (NH_4^+) assimilation, a key aspect of nitrogen cycling in various ecosystems, particularly in soil and composting environments. Microbial communities adeptly assimilate NH_4^+ and convert it into organic nitrogen forms, enhancing nitrogen retention and reducing environmental losses. This process is guided by specific microbial taxa and shaped by environmental conditions (Zhu et al., 2023). Prominent microbial groups, including *Actinobacteria* and *Proteobacteria*, are crucial in assimilating NH_4^+ during composting and soil processes (Liang et al., 2024). In acidic upland soils, bacteria such as *Bacillus* and fungi such as *Penicillium* play essential roles in assimilating inorganic nitrogen (Han et al., 2023). Mechanisms of ammonia assimilation include the glutamate dehydrogenase (GDH) pathway, which plays a major role in ammonia assimilation, contributing to over half of this process in specific environments (Zhu et al., 2023). The interaction between microbial activity and plant nitrogen absorption is vital for sustainable farming, as it enhances soil quality and boosts crop yields (Quadriya et al., 2020). Ammonia assimilation is essential for maintaining nitrogen balance and ecosystem stability; however, environmental conditions such as temperature and soil composition shape its efficiency in nitrogen absorption and transformation.

3.3 Mineralization

Nitrogen is a critical nutrient for plants, but they can only take it up in certain forms. Most of the nitrogen in the soil is indeed in organic forms, such as proteins and amino acids, which plants cannot directly use. The process you are referring to, which converts organic nitrogen into ammonium (NH_4^+), is part of the nitrogen cycle, and it is known as mineralization. Mineralization is done by soil microorganisms, such as bacteria and fungi, which break down organic matter and release nitrogen as ammonium. Once the ammonium is in the soil, it can either be taken up by plants or undergo further transformation into nitrate (NO_3^-) through a process called nitrification, which is carried out by specific soil bacteria. Nitrate is another form of nitrogen that plants can easily absorb.

This whole cycle is essential for maintaining soil fertility and ensuring that plants have access to the nitrogen they need for growth (Kandeler, 1996). Microbes incorporate nitrogen into their biomass during growth, effectively sequestering it and making it temporarily unavailable to plants and other organisms, this is known as immobilization.

The nutrient status of the soil hinges on the balance between these two processes:

- When *mineralization* > *immobilization*: There is a surplus of nitrogen, which benefits plant growth.
- When *immobilization* > *mineralization*: The soil becomes depleted of nitrogen, making it less available for plants and other organisms.

Essentially, this balance is crucial in maintaining soil fertility and influencing plant growth and ecosystem health (Robertson & Groffman, 2007). Through mineralization, organic nitrogen is transformed into ammonium, which plants absorb and utilize for growth. This ongoing process guarantees a steady supply of vital nitrogen for plants. By ensuring the availability of nitrogen, mineralization sustains the entire food chain, from plants to herbivores and predators. Microorganisms play a crucial role in the nitrogen cycle, maintaining the steady supply of this vital nutrient for plants, which ultimately sustains the entire ecosystem.

Increased levels of atmospheric CO_2 can boost nitrogen mineralization rates in forest soils. Research indicates that with CO_2 fertilization, trees direct more carbon to their roots, which stimulates microbial activity and raises net mineralization rates by around 30% (Rumeau et al., 2023). The enhanced availability of ammonium (NH_4^+) is a key outcome, as nitrification and nitrate (NO_3^-) availability remain unaffected, indicating a shift in nitrogen cycling dynamics under elevated CO_2 conditions (Rumeau et al., 2023). Through their metabolic activities, fungal and bacterial communities control nitrogen mineralization. For example, fungal diversity is closely related to soil resistance to acidification and resource utilization, impacting nitrogen cycling under canopy nitrogen deposition. Conversely, bacterial diversity is linked to fast turnover rates and disturbance resilience, significantly contributing to nitrogen cycling in understory nitrogen addition treatments (Yang et al., 2024).

3.4 Nitrification

Nitrification, a key process in the nitrogen cycle, involves the transformation of ammonia (NH_3) into nitrite (NO_2^-) and subsequently into nitrate (NO_3^-). This conversion is vital for making nitrogen accessible to plants and other organisms, while also impacting greenhouse gas emissions and water quality. Recent research has significantly expanded our understanding of the mechanisms, microbial agents, and environmental factors that influence nitrification. Nitrification is an aerobic process consisting of two steps, facilitated by chemolithoautotrophic microorganisms. In the first step, ammonia is oxidized to hydroxylamine (NH_2OH) and then to nitrite (NO_2^-) by ammonia-oxidizing bacteria (AOB) and archaea (AOA). In the second step, nitrite is further oxidized to nitrate (NO_3^-) by nitrite-oxidizing bacteria (NOB) (Daims et al., 2015). In the past, it was believed that different groups of microorganisms were responsible for these steps. However, some findings have revealed the existence of complete nitrifiers, such as Comammox Nitrospira, which can carry out both steps (Yu et al., 2018; Daims et al., 2015). Nitrosomonas and Nitrosospira, which are types of ammonia-oxidizing bacteria (AOB), are vital for the first stage of nitrification. These bacteria, as obligate autotrophs, derive their energy from ammonia oxidation. The activity and community composition of AOB are highly influenced by environmental conditions, such as pH, temperature, and oxygen levels (Raszka et al., 2011). Research has demonstrated that ammonia-oxidizing bacteria (AOB) can adjust to high ammonium levels and stressful environments. For instance, in soils contaminated with hydrocarbons, AOB develops resistance to the inhibitory effects of hydrocarbons, enabling them to sustain nitrification activity. Moreover, ammonia-oxidizing bacteria (AOB) have been observed to prevail in nitrification processes within agricultural soils, where nitrogen deposition boosts their population and activity (Song & Niu, 2022).

In acidic upland soils, active microbial taxa, such as *Bacillus* and *Penicillium*, tend to favor the assimilation of inorganic nitrogen, primarily in the form of NH_4^+ and NO_3^-. Long-term fertilization influences microbial community composition and the rates at which nitrogen is assimilated (Han et al., 2023).

3.5 Denitrification

Denitrification is a vital microbial process that converts nitrates into nitrogen gas, playing a key role in minimizing nitrogen pollution across different ecosystems. This extraordinary process relies on the activities of diverse microbial communities, whose complex interactions and specialized functions are crucial for efficient nitrogen elimination. The subsequent sections gracefully illustrate the significant roles that microbes play in denitrification. Denitrification starts with nitrates (NO_3^-), a type of nitrogen commonly present in soil and water. These nitrates originate from natural processes such as decomposition, as well as human activities such as the use of fertilizers in agriculture and the release of wastewater. Denitrification relies on a variety of microorganisms, including denitrifying bacteria such as *Acinetobacter baumannii* and *Pseudomonas aeruginosa*, which collaborate efficiently to reduce nitrates (He et al., 2024). Ammonia-oxidizing bacteria (AOB) and archaea (AOA) contribute significantly to the nitrification-denitrification process, boosting the overall efficiency of nitrogen removal (Sumitha et al., 2024). Microbial communities carry out enzyme-driven reactions to transform nitrate into nitrogen gas, with certain strains demonstrating the capability for complete denitrification. Pathway partitioning, a process where various microbial species collaborate to divide the tasks involved in denitrification, has been found to improve efficiency while minimizing the accumulation of byproducts such as nitrous oxide (N_2O) (Isokpehi et al., 2024). The process of denitrification generally relies on four essential enzymes—nitrate reductase, nitrite reductase, nitric oxide reductase, and nitrous oxide reductase—that work sequentially to convert nitrate into nitrogen gas (Isokpehi et al., 2024). Certain bacteria, such as *Rhodopseudomonas palustris*, are capable of carrying out partial denitrification, employing only a subset of the required enzymes and depending on interactions within microbial communities to finalize the process (LaSarre et al., 2024). Recent studies have revealed a vast collection of microbial genomes linked to denitrification, with over 62,000 genomes containing the genes for the required enzymes. This underscores the widespread presence of denitrification as an ecological trait across various environments (Isokpehi et al., 2024).

3.6 Anammox

The anammox process holds a pivotal role in the nitrogen cycle by facilitating the direct conversion of ammonium and nitrite into dinitrogen gas, thereby streamlining nitrogen removal, particularly in wastewater treatment systems. This method not only improves the efficiency of nitrogen removal but also significantly reduces energy consumption when compared to conventional techniques. The subsequent sections provide an in-depth discussion on the importance of anammox bacteria and their contribution to the nitrogen cycle. Anammox bacteria play a remarkable role by anaerobically oxidizing ammonium with nitrite serving as the electron acceptor, leading to substantial nitrogen loss in marine ecosystems. This process is also invaluable in wastewater treatment, as it reduces the need for aeration, thereby reducing operational expenses (Bielski & Islam, 2024). Anammox rates demonstrate notable differences across ecosystems, averaging 1.60 ± 0.17 nmol Ng^{-1} h^{-1} in terrestrial settings (Yao et al., 2023). Wetlands display the highest rates at 2.17 ± 0.31 nmol Ng^{-1} h^{-1}, whereas forests and grasslands record comparatively lower rates. These variations are influenced by factors such as soil moisture, organic carbon levels, and nitrogen content, which vary across different ecosystem types (Yao et al., 2023). Anammox serves as a crucial pathway for nitrogen loss, complementing denitrification, though its rates are considerably lower across various soil types. In agricultural soils, anammox contributes to increased N_2 emissions, highlighting its significance in strategies for managing nitrogen effectively (Yan et al., 2024). Although anammox plays a crucial role in nitrogen cycling, its rates are frequently eclipsed by denitrification, which predominantly drives nitrogen loss in numerous soils. Gaining insight into both processes is vital for creating efficient nitrogen management strategies (Figs. 2 and 3).

4 Influence of Climate Change on Soil Nitrogen Availability

Soil nitrogen availability is influenced by various biological, chemical, and physical processes, including mineralization, nitrification, denitrification, and leaching. Climate change, driven by increasing atmospheric CO_2 concentrations, rising temperatures, altered precipitation patterns, and extreme weather events, has profound effects on these processes.

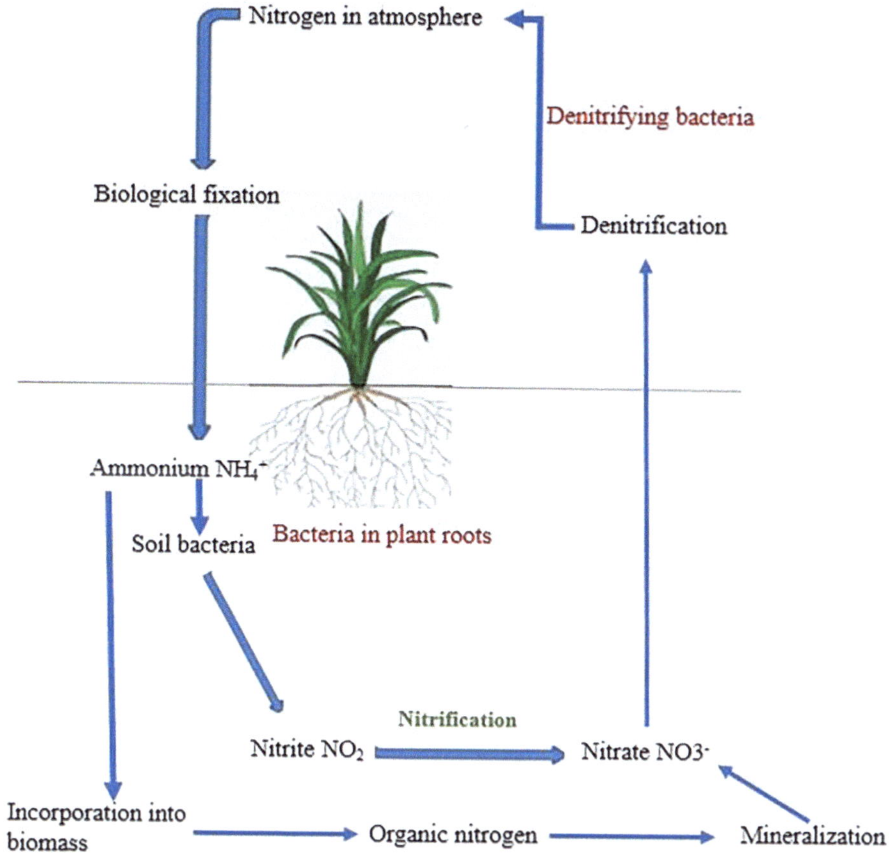

Fig. 2 The nitrogen cycle is a fascinating process. Atmospheric nitrogen (N_2) is converted into ammonium (NH_4^+) by nitrogen-fixing bacteria in the soil or through industrial processes. Ammonium is transformed into nitrite (NO_2^-) and then into nitrate (NO_3^-) by nitrifying bacteria. Plants absorb nitrate from the soil to build proteins and other organic compounds. When plants and animals die or excrete waste, organic nitrogen is converted back into ammonium by decomposing bacteria. Denitrifying bacteria convert nitrate back into nitrogen gas (N_2), releasing it into the atmosphere and completing the cycle (Magray et al., 2022)

4.1 Climate Change Factors Affecting Soil Nitrogen Availability

4.1.1 Rising Temperatures

Increased temperatures generally accelerate microbial activity, enhancing nitrogen mineralization rates (Rustad et al., 2001). However, excessive warming can reduce soil moisture, limiting microbial activity and nitrogen availability (Butler et al., 2012). Faster decomposition of organic matter increases NH_4^+ and NO_3^- release. Higher temperatures may promote denitrification, leading to N loss as N_2O, a potent greenhouse gas (Bai et al., 2013).

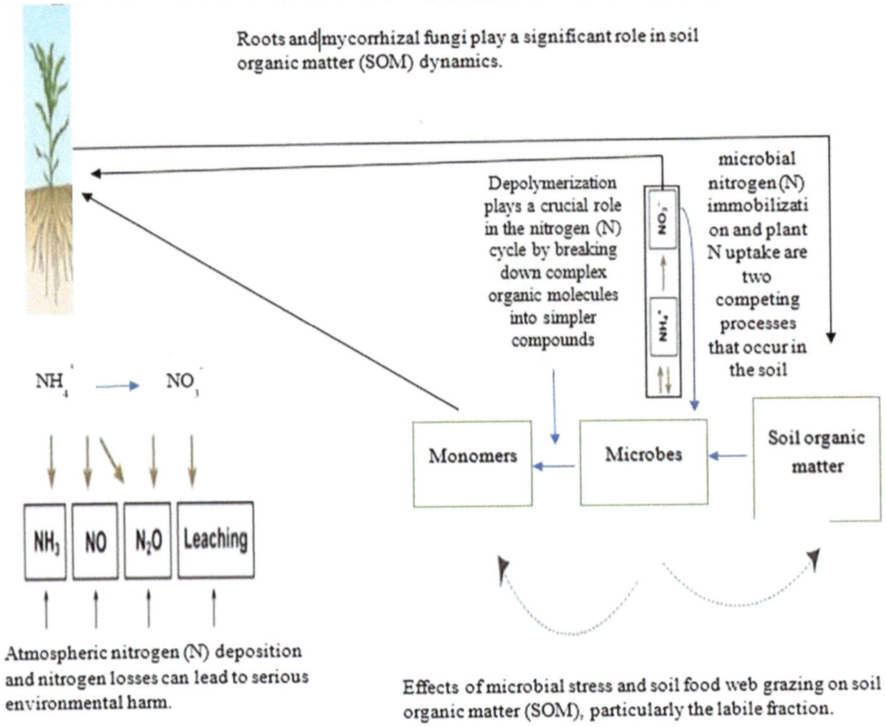

Fig. 3 Nitrogen mineralization is a critical process in the nitrogen cycle. It converts organic nitrogen compounds in soil organic matter into inorganic forms that are accessible to plants

4.1.2 Altered Precipitation Patterns

Changes in rainfall intensity and frequency influence soil moisture, affecting nitrogen cycling: Increased rainfall enhances the leaching of nitrate, reducing plant-available nitrogen (Davidson et al., 2012). Drought conditions can also reduce microbial activity, slowing mineralization and nitrification (Austin et al., 2004).

4.1.3 Elevated Atmospheric CO_2

Higher CO_2 levels stimulate plant growth, increasing nitrogen demand. However, this can lead to nitrogen dilution in plants. Plants may exhibit lower tissue N concentrations, reducing litter quality and slowing decomposition (Luo et al., 2004). Increased C: N ratio in plant residues may lead to microbial N immobilization, temporarily reducing soil N availability (Hu et al., 2001).

4.1.4 Extreme Weather Events

Floods and droughts can disrupt nitrogen cycling. Flooding creates anaerobic conditions, promoting denitrification and N loss (Groffman et al., 2009). Droughts reduce microbial activity, decreasing mineralization and nitrification (Sardans et al., 2008).

Climate change significantly alters soil nitrogen availability through complex interactions between temperature, moisture, and CO_2 levels. Understanding these dynamics is crucial for sustainable agriculture and ecosystem management. Future research should focus on adaptive strategies to mitigate negative impacts on soil fertility and global nitrogen cycles.

5 Paradigm Shift of Global N Cycle Due to Changing Climatic Conditions

A major paradigm shift in the global nitrogen (N) cycle is occurring as a result of heightened human activity and shifting climate conditions. Anthropogenic nitrogen inputs from industry, agriculture, and the burning of fossil fuels have significantly disturbed the nitrogen cycle, which was once a comparatively steady natural process. By changing temperature, precipitation patterns, and the frequency of extreme weather events, climate change makes these disruptions even more difficult and speeds up nitrogen transformation processes, including nitrification, denitrification, and volatilization. These modifications lead to feedback loops that worsen global warming by increasing emissions of nitrous oxide (N_2O)—a powerful greenhouse gas. Furthermore, increased nitrogen deposition promotes plant growth at first but ultimately causes soil acidification, biodiversity loss, and imbalances in the carbon cycle. Reduced biological thresholds due to climate change, along with the emergence of nitrogen hotspots in tropical and sensitive regions, indicate the urgent need to rethink nitrogen management approaches. As a result, the nitrogen cycle is now viewed as an active, dynamic, and climate-sensitive system that needs careful observation and action rather than a passive backdrop.

5.1 Soil Physicochemical Properties Influencing Nitrogen Cycle

Shaw and Harte (2001) showed that nitrogen mineralization increases as the temperature rises to around 30 °C, and a similar pattern is observed for nitrification as well. According to Butterbach-Bahl et al. (2011), fine-textured soils may store more water than sandy soils and they also often contain larger levels of soil organic carbon. After heavy rainfall events, fine-textured soils will exhibit greater soil moisture values and more frequent and intense anaerobic conditions at the same precipitation amount. Because the primarily anaerobic denitrification process is stimulated more often in fine-textured soils, gaseous N losses as well as N_2O and NO emissions are increased. However, nitrate leaching is encouraged by sandy soils' enhanced soil drainage and decreased ion exchange capacity. Apart from this, the C:N ratio, soil pH, and soil organic carbon content (SOC) are other significant variables influencing N cycling (Li et al., 2005; Gundersen et al., 1998; Klemedtsson et al., 2005; Stevens et al., 1998; Kesik et al., 2006).

5.2 Soil Ammonia-Oxidizing Community Influencing Nitrogen Cycle

According to Zhang et al. (2023) nitrogen stimulation of nitrification is maintained by the shifts in soil ammonia-oxidizing community. The researchers compiled 68 field researches from around the globe to assess how N loading affects nitrification and soil ammonia oxidizers. Their investigation depicted climate affects soil pH, N and moisture and eventually regulates ammonia-oxidizing archaea: ammonia-oxidizing bacteria (AOA:AOB) to N loading. Prosser and Nicol (2012) opined that AOB can be favored by alkaline soil and that AOA grows well in acid soil, and this is also corroborated with the data of Zhang et al. (2023). Zhang et al. (2023) also confirmed that as N mineralization rate increases response of AOA also increases, whereas increment of AOB response follows the N loading rate, which was previously proved by Prosser and Nicol (2012). According to Liao et al. (2022) and Yue et al. (2021), soil moisture has a non-significant relationship with AOA and a negative relationship with AOB. The study of Zhang et al. (2023) also shows that interaction between nitrification and AOA:AOB creates a "U" shaped curve, which confirms that in different climatic conditions change in the community structure can shift microbial function.

5.3 Climate Change Influencing the Microbial Abundance

Andrade-Linares et al. (2021) worked on effect of climate change on the microbial abundance in montane grasslands in Europe, which are related to nitrogen turnover. Europe's montane grasslands are subject to temperature increases twice as rapid as the world average. A rise in land use intensity may accompany changes in climate because of the need to boost productivity and the lengthening of the vegetative period. The data of the investigators indicate that, even after a brief 1-year simulation period, there have been noticeable changes in the number of microorganisms that drive the N cycle in soil due to both land use intensification and climate change (Andrade-Linares et al., 2021).

5.4 Factors Influencing Ammonia Oxidation

According to Waqas et al. (2021) nitrification may be impacted in opposing ways by ongoing climate change and increased atmospheric CO_2. The investigators assessed the effects of warming and CO_2 enrichment on the community structure of ammonia-oxidizing archaea (AOA) and bacteria (AOB) by a lengthy 9 years experiment by increasing temperature and CO_2 concentration. According to Schleper (2010), ammonia oxidation is performed by communities of ammonia-oxidizing archaea and ammonia-oxidizing bacteria. Bacteria belonging to the genus *Thaumarchaeota* are members of AOA (Brochier-Armanet et al., 2008; Pester et al., 2011), while AOB includes *Betaproteobacteria* and *Gammaproteobacteria* (Purkhold et al.,

2000). Li et al. (2020) and Nguyen et al. (2019) stated that the composition and characteristics of nitrifiers are controlled by the physicochemical properties such as soil pH, total N content, and soil temperature. Kuzyakov et al. (2018) opined that the nitrification process can be modulated by increased CO_2. Long et al. (2011) in their experiment in North Carolina Forest soil proved that the effect of elevated CO_2 on gene abundance of AOB and AOA is completely different, where it can increase the gene abundance of AOA, but not in AOB. Study of Jia et al. (2021) also stated that the nitrifying community structure of *Robinia pseudoacacia* L. was significantly altered by increased CO_2. A study in Sichuan, China, by Séneca et al. (2020) and Zhang et al. (2017) confirmed that nitrifier can increase in number in elevated temperature. Additionally, some investigations suggested that in dry and arid regions, AOA communities are more susceptible to a brief rise in temperature than AOB groups (Hu et al., 2015; Zhang et al., 2019). In a nutshell, Waqas et al. (2021) proposed that the potential nitrification rate was increased by the tested climate conditions, which will eventually affect the environmental pollution and primary productivity. The investigators further stated that the AOB community can be regulated by the soil physicochemical properties.

5.5 Temperature Influencing Microbial Growth

According to Lee et al. (2010), microbial growth is very much dependent on the temperature. Zhang et al. (2009) stated that survivability of AOB is adapted to 4–45 °C—a wide range and in every 10 °C increase in temperature also resulted doubling of maximum specific growth rate of *Nitrosomonas* (USEPA, 1993). Thus, the growth of AOB can foster the nitrification process. Yang et al. (2007) proved that decreasing pH leads to decreased ammonia. Oldenburg et al. (2002) and Zhang et al. (2021) proved that AOB gets inactivated in higher pH.

6 Climate Change Impact on Soil Microbiome

The soil microbiome, a complex and dynamic community of microorganisms, plays a crucial role in ecosystem functioning, including nutrient cycling, plant health, and carbon sequestration. They mediate critical processes such as organic matter decomposition, nitrogen fixation, and greenhouse gas emissions (Fierer, 2017). Climate change is expected to disrupt these microbial communities, with cascading effects on soil fertility, plant productivity, and climate feedback loops (Classen et al., 2015). Climate change, driven by rising temperatures, altered precipitation patterns, and increased atmospheric CO_2, has profound effects on soil microbial communities. These changes can disrupt soil fertility, carbon sequestration, and agricultural productivity.

6.1 Temperature Rise

Increased soil temperatures alter microbial metabolic rates, community composition, and enzyme activities (Classen et al., 2015). Warmer conditions may favor thermophilic microbes while reducing the abundance of cold-adapted species. Elevated temperatures can also accelerate organic matter decomposition, leading to higher CO_2 emissions (Crowther et al., 2016).

6.2 Altered Precipitation Patterns

Changes in rainfall frequency and intensity influence soil moisture, affecting microbial survival and activity. Drought conditions reduce microbial biomass and diversity, while excessive rainfall can lead to anaerobic conditions, favoring methanogens and denitrifiers (Schimel, 2018). Fungi-dominated systems may become more prevalent under drought, while bacteria thrive in wetter conditions (de Vries et al., 2018).

6.3 Elevated CO_2 Levels

Higher CO_2 concentrations enhance plant growth, increasing root exudates that shape microbial communities. However, shifts in microbial composition may alter nitrogen cycling, potentially limiting plant-available nutrients (van Groenigen et al., 2014).

6.4 Extreme Weather Events

Floods, droughts, and heatwaves cause abrupt shifts in microbial communities, reducing resilience and functional redundancy (Jansson & Hofmockel, 2020).

6.5 Impact of Soil pH

Climate change can also influence soil pH and nutrient availability, which in turn can affect microbial communities. The soil microbial community is the main indicator, which has a crucial role in the remediation of polluted soils. Ecological stability is achieved in a pollution-free environment and pH value. The pH factor has a significant impact on the dynamics of microbes' communities (Naz et al., 2022). Acidification can hinder the growth of certain microorganisms, particularly those that prefer neutral to alkaline conditions (Zhou et al., 2024).

Climate change significantly alters soil microbiomes, with cascading effects on ecosystems and agriculture. Understanding these impacts is critical for developing sustainable soil management practices. Future research should focus on microbial adaptation mechanisms and their role in climate feedback loops.

7 Effect of Shifted N Cycle on Soil Fertility

Natural nitrogen dynamics have been drastically changed by the shifting nitrogen cycle, which is mostly the result of human activities, including industrial fertilizer application, burning fossil fuels, and increased agriculture. This has had detrimental effects on the environment and agriculture. Even while nitrogen is a necessary ingredient for plant growth, adding too much of it to ecosystems upsets the natural equilibrium and frequently. Long-term soil fertility and ecosystem health have been adversely affected by this disturbance, which has resulted in a number of environmental effects such as microbial imbalances, nutrient leaching, and soil acidification (Hu, 2018). According to Karimi et al. (2023), while nitrogen additions initially increase crop yields, overuse or improper management frequently causes organic matter to degrade and vital soil nutrients to be depleted.

7.1 Increased Soil Acidification

Tian and Niu (2015) have done a meta-analysis summing up 106 studies, which sheds light on the soil acidification after N addition. The researchers confirmed that soil pH reduced by 0.26 on average every year. This decrease is more evident in grasslands. More acidity of the soil was caused by the addition of urea and NH_4NO_3 than by NH_4-form fertilizer (Tian & Niu, 2015). The researchers also confirmed that transition of base cations (Ca^{2+}, Mg^{2+}, and K^+) to non-base cations (Mn^{2+} and Al^{3+}) is now occurring in soils all over the world.

Lu et al. (2014) carried out a long-term experiment in a nitrogen-rich tropical lowland forest in Southern China, where they added ammonium nitrate (NH_4NO_3) to study nitrogen deposition. Their results showed that the forest soil had undergone severe acidification, leaving it highly vulnerable due to depleted base cations, reduced base saturation, and a negative acid-neutralizing capacity when tested with water extracts.

7.2 Decline in Soil Organic Matter

Jian et al. (2016) proved that soil organic carbon (SOC) decomposition rate is affected by nitrogen fertilization. The investigators have summed up a total of ten enzymes responsible for extracellular enzyme activities (EEA) from the secondary data. The study demonstrated how nitrogen fertilization affected the activities of soil microbial biomass carbon (MBC), soil nitrogen (TN), hydrolase, oxidase, and SOC. Jian et al. (2016) demonstrated that RRs of the combined hydrolases linked to C acquisition enzymes (i.e., C-acq) and MBC were strongly positively correlated. This indicates that variations in the activity of coupled hydrolases could serve as a stand-in for MBC during N fertilization (Jian et al., 2016). However, this relationship varied between the soils of farmland and forest. The investigators also stated that they witnessed a decline in MBC by 9% upon N fertilization which corroborated the previous study of Treseder (2008).

7.3 Decline in Soil Microbial Diversity

Ramirez et al. (2012) performed a fantastic experiment by collecting soil types from various areas in North America. Over the course of the year-long incubation, N addition reduced microbial respiration rates by an average of 11% and microbial biomass by 35%. After N addition, they observed that relative abundance of *Actinobacteria* and *Firmicutes* increased whereas a decline in *Acidobacteria* and *Verrucinicrobia* occurred.

Fierer et al. (2011) also proved that the prevailing microbial life-history strategies may change as a result of N fertilization, either directly or indirectly, preferring a more active copiotrophic over oligotrophic microbial community. The investigators' analyses of 16S rRNA genes using pyrosequencing showed that N fertilization had no discernible impact on bacterial diversity. However, community composition was influenced significantly—increase of Phyla *Proteobacteria* and *Bacteroidetes* and a decline of *Acidobacteria* upon N addition.

8 Conclusion

In conclusion, the intricate dynamics of the nitrogen cycle are profoundly susceptible to shifts in climatic conditions, with microbial communities emerging as pivotal mediators of these changes. As global temperatures rise and precipitation patterns become more erratic, the delicate balance of nitrogen transformations—including nitrogen fixation, nitrification, denitrification, and ammonification—is significantly altered. Microbes, as the primary drivers of these biogeochemical processes, respond to environmental perturbations in complex ways, affecting the availability of nitrogen for plant growth.

Climate change impacts microbial function, which in turn influences nitrogen cycling, further impacting the climate. Understanding the resilience and vulnerability of these microbial communities is therefore paramount. Future research must continue to unravel the precise mechanisms by which individual microbial groups and their complex interactions respond to multifaceted environmental stressors. Ultimately, a deeper comprehension of these microbial-driven dynamics is indispensable for developing effective strategies to mitigate the ecological consequences of climate change and to sustain the essential ecosystem services provided by the nitrogen cycle.

References

Andrade-Linares, D. R., Zistl-Schlingmann, M., Foesel, B., Dannenmann, M., Schulz, S., & Schloter, M. (2021). Short term effects of climate change and intensification of management on the abundance of microbes driving nitrogen turnover in montane grassland soils. *Sci Total Environ, 780*, 146672. https://doi.org/10.1016/j.scitotenv.2021.146672

Austin, A. T., Yahdjian, L., Stark, J. M., et al. (2004). Water pulses and biogeochemical cycles in arid and semiarid ecosystems. *Oecologia, 141*, 221–235.

Bai, E., Li, S., Xu, W., Li, W., Dai, W., & Jiang, P. (2013). A meta-analysis of experimental 669 warming effects on terrestrial nitrogen pools and dynamics. *The New Phytologist, 199*(2), 441–670.

Batjes, N. H. (1996). Total carbon and nitrogen in the soils of the world. *European Journal of Soil Science, 47*, 151–163.

Bielski, R. G., & Islam, M. A. (2024). Pangenome-scale mathematical modelling of ANAMMOX bacteria metabolism. *SynBio, 2*(1), 70–84. Biochem. 2011;43:1118–29.

Bond-Lamberty, B., Bailey, V. L., Chen, M., Gough, C. M., & Vargas, R. (2018). Globally rising soil heterotrophic respiration over recent decades. *Nature, 560*, 80–83.

Boyd, S. R. (2001). Nitrogen in future biosphere studies. *Chemical Geology, 176*, 1–30.

Brochier-Armanet, C., Boussau, B., Gribaldo, S., & Forterre, P. (2008). Mesophilic crenarchaeota: Proposal for a third archaeal phylum, the Thaumarchaeota. *Nature Reviews. Microbiology, 6*(3), 245–252. https://doi.org/10.1038/nrmicro1852

Butler, S. M., Melillo, J. M., Johnson, J. E., et al. (2012). Soil warming alters nitrogen cycling in a New England Forest: Implications for ecosystem function and structure. *Oecologia, 168*, 819–828.

Butterbach-Bahl, K., Gundersen, P., Ambus, P., Augustin, J., Beier, C., Boeckx, P., Dannenmann, M., Gimeno, B. S., Ibrom, A., Kiese, R., Kitzler, B., Rees, R. M., Smith, K. A., Stevens, C., Vesala, T., & Zechmeister-Boltenstern, S. (2011). Nitrogen processes in terrestrial ecosystems. In *The European nitrogen assessment: Sources, effects and policy perspectives* (pp. 99–125). Cambridge University Press eBooks. https://doi.org/10.1017/cbo9780511976988.009

Bytnerowicz, T. A., Akana, P. R., Griffin, K. L., & Menge, D. N. (2022). Temperature sensitivity of woody nitrogen fixation across species and growing temperatures. *Nature Plants, 8*(3), 1–8.

Chamiedes, W. L., & Perdue, E. M. (1997). *Biogeochemical cycles: A computer-interactive study of earth system science and global change*. Oxford University Press.

Chen, M., Zhu, K., Xie, J., Liu, J., Qiao, Z., Tan, P., & Peng, F. (2023). Ammonium-nitrate mixtures dominated by NH4+-N promote the growth of pecan (*Carya illinoinensis*) through enhanced N uptake and assimilation. *Frontiers in Plant Science, 14*, 1186818.

Classen, A. T., Sundqvist, M. K., Henning, J. A., et al. (2015). Direct and indirect effects of climate change on soil microbial and soil microbial-plant interactions: What lies ahead? *Ecosphere, 6*, art130.

Crowther, T. W., Todd-Brown, K. E., Rowe, C. W., Wieder, W. R., Carey, J. C., Machmuller, M. B., Snoek, B. L., Fang, S., Zhou, G., Allison, S. D., Blair, J. M., Bridgham, S. D., Burton, A. J., Carrillo, Y., Reich, P. B., Clark, J. S., Classen, A. T., Dijkstra, F. A., Elberling, B., Emmett, B. A., Estiarte, M., Frey, S. D., Guo, J., Harte, J., Jiang, L., Johnson, B. R., Kröel-Dulay, G., Larsen, K. S., Laudon, H., Lavallee, J. M., Luo, Y., Lupascu, M., Ma, L. N., Marhan, S., Michelsen, A., Mohan, J., Niu, S., Pendall, E., Peñuelas, J., Pfeifer-Meister, L., Poll, C., Reinsch, S., Reynolds, L. L., Schmidt, I. K., Sistla, S., Sokol, N. W., Templer, P. H., Treseder, K. K., Welker, J. M., & Bradford, M. A. (2016). Quantifying global soil carbon losses in response to warming. *Nature, 540*(7631), 104–108. https://doi.org/10.1038/nature20150. PMID: 27905442.

Daims, H., Lebedeva, E., Pjevac, P., et al. (2015). Complete nitrification by Nitrospira bacteria. *Nature, 528*, 504–509.

Davidson, E. A., Samanta, S., Caramori, S. S., & Savage, K. (2012). The Dual Arrhenius and Michaelis-Menten kinetics model for decomposition of soil organic matter at hourly to seasonal time scales. *Global Change Biology, 18*(1), 371–384.

de Vries, F. T., Griffiths, R. I., Bailey, M., Craig, H., Girlanda, M., Gweon, H. S., Hallin, S., Kaisermann, A., Keith, A. M., Kretzschmar, M., Lemanceau, P., Lumini, E., Mason, K. E., Oliver, A., Ostle, N., Prosser, J. I., Thion, C., Thomson, B., & Bardgett, R. D. (2018). Soil bacterial networks are less stable under drought than fungal networks. *Nature Communications, 9*(1), 3033.

Dubey, R. S., Srivastava, R. K., & Pessarakli, M. (2021). Physiological mechanisms of nitrogen absorption and assimilation in plants under stressful conditions. In *Handbook of plant and crop physiology* (pp. 579–616). CRC Press.

Emerson, S., Stump, C., Johnson, B., & Karl, D. M. (2002). In situ determination of oxygen and nitrogen dynamics in the upper ocean. *Deep-Sea Research Part I, 49*, 941–952.

Fierer, N. (2017). Embracing the unknown: Disentangling the complexities of the soil microbiome. *Nature Reviews. Microbiology, 15*(10), 579–590.

Fierer, N., Lauber, C. L., Ramirez, K. S., Zaneveld, J., Bradford, M. A., & Knight, R. (2011). Comparative metagenomic, phylogenetic and physiological analyses of soil microbial communities across nitrogen gradients. *The ISME Journal, 6*(5), 1007–1017. https://doi.org/10.1038/ismej.2011.159

Follett, R. F., & Hatfield, J. L. (2001). Nitrogen in the environment: Sources, problems, and management. *ScientificWorldJournal, 1*(Suppl 2), 920–926. https://doi.org/10.1100/tsw.2001.269

Fowler, D., Steadman, C. E., Stevenson, D., Coyle, M., Rees, R. M., Skiba, U. M., Sutton, M. A., Cape, J. N., Dore, A. J., Vieno, M., Simpson, D., Zaehle, S., Stocker, B. D., Rinadi, M., Facchini, M. C., Flechard, C. R., Nemitz, E., Twigg, M., Erisman, J. W., Butterbach-Bahl, K., & Galloway, J. N. (2015). Effects of global change during the 21st century on the nitrogen cycle. *Atmospheric Chemistry and Physics, 15*, 13849–13893.

Geisseler, D., Horwath, W. R., Joergensen, R. G., & Ludwig, B. (2010). Pathways of nitrogen utilization by soil microorganisms—A review. *Soil Biology and Biochemistry, 42*, 2058–2067.

Giller, K. E. (2001). Nitrogen Fixation in Tropical Cropping Systems. 2nd edn. *CAB International*, Wallingford, UK, 423 pp.

Greaver, T. L., Clark, C. M., Compton, J. E., Vallano, D., Talhelm, A. F., Weaver, C. P., et al. (2016). Key ecological responses to nitrogen are altered by climate change. *Nature Climate Change, 6*(9), 836–843.

Groffman, P. M., Davidson, E. A., & Seitzinger, S. (2009). New approaches to modeling denitrification. *Biogeochemistry, 93*, 1–5. https://doi.org/10.1007/s10533-009-9285-0

Gundersen, P., Callesen, I., & De Vries, W. (1998). Nitrate leaching in forest ecosystems is related to forest floor ratios. *Environmental Pollution, 102*(1), 403–407. https://doi.org/10.1016/s0269-7491(98)80060-2

Han, C., Jiang, Y., Liu, K., Zhang, J., & Zhong, W. (2023). Active microbial taxa preferentially assimilate inorganic nitrogen in acidic upland soils using a 15N-DNA-SIP approach. *Soil Biology and Biochemistry, 188*, 109226.

He, Y., Yun, H., Peng, L., Ji, J., Wang, W., & Li, X. (2024). Deciphering the potential role of quorum quenching in efficient aerobic denitrification driven by a synthetic microbial community. *Water Research, 251*, 121162.

Hu, J. (2018). Human alteration of the nitrogen cycle and its impact on the environment. *IOP Conf Ser Earth Environ Sci, 178*, 012030. https://doi.org/10.1088/1755-1315/178/1/012030

Hu, S., Chapin, F., Firestone, M., et al. (2001). Nitrogen limitation of microbial decomposition in a grassland under elevated CO2. *Nature, 409*, 188–191.

Hu, H., Macdonald, C. A., Trivedi, P., Anderson, I. C., Zheng, Y., Holmes, B., Bodrossy, L., Wang, J., He, J., & Singh, B. K. (2015). Effects of climate warming and elevated CO2 on autotrophic nitrification and nitrifiers in dryland ecosystems. *Soil Biology and Biochemistry, 92*, 1–15. https://doi.org/10.1016/j.soilbio.2015.09.008

Isokpehi, R. D., Kim, Y., Krejci, S. E., & Trivedi, V. D. (2024). Ecological trait-based digital categorization of microbial genomes for denitrification potential. *Microorganisms, 12*(4), 791.

Jansson, J. K., & Hofmockel, K. S. (2020). Soil microbiomes and climate change. *Nature Reviews. Microbiology, 18*(1), 35–46.

Jia, X., Gao, Y., Li, X., Zhao, Y., Wang, L., & Zhang, C. (2021). Effects of cadmium on soil nitrification in the rhizosphere of *Robinia pseudoacacia* L. seedlings under elevated atmospheric CO2 scenarios. *The Science of the Total Environment, 772*, 145023. https://doi.org/10.1016/j.scitotenv.2021.145023

Jian, S., Li, J., Chen, J., Wang, G., Mayes, M. A., Dzantor, K. E., Hui, D., & Luo, Y. (2016). Soil extracellular enzyme activities, soil carbon and nitrogen storage under nitrogen fertilization: A meta-analysis. *Soil Biology and Biochemistry, 101*, 32–43. https://doi.org/10.1016/j.soilbio.2016.07.003

Kandeler, K. (1996). Nitrogen mineralization. In F. Schinner, R. Öhlinger, E. Kandeler, & R. Margesin (Eds.), *Methods in soil biology*. Springer.

Karimi, S., Soltani, S., & Jasemi, K. (2023). Positive and negative impact of nitrogen fertilizer on soil properties and nutrient dynamic. *Asian Journal of Research in Agriculture and Forestry, 9*(3), 233–240. https://doi.org/10.9734/ajraf/2023/v9i3228

Kesik, M., Blagodatsky, S., Papen, H., & Butterbach-Bahl, K. (2006). Effect of pH, temperature and substrate on N2O, NO and CO2 production by *Alcaligenes faecalis* p. *Journal of Applied Microbiology, 101*(3), 655–667. https://doi.org/10.1111/j.1365-2672.2006.02927.x

Klemedtsson, L., Von Arnold, K., Weslien, P., & Gundersen, P. (2005). Soil CN ratio as a scalar parameter to predict nitrous oxide emissions. *Global Change Biology, 11*(7), 1142–1147. https://doi.org/10.1111/j.1365-2486.2005.00973.x

Knicker, H. (2011). Soil organic N—An under-rated player for C sequestration in soils? *Soil Biology and Biochemistry, 43*(6), 1118–1129.

Kooch, Y., Sanji, R., & Tabari, M. (2018). Increasing tree diversity enhances microbial and enzyme activities in temperate Iranian forests. *Trees, 32*, 809–822. https://doi.org/10.1007/s00468-018-1674-3

Kuzyakov, Y., Horwath, W. R., Dorodnikov, M., & Blagodatskaya, E. (2018). Review and synthesis of the effects of elevated atmospheric CO2 on soil processes: No changes in pools, but increased fluxes and accelerated cycles. *Soil Biology and Biochemistry, 128*, 66–78. https://doi.org/10.1016/j.soilbio.2018.10.005

Ladha, J. K., Peoples, M. B., Reddy, M. R., Biswas, J. C., Bennet, A., Jat, M. J., & Krupnik, T. J. (2022). Biological nitrogen fixation and prospects for ecological intensification in cereal-based cropping systems. *Field Crop Res, 283*, 108541.

LaSarre, B., Morlen, R., Neumann, G. C., Harwood, C. S., & McKinlay, J. B. (2024). Nitrous oxide reduction by two partial denitrifying bacteria requires denitrification intermediates that cannot be respired. *Applied and Environmental Microbiology, 90*, e01741–e01723.

Lee, S., Cho, K., Lim, J., Kim, W., & Hwang, S. (2010). Acclimation and activity of ammonia-oxidizing bacteria with respect to variations in zinc concentration, temperature, and microbial population. *Bioresource Technology, 102*(5), 4196–4203. https://doi.org/10.1016/j.biortech.2010.12.035

Lerman, A., Mackenzie, F. T., & Ver, L. M. (2004). Coupling of the perturbed C-N-P cycles in industrial time. *Aquatic Geochemistry, 10*, 3–32.

Li, C., Frolking, S., & Butterbach-Bahl, K. (2005). Carbon sequestration in arable soils is likely to increase nitrous oxide emissions, offsetting reductions in climate radiative forcing. *Climatic Change, 72*(3), 321–338. https://doi.org/10.1007/s10584-005-6791-5

Li, L., Bebout, G. E., & Idleman, B. D. (2007). Nitrogen concentration δ^{15}N of altered oceanic crust obtained on ODP Legs 129 and 185: Insights into alteration-related nitrogen enrichment and the nitrogen subduction budget. *Geochimica et Cosmochimica Acta, 71*, 2344–2360.

Li, Z., Zeng, Z., Tian, D., Wang, J., Fu, Z., Zhang, F., Zhang, R., Chen, W., Luo, Y., & Niu, S. (2020). Global patterns and controlling factors of soil nitrification rate. *Global Change Biology, 26*(7), 4147–4157. https://doi.org/10.1111/gcb.15119

Li, Z., Zeng, Z., Song, Z., Wang, F., Tian, D., Mi, W., Huang, X., Huang, J., Song, L., Yang, Z., Wang, J., Feng, H., Jiang, L., Chen, Y., Luo, Y., & Niu, S. (2021). Vital roles of soil microbes in driving terrestrial nitrogen immobilization. *Global Change Biology, 27*(9), 1848–1858.

Liang, X., Wen, X., Yang, H., Heng, L., Wang, A., Liu, S., & Li, Q. (2024). Incorporating microbial inoculants to reduce nitrogen loss during sludge composting by suppressing denitrification and promoting ammonia assimilation. *Sci Total Environ, 915*, 170000.

Liao, J., Luo, Q., Hu, A., Wan, W., Tian, D., Ma, J., Ma, T., Luo, H., & Lu, S. (2022). Soil moisture–atmosphere feedback dominates land N2O nitrification emissions and denitrification reduction. *Global Change Biology, 28*(21), 6404–6418. https://doi.org/10.1111/gcb.16365

Liu, L., & Greaver, T. L. (2010). A global perspective on belowground carbon dynamics under nitrogen enrichment. *Ecology Letters, 13*, 819–828.

Long, X., Chen, C., Xu, Z., Oren, R., & He, J. (2011). Abundance and community structure of ammonia-oxidizing bacteria and archaea in a temperate forest ecosystem under ten-years elevated CO2. *Soil Biology and Biochemistry, 46*, 163–171. https://doi.org/10.1016/j.soilbio.2011.12.013

Lu, X., Mao, Q., Gilliam, F. S., Luo, Y., & Mo, J. (2014). Nitrogen deposition contributes to soil acidification in tropical ecosystems. *Global Change Biology, 20*(12), 3790–3801. https://doi.org/10.1111/gcb.12665

Luo, Y., Su, B., Currie, W. S., Dukes, J. S., Finzi, A., Hartwig, U., Hungate, B., McMurtrie, R. E., Oren, R., Parton, W. J., Pataki, D. E., Shaw, R. M., Zak, D. R., & Field, C. B. (2004). Progressive nitrogen limitation of ecosystem responses to rising atmospheric carbon dioxide. *BioScience, 54*(8), 731–739.

Mackenzie, F. T. (1998). *Our changing planet: An introduction to Earth system science and environmental change* (2nd ed.). Prentice-Hall.

Magray, J. A., Zargar, S. A., Islam, T., & Javid, H. (2022). *Factors affecting nitrogen uptake, transport, and assimilation* (pp. 69–85). CRC Press eBooks.

Mandal, D., & Roy, T. (2024). *Climate change impact on soil erosion and land degradation* (pp. 139–161). Springer Nature Singapore.

Mattoo, R., & Suman, B. M. (2023). Microbial roles in the terrestrial and aquatic nitrogen cycle—Implications in climate change. *FEMS Microbiology Letters, 370*, fnad061.

Myrold, D. D. (2021). Transformations of nitrogen. In *Principles and applications of soil microbiology* (pp. 385–421). Elsevier.

Naz, N. M., Dai, Z., Hussain, S., Tariq, M., Danish, S., Khan, I. U., Qi, S., & Du, D. (2022). The soil pH and heavy metals revealed their impact on soil microbial community. *Journal of Environmental Management, 321*, Article 115770.

Nguyen, L. T., Broughton, K., Osanai, Y., Anderson, I. C., Bange, M. P., Tissue, D. T., & Singh, B. K. (2019). Effects of elevated temperature and elevated CO2 on soil nitrification and ammonia-oxidizing microbial communities in field-grown crop. *Sci Total Environ, 675*, 81–89. https://doi.org/10.1016/j.scitotenv.2019.04.181

Oldenburg, P. S., Regan, J. M., Harrington, G. W., & Noguera, D. R. (2002). Kinetics of *Nitrosomonas europaea* INACTIVATION by chloramine. *American Water Works Association, 94*(10), 100–110. https://doi.org/10.1002/j.1551-8833.2002.tb09561.x

Pester, M., Schleper, C., & Wagner, M. (2011). The Thaumarchaeota: An emerging view of their phylogeny and ecophysiology. *Current Opinion in Microbiology, 14*(3), 300–306. https://doi.org/10.1016/j.mib.2011.04.007

Prasad, S., Malav, L. C., Choudhary, J., Kannojiya, S., Kundu, M., Kumar, S., et al. (2021). Soil microbiomes for healthy nutrient recycling. In A. N. Yadav, J. Singh, C. Singh, & N. Yadav (Eds.), *Current trends in microbial biotechnology for sustainable agriculture* (pp. 1–21). Springer. https://doi.org/10.1007/978-981-15-6949-4_1

Priya, A. K., Muruganandam, M., Rajamanickam, S., Sivarethinamohan, S., Gaddam, M. K. R., Velusamy, P., Gomathi, G., Ravindiran, G., Gurugubelli, T. R., & Muniasamy, S. K. (2023). Impact of climate change and anthropogenic activities on aquatic ecosystem—A review. *Environmental Research, 238*, 117233.

Prosser, J. I., & Nicol, G. W. (2012). Archaeal and bacterial ammonia-oxidisers in soil: The quest for niche specialisation and differentiation. *Trends in Microbiology, 20*(11), 523–531. https://doi.org/10.1016/j.tim.2012.08.001

Purkhold, U., Pommerening-Röser, A., Juretschko, S., Schmid, M. C., Koops, H., & Wagner, M. (2000). Phylogeny of all recognized species of ammonia oxidizers based on Comparative 16S RRNA and AMOA sequence analysis: Implications for molecular diversity surveys. *Applied and Environmental Microbiology, 66*(12), 5368–5382. https://doi.org/10.1128/aem.66.12.5368-5382.2000

Quadriya, H., Mir, M. I., Surekha, K., Gopalkrishnan, S., Khan, M. Y., Sharma, S. K., & Bee, H. (2020). *Contribution of microbe-mediated processes in nitrogen cycle to attain environmental equilibrium* (pp. 331–356). Springer.

Ramirez, K. S., Craine, J. M., & Fierer, N. (2012). Consistent effects of nitrogen amendments on soil microbial communities and processes across biomes. *Global Change Biology, 18*(6), 1918–1927. https://doi.org/10.1111/j.1365-2486.2012.02639.x

Raszka, A., Surmacz-Górska, J., Żabczyński, S., & Miksch, K. (2011). The population dynamics of nitrifiers in ammonium-rich systems. *Water Environment Research, 83*(12), 2159–2169.

Robertson G.P. and Groffman P. M. (2007). Nitrogen Transformations, In E. A. Paul (Ed.), *Soil Microbiology, Ecology and Biochemistry* (3rd ed., pp. 341–364). Academic. ISBN 9780125468077, https://doi.org/10.1016/B978-0-08-047514-1.50017-2

Robertson, G. P., & Groffman, P. M. (2015). Nitrogen transformations. In E. A. Paul (Ed.), *Soil microbiology, ecology and biochemistry* (4th ed., pp. 421–446). Academic.

Rumeau, M., Mackenzie, R., Carillo, Y., Sgouridis, F., Reay, M., & Ullah, S. (2023). *Nitrogen cycling in forest soils under elevated CO2: Response of a key soil nutrient to global change.* EGU General Assembly 2023, Vienna, Austria, 24–28 Apr 2023, EGU23-11920.

Rustad, L., Campbell, J., Marion, G., et al. (2001). A meta-analysis of the response of soil respiration, net nitrogen mineralization, and aboveground plant growth to experimental ecosystem warming. *Oecologia, 126*, 543–562.

Sardans, J., Peñuuelas, J., Estiarte, M., & Prieto, P. (2008). Warming and drought alter C and N concentration, allocation and accumulation in a Mediterranean shrubland. *Global Change Biology, 14*, 2304–2316.

Sardar, M. F., Younas, F., Farooqi, Z. U. R., & Li, Y. L. (2023). Soil nitrogen dynamics in natural forest ecosystem: A review. *Frontiers in Forests and Global Change, 6*, 1144930. https://doi.org/10.3389/ffgc.2023.1144930

Schimel, J. P. (2018). Life in dry soils: Effects of drought on soil microbial communities and processes. *Annual Review of Ecology, Evolution, and Systematics, 49*(1), 409–432.

Schleper, C. (2010). Ammonia oxidation: Different niches for bacteria and archaea? *The ISME Journal, 4*(9), 1092–1094. https://doi.org/10.1038/ismej.2010.111

Schmidt, M. W. I., Torn, M. S., Abiven, S., Dittmar, T., Guggenberger, G., Janssens, I. A., et al. (2011). Persistence of soil organic matter as an ecosystem property. *Nature, 478*, 49–56.

Séneca, J., Pjevac, P., Canarini, A., Herbold, C. W., Zioutis, C., Dietrich, M., Simon, E., Prommer, J., Bahn, M., Pötsch, E. M., Wagner, M., Wanek, W., & Richter, A. (2020). Composition and activity of nitrifier communities in soil are unresponsive to elevated temperature and CO2, but strongly affected by drought. *The ISME Journal, 14*(12), 3038–3053. https://doi.org/10.1038/s41396-020-00735-7

Séneca, J., Söllinger, A., Herbold, C. W., Pjevac, P., Prommer, J., Verbruggen, E., Sigurdsson, B. D., Peñuelas, J., Janssens, I. A., Urich, T., Tveit, A. T., & Richter, A. (2021). Increased microbial expression of organic nitrogen cycling genes in long-term warmed grassland soils. *ISME Communications, 1*(1), 69.

Shaw, M. R. & Harte, J. (2001). Response of Nitrogen Cycling to simulated climate change: differential responses along a subalpine ecotone. *Global Change Biology, 7*, 193–210. https://doi.org/10.1046/j.1365-2486.2001.00390.x

Song, L., & Niu, S. (2022). Increased soil microbial AOB amoA and narG abundances sustain long-term positive responses of nitrification and denitrification to N deposition. *Soil Biology and Biochemistry, 166*, 108539.

Sorai, M., Yoshida, N., & Ishikawa, M. (2007). Biogeochemical simulation of nitrous oxide cycle based on the major nitrogen processes. *Journal of Geophysical Research, 112*, 2005JG000109. https://doi.org/10.1029/2005JG000109

Stevens, R., Laughlin, R., & Malone, J. (1998). Soil pH affects the processes reducing nitrate to nitrous oxide and di-nitrogen. *Soil Biology and Biochemistry, 30*(8–9), 1119–1126. https://doi.org/10.1016/s0038-0717(97)00227-7

Sumitha, J., Konwar, T., Raj, P., Nehul, J.-N., Hazarika, P.-P., Gangopadhyay, S., Chattopadhyay, D., Sampathkumar, T., & Sivakumar, K.-K. (2024). Exploring the microbial ecology of nitrification denitrification in wastewater treatment. *Innovations and Future Directions, 45*(2), 92–103.

Tan, X., Machmuller, M. B., Huang, F., He, J., Chen, J., Cotrufo, M. F., et al. (2020). Temperature sensitivity of ecoenzyme kinetics driving litter decomposition: The effects of nitrogen enrichment, litter chemistry, and decomposer community. *Soil Biology and Biochemistry, 148*, 107878. https://doi.org/10.1016/j.soilbio.2020.107878

Tang, Y., Yu, G., Zhang, X., Wang, Q., Ge, J., & Liu, S. (2018). Changes in nitrogencycling microbial communities with depth in temperate and subtropical forest soils. *Applied Soil Ecology, 124*, 218–228. https://doi.org/10.1016/j.apsoil.2017.10.029

Tian, D. Niu, S. (2015). A Global Analysis of Soil Acidification Caused by Nitrogen Addition. *Environmental Research Letter, 10*, Article ID: 024019. https://doi.org/10.1088/1748-9326/10/2/024019

Treseder, K. K. (2008). Nitrogen additions and microbial biomass: A meta-analysis of ecosystem studies. *Ecology Letters, 11*(10), 1111–1120. https://doi.org/10.1111/j.1461-0248.2008.01230.x

U.S. Environmental Protection Agency (USEPA). (1993). *Process design manual: Nitrogen control (EPA/625/R-93/010)*. U.S. Environmental Protection Agency. Retrieved from https://nepis.epa.gov

van Groenigen, K. J., Qi, X., Osenberg, C. W., Luo, Y., & Hungate, B. A. (2014). Faster decomposition under increased atmospheric CO_2 limits soil carbon storage. *Science, 344*(6183), 508–509. https://doi.org/10.1126/science.1249534. Epub 2014 Apr 24. PMID: 24762538.

Viancelli, A., & Michelon, W. (2024). Climate change and nitrogen dynamics: Challenges and strategies for a sustainable future. *Nitrogen, 5*(3), 688–701.

Wallingford, UK., p 423

Wang, X., Wang, C., Cotrufo, M. F., Sun, L., Jiang, P., Liu, Z., et al. (2020a). Elevated temperature increases the accumulation of microbial necromass nitrogen in soil via increasing microbial turnover. *Global Change Biology, 26*, 5277–5289.

Wang, S., Zhuang, Q., Jin, X., Yang, Z., & Liu, H. (2020b). Predicting soil organic carbon and soil nitrogen stocks in topsoil of forest ecosystems in northeastern China using remote sensing data. *Remote Sens, 12*(7), 1115.

Waqas, M. A., Li, Y., Ashraf, M. N., Ahmed, W., Wang, B., Sardar, M. F., Ma, P., Li, R., Wan, Y., & Kuzyakov, Y. (2021). Long-term warming and elevated CO2 increase ammonia-oxidizing microbial communities and accelerate nitrification in paddy soil. *Applied Soil Ecology, 166*, 104063. https://doi.org/10.1016/j.apsoil.2021.104063

Wedepohl, K. H. (1995). The composition of the continental crust. *Geochimica et Cosmochimica Acta, 59*, 1217–1232.

Yan, Z., Chang, B., Song, X., Wang, G., Shan, J., Yang, L., Li, S.-L., Butterbach-Bahl, K., & Ju, X. (2024). A microbial-explicit model with comprehensive nitrogen processes to quantify gaseous nitrogen production from agricultural soils. *Soil Biology and Biochemistry, 189*, 109284.

Yang, J., Harrington, G. W., Noguera, D. R., & Fleming, K. K. (2007). Risk analysis of nitrification occurrence in pilot-scale chloraminated distribution systems. *Journal of Water Supply: Research and Technology-AQUA, 56*(5), 293–311. https://doi.org/10.2166/aqua.2007.007

Yang, A., Zhu, D., Zhang, W., Shao, Y., Shi, Y., Liu, X., Lu, Z., Zhu, Y.-G., Wang, H., & Fu, S. (2024). Canopy nitrogen deposition enhances soil ecosystem multifunctionality in a temperate forest. *Global Change Biology, 30*(3), e17250.

Yao, Y., Han, B., Liu, B., Wang, Y., Su, X., Ma, L., Zhang, T., Nieu, S., Chen, X., & Li, Z. (2023). Global variations and controlling factors of anammox rates. *Global Change Biology, 29*(13), 3622–3633.

Yu, C., Hou, L., Zheng, Y., et al. (2018). Evidence for complete nitrification in enrichment culture of tidal sediments and diversity analysis of clade a comammox *Nitrospira* in natural environments. *Applied Microbiology and Biotechnology, 102*, 9363–9377.

Yue, P., Zuo, X., Li, K., Cui, X., Wang, S., Misselbrook, T., & Liu, X. (2021). The driving effect of nitrogen-related functional microorganisms under water and nitrogen addition on N2O emission in a temperate desert. *Sci Total Environ, 772*, 145470. https://doi.org/10.1016/j.scitotenv.2021.145470

Zhang, Y., Love, N., & Edwards, M. (2009). Nitrification in drinking water systems. *Critical Reviews in Environmental Science and Technology, 39*(3), 153–208. https://doi.org/10.1080/10643380701631739

Zhang, L., Wu, F., Xu, Z., Tan, B., Wang, A., & Yang, W. (2017). Effects of simulated warming on soil ammonia-oxidizing bacteria and archaea communities in an alpine forest of Western Sichuan, China. *Acta Ecologica Sinica, 37*(2), 85–90. https://doi.org/10.1016/j.chnaes.2016.12.004

Zhang, G., Huang, J., Jia, M., Liu, F., Yang, Y., Wang, Z., & Han, G. (2019). Ammonia-oxidizing bacteria and archaea: Response to simulated climate warming and nitrogen supplementation. *Soil Science Society of America Journal, 83*(6), 1683–1695. https://doi.org/10.2136/sssaj2019.05.0134

Zhang, Y., Qiu, J., Xu, X., & Zhou, L. (2021). Disinfection kinetics of free chlorine, monochloramines and chlorine dioxide on ammonia-oxidizing bacterium inactivation in drinking water. *Water, 13*(21), 3026.

Zhang, Y., Cheng, X., Van Groenigen, K. J., García-Palacios, P., Cao, J., Zheng, X., Luo, Y., Hungate, B. A., Terrer, C., Butterbach-Bahl, K., Olesen, J. E., & Chen, J. (2023). Shifts in soil ammonia-oxidizing community maintain the nitrogen stimulation of nitrification across climatic conditions. *Global Change Biology, 30*(1), e16989. https://doi.org/10.1111/gcb.16989

Zhou, X., Tahvanainen, T., Malard, L., Chen, L., Pérez-Pérez, J., & Berninger, F. (2024). Global analysis of soil bacterial genera and diversity in response to pH. *Soil Biology and Biochemistry, 198*, Article 109552.

Zhu, L., Huang, C., Li, W., Weixia, W., Zhurui Tang, Y., & Tian, B. X. (2023). Ammonia assimilation is key for the preservation of nitrogen during industrial-scale composting of chicken manure. *Waste Management, 170*, 50–61.

MIX
Papier aus verantwortungsvollen Quellen
Paper from responsible sources
FSC® C105338

If you have any concerns about our products,
you can contact us on
ProductSafety@springernature.com

In case Publisher is established outside the EU,
the EU authorized representative is:
Springer Nature Customer Service Center GmbH
Europaplatz 3, 69115 Heidelberg, Germany

Printed by Libri Plureos GmbH
in Hamburg, Germany